Handbook of
Teratology

1 General Principles
and Etiology

Handbook of Teratology

Handbook of Teratology

Edited by

JAMES G. WILSON

*The Children's Hospital Research Foundation
and Department of Pediatrics, University of Cincinnati
Cincinnati, Ohio*

and

F. CLARKE FRASER

*McGill University and The Montreal Children's Hospital
Montreal, Canada*

1   General Principles
and Etiology

PLENUM PRESS · NEW YORK AND LONDON

Library of Congress Cataloging in Publication Data

Main entry under title:

Handbook of teratology.

    Includes bibliographies and index.
    CONTENTS: v. 1. General principles and etiology.
    1. Deformities. 2. Teratogenic agents. I. Wilson, James Graves, 1915-   II.
Fraser, F. Clarke, 1920-   [DNLM: 1. Abnormalities. QS675 H236]
QM691. H26   616'.043   76-41787
ISBN 0-306-36241-4 (v. 1)

© 1977 Plenum Press, New York
A Division of Plenum Publishing Corporation
227 West 17th Street, New York, N.Y. 10011

Printed in the United States of America

Contributors

ROBERT L. BRENT, Jefferson Medical College, Thomas Jefferson University, Philadelphia, Pennsylvania

MARSHALL J. EDWARDS, Department of Veterinary Medicine, University of Sydney, Sydney, Australia

F. CLARKE FRASER, Department of Biology, McGill University, and Department of Medical Genetics, The Montreal Children's Hospital, Montreal, Canada

CASIMER T. GRABOWSKI, Department of Biology, University of Miami, Coral Gables, Florida

LUCILLE S. HURLEY, Department of Nutrition, University of California, Davis, California

JEROME E. KURENT, Infectious Diseases Branch, National Institute of Neurological and Communicative Disorders and Stroke, National Institutes of Health, Bethesda, Maryland

H. V. MALLING, Biochemical Genetics Section, Environmental Mutagenesis Branch, National Institute of Environmental Health Sciences, Research Triangle Park, North Carolina

JOHN L. SEVER, Infectious Diseases Branch, National Institute of Neurological and Communicative Disorders and Stroke, National Institutes of Health, Bethesda, Maryland

THOMAS H. SHEPARD, Central Laboratory for Human Embryology, Department of Pediatrics, University of Washington, Seattle, Washington

RAOUL A. WANNER, Department of Veterinary Medicine, University of Sydney, Sydney, Australia

JOSEF WARKANY, Children's Hospital Research Foundation, Cincinnati, Ohio

J. S. WASSOM, Environmental Mutagen Information Center, Information Center Complex/Information Division, Oak Ridge National Laboratory, Oak Ridge, Tennessee

JAMES G. WILSON, Children's Hospital Research Foundation, Cincinnati, Ohio

Preface

In less than 40 years teratology has grown from a little known discipline concerned with studies on the effects of a few physical and chemical stresses on developing fish, amphibians, and birds, to a discipline embracing a vast accumulation of literature on experimental studies in many animal forms—and the results of intensive scrutiny of human development under varied conditions, as well. Emphasis has shifted from preoccupation with descriptions of anatomical defects to concern about subtle and interacting causative factors, to searches for the early reactions to these at the cellular and subcellular levels, and to identification of abnormality in the chemical, the functional, and the ultrastructural realms. These changes in orientation have quite naturally made necessary the recruitment of concepts, methods, and expertise from other disciplines. Hence the foundations of teratology, which once were largely morphological, have extended into genetics, biochemistry, molecular biology, reproductive physiology, epidemiology, and several aspects of veterinary and clinical medicine.

It is not surprising that a student or new investigator approaching the field of teratology may feel some dismay when confronted with the confusing array of cross-disciplinary concepts and principles it encompasses today. One of the aims of this work is to introduce what the editors believe is a logical thread of continuity into a field that may be regarded by some as a welter of disordered information. The major intent of this work, however, has been to assemble for the first time a definitive source of facts, concepts, methods, and references within the broad scope of teratology as it is presently defined. Exhaustive coverage is not possible, even within the generous limits set here, but earnest efforts have been made to provide a working background to most aspects of the subject that are of recent and current research interest. Foremost in the minds of both editors and the authors have been the needs of graduate and other professional students and of beginning investigators. It is hoped, however, that the presentation will prove to be such as to put much of this information within the grasp of intelligent nonspecialists also. Toward that end the inclusion of abstruse and highly technical material has been kept to the minimum need for precise scientific exposition.

Major emphasis has been placed on the causes and mechanisms of abnormal development because recognition and understanding of these aspects are likely to be most helpful in the taking of preventive measures. The quantitative and distributional aspects of epidemiology have been included because, by throwing light on causes and mechanisms, they also contribute to prevention. The clinical manifestations of developmental disease, however, have generally not been presented.

A substantial part of one volume (Vol. 4) has been devoted to discussion of the procedures used in teratological research, in teratological testing in animals, and in certain clinical diagnosis. The purpose is not to provide a step-by-step recipe for using the latest methods, but rather to discuss the relative advantages of available methodologies. In other words, the selection of appropriate methods, with some attention to the design and evaluation of teratological studies, is offered in preference to a listing of the currently preferred method for doing this or that.

The editors are particularly grateful to those authors who were cooperative in promptly submitting well-prepared manuscripts and otherwise making less arduous the labors of putting together these volumes. Gratitude is also extended to Mr. Seymour Weingarten of Plenum Publishing Corporation for his initial encouragement and subsequent assistance in organizing these volumes.

Cincinnati, Ohio   JAMES G. WILSON
Montreal, Quebec   F. CLARKE FRASER

Contents

Introduction

I

History of Teratology

1

JOSEF WARKANY

I. INTRODUCTION

Since ancient times man has taken for granted that "like begets like," and when a child or animal was born that deviated greatly from the parental image it was an impressive event to primitive man. Men and women talked about it and often transmitted their tales through generations, thus developing various myths as the story changed with time. Sometimes they made images of the wondrous birth or perpetuated the event in writing. Attitudes about abnormal infants and children and their parents changed according to the cultural state of the tribe or nation, ranging from admiration or adoration to rejection and hostility. Reactions of different peoples to births of abnormal beings are of general interest because they reflect the population's knowledge, leniency, fears, or cruelty at the time of the unusual incident. Although attitudes changed greatly at different times and in different locations, one finds that there were certain patterns of responses which seem to be inherent in the human mind. Deviations from the normal form of newborn beings vary greatly from minor changes to striking abnormalities. Twins are unusual births in man, but twins, united and inseparable, are most miraculous and impressive apparitions which can not escape notice wherever they occur.

II. CONJOINED TWINS

Conjoined twins are striking biological phenomena which have occurred in man in many locations, races, and time periods. It has been estimated that symmetrical conjoined twins occur once in about 50,000 births (Bartlett,

JOSEF WARKANY · Children's Hospital Research Foundation, Cincinnati, Ohio.

1959). Such monstrous births have impressed primitive peoples as well as civilized nations, and records of their occurrence have come down to us in many forms. Legends, images, and descriptions of conjoined twins, particularly dicephalic monsters, can illustrate the development of ancient teratology to modern scientific recording.

At first the unusual births probably were reported in whispers, talks, or songs. In rather recent times such stories have been collected from natives of some Pacific Islands. Originally oral traditions may have been factual, but as they passed from mouth to mouth they assumed unnatural features. Brodsky (1943) cited such tales transmitted among primitive peoples and told to modern visiting scientists. On Samoa there was a story of Sinalofutu who brought forth twin girls that were joined together by their backs. Their names were Ulu and Ona. After they had grown up "they were startled in their sleep and rushed from the house, each one by a separate door. The door post separated their bodies so that they were parted asunder . . ." And in Papua a legend tells of similar pygopagus twins who "when they would walk, the one must go forwards and the other backwards." Fortunately, these handicapped creatures had a foster son named Iko who remedied the situation: ". . . While they slept at night, he took a bamboo and prised a sliver off it, with an edge sharper than any knife, and so neatly cut away the one from the other that both slept soundly the while, feeling no pain, and in the morning got to their feet, two separate women. . . ." These legends, overheard in recent times, illustrate beautifully how good factual teratologic reports can be combined with miraculous tales which we, not believing in the surgical procedures mentioned, consider myths. Similarly, in the dim past a combination of teratologic observations and cumulative fantasies may have created the mythologies of ancient peoples which have come down to us through written or artistic records. If not depicted or described in durable reports, such legends are lost to posterity. There are many gods who resemble in some ways monstrous or malformed children. The double-headed twin goddess discovered by Mellaart (1963) at Catal Hüyük in Turkey is, to my knowledge, the oldest statue that can be interpreted as a god-image formed after dicephalic conjoined twins (Warkany, 1971). In a Neolithic shrine excavated in 1962 there was found among other figurines a white marble statuette, 16.4 cm high, which shows two heads on one body with two partly folded arms. To the teratologist this is a duplicitas anterior; to the archeologist it is a twin goddess. Since the figurine is thought to date from 6500 B.C., it probably represents the oldest known record of a congenital monstrosity. It is of great interest that a similar sculpture, a chalk carving of a dicephalus dibracchius, was found in New Ireland of the Bismarck Archipelago in the Pacific Ocean. And a huge rock drawing in New South Wales shows a double-headed male figure with six fingers on the right and four on the left hand (Brodsky, 1943). Saint Augustine related that at his time a man was born in the East who was double in his upper half, but single in his lower half, having two heads, two chests, four hands, but only one body and two feet like ordinary man (dicephalus tetrabracchius) and that he lived so long that many had an opportunity to see him. He concluded that it should not seem absurd to us that, as in individual cases there are monstrous births, so in the whole there are monstrous races. If these races are human, they are descended from Adam. Among the many clay figures found in Mexico and Central America, dating from 500 B.C. to 800 A.D., there were some dicephalic specimens (Weisman, 1967) which demonstrate that this subject was of interest to pre-Columbian Americans also. These widely scattered peoples from Asia Minor, Oceania, and America did not have similar religions, but they probably observed and recorded similar dicephalic children, a form of conjoined twinning that is very striking although relatively common (Robertson, 1953; Schwalbe, 1906). The same subject was later treated in pictures and descriptive texts. Monstrous births were often illustrated in broadsheets which were sold at county fairs.

An early record from England concerns the conjoined twin sisters, Mary and Eliza Chulkhurst, the "Biddenden Maids" born in Kent about the year 1100. The sisters certainly had two heads but the areas of their union are doubtful (Ballantyne, 1904). It is said that these girls lived to the age of 34 years; when one of them died in 1134 and separation was suggested, the surviving sister refused, saying:"As we came together we will also go together." The second twin died 6 hours later (Thompson, 1930). The sisters bequeathed 20 acres of land to the church wardens of the parish of Biddenden who distribute cakes bearing the twins' image to strangers and poor people every Easter Sunday. Impressions of the cakes and drawings of the Biddenden Maids have been preserved. Many pamphlets showed fantastic creatures which aroused more interest than objective portraits could. But there exist woodcuts of the sixteenth century which illustrate dicephalic children objectively and with correct anatomical descriptions. Thus a broadsheet of 1512 shows a dicephalus tetrabracchius named Eesbeth and Elisabethen (Hollaender, 1922). Head and body have adult proportions, and the figure is drawn in simple outlines. A masterful drawing by Albrecht Dürer, probably dealing with the same conjoined twins and also dated 1512, is of superior quality. A front and a rear view show the heads and necks separate, and the shoulders and midline humeri joined, while the outer arms are free. The twins are standing, although the umbilical cord is still attached to the common naval. The childhood proportions of head and body and the shading of the muscle groups are admirably rendered. There exists also a broadsheet of 1517 with twins so similar to those just mentioned that it could have dealt with the same birth. The fusion of the upper arms and of the body from the shoulders down is similar to the Dürer drawing. Added is a detailed description of the event which occurred in Landshut in Bavaria. The mother's age was 27 years and the twins were liveborn, but they lived only half an hour; the right twin died first and the left followed. The creature had two heads, two hearts, two livers joined in their cranial aspects, two lungs, two stomachs, two spleens, two gallbladders but only one navel and one vulva. Above the navel they were two perfect children but from the navel down they

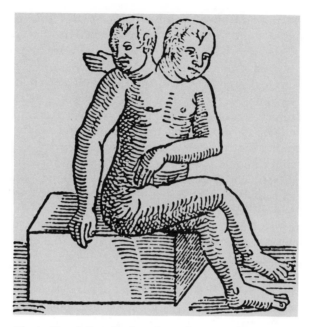

Fig. 1. Dicephalic twins from Lycosthenes, 1557. (From C. J. S. Thompson, *The Mystery and Lore of Monsters.* Williams and Norgate, Ltd., London, 1930.)

were one. These conjoined twins were seen and well described by Wilhelm Rosenzweydt of Landshut, "der sieben freien Kunst und beider Aertzneydoktor" and dissected by a surgeon (Hollaender, 1922). A fairly objective presentation of a dicephalus tribrachius was given by Licetus in 1643 (Gesenius, 1951). One of the most interesting prints dealing with conjoined twins shows a dicephalus, one head affected by a cleft lip, the other lip being normal. This remarkable phenomenon reported in 1620, which illustrates discordance in regard to cleft lip in conjoined twins, has been reported also in several modern publications (Warkany, 1971). But there are also illustrations of less believable occurences. A sheet printed in 1697 shows a double-headed, mustached, adult Turkish archer who had been captured alive in battle. Even less credible is the dicephalic creature with one human and one dog head represented in an Italian broadsheet in 1585 and almost identically reproduced by Aldrovandus in 1642 (Hollaender, 1922).

There exist some interesting reports on functional and emotional behavior of dicephalic twins. Thompson (1930) related that during the reign of Theodosius a child was born with two heads which showed differences in emotions, affections, and appetites. "One head might be crying while the other laughed, or one eating while the other was sleeping. They quarreled sometimes and came to blows. They are said to have lived two years, when one died four days before the other." Another report of emotional dissociation of a conjoined creature of the thirteenth century is mentioned by Thompson (1930) (Fig. 2),

> "one living with two bodies knot together from the top of their heads to their navells, like two grafts in the trunke of a tree, having two heades, two mouthes, two noses, with their faces faire, well-formed and made in every point requisite in a

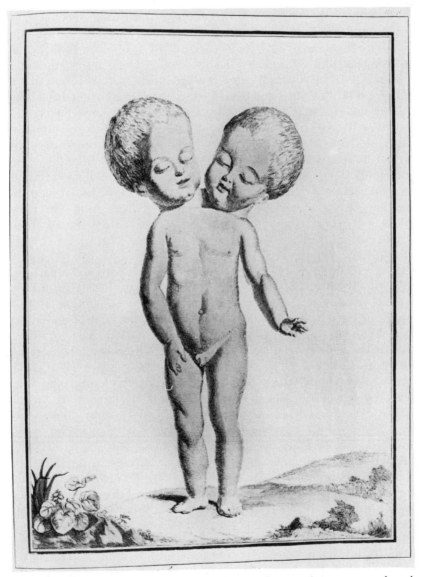

Fig. 2. Dicephalic twins with two well-formed heads. They were born prematurely and did not survive. (From L. J. Moreau de la Sarthe, Paris, 1808.)

nature and from the navell downwardes, it had but the figure and shape of one only, that is to say, two legs, two thighs and one nature. When one wept the other laughed, if the one talked, the other held her peace, as the one eat, the other drank. Living thus a long season till one of them died.

Craniopagus twins are rare. It has been estimated that one craniopagus is born in about 3,000,000 births, or one in 58 conjoined twins (Robertson, 1953) may be a craniopagus. Union can be frontal, parietal, or occipital (Figs. 3 and 4). Some authors limit the term "craniopagus" to double births in which the axes of the two individuals are placed end-to-end (Scholefield, 1929). It is said that the first craniopagus was "observed" by Münster in 1495, but many more cases must have been seen earlier without being recorded. The twin girls born in 1495 in Worms, Germany, were united by their foreheads. They were repeatedly illustrated and described in broadsheets of the fifteenth century (Hollaender, 1922). Münster saw them when they were 6 years old, and they died at 10 years of age. After the death of one of them, an attempt was made to cut the dead twin from the living, but since the head of the survivor could not be closed she too died soon after the surgical attempt. A craniopagus with heads joined at their crowns was born near Bruges (Flanders) in 1682. The girls were perfect except for their union at the top of their heads. Separation was considered but declined by several surgeons. The children were "distinct in life, soul and brains." Their actions have been described and their external appearance was illustrated in a good engraving (Thompson, 1930).

Of special interest are conjoined twins who are born accompanied by a third normally formed infant. Such a "fantastic" report came from England where a double female monster was born in 1664. It is said that at the same delivery, the birth of this monster was followed by the birth of a well-formed female child who survived (Gould and Pyle, 1897). That such an event can actually occur is confirmed by a more recent, well-documented observation reported in 1958 on a triplet pregnancy with craniopagus twins (Franklin *et al.*, 1958). Tan *et al.* (1971) collected nine incidents of conjoined twins in triplet pregnancies. Such observations deserve consideration by those who speculate about the etiology and pathogenesis of double terata. It is also worth noting that one of the craniopagus twins recorded in 1958 developed pyloric stenosis whereas the other did not (Franklin *et al.*, 1958).

An interesting pamphlet, "On a monstrous birth occurring in the ghetto of Venice," was republished by Sapadin (1964). This pamphlet, printed in Bologna in 1576, shows ischiopagus twins (Fig. 5) and discusses many theories about birth defects that range from anomalies of the semen to the shape of the bed used for cohabitation. The author distinguishes immediate, remote, and final causes but stresses that this specific birth was due to the anger of God. Only prayer and penitence could prevent the terrible events which usually follow such monstrous births. In this case the anomalous birth was due to false Jewish beliefs, and the reader was admonished to turn to the right God for salvation.

Fig. 3. The craniopagus or syncephalus frontalis of Worms is seen on the left side of the illustration by Thomas Murner. The center is occupied by the dicephalic Austrian eagle, emblem of the Habsburgs. The conjoined twins were probably interpreted as a sign of one of the unions the emperor Maximilian had achieved at the time (1495). (From E. Hollaender, *Wunder, Wundergeburt and Wundergestalt,* courtesy of Ferdinand Enke Verlag, Stuttgart, Germany, 1922.)

Fig. 4. Occipital craniopagus. The infants were well built; only occipital union
exists. (From L. J. Moreau de la Sarthe, Paris, 1808.)

Enlightenment during the eighteenth century made the life of some conjoined twins tolerable. There are accounts of the "Hungarian Sisters" Helen
and Judith, born in 1701, who were often exhibited and carefully examined
by scientists. They were joined by the sacrum and differed somewhat in appearance, activity, and intelligence from each other. They could read and
write at the age of 9 years and speak three languages. They were well treated
and protected by the Archbishop of Gran and placed in an Ursuline Convent
for their education. In their twenty-second year Judith was seized with a
convulsive fit and remained in a lethargic state thereafter. Helen developed
some fever and fainting spells but she retained her speech and remained

Fig. 5. Ischiopagus twins shown on a broadsheet printed in Bologna, 1576. (From Sapadin, *On a monstrous birth occurring in the ghetto of Venice*, Library of Hebrew Union College, Cincinnati, Ohio.)

conscious. Both sisters died about two weeks after onset of the disease. The care and treatment of the Hungarian Sisters show the tolerant attitude of the eighteenth century toward such an unusual pair of twins (Thompson, 1930).

In the nineteenth and twentieth centuries conjoined twins aroused the interest of pathologists, who supplied detailed anatomical descriptions on this subject. Taruffi (1881–1894) classified symmetrical double terata and Ballantyne (1904) divided those with joined heads into craniopagi (twins united solely by their heads), syncephali thoracopagi (twins with a single head and thorax but two pelves). A second great subdivision included the dicephali who are joined by pelves, thoraces, or spinal columns. The third important subdivision contained the relatively common thoracopagi with their rather numerous variations. In the same year Wilder (1904) wrote a monograph on duplicate twins and double monsters in which a widely used nomenclature was established. Schwalbe in 1906 dealt with "double formations" of human and animal terata in a volume of 410 pages. Symmetrical, as well as asymmetrical, double formations were described and their pathogenesis and etiology discussed. To previous methods of investigation he added roentgenograms of double terata. Other diagnostic measures came into use, such as ventriculograms and angiograms which revealed details of the internal structures in living conjoined twins. Such methods became of increasing importance as some of these "miraculous" births could be kept alive and treatment became possible.

It has been estimated that thoracopagi are the most common conjoined twins (73%); they are followed in frequency by pygopagi (19%) and by ischiopagi (6%). Craniopagi are the rarest type (2%) (Robertson, 1953). Thoracopagi have sometimes severe cardiac problems, but if their degree of union is minor (xiphopagi), they now have a good chance of survival.

The best-known case of survival is that of the Siamese twin brothers Chang and Eng, born in Siam in 1811; they were joined in the sternoxiphoid area and lived together to the age of 63 years. Safe surgical separation was not yet possible, so the rare phenomenon of long and natural lives of two conjoined human beings could be scientifically observed and described (Besse, 1874, Gould and Pyle, 1897; Luckhardt, 1941; Thompson, 1930). Physical and psychological findings which were recorded are extremely interesting, but only a few can be mentioned here. Scientists studying the role of genetic and environmental factors in human development by observation of twins should pay attention to the Siamese and other conjoined person who usually not only have similar genetic endowment, (Aird, 1959) but also similar environments. Some of these united twins are quite alike in physical and behavioral traits while others differ in spite of their common inheritance and postnatal treatment.

The twins of Siam were of Chinese extraction. The family history was of interest because their 35-year-old mother, who had borne four female children prior to Chang and Eng, had afterward several sets of twins, ending up with 14 children in all (Gould and Pyle, 1897). One of the first problems of the united twins was to overcome the belief prevailing in Siam, as elsewhere, that the birth of such unusual creatures portended evil to the kingdom and to escape the king's decision to put them to death to avoid disaster. They were brought up properly and learned to work in unison. They proved also that their lives did not interfere with the prosperous continuation of the kingdom of Siam which still exists as a constitutional monarchy in 1976, i.e., 165 years after the birth of Chang and Eng, at a time when many monarchies have been abolished without the help of conjoined twins.

In 1829 the Siamese twins were brought to Boston and then to London, where they were exhibited and thoroughly examined. Their bodies and limbs were well-formed, but Chang's spine was curved laterally by holding "his arm over the shoulder of Eng." They had great strength and agility; they loved each other and seldom quarreled; they were intelligent and could "engage in different conversations with different persons at the same time upon totally dissimilar subjects." It seems that their dreams differed (Thompson, 1930). After being exhibited all over the globe, they settled down as farmers in North Carolina, where they lived happily until they got married at the age of 44 years to two sisters of English birth. "Domestic troubles, however, soon compelled them to keep the wives at different houses"; they had to visit each wife alternately and produced children who apparently were healthy and strong (Gould and Pyle, 1897; Thompson, 1930). The life and death of the Siamese twins have been well described, including their aging process and development of arteriosclerosis. At the age of 63 years the "intemperate" Chang died, and Eng died a few hours later. A thorough necropsy was performed which showed that their union was relatively minor (Gould and Pyle, 1897; Luckhardt, 1941). They could have been easily separated by modern surgery.

Other conjoined twins who resembled the Siamese also survived birth and were recorded in the nineteenth century. The Sardinian twins, Rita and Christina, born in 1820, and the Tocci brothers, born in Locarno in 1877, had only two legs per pair. The former are of interest because at the time of their birth it was discussed whether they had two souls or one. The latter twins differed in their taste for art—and in development of their feet which made locomotion difficult (Thompson, 1930). Births of conjoined twins became less sensational as time went on and when, in December 1953, a two-headed baby was born in Indiana, the event was recorded under small headlines in the Cincinnati *Enquirer* and a picture of the family, including the dicephalic offspring, was shown in the Cincinnati *Post* a month later without much fanfare.

Reports of the separation of conjoined twins date from early times. Some of these tales are mythical, some dubious, and some well documented. The story from Papua of separation of pygopagus twins has been told before. Several reports deal with separation after the death of one twin. As early as 945 A.D. a pair united by their abdomens was separated after the death of one brother, but the survivor died on the third day after the operation. The famous craniopagi born in 1495 in Worms had a similar fate. When one of them died, aged 10 years, an attempt was made to remove her but the other twin died also. The time was not ripe for neurosurgery. An early successful separation in the seventeenth century has been reported and seems to be confirmed according to Scammon (1926). In modern times several attempts have been successful. In some instances one craniopagus twin could be saved by surgery, and in others both children survived (Baldwin and Dekaban, 1959; Hall *et al.*, 1957; Voris *et al.*, 1957).

Xiphopagi, similar to the Siamese twins, can be separated surgically today. There exists a report in a dictionary of medical sciences published in 1812 which deals with successful operation of twins united at the xiphoid process; both made an early recovery and lived (Gould and Pyle, 1897). In recent times successful xiphopagus operations have been performed with both twins (Holm, 1936; McLaren, 1936; Reitman *et al.*, 1953; Wilson, 1962) or with one twin surviving (Aird, 1959). Thoracopagi were sucessfully separated by Peterson and Hill (1960). In ischiopagi, who shared some vital organs, one twin had to be sacrificed to save the other; the survivor had a complete gastrointestinal tract with malrotation, two uteri with four tubes, four ovaries and three vaginas (Spencer, 1956). Sucessful separation of pygopagus twins has also been reported (Koop, 1961).

Reactions to diploterata and other malformations varied according to the climate of society. It is a common habit of man that he either kills or adores unusual forms of life. Killing of malformed offspring is a rather common habit of animals, whereas adoration of abnormal births is limited to man. It has been said (Martin, 1880) that in general Western peoples have a tendency to kill human monsters, whereas Oriental peoples deify them. From reports coming down to us, we can gather that sometimes the parents and neighbors accepted conjoined twins calmly and admiringly. Some broadsheets of the fifteenth century indicate that the Wormser craniopagi (Hollaender, 1922) were taken as a favorable sign that pointed to harmony of the pope, king, and Christendom. Consequently the girls were well treated and brought up. The double-headed twins found at Catal Hüyük (Mellaart, 1963) point to deification of some of these miraculous births. But more often the "monster" was thought to announce future disaster, and the twins were killed. Thus we are told that the King of Siam at first considered killing the Siamese twins born in 1811 because it was believed that they portended evil. No doubt many cases of diploterata which were never recorded ended in immediate or early death. Some of the illustrated pamphlets mentioned are combined with texts which indicate that the unusual births were sent by God as admonitions to correct the sinful ways of man and as punishment for sins committed. Interpretations of this kind resulted, as a rule, in hostile reception and quick death of the twins. In the centuries of enlightenment more natural and objective explanations were given, and the twins were brought up as well as possible.

Progress in the attitude of society and the medical profession to "monstrous" diploterata is best illustrated by the fact that in 1962 a textbook on pediatric surgery devoted a section to conjoined twins, which had become a recognized subject of surgery (Wilson, 1962). Kiesewetter (1966), writing about surgery of conjoined twins, reviewed 25 operated pairs and listed the results. Of 50 twins, 30 were alive and 20 dead (from any cause within five years of operation). In 1967 a symposium on conjoined twins was published (Bergsma, 1967) in which obstetrical and clinical management, cardiovascular evaluation, surgical separation, and many other medical questions of diploteratology were discussed. The subject had become a chapter in the field of birth defects that could be dealt with in the same manner as single malformations were treated at that time.

Births of conjoined twins reveal man's reactions to teratologic events. Through myths, images, and written records one can follow the changing attitudes to these rare biological phenomena. One can also follow the methods of disposition of children born more or less united. From infanticide to natural death, tolerant rearing or heroic surgery, the attitudes change with changing beliefs and legislations. But even in "enlightened" times like ours, it is curious that there remain ethical and legal questions concerning this subject to which answers cannot be found. These aspects were discussed by Leiter (1932), who was requested by the father of conjoined twins to allow them to die. This was refused, and an attempt at separation was discussed. It was found that only one twin could be saved because of a large bony defect which would remain after operation. But even with the approval of the father, killing one twin to save the other was homicide according to German law. It could be argued, however, that such a double malformation was not two human beings but one monster from which one normal human being could be formed. These fine legal points were solved when both children died after the separation. A year later Hitler came to power in Germany and legal problems of this kind were quickly resolved by the German law of 1933 for the prevention of heritable disorders in offspring (Gütt *et al.*, 1936) which led to speedy disposition of severely handicapped persons. In 1967 in the United States, ethical and moral considerations on the separation of conjoined twins were again discussed in the aforementioned symposium (Bergsma, 1967). Specifically, it was asked what to do about thoracopagus twins who had only one functioning heart. Operation would involve the death of one in order to give the stronger an opportunity for life. The taking of one twin's life in favor of another is contrary to the Hippocratic oath to "maintain life," and different theologians arrived at different answers. Their arguments can be read in the original publication (Pepper, 1967).

III. SINGLE MALFORMATIONS

Among the myths collected in Samoa there is a story of Faga's daughter who was born with an imperforate vagina. This factual legend was told together with more exciting reports about conjoined twins and their fates (Section II). Similar tales of malformations probably were passed on from generation to generation in many communities and countries. There are writers who believe that many figures of mythology originated with unusual and monstrous births since there are similarities between some of these fabulous beings and unusual newborn children and animals. Factual oral reports were gradually changed and assumed unnatural features. Some authors believe that some of the gods of antiquity were shaped according to observations of congenital malformations.

The tendency to consider monstrous infants as divine was called by Ballantyne (1904) euhemeristic after an ancient historian, Euhemerus of Sicily, who wrote a book in which he maintained that deities of Greek mythology were idolized men and women. This idea is said to account for the teratological character of ancient gods and semigods. The description of the deformed child that served as model was often altered by oral tradition; the artist who later produced the god–idol had not seen the monstrous birth. One of the outstanding "euhemerists" was the German gynecologist Schatz (1901), who published a monograph entitled: "Die Griechischen Götter und die Menschlichen Missgeburten." He thought that the Greeks, who loved beauty, had to beautify monsters before they were deified. Thus Polyphemus, a cyclopic god, had a nose below the eye and not a proboscis above the midline eye. Other mythological figures derived from terata were the sirens, who resemble sympodial fetuses. Occipital encephalocele was the model of Atlas; and Janus looked in two directions like a diprosopus. Prometheus, whose liver was torn by a vulture, was suggested by observations of children with omphalocele. Schatz was a man of imagination and originality, but he probably went too far in his interpretation of mythology by teratology. However, there is general agreement that the Egyptian gods Ptah and Bes were achondroplastic dwarfs. Modern teratologists inspecting figurines of these gods may classify Ptah, who has regular facial features, as a case of spondyloepiphyseal dysplasia, whereas Bes, because of his coarse visage, looks more like a typical achondroplastic (Johnston, 1963). Deities of India are typically shown with multiple heads and supernumerary extremities, but it seems less likely that they were sculptured after human malformations than as representing superpowers with unusual physical characteristics.

Stories about one-eyed men are found in many lands, and it seems reasonable to assume that observations of cyclopic children gave rise to the belief that there existed one-eyed persons and peoples. The effects of written records as compared to evanescent oral traditions can be best illustrated by the fame of the Arimaspi who were thought to be a people in the extreme Northeast of Scythia in the Altai Mountains. They were only briefly mentioned in Herodotus' expansive *Histories:* "It is clear that it is the northern parts of Europe which are richest in gold, but how it is procured is another mystery. The story goes that the one-eyed Arimaspians steal it from the griffins who guard it; personally, however, I hesitate to believe in one-eyed men who in other respects are like the rest of us. . . ." This brief and skeptical remark of Herodotus had extraordinary sequences. Pliny took up the story; Saint Augustine knew it; poets from Aeschylus to Milton described the Arimaspi, and many artists pictured them as richly dressed Asiatics, grouped with their griffin foes (*Encyclopedia Britannica,* 1911). The griffins were fabulous beasts with heads and wings of eagles and the bodies of lions. They and their one-eyed enemies, the Arimaspi, represent a remarkable teratologic combination, but we must join Herodotus in his doubts about their existence. Yet there *are* cyclopic infants who may have contributed to the myth, and there *is* gold in the Altai Mountains (*Encyclopedia Britannica,* 1911), so that a core of truth can be discerned in this long-lived and fantastic legend. The same can probably be said of many current teratologic reports which mix truth and imagination in skillful writings in more modern terminology. To those interested in cyclopes and mythology, the article by Esser (1927), which deals with the subject thoroughly, is recommended.

It is less understandable that tales of headless people, with eyes in their breasts, were passed on in many unconnected countries. Arabs had seen them in Java, Sir John Mandeville (in the fourteenth century) alluded to an isle where they dwelt, and Spanish adventurers encountered them on devil-haunted islands. Shakespeare mentioned "men whose heads do grow beneath their shoulders" (Othello, Act 1). Except for acephalic acardii (Warkany, 1971), there are no terata that resemble the headless men of mythology. It is more likely that savage tribes, who used fantastic costumes and masks to scare their enemies and donned covers over their heads with holes for eyes to peep through, could account for the origin of stories about headless people with eyes on their chests.

Other ubiquitous reports deal with men and races that had tails. They were assigned to Ceylon, Formosa, Borneo, Sumatra, Africa, Paraguay, New Guinea (Gould and Pyle, 1897; Thompson, 1930), and other places. It is likely that stories of native tribesmen with tails go back to artificial tails which were and are worn on ritual occasions. But "tails" have been observed in infants and in older persons occurring as caudal appendages in the lumbar or sacrococcygeal region (Warkany, 1971); they could have served as prototypes of mythologic men with tails. Another source of legends of tailed human beings may be the hairy skin seen sometimes over a spina bifida occulta. There is a statue of a faun in the Louvre of Paris that has a hairy appendage resembling those of covered spina bifidas. Fauns and devils may owe their tails to observations of cutaneous manifestations of spina bifida occulta. Thus legends of men with tails may have their origin in teratologic observations.

One can get a good picture of teratologic knowledge and beliefs of the Greeks and Romans before and at the time of Christ, from the *Natural History* of Pliny the Elder (23–79 A.D.). In addition to his official duties, Pliny read and wrote "long before daybreak" on history, geography, and the natural sciences, perusing 2000 books written by hundreds of Roman and Greek authors. He made notes as he read, and passed them on to us, thinking that "there was no book so bad as not to contain something of value (*Encyclopedia Britannica,* 1911). This should be kept in mind when we encounter in his writings fantastic stories which we cannot believe: but Pliny thought that his personal lack of confidence in some of these stories should not keep him from recording them. He explained his point of view by admitting that there are many facts which we do not believe because we haven't witnessed them: "For whoever believed in the Ethiopians before actually seeing them? or, what is not deemed miraculous when first it comes into knowledge? how many things are judged impossible before they actually occur?" These words are certainly understandable to all of us who have witnessed numerous mechanical miracles thought impossible before they actually occurred. As one reads Pliny's unbelievable reports, one must not take them at their face value and reject them as fantastic, but rather accept them as beliefs existing at the writer's time. Many stories are about peoples that lived far away, in Scythia, India, and Africa. In a certain large valley in the Himalayas there were people dwelling in the forests who had their feet turned backward behind their legs and ran extremely fast over the country with the wild animals; one author stated that some of these people had eight toes on each foot. The Monocoli, who had only one leg and moved in jumps with surprising speed, were called the umbrella-foot tribe because in hot weather they lay on their backs on the ground and protected themselves with the shadow of their feet. In India, there were also men with feet 18 inches long, and their women had such small feet that they were called sparrow-feet. The Machlyes were androgynes who performed the function of either sex alternately. Pliny related that according to Aristotle, their left breast was that of a man and their right breast that of a woman.

There followed stories of the "evil eye" and witchcraft, stories which are still believed in modern times. Some of these witches had a double pupil in one eye, and the glance of such women was injurious everywhere. In India there were the biggest animals and trees and many inhabitants were more than seven feet six inches high; the sages of the race stood upright from sunrise to sunset, gazing at the sun with eyes unmoving, standing first on one foot and then on the other in the glowing sand. In some mountains there lived human beings with dog heads whose speech was a bark. There were women in India who bore children only once in their lifetime, and the children began to turn grey directly after birth. In some villages the women had children at the age of seven, and old age came at forty; others conceived at the age of five and did not live more than eight years. There were persistent rumors about people without necks who had their eyes in their shoulders, as mentioned earlier. And a race of nomads in India had only holes in place of the nostrils, like snakes. Near the source of the Ganges there lived people without mouths, who consumed only the air and the scent they inhaled; they could easily be killed by an unusually strong odor. There were pygmies beyond the mountains who did not exceed twenty-seven inches in height; this tribe, known already to Homer, was beset by cranes, but they decimated the aggressive birds by a systematic hunt for their eggs and chicks. "Long-livers" of 100 or 200 years had white hair in their youth that grew black in old age. Some people of India had union with wild animals, and the offspring were of mixed race and half animal. In another area men were born with hairy tails (p. 17), while others were entirely covered by their long ears. These and other varieties of the human race had been produced by the ingenuity of Nature "as toys for herself and marvels for us."

The birth of triplets was an established fact and acceptable, but above that number multiple births were considered portentous, except in Egypt where drinking the water of the Nile caused fecundity. In Pliny's time a woman named Fausta living at Ostia gave birth to quadruplets, two male and two female, which unquestionably portended the food shortage that followed. In the Peloponnesus a woman four times produced quintuplets and in Egypt seven infants were the result of a single birth. Persons born of both sexes combined, hermaphrodites or androgynes, had formerly been considered as portents but later as entertainments. Transformation of females into males was considered possible since a girl at Casinum was changed into a boy under the observation of her parents, but she was deported to a desert island. At Argos, a girl named Arescusa married a man, yet subsequently grew a beard and developed masculine features; later she was named Arescon and took a wife. Pliny himself saw in Africa a person who had turned into a male on the day of marriage to a husband.

As we read about unusual peoples and races in Pliny's *Natural History,* we cannot help asking about the factual background of these wondrous tales. Some of these reports of human pathology can be accepted as facts, worded in ancient terminology. Groups of human beings with polydactyly and severely crippled extremities exist today. Double pupils are rare, but the condition, polycoria, is discussed in modern ophthalmology. Giants and dwarfs occur in our times, and pigmy races have been recently examined and described by endocrinologists. Hermaphroditic conditions are being analyzed and classified, and cases of precocious puberty and progeria are on record. Transformation of females into males occurs in patients with the adrenogenital syndrome; virilization can be congenital or it can take place in postnatal life, resulting in surprises at the time of marriage. Whether androgynes exist, being one half male and one half female, may be doubtful, but Kemp (1951) shows a photograph of such a person. There is no doubt about mixed gonadal dysgenesis, and a uterus is occasionally found in the hernia of a man (hernia uteri inguinalis in the male) (Warkany, 1971). We can believe Pliny's stories about multiple births and match them with stories of our own time.

Many of Pliny's tales cannot be accepted. Whereas we know of cyclopic infants, we do not know of cyclopic tribes; human beings with dog heads and barks are not observed at present. Exaggeration and distortion of actual observations probably account for some of the fantastic tales faithfully recorded by Pliny. Of several works of the Elder Pliny, the *Natural History* is the only one preserved in its entirety. It remained a standard work for many centuries. The Roman Empire collapsed, but the *Natural History* endured. More than 300 years after Pliny's death, he was cited almost verbatim by Saint Augustine (354–430 A.D.), who in *The City of God* (Book XVI,8), discussed the diversity of peoples and the question of monstrous races of men derived from the stock of Adam or Noah: "God sees the similarities and diversities which can contribute to the beauty of the whole. But he who cannot see the whole is offended by deformity . . . ." Saint Augustine himself had observed some malformations. In Hippo there was a man whose hands were crescent-shaped and had only two fingers each; his feet were similarly formed (lobster claw?). Heredity, apparently, was not shown in this case because the author remarked: "If there were a race like him, it would be added to the history of the curious and wonderful." Hermaphrodites were rare, but they did occur.

During the Middle Ages few new contributions were made to teratology. Albertus Magnus (1200–1280) briefly expressed opinions about the origin of monsters, but no original ideas or reports were brought forward. The subject of malformations became popular when printing of woodcuts with texts became possible. The outstanding teratological work of the sixteenth century is the book of Lycosthenes (Conrad Wolffhart, 1518–1557), entitled *Prodigiorum ac Ostentorum Chronicon* which appeared in Basle in 1557 (Thompson, 1930). It is illustrated by hundreds of woodcuts, many of which were used and reproduced by subsequent writers. Lycosthenes apparently was acquainted with the writings of the authors of antiquity whose miraculous reports he collected and illustrated in his standard work. Most of the illustrations are fantastic compositions of earlier descriptions of improbable creatures and events. Figure 6 shows Mother Alcippe feeding an elephant, a transfiguration of a child born with double harelip and a median labial protrusion. The "Satan's birth of Krakow," an often-reproduced monster with fiery eyes, long horns, tail, and fur, exemplifies another sensational report. A dicephalic creature (Fig. 1) represents a more realistic illustration of his work.

Ambroise Paré (1510–1590), devoted a section of his *Chyrurgery* (1579) to "monsters and prodigies" (Ballantyne, 1904; Thompson, 1930). Many unusual children were born during his lifetime, and they were shown or reported to him. He described and depicted them and classified their "causes" coming up with 13 etiologic principles. The first was the glory of God and the last were activities of Satan. Maternal impressions were cited in between and attributed to diabolic forces (principle 13). Thus Paré believed in the multiplicity and interaction of etiologic factors. He can be considered not only the father of French surgery but also the father of multifactorial theories, which have again become fashionable in recent years and are used by puzzled teratologists who cannot fit factual occurrences into their simple systems. Fortunio Licetus followed with *De Monstrorum Natura* in which copper engravings and more natural Aristotelian theories were published. The objective presentation he made of a dicephalus tribracchius has been mentioned before.

Aldrovandus' book *Monstrorum Historia* (1642) combined some fantastic woodcuts with some good factual illustrations. Figure 7 shows a monster with the heads of three animals, but in Fig. 8 a phocomelic child is illustrated that could correspond to an actual observation. The reduction malformations in Fig. 9 and the parasitic twin in Fig. 10 can also be considered fairly authentic. The man with the malformed abdomen (Fig. 11) is of special interest. He is described in detail and illustrated by a woodcut. Among superstitious beliefs about the origin of malformations, Aldrovandus also expressed the belief that abnormal spatial situations within the uterus can be responsible for the birth defects (Gruber, 1964). It can be seen that teratologic works of the seventeenth century already contain some rather objective and credible reports and

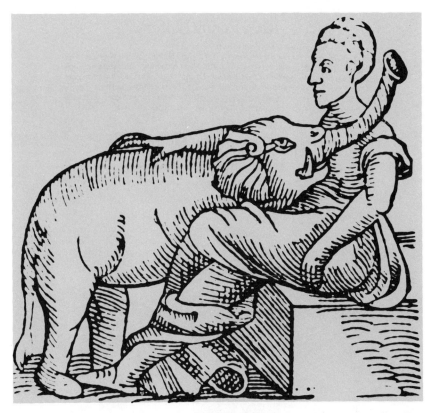

Fig. 6. Alcippe nursing her son. Woodcut from Lycosthenes, 1557. (From Hollaender, *Wunder, Wundergeburt und Wundergestalt*, 1922.)

ideas about congenital malformations. In the same century, William Harvey (1651) introduced the theory of arrested development as a new explanation for congenital malformations. He noted (Harvey, 1651) that fetuses have an oral aperture without lips and cheeks, stretching from ear to ear, ". . . and this is the reason, unless I much mistake, why so many are born with the upper lip divided. . . ." Harvey's thought was not immediately accepted, but during the next century von Haller and C. F. Wolff applied the principle of arrested development to explain ectopia cordis and exomphalos (Ballantyne, 1904). By then comparative anatomy had also been sufficiently developed so that the resemblance of certain malformations, with structures occurring normally in lower animals, made deviations from the normal understandable. Embryologic observations became the basis of teratologic interpretations.

IV. EXPLANATIONS OF THE PAST

Theories of the origin and meaning of congenital malformations were admirably summarized by Ballantyne (1904). Myths and records indicate that from early times unusual births have been considered as portents for events to

Fig. 7. Fantastic tricephalic creature. (From Aldrovandus, 1642.)

come. Hybridization of man and animals supplied another kind of explanation. An alternative theory which has lasted through the ages is that maternal impressions have a formative effect upon the unborn child. Other ideas, more akin to our modern theories of teratogenesis, have their roots in conceptions of the past.

A. Malformations as Portents

The most extensive early records of congenital malformations were kept by Babylonian priests on clay tablets found in a mound near the Tigris River (Ballantyne, 1904; Warkany, 1971). These records were kept for divination,

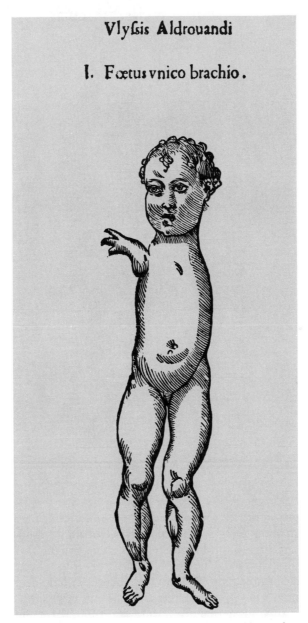

Fig. 8. Fetus with one phocomelic arm. (From Aldrovandus, 1642.)

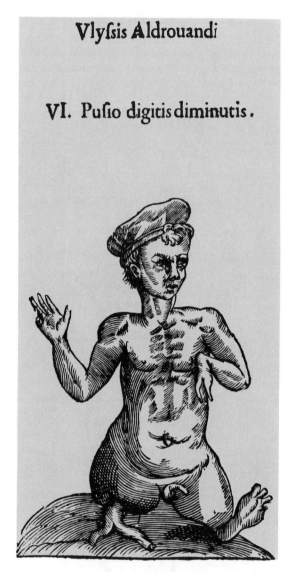

Fig. 9. Man with reduction malformation.

i.e., foretelling the future from observations of children or animals born with more or less striking defects. The appearance of malformations of the ears, nose, mouth, sex organs, and digits had meaning for the future of the king, the land, and the parents. These tablets have been published, translated, and interpreted by teratologists (Ballantyne, 1904; Dennefeld, 1914, Warkany, 1971). Babylonians developed systems of divination which became so firmly established by repetition that they were transmitted for millennia through Egypt, Syria, and Asia Minor, to Greece and Rome, and from there to Europe and America (Warkany, 1971). We do not believe in the auguries of the Babylonian soothsayers and their successors, but we must pay attention to their systematic knowledge of congenital malformations. The attitude of ancient and medieval peoples toward abnormal births, and their acceptance as bad or good omens, can be traced back to the records of the priests of Mesopotamia.

Fig. 10. Epigastric parasite. (From Aldrovandus, 1642.)

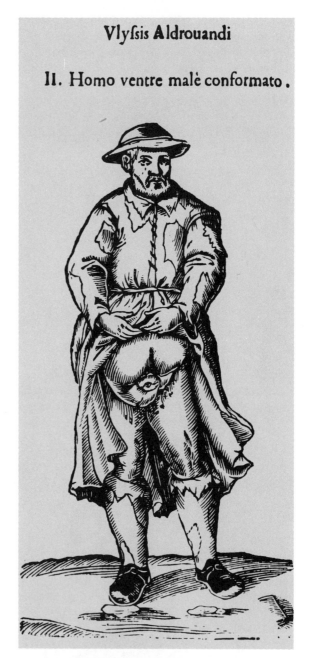

Fig. 11. Man with defect of the anterior abdominal wall.
(From Aldrovandus, 1642.)

Use has been made of monstrous births as portents in Asiatic countries and also in Greece and Rome where philosophers and augurs considered divinations as acceptable and necessary methods. The Stoic philosophers were firm believers in Providence and Fate; by Fate they meant "an orderly succession of causes wherein cause is linked to cause and each cause of future events, necessarily knows what every future event will be" (Cicero; Warkany, 1971). Omens do not cause events, but they foretell what will occur. Cicero and Pliny tell about the widespread beliefs in augury without necessarily accepting them themselves. From the Latin, such words as "portent," "monster," and "prodigy" entered the English and other European languages, thus influencing European thinking. During the Middle Ages, interpretation of monsters as portents continued and played an important role in times of crisis. Many popular pamphlets state that unusual births predict catastrophes, and in Paré's *Chyrurgery* (1579) one finds that monsters are "signs of punishment at hand." Through Europe the Babylonian beliefs came to the New World, where they still can be found in the twentieth century (Warkany, 1971).

The theory of malformations as portents of disaster often led to persecution and even death of affected children in many countries and situations. It is not clear how the Babylonians treated the children with birth defects whom they used for their predictions. If it is true that monsters are merely messengers, not causes, of catastrophes, then their elimination cannot be considered a logical sequitur. Killing the messenger who brings bad news could not prevent the predicted disaster.

B. Hybridization

Through the ages, there was a widespread belief that members of different species might be fertile with one another and thus produce monstrous offspring. According to Ballantyne (1904) the hybridity theory probably originated in India and Egypt where transmigration of souls from man to animal was taught and where animals were thought of as equal or superior to human beings. A cross between different species was not considered repulsive, so that a child who resembled in some way an animal was paid the same respect due to the animal in the official religion.

In a meeting of the Academy of Sciences in Paris on January 9, 1826, Etienne Geoffroy Saint-Hilaire demonstrated a human monster found in a sarcophagus of a burial ground for sacred animals in Hermopolis, Egypt. The child probably had been considered a product of hybridization of a woman and a monkey, an animal that ranked very highly in the Egyptian religion. But Saint-Hilaire recognized the specimen as a human anencephalus, a malformation which was well known to him since he had written several papers on this subject. The find demonstrated that, in contrast to European countries which considered anencephalics as products of evil or signs of coming disaster, Egypt honored these monstrosities as sacred animals (Martin, 1880).

In contrast to these pagan ideas, the Mosaic and Christian laws condemned relations between man and animals as criminal: "And if a man lie with a beast, he shall surely be put to death; and ye shall slay the beast. And if a woman approach unto any beast, and lie down thereto, thou shalt kill the woman and the beast: they shall surely be put to death; their blood shall be upon them." (Levit. 20:15,16). With the acceptance of the hybridity theory, resemblance of a child with an animal made the mother suspect and she and her child were in danger. It is impossible to estimate how many persons became victims of the hybridity belief during the Dark Ages. That punishment was probably not rare may be gathered from the following example. The Danish anatomist, Bartholin, mentions in his writings that a girl who gave birth to a monster with a "cat's head" was burned alive in the public square of Copenhagen "ob lasciviorem cum fele jocum." This happened in 1683 in a civilized country. The fact is reported by Bartholin without criticism or expression of disapproval (Martin, 1880). The event occurred 12 years after Niels Stensen, another Danish anatomist, had demonstrated his teratologic knowledge by an admirable description of a fetus with Tetralogy of Fallot. This is an illustration of the fact that science and coarsest superstition can thrive side by side in the same country at the same time (Warkany, 1971). Comparisons of malformations with structures seen in animals have been continued to our own times in many languages by the use of such expressions as harelip, bec-de-lièvres, gueule-de-loup, Hasenscharte, Wolfsrachen, phocomelia, and others (Warkany, 1959).

There was also danger when an animal was born with a malformation which somehow resembled a person living near the animal's mother. Both the unfortunate person suspected and the animal were treated according to Biblical law. Landauer (1962) has published such cases, the remarkable trial of George Spencer among them. The story is told in the *Records of the Colony and Plantation of New Haven, from 1638 to 1648* (Hoadley, 1857). In 1641, three years after the founding of the colony, a monstrous pig was born that had "but one eye in the midle of the face" and over the eye "a thing of flesh grew forth and hung downe, itt was hollow, and like a mans instrum' of gen'ation." The animal was obviously a cyclopic pig with a proboscis, a malformation not rare in the species (Fig. 12). However, to the good people of New Haven, cyclopia was not known as a spontaneous defect in pigs, and they attributed the birth of this monster to the "unnatureall spell and abominable filthynes" of a servant, George Spencer, who had "butt one eye for vse, the other hath (as itt is called) a pearle in itt" and closely resembled the eye of the miraculous pig. The court procedures, the confessions and retractions, as well as the testimonies, are recorded in great detail. This trial continued from 1641 to 1642, but finally the prisoner was executed on April 8, 1642, after the cyclopic sow had been "slaine in his sight, being run through with a sworde." On reading this trial one realizes that the hybridity theory of India and Egypt was, with

Fig. 12 A cyclopic pig with proboscis. (Engraved by N. F. Regnault in Moreau
de la Sarthe, Paris, 1808.)

modifications, imported to this country—where it had much more serious
consequences than in the lands of its origin.

C. Devils and Witches

Related to the hybridity theory is the belief that malformations are the
result of the association of human beings with demons, witches, and other evil
consorts. Especially in Europe, during the fifteenth and sixteenth centuries, it
was believed that the world was populated by demonic creatures, sorcerers,
witches, incubi, and succubi who plagued human beings waking or asleep.
When a deformed child showed any sign which resembled the imaginary
features of the devil, a demonic origin was suggested. Hairy nevi, ichthyosis,
club feet, shortening of the upper extremities, long and deformed ears, syndactylism, and similar anomalies could be interpreted as signs of satanic ori-

gin. Many pamphlets and broadsheets exist which show such creatures of unmistakable demonic origin (Hollaender, 1922). In these pamphlets the devilish creatures often have wings in place of arms, probably illustrating such malformations as phocomelia, syndactylism, or polydactylism; they may also show tails or trunks. Demonology with all its ramifications had developed into a science and was exacting many sacrifices of human lives.

Midwives were especially liable to be witches either because those who love the black arts have a tendency "to draw towards operations of childbed, or because the Devil is peculiarly anxious to ensnare midwives" (Williams, 1960). Stillbirths could endanger a midwife, who was held responsible if there were too many. Thus in the diocese of Bâle a woman confessed that she had killed more than 40 infants by sticking a needle into their brains; in Strasbourg one confessed that she had killed more than she could count (Williams, 1960). As belief in witchcraft has returned to our present culture—note the "occult" shelves in our modern bookstores—obstetricians, particularly those who practice amniocentesis, may soon be treated like their midwife predecessors.

D. Maternal Impressions

The formative power of visual impressions and emotions of a woman during pregnancy has been an eternal belief, ubiquitous among peoples that were widely separated geographically and culturally. Many authors have written about this subject (Gould and Pyle, 1897; Warkany, 1971), but Ballantyne's (1904) treatment of the history of the psychogenic theories of congenital malformations remains the outstanding work which traces the beliefs from ancient times to the end of the nineteenth century. The idea that the child in the womb is influenced by maternal impressions was not used originally as an explanation of congenital anomalies but was employed for eugenic purposes with the intention of changing and improving the fruit of man or animal. One of the best known examples is found in the Bible in the story of Jacob, who tried to obtain a large number of speckled offspring in Laban's flocks from which all the ring-streaked and spotted animals had been removed. How he did this and succeeded in his breeding experiments is told in the Bible (Genesis, Chapter 30). Methods of mental modification were used in Greece and in Rome, where Pliny wrote that maternal and paternal thoughts at conception can shape the child (Ballantyne, 1904). It is of interest to the teratologist that the idea of maternal impressions was also used in reverse to explain ill-formed children. For instance, the resemblance to monkeys of microcephalic or anencephalic children gave rise to the opinion that it is dangerous for women to look at monkeys during pregnancy. The birth of a black child to white parents was attributed to the picture of a Moor.

It is most interesting to trace this conception through the centuries. The French surgeon, Ambroise Paré, considered a wayward maternal imagination as one of his 13 causes of monsters. The best minds of the sixteenth century adopted the theory of maternal impressions, including the skeptic Montaigne who, in an essay entitled "On the Power of Imagination," explained why this belief was so prevalent. Since man had observed that the mind had definite influence upon such organic functions as yawning, vomiting, potency, and impotence, many unexplained somatic disturbances were attributed to the mind. It was thought that the imagination works not only upon one's own body, but also upon that of others. In the days of Montaigne, transmission of contagious diseases was attributed to visual impressions. "And as an infected body communicates its maladies to those that approach or live near it, as we see in the plague, the smallpox and sore eyes, that run through whole families and cities . . . so the imagination being vehemently agitated, darts out infection capable of offending the foreign object." Thus at a time when neither bacteria nor viruses were known as causative agents of contagious disease, it was believed that communicable diseases could be transferred by visual exposure. Since imagination and fear could influence involuntary functions of the body and since diseases could be transmitted by visual perception, it seemed very likely that maternal impressions could modify the growth of the developing fetus who is a part of the mother's body. Therefore, Montaigne stated, "We know by experience that women impart the marks of their fancy to the bodies of the children they carry in their womb." These thoughts of an enlightened man of the sixteenth century explain why mental factors were and are used in the explanation of disorders for which no visible causes could or can be found (Warkany, 1959, 1971).

In subsequent centuries the theory of maternal impressions continued to flourish, but as knowledge of embryology and animal teratology progressed, objections were raised by some scientists against this primitive explanation of birth defects. One of the best refutations was supplied by William Hunter who in the eighteenth century made a prospective study, asking mothers before their confinement about psychic traumas during pregnancy. In no one instance was there a coincidence between the woman's answer and a subsequent anomaly. But when she knew the type of anomalous structure of her child she frequently named, retrospectively, a cause not previously reported (Warkany, 1971). In spite of Hunter's splendid experimental research, unfortunate experiences during pregnancy of women with deformed children were still cited as causes during the nineteenth century. In 1899 Keating's *Cyclopedia of the Diseases of Children* contained a chapter on maternal impressions which, in scientific language and with modern tabulation, presented 90 instances of defective children recorded together with the "cause or nature of the impression." Another curious publication in the twentieth century explained the teratogenic effects of fright by cortisone release in stress situations (Strean, 1958). An experimental test with strains on mice, in which cortisone administered during gestation can produce cleft palate in the young, failed to produce the malformation by audiogenic stresses (Warkany and Kalter, 1962). This limited experimental approach did not bear out the contention that cortisone releases or injection can be equated with maternal stress.

The theory of maternal impression, although unscientific and unproven, was a rather benign explanation of congenital malformations in comparison with the theories of hybridity and Satanic causation which often resulted in cruel actions against the child, the parents, or undesirable members of the community. Mothers and children who were considered innocent victims of accidental sights could not be held responsible for the unfortunate event and were treated with tolerance.

E. Biological Theories

While superstitions and fantastic explanations of birth defects prevailed, there also existed biological theories which seem rational to us today. The thought that alcohol and improper nutrition could adversely affect the un-born child is suggested in the Bible by an admonition given by an angel of the Lord to the barren wife of Manoah who was promised that she would conceive and bear a son: "Now therefore beware, I pray thee, and drink no wine nor strong drink, and eat not any unclean thing" (Judges 13:4). Modern human and subhuman avian and mammalian teratology still deals with such maternal environmental factors and the proof that they can influence embryonic and fetal development.

Aristotle (1943) had a good knowledge of teratologic facts. He knew of redundance and reduction of fingers, toes, hands, and feet. Imperforate anus, absence of the gallbladder, spleen, or kidney also were among malformations registered by him. Even situs inversus and polysplenia were mentioned in his *Generation of Animals.* He knew that minor malformations are compatible with life and can be found in viable animals, whereas those young which depart more seriously from the normal do not live beyond the newborn period. Aristotle's description of a hermaphrodite or pseudohermaphrodite agrees well with the present-day observations. He did not believe in hybridization of different species if body sizes and gestation periods differed, as they do in man, sheep, dog, and ox. Aristotle's knowledge of malformations and his sound judgments represent the height of ancient teratology. Among Pliny's accounts of malformations are some that are acceptable and others that are fantastic.

Comparative teratology is closely related to observations in chicken eggs. Artificial incubation of eggs was already known in ancient Egypt. Hippocrates and Aristotle early noted variability of embryos during hatching of eggs. Great progress was made along these lines in the eighteenth century when artificial incubation was scientifically established by Réaumur (1683–1757), who experimented with controlled temperature changes. Experimentation with eggs of fowl and production of malformations became possible. In fact it was through the failures of artificial incubation that monstrosities had been obtained. Etienne Geoffroy Saint-Hilaire used these observations for systematic investigations of malformations (Ballantyne, 1904). He developed various teratogenic methods such as shaking and waxing of eggs, and his son Isidore, as well as Allen Thompson in Scotland and Panum in Germany, repeated and extended the experiments of the elder Saint-Hilaire. In the second half of the nineteenth century it was Camille Dareste (1891) who found that different teratogenic methods could result in similar teratologic effects by arrest of development. Thus by changing temperatures he achieved production of numerous and differing malformations. Subsequently Charles Féré used chemicals and microbic toxins which were either applied externally or injected into the albumin of the egg; he found alcohols, nicotine, and many drugs to be teratogenic (1893–1901). Whereas Dareste considered amniotic changes as important mechanisms of teratogenesis, Féré believed that more subtle nutritional or chemical effects upon the embryo were active forces. Experimental teratology was also practiced by investigators who studied developmental embryology in eggs and embryos of reptiles, amphibia, and fishes. They prepared the way for Stockard (1921), who in the first decades of the twentieth century made important contributions to the understanding of structural modifications by environmental influences on the developing embryo. He too stressed the lack of specificity of the inducing agents and the importance of the embryo's developmental stage at the time of interference.

While experimentalists dealt with eggs of lower animals, pathologists of the nineteenth century showed great interest in congenital malformations of man. Many books were entirely devoted to monstrosities and birth defects, and illustrations became realistic and true to life. Johann Friedrich Meckel (1781–1833) gave much space to single and multiple malformations in his *Handbuch der pathologischen Anatomie* (Meckel, 1812–1816). Present-day syndromologists find to their surprise that some "new" disease pictures discovered by them had been described by this great pathologist 150 years before. Etienne Saint-Hilaire (1822) and Isidore Geoffroy Saint-Hilaire (1832) wrote their classical volumes on human monsters and anomalies and attempted comprehensive classifications. W. Vrolik (1849) published his *Tabulae ad Illustrandam Embryogenesin Hominis et Mammalium,* in which such conditions as phocomelia (Fig. 13), osteogenesis imperfecta (Fig. 14), and Kleeblattschädel (cloverleaf skull) (Fig. 15) were illustrated with such perfection that they could not be bettered by modern observations and reproductions. A.Förster's (1865) *Die Missbildungen des Menschen* represents a systematic work and atlas on human malformations which is illustrated by line engravings executed after the author's careful drawings. Förster distinguished between "vitia primae formationis" (monstrosities) and variations which he called "lusus naturae," a term already used by Pliny. Friedrich Ahlfeld, a gynecologist, published a book, *Die Missbildungen des Menschen* (1880–1882), which deals especially with double and cleft formations. The monumental work by C. Taruffi entitled *Storia della Teratologia* appeared between 1881 and 1894 in 8 volumes. It tells not only the history of teratology but is also an encyclopedia of general and special pathology of malformations. Unfortunately the work is not easily available and thus is not used as much as it might be. Gould and Pyle's (1897) *Anomalies and Curiosities of Medicine* contains a great deal of teratology. It represents

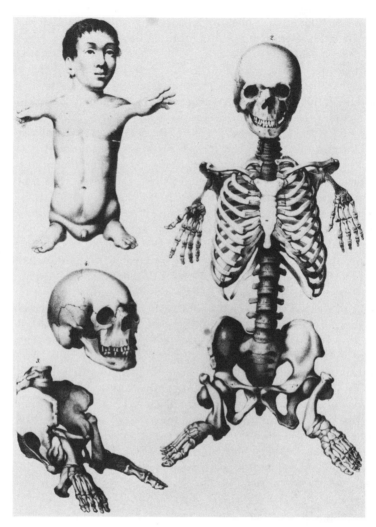

Fig. 13. Reduction malformations (phocomelia). (From Vrolik, 1849.)

"a collection of extraordinary cases derived from an exhaustive research of medical literature abstracted, annotated and indexed." Its title and format made it so popular that the original edition soon became exhausted. A reprint, published many decades later, is read by lay persons and used by scientists in search of old references.

At the beginning of the twentieth century, J. W. Ballantyne's classic *Manual of Antenatal Pathology and Hygiene* was published in two volumes. The first volume (1902), which dealt with disease of the fetus, was probably of prime importance to the author who was an obstetrician and gynecologist. Two years later, in 1904, there appeared the second volume which primarily treated the

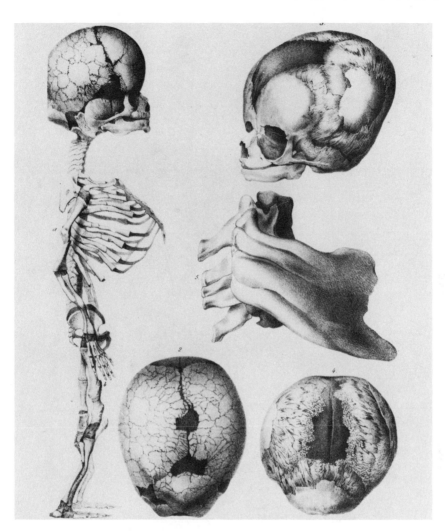

Fig. 14. Osteogenesis imperfecta. (From Vrolik, 1849.)

embryo. This volume, the result of Ballantyne's scholarship and intellectual curiosity, is most appreciated today by teratologists. Ballantyne described the anatomy, physiology, and social implications of monstrosities; he evaluated malformations in the light of their embryology and development. Furthermore he approached the subject from the standpoint of causation, which seemed to him of greatest interest because it promised hope for preventive treatment. Ballantyne's conception of prenatal pathology was far ahead of his time. To teratologists, whether interested in single cases or in general ideas, perusal of Ballantyne's work is a must; some later workers have found that their recent discoveries had been dealt with in that treatise as long-established

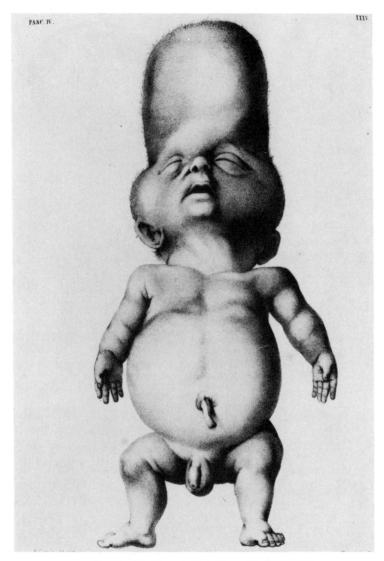

FASC. IV.   XXIV

Fig. 15. Kleeblattschädel. (From Vrolik, 1849.)

facts. Ballantyne's volume on the embryo probably is more popular today than it was in the early years after its publication.

In 1906 and 1907 the first two parts of Schwalbe's *Die Morphologie der Missbildungen des Menschen und der Tiere* were published (Schwalbe, 1906). These volumes represented the beginning of a gigantic handbook which subsequently dealt with many specific areas of teratology. The first volumes summarized the teratologic knowledge up to the beginning of the twentieth century. It contained more material about the earliest phases of normal and abnormal development of the embryo and about experimental developmental mechanics than Ballantyne's work, but it lacked the clarity and unity of the Scottish author's treatise. The second part of Schwalbe's publication was devoted to double formations, conjoined twins, and parasites (asymmetrical duplications). This work, too, should be consulted by present-day teratologists who will find that many modern ideas were anticipated by scientists of the nineteenth century. The encyclopedic knowledge of Ballantyne and Schwalbe cannot be equaled by today's scientists, who may know more about morphologic or developmental details but generally lack the overview of these scholars' writing at the beginning of this century.

At the present time genetic aspects of congenital malformations are in the center of causal explanations and research, and one may ask how the leaders of teratology in the nineteenth century could write meaningfully about causation without the basic knowledge of the principles of heredity. General maxims about transmission of traits from parents to offspring had been known since earliest time. "A bad crow lays a bad egg" was an old Greek maxim about morbid inheritance (Ballantyne, 1904), but there were always observations which contradicted this rule. The Bible states that the Lord, who is slow to anger and forgiving, "will by no means clear the guilty; visiting the iniquity of the fathers upon the children, upon the third and upon the fourth generation" (Numbers 14:18). This passage "explains" not only transmission of diseases and disorders in some families but also implies their cause: Some guilty ancestor had sinned and brought upon the offspring misfortune and malady. A suggestion of dominant inheritance of achondroplasia has been seen in a relief in a temple in Thebes, Egypt, dating back to 1500 B. C.; it shows a queen of Somaliland and her daughter, both dwarfed and possibly chondrodystrophic (Mørch, 1941). Hippocrates knew that children resemble their parents: gray-eyed parents had gray-eyed children, and squinting parents had squinting children. He believed that people after centuries of common environment begin to have traits in common (Petersen, 1946). The Greek philosopher Empedocles (490–430 B.C.), who was regarded by Galen as the founder of Italian medicine, had many modern ideas. Among these was a theory of evolution which accounts for normal and abnormal human development. It assumed an accidental ("mutational") appearance of disjointed organs which formed a variety of combinations and permutations with survival of the fittest. On the earth

> many heads sprung up without necks and arms wandered bare and bereft of shoulders. . . . Eyes strayed up and down in want of foreheads. . . . Solitary limbs wandered, seeking for union. . . . Many creatures with faces and breasts looking in different directions were born. . . . Some, offspring of oxen with faces of men, while others, again, arose as offspring of men with the heads of oxen, and creatures in whom the nature of women and men was mingled, furnished with sterile parts. . . . But as divinity was mingled still further with divinity, these things joined together as each might chance, and many other things besides them continually arose. . . .

The monstrous forms which came into being by the fortuitous union of these parts soon became extinct, but finally the fit formations survived as animals and men (Burnet, 1930). From the fragments of Empedocles' works which are still preserved, one does not learn how he explained the fact that we still see monstrous forms and that some are recurring in certain families (Warkany, 1971).

Pliny mentioned that moles and birthmarks can be transmitted through several generations, and that in one family three children were born with a membrane over their eye. There were, of course, striking examples of inheritance of birth defects which could not be overlooked by alert observers. Trew (1757) described a family in which many members were affected with cleft palate in several generations (Warkany, 1971), and similar pedigrees indicating direct transmission of congenital malformations had been repeatedly recorded before. Ballantyne (1904), who stressed pathologic teratogenic factors acting during embryologic development, did not deny "germinal factors" which determine development in the preembryonic epoch. He admitted that some characteristics, like sex or certain traits transmitted from the father, must be determined in the germinal period of antenatal existence.

> When we meet with the same malformation in two or three generations of the same family, there seems no other reasonable explanation of the occurrence than that either in the spermatozoon or in the ovum there was a teratogenic something which determined the abnormal development of the embryo. There is no lack of instances of such occurrences, as, for example, hereditary hare-lip, fistula in the lower lip, aniridia, coloboma of the iris, microphthalmia, fistula auris, ectromely, ectrodactyly, polydactyly, polymastia, absence of the patella, hypospadias, etc. etc. The occurrence, also, of the same malformation in several members of the same family seems to require a similar explanation.

Thus Ballantyne was fully aware of the existence of germinal teratogenic factors and of inheritance of many congenital malformations. But as an obstetrician, concentrating on antenatal pathology of the embryo, he did not get around to finishing his intended work on pathology of the germ. This would have entailed the writing of a third volume and for such a task Ballantyne did not have the time. In the last three chapters he gave a brief sketch of the plan he had drawn up for the treatment of this subject, and in chapter 33 he showed that he had good knowledge of hereditary disease; however, his knowledge of the pathology of the egg and sperm cells was limited. He wrote his final chapters at the time of the rediscovery of the Mendelian laws (1900), or soon thereafter, and human genetics as it is known today was not known at that time.

Schwalbe (1906) knew a great deal more about embryology, human and experimental. Mechanisms of normal and abnormal development had been thoroughly studied at the time of his writing and were discussed in great detail by him. There were also chapters on regeneration and comparative teratology. Only 14 pages were assigned to inheritance, although Schwalbe dealt with germ cells, chromosomes as carriers of hereditary traits, continuity of the germ plasm, mutation, selection, skipping of generations, and atavism. Reviewing the biological knowledge of riddles of his time, he asked what teratology could do for explanations of biologic inheritance. A year before the appearance of Schwalbe's volume, Farabee (1905) had published a remarkable example of dominant Mendelian inheritance: transmission of brachydactyly in man.

Neither Schwalbe nor Ballantyne were aware of the great contribution made to teratology, almost 40 years before their writing, by a man who had no interest in congenital malformations and who had no reputation among his contemporary fellow scientists, Gregor Mendel (1822–1884). This outsider, working on plant hybrids in the monastery garden of Brünn, Moravia, discovered the fundamental laws of heredity which explain the transmission of normal traits as well as many congenital malformations. His small publication (Mendel, 1866) got little attention until 34 years later when the principles revealed by his work were rediscovered in 1900, and within a few years they were applied to explaining birth defects. Mendel's story shows (Iltis, 1932), that now, as then, great contributions can be made by outsiders who are not in the mainstream of their science and speak a language different from that of the experts: Judgment by one's peers is not always the final judgment.

IV. INFANTICIDE AND THE LAW

Execution of deformed children or of their mothers did not originate in the Middle Ages. Although there are indications that primitive peoples sometimes kept deformed children alive and treated them well (Brodsky, 1943), there were many exceptions to this rule. Infanticide was practiced by many primitive peoples for religious and economic reasons according to the laws of the tribe. In certain populations female infants were endangered; in others males. Twins were sometimes killed together with their mothers. Infanticide was widespread among the early Greeks and Romans. It was practiced, according to Aristotle, when there were too many children, and Plato advocated it for the inferior or deformed in his *Republic*. It should be pointed out that infanticide was not limited to weak or deformed children. Certain cults sacrificed children to their gods for ceremonial reasons. It was the most-loved child, such as a first-born son, who was in danger when a god had to be propitiated. We learn this from the near-sacrifice by Abraham of his only son Isaac, who was already bound on the altar when saved by intervention of the Lord (Genesis 22). Moloch, to whom children were sacrificed, was a god of peoples surrounding the Israelites, but child burnings and offerings to Baal occurred also in Jerusalem, as mentioned and protested by Jeremiah (7:31). The titan Cronos of Greek mythology, who had learned that one of his children would some day dethrone him, swallowed five of his children as soon as they were born, until Zeus his sixth child escaped Cronos when his wife Rhea gave him, instead of the child, a great stone wrapped in swaddling clothes. Thus Cronos was thought to be the ancestor of various gods around the Mediterranean to whom children were sacrificed.

In a different civilization and location child sacrifices have been revealed in recent years. At the Sacred Cenote of the Mayas in Mexico, the Well of Sacrifice of Chichén Itzá, were found human bones of about 1000 victims sacrificed to Chac, the god of rain. It was established that the bones were mostly those of children and not of "virgins" as earlier reported. It seems that the "peaceful" Mayas had developed a child-slaughter cult like many other native inhabitants of the Americas (Ediger, 1971) and like many peoples of the Old World.

In Sparta infanticide was provided for in the constitution and was practiced for the good of the commonwealth. The child was property of the state and not of the parents. The father could not dispose of the child. After a child was born, he was taken before the Elders of the City for inspection at a place called Lesche. If the child was free of blemishes, he was pronounced worthy of being raised, and from that moment on he became owner of a part of the state's land. Plutarch described the Spartan upbringing of young men and women, and their educational system that has become proverbial. To have strong offspring and to have sons who assured rest and well-being of the father's soul in the next life, marriage was a necessity. If a widowed father lost his children—no matter what age he had reached—he had to contract a new union with a young wife. If the father's virility was in doubt, the young wife was entrusted to a man equally young, robust, and sane. Their children belonged to the state, which produced, fed, and educated the offspring, and the sons had to take care of the legal father's soul (Martin, 1880). To us it is of greater importance to learn of the treatment of the weaklings. "If the elders found the child puny and ill-shaped, [they] ordered it to be taken to what was called the Apothetae, a sort of chasm under Taygetus; as thinking it neither for the good of the child itself, nor for the public interest that it should be brought up, if it did not from the very outset, appear made to be healthy and vigorous. Upon the same account, the women did not bathe the newborn children with water as is the custom in all other countries but with wine, to prove the temper and complexion of their bodies; from a notion they had that epileptic and weakly children faint and waste away upon their being thus bathed, while on the contrary, those of a strong and vigorous habit acquire firmness and get a temper by it, like steel" (Plutarch). Plutarch did not describe comparable procedures in ancient Athens, but Martin (1880) thought that malformed children there were not treated better than in Sparta.

In Rome, teratologic laws show that the legal minds of that nation neglected no subject matter that could be legislated. Martin (1880) tells us that Romulus forbade infanticide of boys or of first-born girls unless they had a congenital mutilation. But even when the child was born defective, the parents were not allowed to dispose of it until five neighbors had given their assent. The laws of the Twelve Tables (about 450 B. C.) dealt with monsters and *permitted* the killing of a monstrous child immediately. But according to the code of the decemvirs the deformed *had to be sacrificed,* and it was not necessary to obtain consent of neighbors. Since the father had the absolute right to dispose of the child there may have been some abuse, and sometimes infants were killed that had no evidence of deformity. There is agreement that hermaphrodites were drowned in the Tiber or burned. But even this dreaded anomaly, which in early times foretold disaster, was later considered merely a form of entertainment of Nature (Cicero). There was a slow change from the pitiless period of the Roman law when "portent" became a term for a minor anomaly, whereas "monster" referred to a being that had no human form. Only the latter was deprived of a personality and treated accordingly (Martin, 1880).

The fate of the mother who had given birth to a monstrous child depended on the explanation of the event by the populace. When such a delivery coincided with a public calamity, the mother also was threatened in Greece and in Rome, because the unusual birth was a sign of the gods' anger. Sometimes the woman was stoned by the irate mob to atone for her guilt in connection with the recent disaster, or she had to leave her home, to return when the public wrath had subsided (Martin, 1880). Other theories of the origin of malformations were even more dangerous to the mother. When it was believed that deformities were due to intercourse between a woman and an animal, the mother's life was clearly endangered (p. 28).

All these reactions show that man insists upon an explanation of unusual events such as births of monstrous children. This attitude was present in the past and continues today. Many innocent persons became victims of the false beliefs of former times. Now things have changed. As Brent (1967) has pointed out, in recent times, when injustice occurs, it tends to favor the malformed or their parents. When a family is afflicted by the birth of an abnormal child, there still exists a tendency to blame a person for the misfortune. The mothers or the fathers sometimes feel guilty and blame themselves for the misfortune, but occasionally parents accuse other persons or institutions. If no rational and specific cause can be found, it is often difficult to refute the parents' accusation. Modern teratology has discovered many factors which can be responsible for congenital malformations, some preconceptional and others postconceptional. It is the latter that are best understood by the lay public, and they are often suspected by the parents or their friends. Sometimes these suspicions are justified but much more often they are not. Since environmental events during pregnancy are always complex there is much room for argument and litigation. There are now legal disputes about the right to sue. If the malformed child survives, it can file suit for pain and suffering, and the parents or siblings have the same right. If the malformed infant does not survive, suit can be filed for the estate of the dead in case the fetus was once alive; an abortus cannot sue. It is up to the court to decide whether the fetus was alive or not (Brent, 1967). This question of liveborn vs. stillborn deformed infants has been debated since antiquity. From Ulpian to Justinian the role of deformed children had to be considered in the laws of inheritance because an infant that had breathed was sometimes counted differently from a stillborn (Martin, 1880); and this could decide who was in line to succeed to a hereditary rank or heritage.

In litigation, responsibility or negligence can now be claimed against a physician, a hospital, a pharmaceutical firm, or a lay person who allegedly caused some injury to the mother during pregnancy. It is conceivable that a malformed person can hold his parents legally responsible for inherited or other congenital defects. Many of these questions of medicolegal aspects of teratology have been ably discussed by Brent (1967), who pointed out the difficulties encountered in this field and the injustices caused by scientifically unsound litigations. As in the past, emotions, rather than reason, often decide the case. Legal decisions can be improved by progress in etiologic research but, since causation is always complex, judgments of responsibility will often remain dubious. The most interesting point is a comparison of ancient concepts to those of our time. Whereas in the past innocent parents of deformed children were threatened, now innocent bystanders are sometimes held responsible for an occurrence the cause of which is obscure. But parallels can probably be found in the Middle Ages when innocent neighbors were accused of witchcraft or magic when a "monstrous" child was born (p. 29).

The right to life of fetuses and children with severe physical and mental defects is under serious discussion today as it was in antiquity. Some ethical questions connected with surgical treatment of conjoined twins have been discussed previously. Medical progress, which makes it possible to preserve the life of many children with serious birth defects, poses questions not faced by past generations. Many congenital malformations that were lethal in the past can now be repaired surgically or alleviated by medical procedures. Complicating factors, such as infections which previously eliminated defective children in early years, can now be combated by antibiotic treatment so that the life-span of the affected population is extended. Remarkable successes in correction of birth defects have been achieved by making crippled and handicapped persons useful and functional members of society. But these successes are not complete or universal. There are children born with multiple malformations that are correctable in part but not *in toto*. And there are malformations that are associated with mental defects which are not improved by repair of the physical handicaps. Thus the medical triumphs have resulted in serious questions as the population of handicapped persons grows. Survival of defective children imposes serious responsibilities upon the immediate family and the surrounding community. Normal members of the family and of the community are often put at a disadvantage by the existence of the abnormal child in their midst. Children who are made to survive sometimes suffer because of their anomalies, and their prolonged lives are not always happy ones. Situations like those encountered before with conjoined twins are now experienced daily on infants' wards where physicians, obligated to preserve life by modern methods of treatment, must face parents who will have to live with children who cannot develop properly and without pain.

The situation is further complicated by the ability to diagnose many severe defects before the child is born. Amniocentesis and other procedures available to obstetricians can reveal anomalies of the embryo and fetus months before delivery. Use of prenatal diagnostic findings in decisions on elective abortion is controversial (Fletcher, 1975) on legal and ethical grounds, and those who consider such abortions acceptable are faced with the subsequent question: Do the ethical principles which permit abortion of a defective unborn also permit active or passive euthanasia of defective newborns? Where is the borderline between "genetically indicated" abortion and infanticide (Fletcher, 1975)? And where is the borderline between elimination of genetically undesirable embryos or children and murder of genetically undesirable peoples as practiced by the national socialist government that ruled Germany from 1933 to 1945? Science cannot answer these moral questions which exist today as they existed in the dim past of mankind, in antiquity, and during the Middle Ages. "Enlightened" as man is in regard to questions of etiology, pathogenesis, and treatment of congenital malformations, he does not have simple answers to the ethical questions raised by the birth of children with severe, irreparable anomalies.

REFERENCES

Ahlfeld, F., 1880–1882, *Die Missbildungen des Menschen*, F. W. Grunow, Leipzig.

Aird, L., 1959, Conjoined twins—further observations, *Br. Med. J.* **1:**1313.

Aldrovandus, U., 1642, *Monstrorum Historia*, Bononiae.

Aristotle, 1943, *Generation of Animals*, Harvard Univ. Press, Cambridge, Massachusetts (transl. by A. L. Peck).

Augustine, A., Saint, 1950, *The City of God*, Modern Library, New York.

Baldwin, M., and Dekaban, A., 1959, The surgical separations of Siamese twins conjoined by the head (cephalopagus frontalis) followed by normal development, *J. Neurol. Neurosurg. Psychiatry* **21:**195.

Ballantyne, J. W., 1902, *Manual of Antenatal Pathology and Hygiene. The Foetus*, W. Green and Sons, Edinburgh.

Ballantyne, J. W., 1904, *Manual of Antenatal Pathology and Hygiene. The Embryo*, W. Green and

Bartlett, R. C., 1959, Cephalothoracopagus. Report of a case, *A.M.A. Arch. Pathol.* **68:**292.

Bergsma, D. (ed.), 1967, *Conjoined Twins*, Birth Defects Original Article Series, Vol. III, National Foundation—March of Dimes, New York.

Besse, H., 1874, *Diploteratology*, Gazette Steam Book and Job Office, Delaware, Ohio.

Brent, R. L., 1967, Medicolegal aspects of teratology, *J. Pediatr.* **71:**288.

Brodsky, I., 1943, Congenital abnormalities, teratology and embryology: Some evidence of primitive man's knowledge as expressed in art and lore in Oceania, *Med. J. Aust.* **1:**417.

Burnet, J., 1930, *Early Greek Philosophy*, A. and C. Black, London.

Cicero, M. T., 1938, *De Senectute, de Amicitia, de Divinatione*, W. Heinemann, Ltd., London, and Harvard Univ. Press, Cambridge, Massachusetts (transl. by W. A. Falconer).

Dareste, C., 1891, *Production expérimentale des monstruosités*, C. Reinwald et Cie, Paris.

Dennefeld, L., 1914, *Babylonisch-Assyrische Geburts-Omina*, J. C. Hinrichsche Buchandl., Leipzig.

Ediger, D., 1971, *Well of Sacrifice*, Doubleday, New York.

Encyclopedia Britannica, 11th ed, 1910, Vol. II, p. 491, Vol. XXI, p. 841, University Press, Cambridge, England.

Esser, A. A. M., 1927, Kyklopenauge, Kyklopen und Arimasper, *Klin. Monatsbl. Augenheilkd.* **79**:398.

Farabee, W.-C., 1905, Inheritance of digital malformations in man, *Pap. Peabody Mus. Harvard Univ.* **3**:65.

Féré, C., 1893–1901, Notes, *C.R. Soc. Biol.*

Fletcher, J., 1975, Abortion, euthanasia, and care of defective newborns, *N. Engl. J. Med.* **292**:75.

Förster, A., 1865, *Die Missbildungen des Menschen. Systematisch Dargestellt*, Druck und Verlag von Friedrich Mauke, Jena.

Franklin, A. W., Tomkinson, J. S., and Williams, E. R., 1958, A triplet pregnancy with craniopagus twins, *Lancet* **1**:683.

Gesenius, V. H., 1951, Missgeburten im Wechsel der Jahrhunderte, *Berl. Med. Z.* **2**:359.

Gould, G. M., and Pyle, W. L., 1897, *Anomalies and Curiosities of Medicine*, W. B. Saunders, Philadelphia.

Gruber, G. B., 1964a, Studien zur Historik der Teratologie, *Zentralbl. Allg. Pathol.* **105**:219.

Gruber, G. B., 1964b, Studien zur Historik der Teratologie, *Zentralbl. Allg. Pathol.* **106**:512.

Gütt, A., Rüdin, E., and Ruttke, F., 1936, *Gesetz zur Verhütung Erbkranken Nachwuchses vom 14. Juli 1933 nebst Ausführungsverordnungen*, J. F. Lehmanns Verlag, München.

Hall, K. D., Merzig, J., and Norris, F. H., 1957, Separation of craniopagus: Case report, *Anesthesiology* **18**:908.

Harvey, W., 1651, *Exercitationes de Generatione Animalium*, Typis Du-Gardianis; impensis Octaviani, Pulleyn in Coemetario, Paulino, Amsterdam and London.

Herodotus, 1961, *The Histories*, Penguin Books, Baltimore, Maryland (transl. by A. de Sélicourt).

Hoadly, C. J. (ed.), 1857, *Records of the Colony and Plantation of New Have, from 1638 to 1649*, Case, Tiffany and Co., Hartford, Connecticut.

Hollaender, E., 1922, *Wunder, Wundergeburt und Wundergestalt*, Verlag Von Ferdinand Enke, Stuttgart.

Holm, H. H., 1936, Siamese twins, report of delivery and successful operation, *Minn. Med.* **19**:740.

Iltis, Hl, 1932, *Life of Mendel*, W. W. Norton, New York.

Johnston, F. E., 1963, Achondroplastic dwarfs through history, *Clin. Pediatr.* **2**:703.

Kemp, T., 1951, *Genetics and Disease*, Ejnar Munksgaard, Copenhagen.

Kiesewetter, W. B., 1966, Surgery on conjoined twins, *Surgery* **59**:860.

Koop, C. E., 1961, The successful separation of pygopagus twins, *Surgery* **49**:271.

Landauer, W., 1962, Hybridization between animals and man as a cause of congenital malformations, *Arch. Anat. Histol. Embryol.* **44**:155.

Leiter, K., 1932, Ein Craniopagus parietalis vivens, *Zentralbl. Gynaekol.* **56**:1644.

Luckhardt, A. B., 1941, Report of the autopsy of the Siamese twins together with other interesting information covering their life, *Surg. Gynecol. Obstet.* **72**:116.

Martin, E., 1880, *Histoire des Monstres depuis l'Antiquité jusqu'a Nos Jours*, C. Reinwald et Cie, Paris.

McLaren, D. W., 1936, Separations of conjoined twins, *Br. Med. J.* **2**:971.

Meckel, J. F., 1812–1816, *Handbuch der pathologischen Anatomie*, Redum, Leipzig.

Mellaart, J., 1963, Deities and shrines of neolithic Anatolia, *Archaeology* **16**:29.

Mendel, G., 1866, *Versuche an Pflanzen/Hybriden, Verh. Naturforsch. Ver. Brünn* **4**:3.

Mørch, E. T., 1941, *Chondrodystrophic Dwarfs in Denmark*, Ejnar Munksgaard, Copenhagen.

Montaigne, M. E., 1892, *Essays*, A. L. Burt Co., New York (transl. by C. Cotton and W. C. Hazlitt).

Moreau de la Sarthe, L. J., 1808, *Description des Principales Monstruosités dans l'Homme et dans les Animaux, précédéé d'un discours sur la Physiologie et la Classification des Monstres*, Fournier Fréres, Paris.

Pepper, C. K., 1967, Ethical and moral considerations in the separation of conjoined twins, *in: Conjoined Twins*, (D. Bergsma, ed.), pp. 128–134, National Foundation—March of Dimes, New York.

Petersen, W. F., 1946, *Hippocratic Wisdom*, Charles C Thomas, Springfield, Illinois.

Peterson, C. G., and Hill, A. J., 1960, The separation of conjoined thoracopagus twins, *Ann. Surg.* **152**:375.

Plato, 1937, *Dialogues of Plato,* Random House, New York (transl. by B. Jowett).

Pliny (The Elder), 1939, *Natural History,* Vol. II, Book 7, Harvard Univ. Press, Cambridge, Massachusetts (transl. by H. Rackham).

Plutarch, 1934, *Lives of the Noble Grecians and Romans,* Modern Library, New York (transl. by J. Dryden).

Reitman, H., Smith, E. E., and Geller, J. S., 1953, Separation and survival of xiphopagus twins, *J. Am. Med. Assoc.* **153:**1360.

Robertson, E. G., 1953, Craniopagus parietalis, *Arch. Neurol. Psychiatry* **70:**189.

Saint-Hilaire, Etienne G., 1822, *Philosophie Anatomique des Monstruosités Humaines,* Rignaux, Paris.

Saint-Hilaire, Isidore G., 1832, *Histoire Générale et Particulière des Anomalies de l'Organisation chez l'Homme et les Animaux,* J. B. Bailliére et fils, Paris.

Sapadin, A., 1964, On a monstrous birth occurring in the ghetto of Venice, *in: Studies in Bibliography and Booklore,* Vol. 6. pp. 153–158, Library of Hebrew Union College-Jewish Institute of Religion, Cincinnati, Ohio.

Scammon, R. E., 1926, The surgical separation of symmetrical double monsters, *in: Abt's Pediatrics,* Vol. 6, W. B. Saunders, Philadelphia.

Schatz, F., 1901, *Die Griechischen Götter und die Menschlichen Missgeburten,* J. F. Bergmann, Wiesbaden.

Scholefield, B. G., 1929, Case of craniopagus parietalis, *J. Anat.* **63:**384.

Schwalbe, E., 1906, *Die Morphologie der Missbildungen des Menschen und der Tiere,* Parts 1 and 2, Gustav Fischer, Jena.

Spencer, R., 1956, Surgical separation of Siamese twins: Case report, *Surgery* **39:**827.

Stockard, C. R., 1921, Developmental rate and structural expression: An experimental study of twins, "double monsters," and single deformities, *Am. J. Anat.* **28:**115.

Strean, L. P., 1958, *The Birth of Normal Babies,* Twayne Publishers, New York.

Sugar, O., Grossman, H., Greeley, P., and Destro, V., 1953, The Brodie craniopagus twins, *Trans. Am. Neurol. Assoc., 78th Meeting,* pp. 198–199.

Tan, K. L., Tock, E. P., Dawood, M. Y., and Ratnam, S. S., 1971, Conjoined twins in a triplet pregnancy, *Am. J. Dis. Child.* **122:**455.

Taruffi, C., 1881–1894, *Storia della Teratologia,* 8 vols., Regia Typographia, Bologna.

Thompson, C. J. S., 1930, *Mystery and Lore of Monsters,* Williams and Norgate, Ltd., London.

Trew, C. J., 1757, Sistens plura exempla palate deficientis, *in: Nova Acta Physico-Medica Academiae Caesareae Leopoldino-Carolinae,* Norimbergae, cited in Rischbieth's Hare-lip and cleft palate, *in: The Treasury of Human Inheritance,* 1910, Dulan and Co., London.

Voris, H. C., Slaughter, W. B., Christian, J. R., and Cayia, E. R., 1957, Successful separation of craniopagus twins, *J. Neurosurg.* **14:**54.

Vrolik, W., 1849, *Tabulae ad Illustrandam Embryogenesin Hominis et Mammalium,* G. M. P. Londonck, Amstelodami.

Warkany, J., 1959, Congenital malformations in the past, *J. Chronic Dis.* **10:**84.

Warkany, J., 1971, *Congenital Malformations, Notes and Comments,* Year Book Medical Publishers, Chicago.

Warkany, J., and Kalter, H., 1962, Maternal impressions and congenital malformations, *Plast. Reconstr. Surg.* **30:**628.

Weisman, A. I., 1967, *The Weisman Pre-Columbian Medical Miniature Collection,* University of Cincinnati College of Medicine, Cincinnati, Ohio.

Wilder, H. H., 1904, Duplicate twins and double monsters, *Am. J. Anat.* **3:**387.

Williams, C., 1960, *Witchcraft,* Meridan Books, New York.

Wilson, H., 1962, Conjoined twins, *in: Pediatric Surgery,* Vol. 1, Part IV, (C. D. Benson *et al.,* eds.), Year Book Medical Publishers, Chicago.

Current Status of Teratology

2

GENERAL PRINCIPLES AND MECHANISMS
DERIVED FROM ANIMAL STUDIES

JAMES G. WILSON

I. INTRODUCTION

Experimental teratology in the modern sense can be said to have begun in the 1940s when Warkany and his associates (Warkany and Nelson, 1940; Warkany and Schraffenberger, 1947; and others) first forcefully called attention to the fact tht environmental factors such as maternal dietary deficiencies and X-irradiation could adversely affect intrauterine development in mammals. Earlier studies in which amphibian, fish, and chick embryos had been subjected to altered environments had shown these forms also to be quite susceptible to unfavorable influences during development, but experiments of this type were not generally accepted as purporting similar vulnerability for higher animals. It was widely assumed in biology and in medicine that the mammalian embryo developed within the virtually impervious shelter of the uterus and the maternal body where it was protected from extrinsic factors. This view was consistent with the generally held opinion that most aspects of normal as well as abnormal development were genetically determined, after the rediscovery of the Mendelian principles of inheritance at the beginning of the twentieth century. Although there were published accounts that terata had been observed after exposure of pregnant women and animals to ionizing radiations (Goldstein and Murphy, 1929), the fundamental implication that mammalian embryos were susceptible to environmental factors was not immediately grasped.

JAMES G. WILSON · Children's Hospital Research Foundation, Elland and Bethesda Avenues, Cincinnati, Ohio 45229.

In fact, the precise demonstrations by Warkany and his associates that predictable numbers and types of defective offspring could be produced by exposing pregnant mammals to environmental stresses surprisingly did not arouse much concern about risks of that nature to the human embryo. Even the revelation by Gregg (1941) and others that the rubella virus could indeed divert the course of normal development in the human embryo did not immediately alert biomedical scientists to the possibility that other extrinsic agents might also penetrate the protective mechanisms of the mother to damage the conceptus. The real implications of this possibility were realized only after Lenz (1961) and McBride (1961) related the appearance of an epidemic of limb-reduction malformations in newborn infants to the taking by the pregnant mothers of a presumably harmless sedative–hypnotic drug, thalidomide.

Thus, despite its cost in human suffering, the thalidomide catastrophe can be regarded as having long-range benefits for mankind. It called attention to the fact, as had no other prior event, that human and other mammalian embryos can be highly vulnerable to certain environmental agents even though these have negligible or no toxic effects in postnatal individuals. This realization has, of course, given great impetus to the science of teratology; but it has also spread wavelike through other scientific disciplines, through the chemical and pharmaceutical industries, and into many governmental regulatory agencies. It has resulted in the formulation of safety evaluation procedures to estimate risks to the unborn where none existed previously. It has raised questions about possible adverse effects on intrauterine development from new environmental changes, as well as from conditions in man's surroundings that have long been accepted without concern. It has even introduced caution into what at times has surely been indiscriminate use of medication during human pregnancy.

To say that the thalidomide catastrophe was a blessing in disguise would be callous as well as trite, but it has undeniably heightened awareness of the vulnerability of the intrauterine occupant to outside influences. A natural by-product has been a tremendous increase in the number of animal studies designed to reveal the effects of a great variety of environmental stresses, excesses, and deficiencies during pregnancy. This is not to say that such studies were not done before thalidomide. A sizable literature on experimental teratology in mammals already existed (see reviews by Giroud and Tuchmann-Duplessis, 1962; Kalter and Warkany, 1959; Wilson, 1959); but the number of papers published annually has more than doubled since 1961 (Larsson, 1974).

This accumulated literature on experimental teratology represents a considerable resource of information and experience. It is no longer possible to review it in any comprehensive way, and if it were, the practical value of an overall review is questionable. Regardless of the welter of detail which comprises this literature, however, undoubtedly there are within it important generalizations which can be distilled out. Accordingly a list of five such generalizations was tentatively formulated as early as 1959 (Wilson, 1959) as "principles" of teratology. Subsequent authors (Nishimura, 1964; Beck and Lloyd, 1965) have accepted these, with some modifications. But as the field of teratology has continued to expand, emphasis has extended the largely theoretical interests that prevailed during the 1950's and early 1960's to a greater concern about human safety. For example, the original list of principles did not contain one dealing with dose–response relationships, but recent events have made mandatory the addition of generalizations covering this practical aspect of the subject (Wilson, 1973a). It is probable that further additions and modifications will be necessary as expansion of the field continues, particularly as the little-studied subject of mechanisms is explored.

At present six teratological generalizations are sufficiently well defined to justly being called principles. These are stated below together with a brief discussion of the main corollary conditions associated with each. It is neither feasible nor desirable in the present context to try to catalog all of the ramifications and exceptions that could be found for each principle. Furthermore, since previous discussions (Wilson, 1973a,b) have attempted to provide full documentation, with many illustrative examples from published literature, references citations have for the most part not been repeated here in the interest of brevity and an easier expository style.

II. GENERAL PRINCIPLES OF TERATOLOGY

1. Susceptibility to teratogenesis depends on the genotype of the conceptus and the manner in which this interacts with environmental factors.

In simple terms, genetic makeup of the developing organism is the setting in which induced teratogenesis occurs, the genes and the extrinsic influences interacting to varying degrees. Differences in the reaction to the same potentially harmful agent by individuals, strains, or species are presumed to depend on variations in biochemical or morphological makeup that are determined by genes. For example, the fact that mouse embryos are usually susceptible to cleft palate induction by glucocorticoids, whereas most other mammalian embryos are resistant to this effect, can be interpreted to mean that mice possess inborn chemical or anatomical features which make them more vulnerable (less resistant) to these agents than are other animals. This susceptibility could relate to such features as the rates at which the hormone is absorbed, eliminated, or transformed by the maternal animal; its rate of passage across the placenta; or the nature of its interactions within the cells and tissues of the embryo. Whatever it is that allows the hormone to get to the mouse embryo in larger amounts, for longer periods, or to have a more disruptive effect, it is at least to some extent genetically determined. Thus, the same sorts of determinants that give individuals, strains, and species their distinctive similarities and dissimilarities in normal structure and function probably also give them varying degrees of susceptibility to adverse influences.

An interesting species-to-species gradation of the factors that determine susceptibility may have been illustrated with thalidomide: Man and other higher primates are extremely sensitive, reacting with a characteristic series of reduction defects of limbs and face; some rabbit and a few mouse strains react in a less characteristic way to much larger dosage, whereas most other mammalian forms are quite resistant. Other intra- as well as interstrain differences in sensitivity to teratogenic agents, not to mention interspecies differences, are so well known as to require no further discussion here. It only need be emphasized that most such differences probably have an hereditary basis, although its precise nature may not be known because the metabolic and structural sites of action of teratogenic agents are often not known.

Individuals carrying mutant genes that destabilize developmental pathways, as postulated by Goldschmidt (1957) and Landauer (1957), would be most easily diverted from the course of normal development by unfavorable environmental influences. This is thought to explain the frequently observed situation in teratology in which a teratogenic agent increases the frequency of a naturally occurring defect above its usual low incidence. It is unlikely, however, that all abnormal development can be explained in this way. Many defects that rarely or never occur spontaneously can be induced in predictable numbers by applying specific tetratogenic agents at critical times in embryogenesis. These can often be shown to be the direct result of specific types of injury, e.g., selective damage to a particular population of cells, or perhaps physical distortion resulting from hemorrhage in an appendage. In such situations vulnerability to extrinsic influences may be more easily related to observable interference with normal embryogenesis, e.g., morphogenetic movements, proliferative rates, etc., than to interaction between genes and outside influences. Thus the proportion between genetic and extrinsic influences which interact to produce a given defect probably varies widely, as Fraser (1959) proposed, ranging from a minority that have major genetic causation, through all degrees of interaction between intrinsic and extrinsic factors, to a minority that have major environmental causation. Many examples of varying degrees of interaction between genetic and environmental influences on development will be given in the chapters in *Causes of Maldevelopment* (Part II of this volume).

2. *Susceptibility to teratogenic agents varies with the developmental stage at the time of exposure.*

It is a basic precept of biology that immature or developing organisms are more susceptible to change than are mature or fully developed ones. This implies that heightened susceptibility extends throughout the developmental span, that is, from fertilization to postnatal maturation, although not necessarily at a constant level. Historically a large share of the total work in experimental teratology has been aimed at studying a known period of high sensitivity, namely, that part of intrauterine development encompassing germ layer formation and organogenesis. It was soon determined that specific organs could often be rendered abnormal by administering a teratogen during or before the early formative stages of that organ, the "critical moments" concept of Stockard (1921), which is today more generally known as "critical periods" in organogenesis. This has subsequently been confirmed to be the time of greatest susceptibility, i.e., when the highest frequency of defects occurs, as far as the induction of gross anatomical defects is concerned, although it is now realized that such defects are only one of several types of deviant development and that other types may have peaks of sensitivity at other times.

The brief survey of the total developmental span which follows attempts to identify times when various types of developmental toxicity can be induced. If development is considered to begin at fertilization, the first logical subdivision is that beginning with cleavage of blastomeres and continuing through completion of blastocyst. Cleavage involves little morphological differentiation of cells except that as numbers increase, relationships to the surface change, i.e., some cells retain a position at the external surface of the morula while others are removed from it to varying degrees by proliferation. After an interior surface in the form of the segmentation cavity or blastocele appears, this permits further positional relationships: Cells that make up the blastocyst wall have exposure on both the interior and exterior surfaces, others remain clustered in the inner cell mass with no free surface, while others are exposed on one or the other of the surfaces. Alterations in these positional relationships are not known to play any part in teratogenesis, although it can be surmised that the type of surface exposure could affect susceptibility to chemical or physical agents in the vicinity of the early embryo and hence lead to selective damage to select groups of cells.

Since mammalian blastomeres probably have not undergone true differentiation, however, except in the positional sense described above, it is doubtful that they have achieved specific developmental roles, and injury to one or a few would not be expected to lead to specific developmental deviations. If all cells within early embryos, blastocysts, have the same metabolic requirements, localized damage would most likely reflect only surface relationships. In fact, adverse influences applied during this early predifferentiation stage have repeatedly been reported not to cause localized defects. Typically, if the dosage of a harmful agent at this time is high, it leads to death of the embryo, otherwise the embryo survives with no greater abnormality than a slight delay in the overall developmental schedule.

Embryonic differentiation, in the sense of determination that a particular group of cells will have a specific role in subsequent development, is of crucial importance in teratogenesis. This is first indicated by the appearance of the embryonic germ layers, ectoderm, entoderm, and mesoderm, a gross segregation of cells into groups with vaguely defined but different potentialities in future development. In lower forms such as chick, regions of the three-layered embryonic disk can be identified with future organs, e.g., presumptive heart, brain, or eye areas, although damage to presumptive organ regions are not necessarily later represented by circumscribed defects of that organ. In mammals the association of early germ layers with future organs is less definite than in chicks, although certain broad generalities can be made. The central nervous system is derived from the ectoderm and the cardiovascular system from the mesoderm. Nevertheless, it is toward the end of the germ-layer stage that specific organ differentiation begins. Even before structural primordia are apparent, experimental embryology and teratology have demonstrated that chemical differentiation (induction) has begun because removal of or damage to localized regions at this time may later be associated with specific organ defects.

The nature of induction or early chemical determination is still not understood, after more than half a century of study. It is, nevertheless, known to precede structural differentiation by several hours to a few days. Whether differentiation is at the chemical or structural level is largely an academic issue because once groups of cells are destined to have specific roles in organ formation, whether visibly different from other cells or not, such cells probably acquire metabolic individuality. It is assumed that the more specialized the metabolic requirements of differentiated cells, the more liable they are to have specific sensitivities to deprivation or damage. Although the association between assigned developmental role and special sensitivity is largely conjectural, it has been repeatedly demonstrated in experimental teratology that localized defects cannot be predictably induced until tissue and organ differentiation has begun.

Thus the period of organogenesis, characterized by the segregation of cell groups and tissues into primordia that will form future organs, is a time of particular vulnerability for induction of structural defects in organs and systems. The early events in organ formation have generally been found to be more sensitive to interference from extraneous influences than later ones, but "critical periods" are known to occur during relatively late stages in the formation of some organs and systems. As the more basic organizational patterns of organ formation are completed, the likelihood of major structural deviation naturally diminishes, and when definitive form and relationships are achieved, maldevelopment in the sense of gross defects is no longer of concern.

The gross modeling that occurs during organogenesis, however, does not equip the organ to assume its ultimate functional role. The processes of cellular and tissue differentiation continue on more refined scales through histogenesis. In other words, histological organization must now proceed so that some elements will become stroma for mechanical support, while others form parenchyma to fulfill specialized functional needs. Histogenesis generally begins before organogenesis is completed and continues well into the subsequent growth phase of most organs. It is hardly completed in the human kidney at birth, and in brain and certain endocrine organs it is not completed until months or years after birth. Harmful influences during histogenesis would not result in gross malformations but could certainly result in finer structural defects at the microscopic level.

In close sequence with the progress of histogenesis is the acquistion of function. In fact a level of functional activity begins in many organs while histogenesis is in progress, but definitive function is usually not achieved until histological differentiation is completed. It is thus not always possible to determine whether an agent causing functional abnormality has acted by interfering with ongoing histogenesis or has prevented the final step in functional maturation. For example, diabetes may result from a deficiency of islet tissue in the pancreas or from reduced secretion of insulin by what appears to be ample numbers of islet cells. In any event, failure to attain full functional competence may be as much a developmental abnormality as is the gross structural defects that are traditionally thought of as terata, because they may be equally the result of interference with developmental processes.

In the overall developmental span birth is hardly more than a landmark. It does not represent a sharp boundary between periods or prenatal susceptibility and postnatal refractoriness, as is sometimes assumed. There is little evidence that systems undergo significant change in their finer functional activities at birth, although gross rearrangements do occur in the heart and lungs to accommodate respiration by breathing air. The processes of growth, histogenesis, and functional maturation continue with only momentary, if any, interruption in transition from the fetal to the infantile periods at birth. The infant, however, may for a time be more vulnerable because he is more directly exposed to environmental changes than was the fetus.

Structural and functional maturation continue after birth in several systems. Myelinization and glial proliferation in the nervous system continue for approximately two years after birth. Attainment of immunologic competence is probably accelerated soon after birth, but this can be attributed to the greater degree and variety of challenges presented by extrauterine life, and it does continue to undergo adjustments throughout most of childhood. The postnatal development of general metabolic capabilities, in terms of detoxification of foreign chemicals, has not been extensively studied in human infants, but it is known to be incomplete at birth. This was illustrated a few years ago by the "gray-baby syndrome" which resulted from inability of newborn infants to conjugate the antibiotic chloramphenicol as do older children. Endocrine functions, particularly those associated with sex and somatic growth, do not become stabilized until after puberty. The extent to which these and other late-maturing functions remain vulnerable in infancy and childhood has by no means been fully evaluated, but there is growing concern that postnatal development can be diverted in significant ways by adverse environmental factors. For example, grossly inadequate nutrition during infancy may stunt later mental as well as physical development in animals and in man (see Chapter 7).

3.   Teratogenic agents act in specific ways (mechanisms) on developing cells and tissues to initiate abnormal embryogenesis (pathogenesis).

Early studies in experimental teratogenesis were often thought to show that particular agents produced characteristic types of malformations. Such associations of patterns of defects with causative factors were sometimes referred to as "agent specificity," implying that each agent had the capability to elicit specific defects. As more and more causative factors were described, however, it became apparent that the composition of the patterns of defects overlapped, i.e., that different agents often produced some of the same types of defects, or the same range of defects in different proportions, or sometimes even the same defects in similar incidences. If each causative factor did not produce a distinctive array of defects, there must be some commonality in the developmental pathways between induction, by whatever means, and the final expression of abnormality. With this in mind, much of the available teratological literature was examined for clues as to how causative agents might act upon developing cells or tissues to initiate abnormal development. These initial effects were designed as *mechanisms* because they were thought to some degree to be determinative of subsequent events.

The term "mechanisms" requires definition because it is variously applied to any or all of a series of events intervening between a cause and an effect. Here the term is restricted to the *early*, if not the *first*, events in such a series. The initial event is possibly the most imporant of the series, not only because it is the connecting link between the cause of the several pathogenetic changes which follow, but also because it very likely influences the nature of these later changes. Athough to assign major importance to the initial change seems justified on logical grounds, it cannot be proven until all events in pathogenesis are known. Since relatively little is presently known about the earliest reactions to most teratogenic causes, the designation of any as being more significant than the others is largely empirical. Furthermore, available information indicates that many of the early changes induced in developing systems are at the molecular or subcellular level and consequently are difficult to discern by present methods.

Nevertheless, evidence can be assembled from the experimental literature to support the existence of some nine ways or mechanisms by which teratogenic causes seem to impinge on developing systems to initiate abnormal development. These proposed mechanisms are presented in Fig. 1. The evidence, or in some instances logical presumption, favoring each is discussed later in this chapter and elsewhere in this volume. For the time being attention is focused on the general concept that unfavorable factors in the environment

Mechanisms	Pathogenesis	Common pathways	Final Defect
Initial types of changes in developing cells or tissues after teratogenic insult: 1. Mutation (gene) 2. Chromosomal breaks, nondisjunction, etc. 3. Mitotic interference 4. Altered nucleic acid integrity or function 5. Lack of normal precursors, substrates, etc. 6. Altered energy sources 7. Changed membrane characteristics 8. Osmolar imbalance 9. Enzyme inhibition	Ultimately manifested as one or more types of abnormal embryogenesis: 1. Excessive or reduced cell death 2. Failed cell interactions 3. Reduced biosynthesis 4. Impeded morphogenetic movement 5. Mechanical disruption of tissues	1. Too few cells or cell products to effect localized morphogenesis or functional maturation 2. Other imbalances in growth and differentiation	

Fig. 1. Diagram of the successive stages in the pathogenesis of a developmental defect, beginning with the initial types of changes in developing cells or tissues (the mechanism) and continuing to the final defect. One or more mechanisms is initiated by the teratogenic cause from the enviroment. This leads to changes in the developmental system which become manifested as one or more types of abnormal embryogenesis. This in turn leads into pathways that seem often to be characterized by too few cells or cell products to effect morphogenesis or functional maturation, but the suggestion that this is a single or common pathway for all developmental defects is conjecture.

are able to trigger changes in developing cells or tissues so as to alter the course of subsequent development. These changes are not necessarily specific to the type of causative factor. For example, ionizing radiations may initiate mutations, chromosomal aberrations, mitotic interference, or enzyme inhibition, but other teratogens are also known to be capable of causing one or more of these changes. Nutritional deficiency would most often lead to lack of precursors and substrates for biosynthesis, but it could also conceivably result in altered energy sources or in osmolar imbalance. Thus a causative agent may initiate one or more mechanisms, and any one mechanism may be set off by different types of causes.

As already noted, these early changes in a developing system are often not readily apparent because they are at subcellular or molecular levels. To become manifest the postulated mechanisms must lead to grosser, more readily demonstrable cellular and tissue alterations that constitute recognizable events of abnormal embryogenesis. These sequentially constitute the pathogenesis of abnormal development.

Pathogenesis is said to have begun when demonstrable signs of cellular and tissue damage are present in an embryo (Fig. 1). The most frequently observed sign of abnormal embryogenesis is cell death in excess of that normally associated with development (see chapter by Scott, Vol. 2), and some investigators have suggested that the teratogenic process always involves some degree of focal cell necrosis, although this has not always been demonstrated. Cellular interaction in the time-honored sense of induction is an important part of normal embryogenesis; when it fails to occur in proper amount or sequence, it may be equally critical in the pathogenesis of abnormal development (see chapter by Saxén, Vol. 2). The role of reduced biosynthesis of the macromolecules needed in growth and differentiation, e.g., DNA, RNA, proteins, mucopolysaccharides, etc., has not been widely studied, but sufficient information is available to indicate that this may sometimes be a major manifestation of abnormal embryogenesis (see chapter by Ritter, Vol. 2). Impairment of morphogenetic movements, that is, interference with the migration or translocation of cells or groups of cells, has not often been invoked to explain maldevelopment, but a few conspicuous examples are known, most involving failure of groups of neurons to follow a prescribed course of migration (see chapter by Yamada, Vol. 2). Finally, tissues that were initially normal may be secondarily disrupted as a result of incursion of foreign materials or abnormal accumulation of tissue fluids or blood, such as in edema or hemorrhage (see chapter by Grabowski, Vol. 2).

Thus abnormal embryogenesis becomes manifested in these relatively few ways, as a result of the changes wrought in developing cells and tissues by the initiating mechanisms. Is it possible that the processses of pathogenesis are further simplified into one or two common pathways as suggested in Fig. 1? All five types of abnormal embryogenesis could conceivably lead to a situation characterized by too few cells or cell products to effect localized morphogenesis or functional maturation. This possibility becomes more plausible when it is considered that a majority of developmental defects do in fact involve deficiencies of tissue elements or of their biochemical products. Even the final defects which do not overtly involve tissue deficiency may, nevertheless, include in their early pathogenesis a phase in which cell death plays a part, as was recently demonstrated by Scott *et al.* (1975). It was noted that certain cytotoxic agents, which by killing cells and initially causing localized cellular depletion, may ultimately result in a tissue-excess abnormality such as polydactyly. Although this paradoxical situation was not fully resolved, it was hypothesized that cellular depletion was greater in limb bud mesoderm than in ectoderm, resulting in an imbalance in the inductive influence thought to be exerted by the apical ectodermal ridge in determining the number of digital rays in the mesoderm. Whether this explanation is valid in all particulars is incidental; the possibility remains that localized deficiencies of cells may result not only in imbalances between interacting groups of cells but may also be followed by a rebound in proliferative rate. It is, however, premature to postulate that a single common pathway could lead to the broad array of structural and functional defects now known. This does not diminish the likelihood that an ordered sequence of events lies between the initial change induced by an environmental influence, the mechanism, and the end expression of ensuing pathogenesis, the final defect.

4. *The final manifestations of abnormal development are death, malformation, growth retardation, and functional disorder.*

These possible consequences of abnormal development are not equally likely to occur after exposure to adverse influences at all times in the developmental span. Although any one or all could probably be induced by embryotoxic insults of sufficient dosage at times of generally high sensitivity, such as during organogenesis, certain manifestations are most likely to follow exposure at particular stages. As already noted, before differentiation begins the early embryo is refractory to most stimuli; however, if dosage is sufficiently high, death of the conceptus can often be induced before maternal toxicity intercedes. After organogenesis begins malformation of specific organs or systems is produced with relative ease, reflecting special sensitivities and needs of rapidly differentiating and growing tissues. The generally elevated sensitivity of most tissues during organogenesis also makes the embryo vulnerable to death. Furthermore, a general cell necrosis less than the critical amount causing death of the embryo could slow the overall rate of growth proportionate to the time required to replenish the lost cells; hence some general growth retardation might also occur. Functional deficit would not ordinarily be expected following damage inflicted during organogenesis *per se* because histogenesis and functional maturation are not conspicuous aspects of development at this time. Nevertheless, such has been reported (Spyker, 1975), and it could be explained as being the result of unreplenished cellular depletion or other derangements in tissues critical to later functions.

Growth retardation in particular would be expected after damage inflicted during the fetal period which is overtly characterized by growth. Less conspicuously, but of equal or greater importance for postnatal life, the fetal period is the time of greatest progress in histogenesis and functional maturation. Accordingly, unfavorable factors at this time are likely to result in structural defects at the tissue level, or in functional deficits, or both. Since relatively little histogenesis continues beyond birth, structural defects even at the histological level are not likely to be induced after birth. Certain aspects of functional maturation, however, can be interfered with throughout infancy and into childhood (see principle 3).

Superficially these manifestations appear to be discrete, but in fact they may be interrelated to varying degrees. Death of the organism is seemingly unequivocal, but when it occurs in an embryo or fetus it often is associated with and may be the consequence of severe malformation, overall cessation of growth, or general breakdown of essential functions. Growth retardation at first thought might be attributed simply to slowed proliferation, but considering the complex metabolic and transport activities essential to support normal growth, failure to grow could reflect a variety of functional deficits. The dependence of normal function on gross as well as minute structural integrity has been emphasized, but it should also be recalled that functions may be interdependent, and failure of one may result in breakdown of another.

Of the four different manifestations of abnormal development, malformation may be the best understood, in terms of types of causation, time of induction, and perhaps also of developmental history, i.e., its pathogenesis. Even this most apparent sequence of events between cause and final effect in malformation, however, is still not simple in most instances, contrary to the oversimplification depicted in Fig. 1. For a further example, from the reverse point of view, some cardiovascular defects appear to involve three or four different developmental errors (as in Tetralogy of Fallot) but in fact represent a collection of secondary and tertiary adjustments, some structural, others hemodynamic, to what was originally a single abnormality (pulmonary stenosis).

Whatever the differences and interrelations between death, malformation, growth retardation, and functional deficit, they can all result from developmental aberrations. There is no logical basis for giving undue emphasis to one over the other in evaluating adverse influences on developmental processes. Traditionally malformation has been used as the main criterion in estimating adverse effects, probably because structural defects were more conspicuous, but this does not justify ignoring changes in mortality, growth rate, or functional capacity. Instances will be cited in subsequent chapters in which the main, sometimes the only, sign of aberrant development is not morphological. Therefore, no estimation of adverse effects on development can be considered complete unless it includes all manifestations of abnormal development.

5. *The access of adverse environmental influences to developing tissues depends on the nature of the influences (agent).*

Basically there are two routes of access by which environmental influences may reach developing tissues *in utero:* by directly traversing the maternal body, or by being indirectly transmitted through it. Examples of agents that pass directly through maternal tissues without modification, except dosage reduction, are ionizing radiations, certain microwaves, and ultrasound. Of these only ionizing radiations appear to have sufficient selectivity of action to cause localized damage so that the embryo or fetus can survive in spite of the damage (see Chapter 5). Other physical agents such as extremes of heat and cold are not directly transmitted to the conceptus because the homeostatic processes of the maternal body are, at least initially, able to counteract the effects. Mechanical impact, short of major trauma, also is to a large extent absorbed by the maternal body and the hydrostatic cushions provided by the chorion and amnion.

Chemical agents or their degradation products usually reach the embryo or fetus in some fraction of their concentration in maternal blood when this is more than negligible. Whether or not they reach effective concentrations in the conceptus depends upon many sets of variables. The first of these are maternal dosage and rate of absorption into the blood stream, which in turn depend upon physical form of the agent (whether solid, liquid, or gas), route of entry, etc. Dosage and absorption sufficiently high to result in significant entry into maternal blood do not necessarily mean that the conceptus will receive a proportionately high dosage. Several homeostatic devices in the maternal organism promptly begin to reduce the blood concentration, e.g., by excreting, detoxifying, or storing the compound so that the number of molecules free in the plasma and therefore free to cross the placenta are considerably less than if they accumulated.

The placenta is often given credit for serving as a barrier behind which the embryo or fetus is protected from foreign chemicals. Actually this is not the case, as evidence is accumulating to indicate that virtually all unbound chemicals in maternal plasma have access to the conceptus across the placenta. Many molecules of small size (less than 600 mol. wt.) and low ionic charge cross by simple diffusion, others by facilitated diffusion, active transport, pinocytosis, or perhaps also by leakage. Lipophilic chemicals are known to cross the placenta and other membranes more readily than other compounds. The question then is not whether a given molecule or ion crosses the placenta but at what rate, as determined by its size, charge, lipid solubility, affinity to complex with other chemicals, etc.

The total dose of a chemical reaching the conceptus is a product of the interaction of many variables, some relating to maternal functional capacity, others dependent on the nature of the chemical itself, and yet others undoubtedly reflecting little understood characteristics of the placenta. No equation has yet been devised to express these complex interactions, in part because all of the variables are not known, much less readily measured. What constitutes an effective dose for the embryo is equally uncertain. As will be discussed under principle 6, the embryo probably has a threshold for most chemical agents, that is, a dose below which no effects occur and above which persistent changes may be induced. Little is known about how an effective dose has to be delivered to the embryo to be effective, e.g., which is more important in producing embryotoxicity—peak concentration or duration of concentration above a given level?

Infectious agents that are teratogenic (see Chapter 6) also must enter the conceptus to be effective, presumably by crossing the placenta, although little is known about their manner of crossing. Most teratogenic infections are viruses indigenous to man and other mammals, and those which have been shown to be teratogenic seem not to be the more virulent strains. It may be that virulent strains reaching the conceptus cause prompt death of the embryo, while less virulent strains are able to cause focal damage but permit the embryo to survive with localized defects. The conditions controlling access of viral infections to the conceptus are much in need of study.

6. Manifestations of deviant development increase in degree as dosage increases from the no-effect to the totally lethal level.

This principle concerns dosage effects generally but is of critical importance to the current arguments about the existence of thresholds for various toxicologic effects. In the safety evaluation of new drugs and chemicals a major policy decision hinges on whether a range of dosage exists below which no adverse effect occurs. If it does, it permits compounds which may be deleterious at high dosage to be used with relative impunity at subthreshold dosage. If it does not, there is no entirely safe level, for as dosages decreases the probability of adverse effect also decreases, but theoretically does not reach zero until dosage is nil.

In typical teratological studies in which intrauterine death and malformation were the criteria of adverse effect, a no-effect level has usually been found when a suitable range of dosage has been used. The same seems to hold true as regards growth retardation, although here the matter has been less extensively explored. The only embryotoxic manifestation about which uncertainty exists is postnatal functional abnormality; this uncertainty is owing to lack of information rather than to evidence refuting the existence of a threshold. Thus, past experience and present indications are that all manifestations of abnormal development begin to be expressed only when dosage exceeds a demonstrable threshold or, conversely, that a lower range of dosage exists at which no embryotoxic effects occur. Different thresholds have been shown to exist for different types of embryotoxicity, even when caused by the same agent. In fact, a low threshold for one manifestation, e.g., embryolethality during early stages, might tend to preclude the recognition of other embryotoxicity; or postnatal functional deficit might occasionally occur at low doses causing no prior developmental abnormality.

The unfounded assumption that carcinogenicity and mutagenicity bear more or less close similarity to teratogenicity has resulted in erroneous assumptions about dose–response relationships in the latter. Mutations and cancers often appeared to have a straight-line dose–response relation, i.e., incidence is proportional to dosage at all levels above zero, although this also has been debated. Nevertheless, if there are no-effect levels for carcinogenesis and mutagenesis, they are sufficiently low to make the matter arguable, which seems not to be the case for teratogenesis. A logical explanation for this apparent difference may relate to the number of cells needed to initiate each of these three pathologies: Carcinogenesis and mutagenesis depend basically on alteration of a single cell from which similarly altered cells are proliferated; teratogenesis seems usually to depend upon damage, perhaps death, to a critical number of cells in a population of many destined to form a tissue, organ, or organism. Many situations are known in which damage to a moderate number of cells by low dosage with a teratogen produces no persistent effect, whereas damage to larger numbers of cells by larger dosage leads to embryotoxicity (see by Scott and Ritter, Vol. 2). Teratogenesis in simplest terms depends upon destruction of a critical number of cells in excess of that which the embryo is able to restore by later proliferation. Thus, the repair and regenerative capacities of an organized, multicellular embryo appear to be greater than those of the individual cells on which the induction of carcinogenesis and mutagenesis depends.

A practical consideration in the dose–response relationships of teratogenesis requires special emphasis. It has become axiomatic in experimental teratology that agents capable of causing any adverse biological effects can usually also be shown to be embryotoxic under the right conditions of dose, developmental stage, and species susceptibility, unless maternal toxicity intercedes. Whether this generalization is true in every instance may be debatable, but the fact remains that virtually all drugs and a great range of chemicals can indeed be shown to be embryotoxic under appropriate laboratory conditions. In view of this, to apply arbitrary rules that would eliminate the use of drugs and chemicals because they can be shown to be embryotoxic at high dosage would be unacceptable. Thus, to apply to the so-called "Delaney clause," which prohibits use of food additives shown to be carcinogenic at any dose, to those that are embryotoxic at any dose would eliminate most drugs and many useful chemicals upon which modern society depends heavily.

Typically the dose–response curve for embryotoxic effects has a steep slope, sometimes going from minimal to maximal effect levels merely by doubling the dose. An exception, however, is the most potent of known human teratogens, thalidomide, which is teratogenic in higher primates at a relatively small dose but can be tolerated without embryolethality at several multiples of this dose. Other potent embryotoxic agents, e.g., actinomycin D, have a very narrow teratogenic range of dosage; they tend to become embryolethal at the same dose that begins to cause malformation. The presence or absence of maternal toxicity is not a reliable indicator of embryotoxicity. Alkylating compounds often cause embryotoxic and maternal toxic effects at similar doses; thalidomide on the other hand is virtually devoid of maternal toxic effects at any dose.

III. MECHANISMS OF TERATOGENESIS

As indicated in principle 3 above, mechanisms are thought to occupy a pivotal position in the series of events between the causative factor in the environment and the ultimate expression of developmental abnormality, the final defect (Fig. 1). They represent the earliest identifiable reaction of the developing system to the environmental cause. They usually are at the subcellular or molecular level, hence are not readily apparent by ordinary means. They lead to the more familiar and demonstrable signs of abnormal embryogenesis (pathogenesis) such as excessive cell death or tissue disruption.

The accumulated literature of experimental embryology and teratology provides evidence and clues that there may be eight or ten such mechanisms by means of which causative agents impinge upon developing cells to change their prescribed course in embryogenesis. Some evidence and logical presumption supporting the existence of the mechanisms identified to date follows. A more extensive review has been presented elsewhere (Wilson, 1973a,b).

A. Mutation

This is probably the most firmly established mechanism of teratogenesis. It is the basis of all heritable developmental defects and consists essentially of a change in the sequence of nucleotides on the DNA molecule. The nature of the change and the number of nucleotides involved varies, as indicated in Fig. 2; but whatever its nature, the end result of the change may be that the information pertaining to development, which is normally encoded in the DNA, will now be erroneously transcribed into RNA, and ultimately into proteins. It is estimated that some 20–30% of human developmental errors can be attributed solely or primarily to mutation in a prior germ line. If the effect is in a germinal cell, the mutation is likely to be hereditary; if it occurs in a somatic cell, it will be transmitted to all descendants of that cell but it cannot be hereditary. When somatic mutations occur in an early embryo, enough

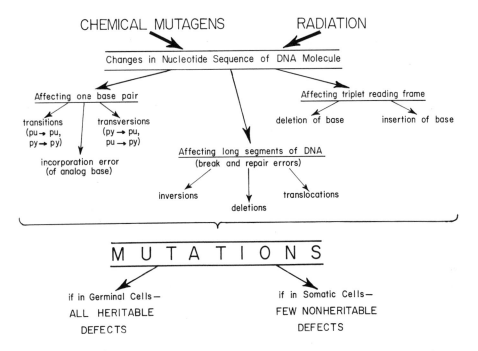

CHEMICAL MUTAGENS   RADIATION

Changes in Nucleotide Sequence of DNA Molecule

Affecting one base pair   Affecting triplet reading frame

transitions   transversions   deletion of base   insertion of base
(pu → pu,   (py → pu,
py → py)   pu → py)

incorporation error   Affecting long segments of DNA
(of analog base)   (break and repair errors)

inversions   translocations

deletions

M U T A T I O N S

if in Germinal Cells—   if in Somatic Cells—
ALL HERITABLE   FEW NONHERITABLE
DEFECTS   DEFECTS

Fig. 2. Illustration of the various ways in which nucleotide sequences in the DNA molecule may be changed in the production of mutations. All of these changes could lead to what are loosely called point mutations, in distinction to chromosomal aberrations which involve whole chromosomes or major parts of a chromatid, although chromosomal abnormalities could also be included under "long segments of DNA." Frame-shift mutations affect one, two, or more than three nucleotides. (From Wilson, 1973.)

progeny cells may be affected to produce a demonstrable structural or functional defect, but it should be emphasized that these are infrequent and random occurrences and certainly cannot explain the specific arrays of defects produced by teratogens that may also be mutagenic. Mutations are caused by ionizing radiations, a number of chemical mutagens such as nitrous acid, alkylating agents, and many carcinogens, and such other factors as might lead to chromosomal breaks or crossovers (Freese, 1971). They may also result from interference with the normal processes of DNA repair.

B. Chromosomal Nondisjunction and Breaks

These give rise to microscopically visible excesses, deficiencies, or rearrangements of chromosomes, chromatids, or parts thereof. They have been associated causally with only a small portion (~3%) of human developmental defects. Their true status as a mechanism is uncertain because the visible changes in the chromosomes are not known to be the first event in response to the environmental stimulus which is the cause. They differ from the point mutations discussed above at least in quantitative terms, and they are not hereditary in the usual sense, although translocations of chromatid parts may be transmitted to half of the offspring. Whole-chromosomal anenploidy originates during meiotic division in the maturation of germ cells or during ordinary mitosis when newly divided chromosomes fail to separate, resulting in nondisjunction which has no counterpart in mutagenesis. Nondisjunction is thought to occur with greater frequency in the germ cells of aged parents and also after aging of germ cells in the genital tract before fertilization. There is also evidence that nondisjunction, as well as other gross chromosomal aberrations, may result from viral infections, irradiation, and certain chemical agents, but much remains to be learned about the causation of all types of chromosomal abnormality (see chapter by Hsu and Hirschhorn, Vol. 2).

Deficiency of whole chromosomes is poorly tolerated and is usually lethal to the cell or the organism, although absence of one sex chromosome in Turner's syndrome is tolerated with relatively slight effect on development. Excess of chromosomes is also usually detrimental. Trisomies of several smaller autosomes are compatible with survival beyond birth, although they are associated with moderate to severe developmental disorders except in the case of sex chromosomes, excess of which usually entails only mild defects (see chapter by Hsu and Hirschhorn, Vol. 2).

C. Mitotic Interference

This designation very likely includes more than one type of primary effect because the mitotic process apparently can be interfered with in a number of ways (Table 1). Many cytotoxic agents are known to act by inhibiting synthesis of DNA, thereby slowing or arresting mitosis, since the process cannot progress beyond the S phase. The mitotic spindle can be prevented from forming, or be dissolved after formation, by several chemical agents which interfere with the polymerization of tubulin into the microtubules of the spindle (Borisy and Taylor, 1967; Malowista et al., 1968). Without a spindle, chromosomes do not separate at anaphase. Finally, even when DNA is synthesized and a spindle is formed, chromosomes may not be able to separate owing to an apparent "stickiness" or physical continuity known as bridges. These conditions occur after high dosage with radiations or radiomimetic chemicals (Hicks and D'Amato, 1966). Since these three different types of interference with mitosis seem highly unlikely to be traceable to a single primary effect, it is probable that more than one mechanism is involved. They are associated here for the arbitrary reason that they have a common visible endpoint, namely mitotic interference.

Table 1. Means of Interference with Mitosis

Apparent mechanisms	Some known causes
1. Slowing or arrest of DNA synthesis	Cytosine arabinoside, hydroxyurea, irradiation
2. Failure of mitotic spindle by preventing microtubule formation	Colchicine, vincristine, griseofulvin, some anesthetics
3. Improper formation or separation of chromatids ("stickiness," bridges, etc.)	Irradiation, radiomimetic chemicals

D. Altered Nucleic Acid Integrity or Function

In addition to mutations, this is the mechanism or mechanisms by which many antibiotic and antineoplastic drugs are teratogenic. Biochemical changes that interfere with nucleic acid replication, transcription, natural base incorporation, or RNA translation (protein synthesis), without producing heritable changes in the DNA of germ cells, are included here. The simplest way to illustrate these alterations in nucleic acid integrity or function is to list some of the chemical agents thought to act in each of the four ways, as presented in Table 2. Agents in each of the categories listed have been demonstrated to be frankly teratogenic in mammals, except the last which concerns interference with information translation in the form of protein synthesis. This in no sense means that protein synthesis is not essential to developmental processes; on the contrary it probably means that any appreciable reduction is incompatible with continued development or with the survival of the embryo or fetus. In fact, several attempts to produce embryotoxicity with the agents listed in item 4 of Table 2 have resulted mainly in embryolethality at higher doses, with some growth retardation but little or no malformation at lesser doses (unpublished data in author's laboratory). Presumably protein synthesis is so essential and so ubiquitous to all embryonic cells that interference cannot be localized sufficiently to produce specific malformation, but instead is generalized and usually causes death.

This is clearly not a well-circumscribed mechanism, in the sense of being a discrete biochemical effect on developing cells, probably ranging from competitive inhibition of enzymes to cross-linking of DNA strands (Fig. 3). Further studies on molecular pathways will be necessary to clarify these relationships. Although some of these agents are known to be mutagenic, certainly not all agents that affect nucleic acid metabolism and function are mutagenic. The question of the extent of overlap between mutagenic and teratogenic chemicals has been critically examined in at least two studies (Kalter, 1971;

Table 2. Types of Interference with Nucleic Acid Metabolism or Function, as Studied Largely in Cell Cultures and Microorganisms[a]

1. Interference with replication or integrity of DNA

Cytosine arabinoside[b]	Inhibits DNA polymerase
Hydroxyurea[b]	Blocks conversion of deoxyribonudeotides to deoxynucleotides
Mitomycin C[b]	Cross-links complementary strands of DNA
Streptonigrin[b]	Selectively binds DNA, inhibits incorporation of adenine
6-Mercaptopurine[b]	Blocks precursors of adenylate and guanylate

2. Interference with RNA synthesis (transcription)

Actinomycin D[b]	Intercalates DNA, binds deoxyguanosine
Nogalamycin	Binds deoxyadenosine and/or thymidine of DNA
Acridine orange	Binds euchromatin DNA, is mutagenic

3. Erroneous incorporation into DNA or RNA

Tubercidin	Acts as analog of adenosine
8-Azaguanine[b]	Acts as analog of guanine
5-Bromouracil	Replaces thymine in synthesis, is mutagenic
5-Fluorouracil[b]	Incorporated mainly in RNA

4. Interference with RNA translation (protein synthesis)

Puromycin	Complexes with and aborts incipient protein
Cycloheximide	Interferes with transfer of tRNA to ribosome
Streptomycin	Binds with and causes misreading of mRNA
Lincomycin	Inhibits tRNA attachment to ribosome

[a]From numerous sources, particularly Balis (1968). In several instances the suggested mode of action is uncertain, although some evidence is available in support of the pathway cited.
[b]Known to be teratogenic in mammals.

Wilson, 1972), and both concurred that there appears to be little overlap, this consisting principally of the polyfunctional alkylating agents and a few antibiotics and antimetabolites. Many of the mechanisms of teratogenesis proposed here simply do not apply to mutagenesis (e.g., several of the succeeding mechanisms discussed in this section have little to do with the sequence of DNA nucleotides).

E. Lack of Precursors and Substrates Needed for Biosynthesis

This is probably one of the better-established mechanisms. The materials essential for biosynthesis and maintenance of growth and differentiation can be withheld from the sites at which they are utilized in the embryo by four means which are summarized in Table 3. One of these, specific dietary deficiency, particularly of vitamins and minerals, has been shown repeatedly over the past 35 years to be teratogenic, embryolethal, and growth inhibiting to the offspring of pregnant mammals (see chapter 7). In fact these studies made up a substantial part of early experimental teratology in mammals. Contrary to the old adage that the embryo lives as a favored parasite upon the body of

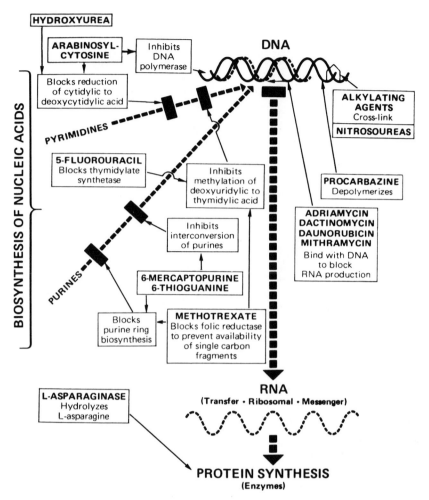

Fig. 3. Pathways of nucleic acid metabolism and function that are thought to be interfered with by several anticancer and antibiotic drugs. (From Irwin H. Krakoff, Memorial Hospital for Cancer and Allied Disease, New York City. Published in *Science* **184,** May 31, 1974.)

the mother, receiving ample supplies of scarce materials at her expense, the embryo has often been shown to manifest deficiency by death or abnormal development before the mother showed serious signs of deprivation.

Deficiency of essential materials in the embryo, however, can occur in spite of adequate supplies of such materials in the maternal diet. The presence of analogs or antagonists to vitamins, essential amino acids, purines, pyrimidines, etc., may result in the utilization of these spurious metabolites in biosynthesis instead of the normal precursors. Although analogs and antagonists may be incorporated at early steps in metabolic pathways, they rarely fulfill the ultimate biological role of the natural metabolite and consequently

**Table 3. Lack of Precursors, Substrates, etc.,
Needed in Biosynthesis**

Cause of deprivation	Examples associated with teratogenesis
1. Specific dietary deficiency	Riboflavin, vitamin A, vitamin E, folic acid, pantothenic acid, Zn, Mg, Mn
2. Specific analogs or antagonists	Purines, pyrimidines, glutamine, adenine, leucine, folic acid, riboflavin, nicotinic acid
3. Failed absorption from maternal gut	Copper due to excess Zn or SO$_3$ (mothers of swayback lambs?); iodine in presence of high Ca (endemic cretinism?)
4. Failure of placental transport caused by	Azo dyes, tissue antisera

often cause abnormal development or death of early embryos (Seegmiller and Runner, 1974).

Even in the event of adequate supplies of normal precursors and substrates in maternal diet, and without intervention of antagonists, normal materials may fail to be absorbed from the maternal digestive tract or to be transported in maternal blood and across the placenta. Well-documented instances of failed absorption as a mechanism of teratogenesis are not known, although possible examples are the situations mentioned in Table 3 (Volpin-testa, 1974).

Failure of placental transport of essential metabolities is now generally accepted as the underlying fault in the teratogenic effects associated with azo dyes (Lloyd et al., 1968) and tissue antisera (Brent, 1971). The physical presence of these substances in the visceral yolk sac epithelium of early rodent embryos is though to interfere with either absorption and/or transfer of metabolites essential for normal embryonic growth; thereby creating conditions tantamount to nutritional deficiency. It should be emphasized that these instances of apparent failed placental transport are known only in rodents and rabbits which are almost unique among mammals in being dependent for early placental transport on the inverted yolk sac placenta. No instance of teratogenesis associated with failed placental transport has been demonstrated in higher mammals.

F. Altered Energy Sources

On logical grounds alone this would have to be postulated as a teratogenic mechanism, in view of the need for uninterrupted high levels of energy by proliferating, rapidly synthesizing tissues. Evidence is now accumulating to indicate that four pathways of energy supply may be interfered with by known teratogens (see chapter by Krowke and Neubert, Vol. 2; Bowman, 1967; Cox and Gunberg, 1972; Mackler, 1969; Netzloff *et al.*, 1968; Shepard *et al.*, 1970). Table 4 shows these pathways and some causes thought possibly to interfere with energy (ATP) generation. It must be emphasized that the establishment of precise cause–effect relationships in this area has been elusive and much research will probably be required to clarify the details of these pathways which at present are to a considerable degree hypothetical as far as teratogenesis is concerned.

G. Enzyme Inhibitions

Enzymatic functions are probably ubiquitous in all aspects of differentiation and growth, hence they would almost certainly be interfered with by factors adversely affecting these developmental processes. The question here is whether enzymes are ever primarily and specifically the sites of teratogenic action. This subject has not been extensively studied, but in a few instances, such as those listed in Table 5, agents shown to be teratogenic in mammals are either known, or can be assumed from studies in other growing systems, to act on specific enzymes (Jaffe, 1974; Johnson, 1974; Runner, 1974). In addition to the enzymes listed in Table 5 which are known to be inhibited by agents that are also known teratogens, there undoubtedly are enzyme inhibitions involved in some of the mechanisms discussed above. For example, the repair of spontaneous or induced breaks in the DNA strands is dependent on specific enzymes and is important in keeping mutations to a minimum. It is now recognized that some mutagenic agents act by inhibiting these repair enzymes (Freese, 1971). Furthermore, the pathway of action for some agents that interfere with formation of the mitotic spindle is thought to be inhibition of the polymerase that converts tubulin into the microtubules to form the

Table 4. Reduced Energy (ATP) Sources in the Embryo

Affected pathways	Causes associated with teratogenesis
1. Inadequate glucose sources	Dietary deficiency, induced hypoglycemia
2. Interference with glycolysis	6-Aminonicotinamide, iodoacetate
3. Interference with critic acid cycle	6-Aminonicotinamide, riboflavin deficiency (w/wo galactoflavin)
4. Impairment of terminal electron transport system	Hypoxia, dinitrophenol, cyanide

**Table 5. Some Specific Enzyme Inhibition Thought
to Be Involved in Teratogenesis**

Enzymes inhibited	Teratogenic agent
Dihydrofolate reductase	Folic acid antagonists
Thymidylate synthetase	5-Fluorouracil
Ribonucleoside diphosphate reductase	Hydroxyurea
Carbonic anhydrase (?)	Acetazolamide
DNA polymerase	Cytosine arabinoside
Glucose-6-P dehydrogenase	6-Aminonicotinamide

astral rays of the spindle. Thus, some of the teratogenic mechanisms discussed above as separate entities may in fact involve enzyme inhibitions as the primary or initiating event in abnormal embryogenesis.

H. Osmolar Imbalance

The action of this teratogenic mechanism is one of the few which can be traced stepwise through pathogenesis, beginning with the first reaction after an environmental cause impinges on an embryo and continuing to the appearance of the final defects. Based on the "edema syndrome" as studied by Grabowski (see chapter by Grabowski, Vol. 2) in chick embryos subjected to hypoxia, it is possible to diagram in some detail the successive events in the teratogenesis of the characteristic head, limb, and rump malformations, as shown in Fig. 4. The first recognized reaction in the embryonic system after the cause is applied in hypoosmolarity in certain extraembryonic compartments and, as defined earlier, this would technically be designated as the *mechanism*. Almost simultaneously, however, would be an inrush of fluid from these compartments into the embryo to result in hypervolemia and increased blood pressure within the embryo. Thus, all of these early fluid–osmolyte changes could be regarded as mechanistic in the sense that they are all necessary to cause the first overt damage in the form of edema, hematomas, and blisters. These in turn lead to other pathogenetic events such as mechanical distortion and ischemia in the tissues, which reaches severe proportions in the extremities and the surface structures of the head where abnormal embryogenesis eventually renders the affected parts into the final defects.

Similar pathogenetic events are thought to follow osmolyte imbalances resulting from such agents as trypan blue, hypertonic solutions, and adrenal cortical hormones. Malformations in the tail and extremities of mice, induced by injecting pregnant females with hypertonic saline solutions, have been attributed to tissue damage following edema, blisters, and hemorrhages (Tanaka *et al.*, 1968), providing a mammalian example of the edema syndrome. Likewise, the orofacial malformations seen in the offspring of rats given benzhydryl piperazine compounds during pregnancy have been attrib-

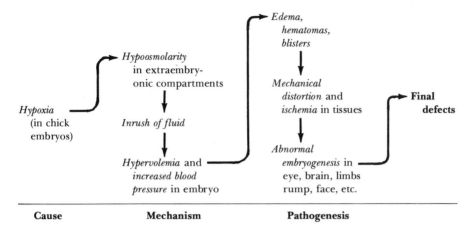

Fig. 4. The "edema syndrome" illustrates the sequence of events between the application of a teratogenic dose of hypoxia (the cause) from the initial changes in the developing organism (the mechanism), through the appearance of various structural derangements (pathogenesis) to the final defect.

uted to the edema known to have occurred in the embryos after treatment (Posner, 1972). Fluid accumulations, hemorrhages, and blisters have often been reported in rodent embryos after trypan blue treatments. Thus there seems little doubt that abnormal fluid accumulations can cause tissue distortions sufficient to lead to malformations. The critical question is whether osmotic disturbances are actually the primary reactions to the environmental cause or secondary to still earlier, unrecognized changes.

I. Altered Membrane Characteristics

Abnormal membrane permeability can lead to osmolar imbalances as described above, but in other cases the altered membrane rather than the fluid compartments which it separates would be the primary site of teratogenic action. Good examples of primary effects on membranes in relation to teratogenesis are not known to the writer, although a few experimental situations are possibly applicable. Solvents such as DMSO have been noted to produce swellings and blisters associated with ionic shifts between compartments in chick embryos, and it was postulated that the solvent had altered the permeability of cell and other membranes (Browne, 1968). Teratogenic doses of vitamin A, leading to hypervitaminosis A, have been reported to cause ultrastructural damage to cellular membranes in rodent embryos (Morriss, 1973; Nakamura *et al.*, 1974). At present, however, altered membrane properties in the embryo can only be regarded theoretically as a teratogenic mechanism. To establish it as a clearcut mechanism will be complicated by the need to distinguish it from osmotic imbalance, which also requires maintenance of prescribed osmotic differences on two sides of a membrane.

IV. WHY BE CONCERNED WITH MECHANISMS?

It is clear from the above that information on teratogenic mechanisms is limited, but it may not be apparent why a knowledge of mechanisms is important. Pragmatists have argued that a machinist is able to repair an engine without understanding the theory behind its operation. But teratology is concerned with more than repair. One of the major objectives is to anticipate risks before they materialize. The anticipation of teratic risks in today's rapidly changing environment becomes an endless succession of screening tests unless a knowledge of mechanisms can lead to extrapolations, generalizations, and shortcuts that will simplify the task. Furthermore, the use of animal tests for evaluation of human risk will become more than empirical only when the degree of comparability of mechanisms between test animal and man is understood. Finally, with a better knowledge of mechanisms, unknown causes may be more easily recognized.

Correction and treatment of developmental defects should not be limited to searching for palliatives at the symptomatic level. Prevention of developmental disease can be attempted at many points prior to their final manifestation, if not by diverting the initiating mechanism, at least by intervention at some point in pathogenesis between the mechanism and the final manifestation. For example, a deficient enzyme may be supplemented, an excessive accumulation of a substrate avoided by dietary restriction, or an inadequate transport system augmented before irreparable damage is done.

The prevention of "spontaneous" defects is said to be a matter that lies in the future (Lederberg, 1970). This is not necessarily so if a defect without apparent single causation can be shown or presumed to result from interacting or multiple causes. There is at present no reason to believe that multiple causes act through mechanisms different from those that are triggered by single causes. If this is the case, better understanding of mechanisms could simplify the complex task of identifying the multiple variables that combine to activate known mechanisms.

REFERENCES

Balis, M. E., 1968, *Antagonists and Nucleic Acids,* North-Holland Publishing, Amsterdam.

Beck, F., and Lloyd, J. B., 1965, Embryological principles of teratogenesis, *in: Embryopathic Activity of Drugs* (J. M. Robson, F. Sullivan, and R. L. Smith, eds.), pp. 1–20, Churchill, London.

Borisy, G. G., and Taylor, E. W., 1967, The mechanism of action of colchicine. I. Binding of colchicine-³H to cellular protein, *J. Cell Biol.* **34**:525–534.

Bowman, P., 1967, The effect of 2,4-dinitrophenol on the development of early chick embryos, *J. Embryol. Exp. Morphal.* **17**:425–431.

Brent, R. L., 1971, Antibodies and malformations, *in: Malformations Congénitales des Mammifères* (H. Tuchmann-Duplessis, ed.), pp. 187–220, Mason et Cie, Paris.

Browne, J. M., 1968, Physiological effects of dimethylsulfoxide (DMSO) on the chick embryo, *Teratology* **1**:212.

Cox, S. J., and Gunberg, D. L., 1972, Energy metabolism in isolated rat embryo hearts: Effect of metabolic inhibitors, *J. Embryol. Exp. Morphal.* **28**:591–599.

Fraser, F. C., 1959, Causes of congenital malformations in man, *J. Chronic Dis.* **10**:97–110.

Freese, E., 1971, Molecular mechanisms of mutations, *in: Chemical Mutagens. Principles and Methods for Their Detection*, Vol. 1 (A. Hollaender, ed.), pp. 1–56, Plenum Press, New York.

Giroud, A., and Tuchmann-Duplessis, H., 1962, Malformations congénitales. Role des facteurs exogènes, *Pathol.-Biol.* **10**:119–151.

Goldschmidt, R. B., 1957 Problematics of the phenomenon of phenocopy, *J. Madras Univ.* **B27**:17–24.

Goldstein, L., and Murphy, D. P., 1929, Etiology of ill-health in children born after maternal pelvic irradiation. II. Defective children born after postconception pelvic irradiation, *Am. J. Roentgenol. Radium Ther. Nucl. Med.* **22**:322–331.

Gregg, N. McA., 1941, Congenital cataract following German measles in the mother, *Trans. Ophthalmol. Soc. Aust.* **3**:35.

Hicks, S. P., and D'Amato, C. J., 1966, Radiosensitivity at various stages of the mitotic cycle and cellular differentiation, *in: Advances in Teratology* (D. H. W. Woollam, ed.), pp. 213–227, Logos Press, London.

Jaffe, N. R., 1974, Specific activity and localization of enzymes associated with osteochondrodysplasia in the developing rat, *Teratology* **9**:A-22.

Johnson, E. M., 1974, Organ specific patters of abnormal molecular differentiation, *Teratology* **9**:A-23–24.

Kalter, H., 1971, Correlations between teratogenic and mutagenic effects of chemicals in mammals, *in: Chemical Mutagens* (A. Hollaender, ed.), pp. 57–82, Plenum Press, New York.

Kalter, H., and Warkany, J., 1959, Experimental production of congenital malformations in mammals by metabolic procedures, *Physiol. Rev.* **39**:69–115.

Landauer, W., 1957, Phenocopies and genotype, with special reference to sporadically-occuring developmental variants, *Am. Nat.* **91**:79–90.

Larsson, K. S., 1974, Data banks, *in: Birth Defects* (A. G. Motulsky and W. W. Lenz, eds.), pp 187–190, Excerpta Medica, Amsterdam.

Lederberg, J., 1970, Genetic engineering and the amelioration of genetic defect, *Bioscience* **20**:1307–1310.

Lenz, W., 1961, Kindliche Missbildungen nach Medikament wahrend der Draviditat?, *Deutsch. Med. Wochenschr.* **86**:2555–2556.

Lloyd, J. B., Beck, F., Griffiths, A., and Parry, L. M., 1968, The mechanism of action of acid bisazo dyes, *in: The Interaction of Drugs and Subcellular Components on Animal Cells* (P. N. Campbell, ed.), pp. 171–202, Churchill, London.

Mackler, B., 1969, Studies of the molecular basis of congenital malformations, *Pediatrics* **43**:915–926.

Malowista, S. E., Sato, H., and Bensch, K. G., 1968, Vinblastine and griseofulvin reversibly disrupt the living mitotic spindle, *Science* **160**:770–771.

McBride, W. G., 1961, Thalidomide and congenital abnormalities, *Lancet* **2**:1358.

Morriss, G. M., 1973, The ultrastructural effects of excess maternal vitamin A on primitive streak stage rat embryo, *J. Embryol. Exp. Morphal.* **30**:219–242.

Nakamura, H., Yamawaki, H., Fujisawa, H., and Yasuda, M., 1974, Effects of maternal hypervitaminosis A upon developing mouse limb buds. II. Electron microscopic investigation, *Cong. Anom.* **14**:271–283.

Netzloff, M. L., Johnson, E. M., and Kaplan, S., 1968, Respiratory changes observed in abnormally developing rat embryos, *Teratology* **1**:375–386.

Nishimura, H., 1964, *Chemistry and Prevention of Congenital Anomalies*, Thomas, Springfield, Illinois.

Posner, H. S., 1972, Significance of cleft palate induced by chemicals in the rat and mouse, *Food Cosmet. Toxicol.* **10**:839–855.

Punner, M. N., 1974, A specific developmental defect related to an ubiquitous enzyme: Induced micromelia in the chick embryo traceable to defects in dehydrogenase and molecular assembly, *in: Congenital Defects—New Directions in Research* (D. T. Janerich, R. G. Skalko, and I. H. Porter, eds.), pp. 275–281, Academic Press, New York.

Scott, W. J., Ritter, E. J., and Wilson, J. G., 1975, Studies on induction of polydactyly in rats with cytosine arabinoside, *Dev. Biol.* **45:**103–111.

Seegmiller, R. E., and Runner, M. N., 1974, Normal incorporation rates for precursors of collagen and mucopolysaccharides during expression of micromelia induced by 6-aminonicotinamide, *J. Embryol. Exp. Morphal.* **31:**305–312.

Shepard, T. H., Tanimura, T., and Robkin, M. A., 1970, Energy metabolism in early mammalian embryos, *Dev. Biol.* **4:**42–58.

Spyker, J. M., 1975, Behavioral teratology and toxicology, *in: Behavioral Toxicology* (B. Weiss and V. G. Laties, eds.), pp. 311–349, Plenum Press, New York.

Stockard, C. R., 1921, Developmental rate and structural expression: An experimental study of twins, "double monsters" and single deformities, and the interaction among embryonic organs during their origin and development, *Am. J. Anat.* **28:**115–227.

Tanaka, S., Ihara, T., and Mizutani, M., 1968, Apical defects in rat fetuses observed after intraperitoneal injection of high concentration of sodium chloride, *Cong. Anom.* **8:**197–209.

Volpintesta, E. J., 1974, Menkes kinky hair syndrome in a black infant, *Am. J. Dis. Child.* **128:**244–246.

Warkany, J., and Nelson, R. C., 1940, Appearance of skeletal abnormalities in the offspring of rats reared on a deficient diet, *Science* **92:**383–384.

Warkany, J., and Schraffenberger, E., 1947, Congenital malformations induced in rats by roentgen rays, *Am. J. Roentgenol. Radium Ther.* **57:**455–463.

Wilson, J. G., 1959, Experimental studies on congenital malformations, *J. Chronic Dis.* **10:**111–130.

Wilson, J. G., 1972, Interrelations between carcinogenicity, mutagenicity, and teratogenicity, *in: Mutagenic Effects of Environmental Contaminants* (H. E. Sutton and M. I. Harris, eds.), pp. 185–195, Academic Press, New York.

Wilson, J. G., 1973a, *Environment and Birth Defects,* Academic Press, New York.

Wilson, J. G., 1973b, Principles of teratology, *in: Pathology of Development* (M. J. Finegold and E. V. Perrin, eds.), pp. 11–30, Williams and Wilkins, Baltimore.

Relation of Animal Studies to the Problem in Man

3

F. CLARKE FRASER

I. INTRODUCTION

By "the problem" in man we refer to the fact that some 3–7% of human babies are born with malformation serious enough to require treatment. Etiologically these can be broken down into five categories: those caused by a mutant gene (roughly 5%), those in which there is evidence for a multifactorial basis (roughly 20%), those associated with a chromosomal aberration, either autosomal (2%) or sex-chromosomal (8%), those caused by an identifiable environmental agent (roughly 5%), and those in which no cause has been identified (roughly 60%).

Since the "unknown" category is the largest, one might reason that the most important problem in man is that of identifying the causal factors in this group, since recognition is usually necessary for prevention. But there is much that needs to be learned also about the nature and mode of action of the known causes of malformations. The problem in man has many aspects and the relations of animal studies to these aspects can be illustrated by a series of questions. Each of these is discussed in detail elsewhere in these volumes.

1. What environmental agents are teratogenic in man, and under what circumstances? (Chapters 6–9, this volume.)
2. How do mutant genes or teratogens causing malformations act to produce their particular alteration in development? (All Vol. 2.)

F. CLARKE FRASER · Department of Biology, McGill University, and Department of Medical Genetics, The Montreal Children's Hospital, Montreal, Canada.

3. What are the causes of chromosomal aberrations? (See chapter by Hsu and Hirschhorn, Vol. 2.)
4. What role is played by the interaction of an environmental agent, in otherwise harmless amounts, with a mutant gene, in otherwise harmless dosages? (Chapter 13, this volume.)
5. What are the biological attributes of genetically determined susceptibility? That is, what differences in developmental pattern make an individual more prone to have a malformation, how do these bring about the increased susceptibility, and how can they be recognized postnatally? (See chapter by Trasler and Fraser, Vol. 2; elsewhere.)

The better the answers we have to these questions, the better our chances of reducing the frequency of malformations. Since man is in many ways unsuitable for teratologic studies, experiments in animals are often the best way, and sometimes the only way, to search for answers. The following discussion will develop some general themes, consider the advantages of animal studies in helping to solve, or at least ameliorate, the human problem, and point out some of the limitations in extrapolating from animal studies to the human problem.

II. TESTING DRUGS AND OTHER ENVIRONMENTAL AGENTS FOR TERATOGENICITY

The discovery that a drug (cortisone) was teratogenic in the mouse (Fraser and Fainstat, 1951) aroused a significant increase in concern about the possible teratogenicity of drugs in man and a sharp increase in experimentation on their effects on development of the embryo/fetus in experimental animals. The results were at first misleading, or at least misinterpreted. Although marked differences between species, and between genotypes within a species, were noted, the implication that man might be highly sensitive to a drug that seemed relatively nonteratogenic in mice and rats was not appreciated or at least underrated. One teratological pundit, in his wisdom, wrote that

> the agents known to be teratogenic in the lower animals are effective only in or close to the range that causes abortion or resorption of some embryos in the litter. If you suffocate a pregnant animal to the point of prostration, or give her amounts of cortisone that might kill her if she were not pregnant, or starve her till she loses 20 per cent of her body weight, or batter her with sublethal doses of roentgen rays, or create in her an acute vitamin deficiency, the offspring are likely to be malformed. These are sledgehammer blows to the embryo. They are useful tools in the analysis of abnormal development, but I doubt that they have very much to do with the bulk of human malformations where no such violent maternal insults are evident. (Fraser, 1959)

Quantitatively the statement was true, but it may have reduced unduly our concern about the possible exceptions. Then came thalidomide, followed by a spate of symposia, conferences, and workshops on how to be sure a drug would not be teratogenic in man before releasing it for trial in man, and the same author, using his retrospectoscope, concluded that: "a drug that is not demonstrably teratogenic in experimental animals may be so in man. A drug that is demonstrably teratogenic in animals may not be so in man. Therefore, the final proof of whether a drug is likely to be teratogenic in man must be sought in man" (Fraser, 1964). Unfortunately, the practical problems involved in setting up programs capable of testing the effects of the new drugs on the embryo/fetus are so great that most of our present information about human drug teratogenicity has been derived retrospectively.

Of course, if we knew all about the biochemistry and physiology of the developing embryo, and about the relationship of chemical structure to pharmacologic action of a drug, we could predict which drugs would be teratogenic and there would be no need for animal testing. But we do not, and since testing for teratogenicity in man without preliminary screening would be both impractical and immoral (except in very special circumstances) we must resort to animal testing and resolve to do the best we can in spite of its limitations. The demonstration that a drug or other agent is nonteratogenic at well above therapeutic doses in a variety of animal species is reassuring even though one cannot assume that man will conform. If an agent *is* teratogenic in lower animals one must not jump to the conclusion that it will be so in man (as was done, so uncritically, in the case of the herbicide 2,4,5-T and its dioxin contaminants) but must extend the range of testing. The important questions of what species to use, how many animals are needed, how they should be examined, what agents should be tested, and how the agent should be given are dealt with in Volume 3.

We will mention here two areas where animal models have a particularly important role to play in testing for teratogenicity. The first is in searching for postnatal effects of prenatal exposure. Relatively little work on behavioral effects of embryo/fetus exposure to drugs and other agents has been done, particularly in animals with behavioral attributes similar to those of man, but progress in beginning to be made (see chapters by Rodier and Collins, Vol. 3). Here, animal studies are mandatory, as human studies cannot be done, except retrospectively. Another example is that of transplacental carcinogenesis (Rice, 1973). It would have been nice to be able to say that the production of vaginal adenocarcinoma in postpubertal females exposed *in utero* to diethylstilbestrol had been predicted from the animal studies of transplacental carcinogenesis, but in fact the effect was discovered by retrospective studies in man. Nevertheless, animal studies are important in elucidating mechanisms and may alert us to further hazards (see chapter by Bolande, Vol. 2).

Second, it might be possible to circumvent species differences in teratogenicity due to differences in metabolism, transport, and maternal–fetal membrane relationships to some degree by the use of special test systems whereby the effect of an agent could be observed on specific organs rather than on the maternal/fetal complex as a whole. Culture of whole embryos or organs during the period of organogenesis removes confounding factors due to the mother and membranes (see chapters by Shepard and Pious, and New, Vol. 3). One might also look at effects on specific developmental processes, such as the inductive relations of neural tube and somites, the process of cell sorting and aggregation, the interaction of neuroblast and myoblast, epithelial–mesenchyme interactions, induction of hormones and enzymes, and neural crest cell migration, by using experimental systems already well known to experimental embryologists (see chapters by Saxén and Yamada, Vol. 2). These approaches to dissecting the complexities of teratogenicity will, no doubt, play a useful role in aiding our understanding of how the various parts work and remind us that the key to many malformations may lie in the interactions between the parts, both maternal and fetal.

III. USE OF TERATOGENS AND GENES TO ANALYZE MECHANISMS OF DEVELOPMENT AND THEIR ERRORS

A more intellectually rewarding activity than testing compounds for teratogenicity with a view to protecting the fetus from environmental hazards is the use of teratogens to create malformations under sufficiently controlled conditions that one can analyze how the malformations came about. Using appropriate teratogens and test animals, it may be possible to produce a particular malformation in virtually all the exposed embryos, providing the opportunity to study fixed embryos between the time of exposure and the time the malformation appears, knowing that the observed animal *would have had the malformation* if allowed to continue development. By identifying a point in the developing system where something first goes wrong after exposure and following, in a series of embryos, the series of pathogenetic changes leading to the full-blown malformation, one may be able not only to demonstrate one mechanism by which the malformation can arise, but to gain insight into the normal process. One can do this either at the "structural" level—identifying deficient or aberrant growth of a tissue, failure of fusion, mechanical obstruction to morphogenetic movements, and so on; or at the "biochemical" level—identifying the molecular and metabolic basis for the morphogenetic errors. There has been considerably more progress at the structural than the biochemical level (presumably because it is easier to see morphological differences than to identify the *relevant* underlying biochemical changes), although the latter probably has a better chance of leading to preventive methods.

Malformations caused by mutant genes, or occurring with characteristic frequencies in inbred strains, do not usually occur in all the offspring (unless they are trivial), so identification of the beginning of a dysmorphogenetic process may be more difficult than for environmental teratogens. Nevertheless, with the use of proper controls, the analysis of genetically determined errors of morphogenesis can also provide insight into mechanisms of development, both normal and abnormal (see chapter by Glucksohn-Waelsch, Vol. 2).

There are now a large number of examples of teratogen-induced malformations, but in spite of the advantage mentioned above, disappointingly few have been worked out in terms of pathogenetic mechanisms and even fewer have been related to corresponding problems in man. For mutant genes the situation is somewhat better, but numerous additional opportunities are begging for attention. In this chapter we will point out some examples of how the experimental study of malformations has elucidated developmental problems and how this knowledge has been, or could be, related to the problem in man.

A. The Multifactorial/Threshold Concept

1. History

One animal model that has had a significant influence on thinking about human malformations is the multifactorial distribution/developmental threshold model. I apologize to Dr. Josef Warkany, who has been my teacher and source of teratological wisdom ever since his writings first inspired my interest in teratology, for the use of the word "multifactorial." As he points out, all malformations are multifactorial, since a mutant gene or teratogen does not produce its effects in isolation, but in conjunction with many other factors. Nevertheless, usage seems to have entrenched the term firmly in the literature; we use it to mean "resulting from a combination of several genetic and/or environmental factors." The term should not be used as a synonym for polygenic, which is taken to mean "resulting from the action of many genes, each with a small effect, acting additively." The terminology is discussed in more detail elsewhere (Fraser, 1976).

The concept was probably first introduced by Sewall Wright (1934) on the basis of his study on polydactyly in the guineapig. He made crosses between a polydactylous and a normal inbred strain and observed that the frequency of polydactyly in the F_1, F_2, and first back-cross generations fitted the expectation for a Mendelian recessive gene. A less careful worker would have stopped there, but he proceeded to do a second back-cross, which did not fit the Mendelian expectation. Further extensive breeding led him to conclude that the trait was determined by a polygenic system with a physiological threshold. However, he did not express any idea of the nature of the threshold.

Gruneberg (1952) developed the concept further, coined the term "quasicontinuous variant," described many genetic properties of these variants in the mouse, and pointed out that some common malformations in man also show the "stigmata" of quasi-continuity. Although he was dealing with "spontaneous" variants, his writings on this subject are still rewarding reading for those interested in teratogenic mechanisms. One of the first examples that led him to the concept was absence of the third molar tooth in the mouse.

He showed, for example, that frequency of missing third molar in various strains and crosses was correlated with size of the tooth when it was there, and suggested that a critical mass of the tooth anlagen might be the developmental threshold in this case. The "stigmata" of quasi-continuous variants included: (1) different stable frequencies of the trait in different strains expressing the trait, (2) widely different frequencies of the trait in outcrosses of a strain expressing the trait to different "normal" lines (3) a correlation between penetrance and expressivity of the underlying genetic system, (4) sensitivity to environmental differences such as diet and maternal physiology, and (5) sensitivity to genetic differences, such as sex, or major mutant genes.

The McGill group found that cleft palate can be produced by maternal treatment with cortisone (Fraser and Fainstat, 1951), that there were strain differences in frequency of induced cleft palate, and that the movement of the palate shelves from vertical to horizontal during closure was delayed by cortisone (Walker and Fraser, 1957). Noting that the palate shelves, in untreated embryos, tended to move to the horizontal earlier in those strains and crosses that had the lower frequencies of cleft palate when treated by cortisone (Walker and Fraser, 1956; Trasler, 1965), they suggested that cleft palate in this case might be a multifactorially determined threshold character. If the shelves were delayed by more than a certain critical amount, a cleft palate would result. The latest stage at which the shelves could come up and still meet would be the threshold between normality and abnormality.

Carter showed that, for certain common congenital malformations in man, the frequency of the condition in various categories of relatives of affected probands fitted the expectation for a multifactorial threshold character in several ways (Carter, 1969, 1976; Fraser and Nora, 1975). In particular: (1) If the malformation affects one sex more often than the other, the probands of the least often affected sex have relatives with the higher risk; (2) The frequency in the relatives decreases sharply between first- and second-degree relatives, and much less sharply between second- and third-degree relatives, rather than linearly; (3) The frequency in the first-degree relatives approximates the square root of the frequency of the malformation in the general population (Edwards, 1969; Newcombe, 1963); (4) The recurrence rate in the first-degree relatives is higher when the defect is more severe in the proband.

Quite a large number of common congenital malformations and other diseases have now been shown to fulfil these criteria more or less fully. These include the classical example congenital hypertrophic pyloric stenosis (Carter, 1961), cleft lip and palate (Fraser, 1963, 1970), various congenital heart malformations (Nora, 1971), pes equinovarus, (Wynne-Davies, 1965), congenital dislocation of the hip (Wynne-Davies, 1970), anencephaly/spina bifida (Carter and Evans, 1973), Legg–Perthe disease (Gray et al., 1972), and Hirschsprung disease (Passarge, 1973).

The demonstration that various human malformations (and other diseases) could be multifactorial/threshold characters encouraged use of the method of quantitative genetics for estimating the heritability of these conditions and for testing possible models (Falconer, 1967; Edwards, 1969; Morton *et al.*, 1970; Smith, 1974; Cavalli-Sforza and Bodmer, 1971; Elston and Yelverton, 1975), and, conversely, for predicting recurrence risks in categories of relatives for which empirical data were not available (Bonaiti-Pellié and Smith, 1974; Mendell and Elston, 1974).

The next stage is to extrapolate our insights gained from the study of animals to the problem in man. A beginning has been made in the case of cleft lip and palate, where the idea of embryonic face shape as a component of susceptibility to cleft lip in the mouse (Trasler, 1968) is being tested in human families (Fraser and Pashayan, 1970) (see Section III. A. 4). Conversely, few attempts have been made to test in animal models some of the characteristics of multifactorial/threshold characters in man, such as higher frequency in offspring of parents of the least affected sex for characters showing a sex ratio difference. One such attempt fell through when the sex difference in frequency of microphthalmia in the C57BL/6 strain disappeared entirely in the offspring of sublines derived from outcrosses! (F. C. Fraser, unpublished results). It would also be interesting to test the models underlying the proposed methods of estimating heritability, from frequencies of the defect in various categories of relatives, in systems where the genotypes were known and the underlying distributions and threshold could be observed, rather than deduced as is usually necessary in man.

2. Cleft Palate as a Multifactorial/Threshold Model

The malformation most extensively studied from a teratological point of view is cleft of the secondary palate, probably because it happens relatively late in development and is therefore easier to observe. Study of the process of closure and the effects of teratogens upon it has identified many factors that must integrate with one another to bring about successful formation of the intact palate (see chapter by Trasler and Fraser, Vol. 2). There appears to be an intrinsic force in the palatine shelves that promotes their movement from a vertical orientation, lying along the sides of the tongue, to a horizontal orientation about the tongue, where their distal borders fuse to form the completed structure. Opposing this force is the bulk of the intervening tongue. Various factors impinge on the struggle of the shelves to force their way between the tongue and the floor of the skull. The tongue becomes mobile, and moves forward, the mandible lengthens, and the craniofacial complex enlarges, increasing the space between the shelves. Perhaps even swallowing may help, by pushing the posterior tips of the shelves medially and driving the thin end of the wedge into the gap. Once the shelves get above the tongue their subsequent flattening and extension towards the midline may be aided by pressure of the tongue from below and of the cheeks from the sides. The flexion of the neck diminishes, perhaps at first spasmodically, so that the mandible is no longer in contact with the anterior chest wall, and is therefore more free to move and allow room for the tongue as the shelves move in above it. Finally, the epithelia of the shelves meet, fuse, and break down so that a mesenchymal bridge is established across the midline. Thus many factors appear to contribute to successful closure.

The multifactorial/threshold model for cleft palate is illustrated in Fig. 1. It postulates that the time at which the palate shelves become horizontal and thus able to extend towards the midline, meet, and fuse, is continuously distributed. (In the hypothetical diagrams usually shown this continuous variable is also represented as more or less normally distributed, but see p. 86). Thus in some embryos the shelves become horizontal relatively early and in others relatively late. Within inbred strains, or F_1 crosses between them, all the variation would be due to environmental factors or to "random" fluctuations. Head growth is assumed to continue during the period of closure so that head width continuously increases. If shelf movement is delayed beyond a certain critical point the shelves will be too far apart to accomplish fusion, and a cleft palate will result. Thus a discontinuous variable (cleft palate vs. normal palate) is determined by whether a continuous variable (stage at which shelf becomes horizontal) puts the embryo on one side or the other of a developmental threshold (latest stage at which shelves that have reached the horizontal can fuse). Both the distribution and the threshold can be influenced by both genetic and environmental factors.

The diagram illustrates the position of the distribution (relative to the threshold) as being determined primarily by the interaction of shelf force (promoting shelf movement) and resistance of the tongue (delaying it). The tongue moves forward during closure, which may aid the shelves in their struggle to move into the space above it. Both shelf force and tongue resistance can be influenced by other factors. The rate of growth of the mandible may influence forward movement of the tongue and alter its resistance. Motility of the tongue may also promote success of the shelves in invading the space above it. The shelf force presumably depends on the physical integrity of its structure, which may vary with rate of mucopolysaccharide synthesis, the development of contractile elements, hydration, etc., and may also vary with the rate of extension of the cranial base. The probability of an embryo falling beyond the threshold also varies with the position of the threshold, and this varies with shelf width and shelf length. A relatively wide shelf would allow successful closure even if the shelf moved relatively late (i.e., the threshold is moved to the right, relative to the distribution) and a relatively wide head would require the shelves to become horizontal relatively early to close successfully (the threshold is moved to the left). Thus the position of the threshold is also depicted as a continuous variable. This helps to emphasize that the probability of sucessful closure is assumed to depend not only on when the shelves become horizontal but how far apart their medial borders are when they do.

Susceptibility to cleft palate depends on the relation of distribution to threshold. Thus the shelves move to the horizontal later in development in

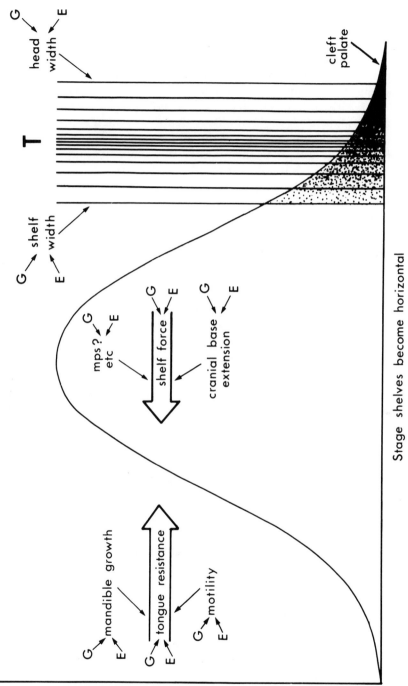

Fig. 1. The multifactorial/threshold model for cleft palate.

A/Jax than in C57BL embryos suggesting that the genes of the A/Jax embryo put it relatively closer to the threshold and make it relatively susceptible to cleft palate induced by any environmental factor that delays shelf movement. Furthermore, the correlation between stage of shelf movement in the untreated embryo and resistance to cortisone treatment (as measured by cleft palate frequency) is reasonably good for six crosses (Trasler and Fraser, 1958; Trasler, 1965). Although the evidence that one is causally related to the other is only inferential, it seems highly unlikely that this relation is coincidental.

The use of developmental stage as the underlying variable is obviously an oversimplification. It may not be the time of approximation of the shelves that is critical but some other variable, such as the strength of the forces that keep them approximated (e.g., shelf growth vs. head growth) while they are undergoing fusion or those that attempt to pull them apart (e.g., head growth). The stage at which the shelves move is a reflection of many interacting underlying variables.

Both genes and environmental factors can influence the position of a given embryo relative to the threshold by acting at each of the points depicted or, indeed, at others not depicted. (For instance we have omitted rate of longitudinal maxillary growth which might well influence the rate of forward movement of the tongue.) A number of examples are presented in the chapter by Trasler and Fraser (Vol. 2).

If the correlation between normal stage of shelf movement and resistance to cortisone is not coincidental (and we think it is not), it illustrates how *the susceptibility of an embryo to a teratogen can be influenced by the embryo's normal developmental pattern.* The fact that exposure to acetylsalicylic acid increased the frequency of lateral cleft lip in A/J embryos, but caused median cleft lip in C57BL embryos, illustrates the same principle if one accepts that the difference in susceptibility to the teratogen is related to the difference in embryonic face shape (Trasler, 1968; Juriloff and Trasler, 1976).

Another possible example, although less well documented, is the differential response of the same two strains of dextroamphetamine. There is a low spontaneous frequency of atrial septal defect in the A/J strain and of ventricular septal defect in the C57BL/6 strain. Maternal treatment with dextroamphetamine increases the frequency of the type of defect that occurs spontaneously (Nora *et al.,* 1968). The atrial septum secundum in the A/J strain completes closure of the ostium primum just around the time of birth, so a relatively small amount of delay will cause a septal defect (Fraser and Rosen, 1975). Comparable observations have not yet been made in the C57BL strain; it would be interesting to see if the atrial septum grew more rapidly and the ventricular septum more slowly in this strain.

3. Implications for Animal and Human Studies

In considering the implications of the model one must keep in mind that it represents a very simplified view of reality, and although it may provide a useful conceptual framework to aid one's thinking about the problem, it must not be accepted glibly. The following implications and caveats are presented.

a. The Contribution of Normal Developmental Patterns to Teratogenic Susceptibility. One of the most significant features suggesting that cleft palate induced by teratogens was an example of the multifactorial/threshold model was the correlation between time of palate closure in untreated embryos and frequency of cleft palate in embryos exposed to cortisone or 6-aminonicotinamide. By and large, the earlier the palate shelves normally closed, the more resistant the embryo was to the teratogen (Walker and Fraser, 1956.) This illustrates, as we have mentioned, that the embryo's teratogenic susceptibility can be influenced by its normal developmental patterns and implies that such a developmental difference would impose a difference in susceptibility to any teratogen that retarded shelf closure.

If this is so, it implies that a strain difference in frequency of an induced malformation could occur even though the primary effect of the teratogen was the same in the two strains. For instance Larsson (1962) did not find a difference in degree of embryonic acid mucopolysaccharide synthesis produced by cortisone treatment in the A/J and C57BL strains. This does not rule out the hypothesis that cortisone causes cleft palate through its action on acid mucopolysaccharide synthesis, since the strain difference in cleft palate frequency could have resulted simply from the strain difference in palate closure pattern, and the same amount of delay in both strains would still result in a strain difference in cleft palate frequency.

Furthermore, the fact that one strain closes its shelves earlier than another does not mean that the critical period of inducing a cleft by a teratogen is earlier in that strain or that one could increase the frequency in the early-closing (and thus resistant) strain by earlier treatment (Verrusio, 1966; Fraser, 1969). The difference in closure time is presumably because one or more of the factors in the system is more "efficient" or "stronger" in the early-closing strain, resulting both in the early shelf closure and in the greater resistance to the teratogen.

b. Penetrance and Expressivity. If the delay in palate closure were caused by a major mutant gene, rather than cortisone, one would say that the gene had full penetrance in the A/J strain and reduced penetrance in the C57BL strain (Fraser, 1965). Thus the multifactorial/threshold model provides one way of illustrating how "modifiers" in the "genetic background" can affect the penetrance of a mutant gene.

A third implication is that penetrance would be positively correlated with expressivity, a well-recognized phenomenon (Gruneberg, 1952; Landauer, 1955; Sang, 1963). In a series of populations carrying a mutant gene, and differing in the relation of distribution to threshold, the closer the mean to the threshold, the greater the deviation from normality at the right-hand tail of the distribution (increased expressivity of the defect) and the more individuals would fall beyond the threshold (increased penetrance). Related to this is the fact that the recurrence risk for certain human malformations increases with the severity of the malformation in the proband (Carter, 1969; Fraser, 1970).

Finally, identifying the basis for a genetic (strain) difference in susceptibility to a teratogen does not tell us anything about the cause of the cleft palate induced, i.e., about the mode of action of the teratogen. The difference between the A/J and C57BL strains might, for instance, stem from a difference in shelf force, and the teratogen might act by paralyzing the tongue equally in the two strains. Yet the C57BL strain, being farther away from the threshold, would be more resistant.

Furthermore, the above model does not include all possible causes of cleft palate. It omits, for instance, structural abnormality of the shelf, and interference with the process of fusion itself. In the latter case, however, the strain difference in palate closure time could result in a strain difference in cleft palate frequency; the C57BL embryo, closing its shelves earlier, would be less affected by growth of the head that might tend to draw the shelves apart and thus would be less affected by a teratogen that impaired fusion.

c. Effects of a Threshold on Frequency of a Quasi-Continuous Trait.
1. If a given malformation is a threshold character with an underlying continuous distribution of some variable, then the frequency of affected individuals does not vary directly with the distance of the threshold from the mean, but with the area of the distribution curve falling beyond the threshold. Therefore, statistical comparisons of frequencies produced by different doses of teratogen, genotypes, etc., should be based on a probit, or other appropriate transformation, of the data, rather than the means and standard errors of the frequencies.

Thus, for a threshold character, with the underlying variable normally distributed, a *linear* dose–response curve, where response is measured by how far a given dose shifts the mean of the distribution, would appear to be *sigmoidal,* as measured by frequencies of induced malformations, unless these were converted to probit units.

2. Unfortunately, however, the underlying distribution may not be normal, even though it is usually depicted as such in diagrams illustrating the model and is assumed to be so in the mathematical treatments of the properties and implications of the model. Normality is clearly not the case in some examples where distribution can be observed. For instance, the distribution of palate closure in embryos with respect to chronological age is quite different for A/J and C57BL embryos. A/J embryos begin shelf movement later than C57BL embryos but complete the process more rapidly (Walker and Fraser, 1956). We interpret this to mean that the shelf force does not develop as strongly in A/J embryos as in C57BLs, so that movement towards the horizontal does not begin until the other relevant factors (tongue position and motility, maxillary size, etc.) are optimal. At this point closure is quickly accomplished. In the C57BL embryos, on the other hand, the shelf force is stronger, and allows the shelves to begin movement before the optimal time; however, because conditions are not optimal, they have a harder time reaching the horizontal and therefore take longer. If this is the correct interpretation, it provides indirect support for the idea that the genetically determined dif-

ference in time of palate closure between the two strains resides in the shelf force.

If the distribution is not normal in situations where it can be observed, we must be cautious in assuming normality where the distribution cannot be observed (as in most situations involving congenital malformations), which makes the choice of transformation difficult. This (as well as the one discussed next) is a problem that plagues, for instance, studies attempting to estimate the heritability of malformations from the frequencies in relatives (Falconer, 1967; Edwards, 1969; Cavalli-Sforza and Bodmer, 1971), or the number of genes underlying a strain difference in malformation frequency, either spontaneous or induced (Dagg *et al.*, 1965; Davidson *et al.*, 1969). A transformation, such as arcsin, if it renders the data linear, provides some reassurance (Biddle, 1975), but caution is recommended in the appraisal of conclusions reached by such statistical manipulations.

3. In a multifactorial threshold model an increase in frequency of the malformation can result from either a shift of the distribution to the right (in the conventional diagram) or an increase in variance, which in effect puts more individuals in the tails of the distribution. The latter possibility is almost consistently ignored in the interpretation of teratological data. It was pointed out by Newcombe (1963) in relation to the effects of inbreeding in man, and it can also result from assortative mating. This could be misleading when conclusions are based on the assumption that changes in frequency of the malformation are due entirely to shifts in the mean of the distribution.

4. In a multifactorial/threshold model, if two teratogens interact additively—that is, together they shift the distribution mean by the sum of their separate effects—the increase in the frequency of the malformation will be greater than the sum of the individual effects. In other words the teratogens will appear to be synergistic (at least when the individual teratogens are used at doses producing less than 50% malformations), unless an appropriate transformation is used (Fraser, 1965). Thus additive interaction could be misinterpreted as potentiation (see Chapter 13).

Similarly, it has sometimes been assumed that if two recessive mutant genes each produce a similar phenotype when homozygous, and the F_1 between the two homozygotes also show the mutant phenotype, then the two mutant genes must be allelic; this is not necessarily so. Each mutant could be having an intermediate (additive) effect in combination with its normal allele, which would result in a normal phenotype, but in combination the two mutants, acting at different points of the system, could shift the distribution so that the compound heterozygote fell beyond the threshold and had the mutant phenotype (Harris and Fraser, 1968).

Thus the model provides a number of useful insights into some of the properties of congenital malformations, and also points up some possible pitfalls in interpreting teratological data. Perhaps the most useful aspect of the model is to teach us respect for the complexities of developmental systems and the causes of their failures.

4. Significance for the Human Problem

As for the significance of this approach for the human problem, probably its greatest value has been to provide a substantive model of a multifactorial/threshold system underlying a common congenital malformation. Indirect evidence for such a system was first provided by Sewall Wright, from his studies of the genetics of polydactyly in the guineapig (1934), elaborated by Gruneberg (1952) in the mouse into the concept of quasi-continuous variation, and extrapolated to man by Carter and others (Carter, 1969; Fraser and Nora, 1975). The cleft palate model provided an experimental illustration, identifying the threshold, the nature of various components of the system, and the complex and subtle interrelationships referred to as "gene–environment interactions" (Fraser, 1969).

The model is not likely to provide us with any immediate means to lower the cleft palate frequency in man, but one hopes that the better the understanding of the system, the better the chances of solving its problems. It is clear that there is no one cause of cleft palate, nor one means of prevention, but this should not discourage us from trying to sort out the various components of the system. Most cleft palates in man are the result of the interaction of many factors, but the relative weights of the factors will vary from case to case, and perhaps it will be possible by closer examination of the palate shelves, maxilla, mandible, and tongue in man, and by comparison with those induced by teratogens in animals, to separate out some categories where one component or another is the major culprit. This may make a difference to management and lead to etiological clarifications and improved predictions of recurrence risks.

For instance, Trasler (1968) showed that maternal treatment with salicylate caused lateral cleft lip in the A/Jax inbred strain (with a relatively narrow, pointed face) and median cleft lip in the C57BL strain (with a relatively broad, square face). This led to the hypothesis that the human embryo's liability to cleft lip might be altered by the embryonic face shape. Thus if embryonic face shape was reflected in postnatal face shape, and if the relevant changes in face shape were genetically determined, then the near relatives of children with cleft lip ought to show significant differences in face shape from the population average. Observations on face shape in the affected children (of which there are many) are confounded by the presence of the anatomical defect and, prior to the experimental evidence, the idea of looking for relevant changes in the unaffected near relatives did not seem to have occurred to those studying the genetics of cleft lip in man. Preliminary studies on the external facial dimensions supported the hypothesis (Fraser and Pashayan, 1970), and cephalometric studies have also provided some support (Coccaro *et al.*, 1972; Kurisu *et al.*, 1974), although the evidence is not entirely consistent. (Incidentally, it is possible that skeletal dimensions may not be as good a reflection of embryonic face shape as is the external topography, and this would be an interesting topic for study.) Other possible approaches to testing the hypothesis would be studies of face shape in population groups with high frequencies of cleft lip (certain American Indian tribes; the Japanese) and family studies where dominant mutant genes causing cleft lip or palate are sometimes transmitted by unaffected persons (reduced penetrance). Is the lack of penetrance or resistance to the cleft-producing property of the mutant gene related in part to the person's face shape? For instance in the family described by Weinstein and Cohen (1966) cleft palate is transmitted in an apparently sex-linked recessive pattern, but the affected males also have hypertelorism, and so do their mothers. The pedigree is also compatible with inheritance of an autosomal gene predisposing to cleft palate (because it increases the width of the embryonic maxilla?) with male embryos being more susceptible than female (because they have wider maxillae?). These speculations serve to illustrate how concepts derived from animal teratology can suggest hypotheses—some of them testable—to account for dysmorphogenic problems in man. Other examples will be found throughout this volume.

On the other hand, animal models may caution us against drawing unjustified conclusions from end results. For instance the occurrence of micrognathia with cleft palate does not necessarily mean that the cleft palate was caused by the inhibition of mandibular growth; in the case of cleft palate caused by maternal treatment with vitamin A the micrognathia appears after the palate has failed to close (Shih *et al.*, 1974).

Finally, the model implies that for threshold characters the dose–response curve for a particular teratogen may (paradoxically) have no threshold. If the tail of the untreated test population, be it inbred mouse strain or people, extends beyond the threshold (i.e., some individuals are affected in the absence of the teratogen) the dose–response curve would be expected to taper off gradually with decreasing doses to the point where the effects of the teratogen merge with the physiological "noise"; it would not descend to a certain point and then drop suddenly to the background (population) level.

As a corollary to this one should not assume that there is a "safe" dose for a drug, at least in the range where there is any pharmacological response at all. Any maternal treatment that alters the maternal–fetal system, if it occurs at a critical period, to *an embryo at the tail of the distribution of susceptibility*, may change the relation of distribution to threshold and result in a developmental error. Thus no drug should be given during pregnancy unless there is a need for it that is urgent enough to justify the unknown but probably finite risk of malforming the embryo.

B. Pathogenetic Clues to Syndromes

One of the most puzzling features of many human dysmorphogenic syndromes is the fact that many of the component anomalies have no obvious pathogenetic relationship to one another, and an important use of teratological studies in animals is to provide models that may explain these curious associations.

For example, a plausible model for a group of syndromes included in the so-called first and second branchial arch syndrome has been provided by Poswillo (1973). The affected persons shown varying degrees of under- or maldevelopment of the external ear, the middle ear ossicles, the condyle and ramus of the mandible, the zygomatic arch, the malar bone, the temporal bone, and the muscles of mastication. Why do the associated defects usually occur only on one side; why are they so variable; and why do they involve structures not arising from the branchial arches, such as the temporal bone, whereas they spare some structures that do (the cranial part of the hyoid bone, the body of the mandible)? A combination of astute anatomical observation and embryological knowledge led to the suggestion that the basic cause was ischemic necrosis resulting from anomalous development of the stapedial artery (Braithwaite and Watson, 1949). The suggestion was borne out by experimental demonstration of this mechanism in mice, using the folic acid antagonist triazene, which was shown to cause hemorrhage in the area of the forming stapedial artery, with subsequent deformities in the adjacent structures (Poswillo, 1973). In contrast, the characteristic features of mandibulofacial dysostosis (Treacher–Collins syndrome) can be produced in rats by maternal treatment with vitamin A, which causes focal death of preotic neural crest cells and consequent anomalous development of the ear, as well as reduction of ectomesenchyme in the first and second branchial arches (Poswillo, 1975).

Similarly, studies on the results of experimentally induced oligohydramnios have demonstrated that constriction of the embryo may produced cleft palate, presumably by increasing neck flexion and making it more difficult for the shelves to move in above the tongue (Trasler *et al.,* 1956; Poswillo, 1966). It may also produce deformitites of the limbs (flexion contractures, "amputations") similar to those reported in association with putative amniotic bands in human babies (Jones *et al.,* 1974). This suggests that the bands are secondary to local necrosis and the formation of adhesions. Further work on experimental animals with longer gestation periods than mice might be revealing.

Another attempt to provide an animal model for a human syndrome involves the recessively inherited combination of goiter and nerve deafness known as Pendred syndrome. Maternal hypothyroidism was induced in an inbred strain of mice by maternal treatment with propylthiouracil (Deol, 1973). The offspring were found to be deaf and had structural anomalies of the cochlea. Unfortunately, the state of the thyroid in the offspring is not mentioned. Addition of thyroxine to the drinking water containing the propylthiouracil prevented the effects of the propylthiouracil. Although this experiment cannot be considered a valid model for Pendred syndrome (since patients are usually euthyroid, and congenital cretins are not usually deaf), nevertheless it reveals an interesting relationship between development of the inner ear and the thyroid, and it should be further exploited.

A neglected area that might throw some light on the bases for *associations of features* of syndromes is the *association of features* in embryos treated with teratogens. Almost all data on the results of experimental treatment with teratogens present the frequencies of induced anomalies in a form that would not reveal associations of features. This may be partly due to the fact that the frequencies of skeletal anomalies are usually measured in cleared, alizarin-stained specimens, and the frequencies of soft-tissue anomalies from (other) dissected or sectioned embryos. This is unfortunate, as associations of anomalies may be a clue to developmental relationships and to the underlying nature of counterpart human syndromes.

The situation is somewhat better with pleiotropic dysmorphogenetic genes, but their existence tends to be ignored by experimental teratologists and syndromologists alike. Here we will mention only a few examples to illustrate their potential relevance to syndromes. Most of them are drawn from Gruneberg's informative book, *The Pathology of Development* (1963). Further examples of how the study of mutant genes has helped to elucidate developmental mechanisms will be found in the chapter by Gluecksohn-Waelsch (Vol. 2).

Let us consider some of the problems of extrapolating from animal models to human syndromes. Firstly there is the problem of whether the animal mutant is really a counterpart of the human syndrome. For instance the mutant gene *Naked* in the mouse produces a phenotype which is superficially quite similar to that of the hydrotic type of ectodermal dysplasia (Williams and Fraser, 1974), but the biochemical defects in the two mutants turn out to be quite different (Gold and Scriver, 1972; Tenenhouse *et al.,* 1974). Secondly, some of the features useful for the delineation of human syndromes are minor anomalies that have no counterpart in experimental animals, and *vice versa.* What is the mouse counterpart of synophris (confluent eyebrows), or epicanthic folds, and what is the human counterpart of crooked tail? (E. Perrin's suggestion of "dishonest prostitute" is not acceptable.) On the other hand, the basic developmental defect underlying the association of malformations in a syndrome may be *transient* (a maturational delay in an enzyme, for instance), so that it may be impossible to detect in the postnatal infant, and animal models provide the only available approach to the problem.

The mutant gene for "myelencephalic bleb" causes a variety of malformations including exencephaly, microphthalmia, small pinna, club hands and feet, preaxial polydactyly, occasional tibial hemimelia, syndactyly, renal agenesis or hypoplasia, ureteral anomalies, and alopecia areata. The mutant has three major gene effects—on neural tube closure, on the morphogenetic field of the hind limb and ureter, and the production of blebs of mesenchymal intercellular fluid (Carter, 1959). These blebs may lodge under the skin (hair defects), over the eye (microphthalmia), or in the limbs (malformations of the extremities). What the basic effect of the gene is remains tantalizingly unclear, but if found might throw much light on developmental mechanisms. It is interesting that the frequencies of different features produced by the mutant gene vary with the genetic background; it is possible to select a subline with a high frequency of forelimb defects, or of eye defects, and so on, and these changes in "expressivity" of the gene appear to be related to changes in the surface contours of the embryo, affecting the paths of migration of the sub-cutaneous blebs. The lesson for syndromologists is that the same basic cause may produce what might be classified on the basis of phenotypes as two different syndromes.

To turn to the chick, for a change, consider the mutant gene *talpid*[3] (Ede and Kelly, 1964a,b) which causes hypotelorism, absence of midface, small cerebral hemispheres with abnormal segregation—all reminiscent of the holoprosencephaly defect (deMyer *et al.*, 1964)—as well as short limbs, spadelike feet, polydactyly, vertebral anomalies, and subcutaneous hemor-rhages. The picture is not unlike that of the D trisomy syndrome. The basic cause appears to be increased adhesiveness of mesenchymal cells, producing decreased migrational motility; smaller, denser condensations of cartilage; and more rapid aggregation (Ede and Agerback, 1968). Because of this the long-bone chondroblasts are poorly organized, the prechordal mesoderm fails to split into the lateral maxillary rudiments (mid-face absence), and in the distal limb reduced motility in the dividing zone results in a short wide plate, which allows room for more centers of condensation (polydactyly). Thus widely different kinds of malformation can result from the same basic cellular defect, which is not obvious at birth. But one wonders if the increased adhe-siveness would still be demonstrable in fibroblasts or chondroblasts after hatching.

Another educational mutant is short ear (*se*) in the mouse. This mutant gene causes short ears, a bifurcated or absent xiphoid process, missing ribs, and diminution of all bones that pass through a cartilaginous phase (Green, 1968). The cartilaginous condensations are reduced in size whenever they appear, although in this case cell adhesiveness does not seem to be involved, and the defect does extend into postnatal life. The reduction in size causes visceral crowding that plausibly accounts for several visceral defects, including ectopic left gonad and right renal artery, and hydroureter. The lesson here is that mutant genes do not necessarily act at a particular stage of development—this one affects a *process*, wherever and whenever it occurs.

The phocomelic mutant causes short limbs, cleft palate, a small maxilla and mandible, hypertelorism, a pointed narrow nose, and preaxial polydac-tyly. In this case the blastemal precartilaginous condensations are normal in size and appearance, but appear about a day later than usual. The skeletal abnormalities derive from the resulting asynchronies (Sisken and Gluecksohn-Waelsch, 1959).

In "tail short" the nose is also small and short, there are hypertelorism, vertebral fusions, spina bifida occulta, extra ribs, fused ribs, a shortened left forelimb and/or right hindlimb, occasional reduction deformities of the radius and/or a triphalangeal thumb, and fusions of carpal and tarsal bones. The basic cause, surprisingly, is not in the skeletal anlagen at all. It is an anemia, at the stage when the yolk sac is producing the blood cells, leading to general retardation, which affects the development of specific organs differentially, according to their susceptibility at this particular stage. The defect is *transient,* disappearing when the liver takes over hematopoiesis, and would be difficult to deduce from postnatal observations (Deol, 1961).

Finally, we mention the mutant gene for syndactylism, which causes fusion of digits 3 and 4 with or without fusion of 1 and 2 on the hindfoot. In the hindfoot the phalanges are fused, in the forefoot the soft tissue only. There are fusion of the navicular and cuneiform bones, tail anomalies, and epidermal excrescences. These features are reminiscent, in some respects, of the focal dermal hypoplasia syndrome (Goltz *et al.,* 1970). The cause traces back to premature maturation and keratinization of the epidermis, so that the apical ectodermal ridge is narrow, the foot plate is constricted, and the digits squeezed together (Gruneberg, 1960). Thus again the skeletal defect has a systemic cause, a transitory skin disease, that would be impossible to identify postnatally.

C. Prenatal Prevention of Malformations

Finally, the use of animal models to suggest possible methods of preventing malformation should be mentioned. An outstanding example is the "pallid" mutant in the mouse, in which defects in balance were traced to a deficiency of otoliths in the ear. It was observed that maternal manganese deficiency in the rat suppressed formation of the otoliths in the embryo, leading to similar postnatal ataxia. When the maternal diet was supplemented with large amounts of manganese, the genetically "pallid" embryos formed otoliths and the ataxia was corrected (Erway *et al.,* 1971). Another example is the prenatal prevention of the "open eye" defect in the mouse by maternal treatment with cortisone, although this was discovered as an unexpected result of an attempt to increase the penetrance of the "lid-gap" gene (Watney and Miller, 1964).

Thus there are many lessons about human dysmorphogenesis to be learned from experimental teratology. The lessons may serve either as cautions or as signposts, and we hope the examples cited in this chapter will inspire further use of animal models to elucidate the mechanisms underlying human malformations.

REFERENCES

Biddle, F. G., 1975, Teratogenesis of acetazolamide in the CBA/J and SWV strains of mice II. Genetic control of the teratogenic response, *Teratology* **11**:37.

Bonaiti-Pellié, C., and Smith, C., 1974, Risk tables for counselling in some common congenital malformations, *J. Med. Genet.* **11**:374.

Braithwaite, F., and Watson, J., 1949, A report on three unusual cleft lips, *Br. J. Plast. Surg.* **2**:38.

Carter, C. O., 1961, The inheritance of congenital pyloric stenosis, *Br. Med. Bull.* **17**(3):251.

Carter, C. O., 1969, Genetics of common disorders, *Br. Med. Bull.* **25**(1):52.

Carter, C. O., 1976, Genetics of common single malformations, *Br. Med. Bull.* **32**:21.

Carter, C. O., and Evans, K., 1973, Spina bifida and anencephalus in Greater London, *J. Med. Genet.* **10**(3):209.

Carter, T. C., 1959, Embryology of the Little and Bagg X-rayed mouse stock, *J. Genet.* **56**(3):401.

Cavalli-Sforza, L. L., and Bodmer, W. F., 1971, *The Genetics of Human Populations*, W. H. Freeman, San Francisco.

Coccaro, P. J., D'Amico, R., and Chavoor, A., 1972, Craniofacial morphology of parents with and without cleft lip and palate children, *Cleft Palate J.* **9**:23.

Dagg, C. P., Schlager, G., and Doerr, A., 1965. Polygenic control of the teratogenicity of 5-fluorouracil in mice, *Genetics* **53**(6):1101.

Davidson, J. G., Fraser, F. C., and Schlager, G., 1969, A maternal effect on the frequency of spontaneous cleft lip in the A/J mouse, *Teratology* **2**(4):371.

deMyer, W., Zeman, W., and Palmer, C. G., 1964, The face predicts the brain: Diagnostic significance of median facial anomalies for holoprosencephaly (arrhinencephaly), *Pediatrics* **34** (2):256

Deol, M. S., 1961, Genetical studies on the skeleton of the mouse XXVIII. Tail-short, *Proc. R. Soc. (London), Ser. B* **155**:78.

Deol, M. S., 1973, An experimental approach to the understanding and treatment of hereditary syndromes with congenital deafness and hypothyroidism, *J. Med. Genet.* **10**(3):235.

Ede, D. A., and Agerback, G. S., 1968, Cell adhesion and movement in relation to the developing limb pattern in normal and talpid³ mutant chick embryos, *J. Embryol. Exp. Morphol.* **20**(1):81.

Ede, D. A., and Kelley, W. A., 1964a, Developmental abnormalities in the head region of the talpid³ mutant of the fowl, *J. Embryol. Exp. Morphol.* **12**:161.

Ede, D. A., and Kelley, W. A., 1964b, Developmental abnormalities in the trunk and limbs of the talpid³ mutant of the fowl. *J. Embryol. Exp. Morphol.* **12**:339.

Edwards, J. H., 1969, Familial predisposition in man, *Br. Med. Bull.* **25**(1):58.

Elston, R. C., and Yelverton, K. C., 1975, General models for segregation analysis, *Am. J. Hum. Genet.* **27**:31.

Erway, L. C., Fraser, A. S., and Hurley, L. C., 1971, Prevention of congenital otolith defect in pallid mutant mice by manganese supplementation, *Genetics* **67**:97.

Falconer, D. S., 1965, The inheritance of liability to certain diseases estimated from the incidence among relatives, *Ann. Hum. Genet.* **29**:(1):51.

Falconer, D. S., 1967, The inheritance of liability to diseases with variable age of onset, with particular reference to diabetes mellitus, *Ann. Hum. Genet. Lond.* **31**(1):1.

Fraser, F. C., 1959, Antenatal factors in congenital defects. Problems and pitfalls, *N.Y. State of J. Med.* **59**(8):1597.

Fraser, F. C., 1961, The use of teratogens in the analysis of abnormal development mechanisms, *in: First International Conference on Congenital Malformation Genetics* (M. Fishbein, ed.), pp. 179–186, J. B. Lippincott, Philadelphia.

Fraser, F. C., 1963, Hereditary disorders of the nose and mouth, *in: Proceedings of the IInd International Conference on Human Genetics* (L. Gedda, ed.), Vol. 2, pp. 1852–1855, Instituto G. Mendel, Rome.

Fraser, F. C., 1964, Experimental teratogenesis in relation to congenital malformations in man, *in: Proceedings of the 2nd International Conference on Congenital Malformations* (M. Fishbein, ed.), pp. 277–287, International Medical Congress Ltd., New York.

Fraser, F. C., 1965, Some genetic aspects of teratology, *in: Teratology, Principles and Techniques* (J. G. Wilson and J. Warkany, eds.), pp. 21–38, University of Chicago Press, Chicago.

Fraser, F. C., 1969, Gene–environment interactions in the production of cleft plate, *in: Methods for Teratological Studies in Experimental Animals and Man* (H. Nishimura and J. R. Miller, eds.), pp. 34–49, Igaku Shoin Ltd., Tokyo.

Fraser, F. C., 1970, The genetics of cleft lip and cleft palate, *Am. J. Hum. Genet.* **22**(3):336.

Fraser, F. C., 1976, Uses and misuses of the multifactorial/threshold concept, *Teratology* **13** (in press).

Fraser, F. C., and Fainstat, T. D., 1951, The production of congenital defects in the offspring of pregnant mice treated with cortisone. A progress report, *Pediatrics* **8:**527.

Fraser, F. C., and Nora, J. J., 1975, *Genetics of Man,* Lea and Febiger, Philadelphia, Chapter 11.

Fraser, F. C., and Pashayan, H., 1970, Relation of face shape to susceptibility to congenital cleft lip. A preliminary report, *J. Med. Genet.* **7**(2):112.

Fraser, F. C., and Rosen, J., 1975, Association of cleft lip and atrial septal defect in the mouse, *Teratology* **11:**321.

Gold, R. J. M., and Scriver, C. R., 1972, Properties of hair keratin in an autosomal dominant form of ectodermal dysplasia, *Am. J. Hum. Genet.* **24:**549.

Goltz, R. W., Henderson, R. R., Hitch, J. M., and Ott, J. E., 1970, Focal dermal hypoplasia syndrome, *Arch. Derm.* **101**(1):1.

Gray, I. M., Lowry, R. B., and Renwick, D. H. G., 1972, Incidence and genetics of Legg–Perthes disease (osteochondritis deformans) in British Columbia: Evidence of polygenic determination, *J. Med. Genet.* **9:**197.

Green, M. C., 1968, Mechanism of the pleiotropic effects of the short-ear mutant gene in the mouse, *J. Exp. Zool.* **167**(2):129.

Gruneberg, H., 1952, Genetical studies on the skeleton of the mouse. IV. Quasi-continuous variations, *J. Genet.* **51:**95.

Gruneberg, H., 1960, Genetical studies on the skeleton of the mouse. XXV. The development of syndactylism, *Genet. Res.* **1**(2):196.

Gruneberg, H., 1963, *The Pathology of Development,* Blackwells, Oxford, U.K.

Harris, M. W., and Fraser, F. C., 1968, Lid gap in newborn mice: A study of its cause and prevention, *Teratology* **1**(4):417.

Johnston, M. C., and Listgarten, M. A., 1972, The migration, interaction and early differentiation of oro-facial tissues, *in: Development Aspects of Oral Biology* (H. S. Slavkin and L. A. Baretta, eds.) p. 53, Academic Press, New York.

Jones, K. L., Smith, D. W., Hall, B. D., Hall, J. G., Ebbin, A. J., Massoud, H., and Golbus, M.S., 1974, A pattern of cranio-facial and limb defects secondary to aberrant tissue bands, *J. Pediatr.* **84:**90.

Juriloff, D. M., and Trasler, D. G., 1976, Test of the hypothesis that embryonic face shape is a factor in genetic predisposition to cleft lip in mice, *Teratology* **13** (in press).

Kurisu, K., Niswander, J. D., Johnston, M. C., and Mazaheri, M., 1974, Facial morphology as an indicator of genetic predispositions to cleft lip and palate, *Am. J. Hum. Genet.* **26:**702.

Landauer, W., 1955, Recessive and sporadic rumplessness of fowl: Effects on penetrance and expressivity, *Am. Nat.* **89:**35.

Larsson, K. S., 1962, Studies on the closure of the secondary palate. IV. Autoradiographic and histochemical studies of mouse embryos from cortisone-treated mothers, *Acta Morphol. Neerl.-Scand.* **4:**369.

Mendell, N. R., and Elston, R. C., 1974, Multifactorial qualitative traits: Genetic analysis and prediction of recurrence risks, *Biometrics* **30:**41.

Morton, N. E., Yee, S., Elston, R. C., and Lew, R., 1970, Discontinuity and quasi-continuity: Alternative hypotheses of multifactorial inheritance, *Clin. Genet.* **1:**81.

Newcombe, H. B., 1963, The phenodeviant theory, *in: Congenital malformations* (M. Fishbein, ed.), pp. 345–349, International Medical Congress Ltd., New York.

Nora, J. J., 1971, Etiologic factors in congenital heart disease, *Pediatr. Clin. North Am.* **18**(4):1059.

Nora, J. J., Sommerville, R. J., and Fraser, F. C., 1968, Homologies for congenital heart diseases: Murine models, influenced by dextroamphetamine, *Teratology* **1**(4):413.

Passarge, E., 1973, Genetics of Hirschsprung's disease, *Clin. Gastroenterol.* **2**(3):507.

Poswillo, D., 1966, Observations of fetal posture and causal mechanisms of congenital deformity of palate, mandible, and limbs. *J. Dent. Res.* **45**(Suppl. 3):584.

Poswillo, D., 1973. The pathogenesis of the first and second branchial arch syndrome, *Oral Surg., Oral Med., Oral Pathol.* **35**:302.

Poswillo, D., 1975, The pathogenesis of the Treacher Collins syndrome (mandibulofacial dysostosis), *Br. J. Oral Surg.* **13**:1.

Rice, J. M., 1973, An overview of transplacental chemical carcinogenesis, *Teratology* **8**:113.

Sang, J. H., 1963, Penetrance, expressivity and thresholds, *J. Hered.* **54**(4):143.

Shih, L. Y., Trasler, D. G., and Fraser, F. C., 1974, Relation of mandible growth to palate closure, *Teratology* **9**:191.

Sisken, B. F., and Gluecksohn-Waelsch, S., 1959, A developmental study of the mutation "Phocomelia" in the mouse, *J. Exp. Zool.* **142**(1,2,3,):623.

Smith, C., 1971, Recurrence risks for multifactorial inheritance, *Am. J. Hum. Genet.* **23**(6):578.

Smith, C., 1974, Concordance in twins: Methods and interpretation, *Am. J. Hum. Genet.* **26**(4):454.

Tenenhouse, H. S., Gold, R. J. M., Kachra, Z., and Fraser, F. C., 1974, Biochemical marker in dominantly inherited ectodermal malformation, *Nature* **251**:431.

Trasler, D. G., 1965, Strain differences in susceptibility to teratogenesis, Survey of spontaneously occurring malformations in mice, *in: Teratology* (J. G. Wilson and J. Warkany, eds.), pp. 38–55, Univ. of Chicago Press, Chicago.

Trasler, D. G., 1968, Pathogenesis of cleft lip and its relation to embryonic face shape in A/J and C57BL mice, *Teratology* **1**(1):33.

Trasler, D. G., and Fraser, F. C., 1958, Factors underlying strain, reciprocal cross, and maternal weight differences in embryo susceptibility to cortisone induced cleft palate in mice, *Proc. X Int. Cong. Genet.* **2**:296.

Trasler, D. G., and Fraser, F. C., 1963, Role of the tongue in producing cleft palate in mice with spontaneous cleft lip, *Dev. Biol.* **6**:45.

Trasler, D. G., Walker, B. E., and Fraser, F. C., 1956, Congenital malformations produced by amniotic-sac puncture, *Science* **124**:439.

Verrusio, A. C., 1966, Biochemical basis for a genetically determined difference in response to the teratogenic effects of 6-aminonicotinamide, Ph. D. thesis, McGill University, Montreal, Canada.

Walker, B. E., and Fraser, F. C., 1956, Closure of the secondary palate in three strains of mice, *J. Embryol. Exp. Morphol.* **4**:176.

Walker, B. E., and Fraser, F. C., 1957, The embryology of cortisone-induced cleft palate, *J. Embryol. Exp. Morphol.* **5**(2):201.

Watney, M., and Miller, J. R., 1964, Prevention of a genetically determined congenital eye anomaly in the mouse by the administration of cortisone during pregnancy, *Nature* **202** (4936):1029.

Weinstein, E. D., and Cohen, M. M., 1966, Sex-linked cleft palate—report of a family and review of 77 kindreds, *J. Med. Genet.* **3**(1):1.

Williams, M., and Fraser, F. C., 1966, Hydrotic ectodermal dysplasia—Clouston's family revisited, *Can. Med. Assoc. J.* **96**:377.

Wright, S., 1934, The results of crosses between inbred strains of guinea pigs, differing in number of digits, *Genetics* **19**:537.

Wynne-Davies, R., 1965, Family studies and aetiology of club foot, *J. Med. Genet.* **2**(4):227.

Wynne-Davies, R., 1970, A family study of neonatal and late-diagnosis congenital dislocation of the hip, *J. Med. Genet.* **7**(4):315.

Causes of Maldevelopment

II

Action of Mutagenic Agents

4

H. V. MALLING AND J. S. WASSOM

I. INTRODUCTION

The biochemical processes which ensure the stability of genetic material transmitted from one generation to another are complex. A delicate balance exists between the fidelity and stability of this material which enables organisms to proliferate in their environment. Unfortunately, widespread use has been made of both man-made and naturally occurring chemical formulations which can interfere with these genetic processes. There is growing concern that these agents may be significantly contributing to the health burden of man by increasing the frequency of genetic diseases.

In the present chapter, we have attempted to provide some insight into this issue by reviewing what is currently known about the mammalian gene and how mutations may occur in DNA, the genetic material common to all organisms excluding, of course, RNA viruses. We have also catalogued a number of widely used compounds according to their mode of action and genetic effects. In order to give some perspective about the occurrence of

H. V. MALLING · Biochemical Genetics Section, Environmental Mutagenesis Branch, National Institute of Environmental Health Sciences, P. O. Box 12233, Research Triangle Park, North Carolina 27709. J. S. WASSOM · Environmental Mutagen Information Center, Information Center Complex/Information Division, Oak Ridge National Laboratory, P.O. Box Y Bldg. 9224, Oak Ridge, Tennessee 37830. This work was supported by the National Institute of Environmental Health Sciences, the National Cancer Institute, and the Energy Research and Development Administration. Oak Ridge National Laboratory is operated by Union Carbide Corporation for the Energy Research and Development Administration.

potential mutagens in the environment, we have provided the reader with a listing of the various uses and/or applications which several chemicals have in our environment.

II. ENVIRONMENTAL MUTAGEN INFORMATION CENTER (EMIC)

Since the primary source of all the information used and discussed in this chapter came from the Environmental Mutagen Information Center, we thought it useful to provide the reader with a concise description of this specialized information center. The Environmental Mutagen Information Center (EMIC) is a computerized informational facility which was organized by the American Environmental Mutagen Society in 1969 and is located at the Oak Ridge National Laboratory. The society's commission to the center was to assist in the collection and dissemination of chemical mutagenesis information. As of July 1, 1976, the center had 17,500 references in its computer file, most of which (89%) were published during the years 1968–1976. The operational concept and the programs used by the center are designed primarily for the benefit of scientists and health administrators but can be used by any individual knowledgeable in the field.

EMIC screens the scientific literature for information to add to its data base by searching several large multidisciplinary resources for chemical mutagenesis information. (For a complete review of EMIC's literature screening program, see Wassom, 1973). Also, many (112) key genetic, cytological, and biochemical journals are continuously monitored for information. Extreme care is taken to ensure that data obtained from papers selected to become a part of the EMIC files are accurately recorded. Several checks are built into the system to guarantee this. The following scheme summarizes the operational format EMIC uses in selecting data.

Screening and Selection of Information from the International Scientific Literature.

A. *Search Methods*
1. Computerized searches of large data bases (e.g., *Chemical Abstracts*)
2. Manual screening of key source journals
3. Correspondents*

B. *Selection Criteria*
Articles are selected which report on or review the testing of chemicals and/or biologicals for the induction of
1. All types of mutations (such as single base-pair changes, chromosome aberrations, changes in ploidy, etc.)

*Individuals are located in various countries throughout the world who screen the literature of their respective countries for material to be considered for the EMIC data file.

2. Effects on DNA (such as binding, breaks, base modification, repair, etc.)
3. Mitosis
4. Meiosis
5. Oogenesis
6. Spermatogenesis

A typical entry, after it has been initially processed, would look like the example at the bottom of the page.

This information center was created and organized to benefit those individuals engaged in work in the field of chemical mutagenesis, and we believe it serves an important role in this area. Even though considerable information is available to the teratologist from this center, it cannot serve all his needs. Therefore, an information center designed along the same operating principles as EMIC to handle teratological information is now being organized by Dr. Robert E. Staples of the National Institute of Environmental Health Sciences in collaboration with EMIC.

III. TYPES OF MUTATIONS

A. Introduction

A mutation can be defined as a macromolecular or micromolecular change in cellular DNA. The central dogma which is accepted as the basis of our understanding of how DNA functions is that coded messages in this macromolecule are transcribed in RNA which then translates them into the

---

Author(s):   Leonard, A.; Linden, G.

Title: Observation of dividing spermatocytes for chromosome aberrations induced in mouse spermatogonia by chemical mutagens

Source: Mutation Research 16:297–300, 1972

Key Index Terms:
  Organism (common classification):   Mammal, mouse
  Organism (specific classification):   Mus, BALB/C
  Sex of Test Animal(s):   Male
  Type of Study: *In vivo*
  Biological Endpoint of Test: Chromosome aberrations; translocations
  Agent(s) and Chemical Abstracts Service Registry No(s):
    Trenimon (68-76-8); Endoxan (50-18-0); MMS (66-27-3);
    Propyl methanesulfonate (1912-31-8); Isopropyl methane-
    sulfonate (926-06-7); Myleran (55-98-1).

---

primary structure of proteins. There are, however, several types of DNA which do not follow this precept. For example, there is one class of DNA in which the triplet code is only transcribed but is never translated as in the case of the DNA triplets which code for transfer-RNA. In another class, the triplet code may not even be transcribed at all, as in the case of repetitive DNA. A mutation in one of these classes will have different effects on the organism than a mutation occurring in DNA which follows the central dogma mentioned above. In order to obtain a better understanding of the processes which lead to a mutation, one must first become familiar with the structure of genetic material. This understanding should begin with the primary structure of DNA and eventually reach the complexity of the chromosome in higher organisms.

B. The Basic Structure of the Inherited Material

1. DNA and Unique Coding

It is now well known that most DNA is composed of two nucleotide chains which are wound around each other to form a double helix. The backbone of this helix consists of a sugar moiety, 2-deoxyribose, which is linked to phosphate at the 3' and 5' positions. This arrangement induces polarity in the chain; if in one chain the phosphate is attached to the 3' position of the 2-deoxyribose below it and to the 5' position of the 2-deoxyribose above it (deoxyribose-3'-P-5'-deoxyribose), then in the other strand of the double helix, the lower deoxyribose will have the phosphate attached at the 5' position and to the 3' position of the deoxyribose located above it (deoxyribose-5'-P-3'-deoxyribose). Also attached to each deoxyribose moiety is one of four bases which may either be adenine (A), thymine (T), guanine (G), or cytosine (C). These bases are arranged into the middle of the double helix and form the core of the molecule with the sugar–phosphate backbones on the periphery.

These four bases pair in a unique manner—A normally pairs with T and G normally pairs with C. Each single strand of the double-stranded DNA helix therefore complements the other (Fig. 1). The sequence of the bases which are arranged along one DNA strand of a gene determines the amino acids which will compose the specific protein coded by this gene. The code consists of triplet combinations made from the four bases. Arrangements of these four bases into three-letter groups can be done for 64 different codes. There are, however, only 20 amino acids. This gives room for redundancy; in fact, one amino acid may be coded by several three-letter DNA words (e.g., leucine is coded by six different codons). During the transcription of the DNA triplet code into mRNA, guanine is paired with cytosine, but adenine is paired with uracil (U). Thymine in DNA corresponds to uracil in RNA, and therefore uracil has the same pairing properties as thymine. At the time of translation, when the tRNA locates its proper code in the mRNA, guanine pairs with cytosine and uracil with adenine. Since chromosomal DNA is one long strand, there has to be a starting sign to initiate reading in the mRNA and a stop sign to terminate it. There is therefore an initiation codon (possibly AUG which is the code for methionine and/or GUG which is the code for valine or formyl-methionine) and a termination codon (UAA) in mRNA.

2. Physical Structure of the Chromosome

The structural model for chromosomes has changed considerably during the last 20 years (for a review see DuPraw, 1970). An early model visualized the chromosome in the shape of a lampbrush with an inner core to which the separated genes were attached. Electronmicrographs did not, however, reveal this inner core but showed that chromosomes consisted of fibers that are at least 500 microns in length (Cairns, 1966). The inner core of these fibers seems to be composed of DNA. The diameter of the fibers varies from 100 Å to 500 Å in diameter, depending on the amount of protein (histones) complexed with the DNA (DuPraw, 1973).

3. Repetitive DNA

During evolution from microorganisms (excluding viruses) to higher animals, there has been an approximate 1000-fold increase in the amount of

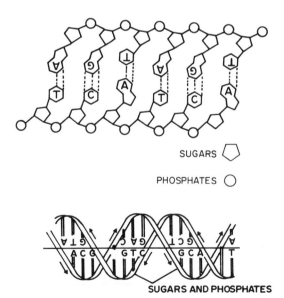

Fig. 1. Schematic model of the DNA double helix.

DNA per haploid genome (Sparrow *et al.*, 1972). Fungi such as *Saccharomyces cerevisiae* have the lowest haploid DNA content (4.5×10^7 nucleotides) found in any eukaryotic cell. The average size of a gene is estimated to consist of only 1500 base pairs (Watson, 1965). The yeast cell contains enough DNA to code for 13,000 genes. Crow and Kimura (1970) on the other hand, have estimated the number of genes in a mammal to be about 30,000. The mammalian haploid genome contains 6×10^9 nucleotides. That means that only 6–9% of the available DNA functions as genes (Kimura and Ohla, 1971). Most of the remaining DNA is not transcribed. The genes which are translated are likely to have unique sequences. Melting-point analysis of mammalian DNA, however, indicated many different classes of repetitive DNA. In yeast, the amount of repetitive DNA is low, indicating that the number of genes necessary to carry out the functions of a simple eukaryotic cell is close to 10^4. Presumably the evolution from yeast to man has only required a threefold increase in the number of functional genes. The main increase in the amount of DNA has been in the different classes of repetitive DNA.

4. The Organization of the Gene

The base-sequence triplet code in the gene is transcribed into an anticode in the RNA (HnRNA, heterogeneous nuclear RNA). The code in the RNA is then translated into the amino acid sequence of proteins. The gene transcribes, however, a much bigger piece of RNA than the RNA (mRNA) which is translated in the cytoplasm into proteins. The HnRNA undergoes extensive posttranscriptional modifications, including the addition of a segment of polyadenylic acid [poly (A)] (200 nucleotides) at the 3' terminal position. The poly(A) is not transcribed from the DNA but originates in the nucleus as a posttranscriptional addition product to HnRNA. The studies on poly(A) have been summarized recently (Darnell *et al.*, 1973). The HnRNA with poly(A) seems to contain 5000–50,000 nucleotides, of which only 3000–4000 nucleotides become mRNA (Crippa *et al.*, 1973). This mRNA is approximately two to three times bigger than the piece which is translated into the protein. Between each gene there seems to be a piece of repetitive DNA approximately 300–500 nucleotides long. A general, but too simple, view of the mammalian gene, is given in Fig. 2. This model does not account for the variation in size of HnRNA and how series of duplicated genes are arranged in, for instance, the complex locus for β-hemoglobin. In this complex locus, closely related genes are placed end-to-end on the DNA in the following sequence: gamma, alpha, delta, and beta. There is so little difference in the amino acid composition of the polypeptides specified by these genes that it appears they have arisen through tandem duplications. The ratio of the three types of hemoglobins found in the human body varies with the stage of development. Gamma and alpha chains are the fetal hemoglobins and are produced by the fetus, whereas alpha, beta, and delta chains are produced by adults. The relation-

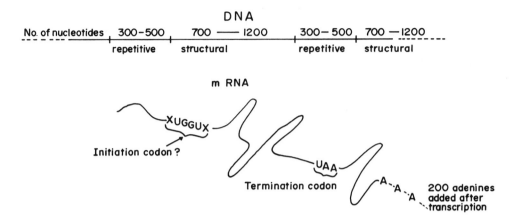

Fig. 2. Simplified model of the mammalian gene and mRNA.

ship between, and the control of, the tandem array of genes responsible for the production of hemoglobins is not known. The subject of human hemoglobins is explored further in the section on small deletions (Section III. E).

It can be concluded that the complexity of the mammalian gene surrounded by the repetitive DNA is likely to give a different spectrum of mutations than that obtained in microorganisms using the same mutagen.

5. Extranuclear DNA

DNA, which is transcribed and translated, occurs not only in the nucleus but also in the various cytoplasmic organelles (Borst, 1972). The mitochondria of eukaryotes contain closed circular duplexes (mtDNA) with a molecular weight of approximately 10^7 daltons, corresponding to 5–6 microns in contour length and 15,000 bases in each mitochondrion.

A mammalian cell may contain a few hundred mtDNA molecules. This DNA specifies at least mitochondrial, ribosomal, and mitochondrial transfer RNAs. In yeast, the mtDNA is susceptible to various intercalating dyes such as acridines and ethidium bromide (for additional information see Table 2). During treatments with these compounds, mtDNA synthesis stops and degradation of preexisting DNA starts (Goldring *et al.,* 1970; Perlman and Mahler, 1971). These actions result in the induction of mutations, referred to as petite mutations, which make it necessary for cells to rely on anaerobic fermentation for growth and therefore form smaller colonies. DNA synthesis is also inhibited by these dyes in mammalian cells (Smith *et al.,* 1971). At least two types of mutants induced in mitochondrial DNA have been observed: (1) mutants resistant to inhibitors of mitochondrial synthesis, and (2) mutants resistant to inhibitors of uncouplers of oxidative phosphorylation (Bunn *et al.,* 1970). Cytoplasmic mutations such as these show maternal inheritance.

C. Base-Pair Substitutions

The four bases (adenine, guanine, cytosine, and thymine) of DNA are paired in the unique manner described earlier: adenine always pairs with thymine, and guanine always pairs with cytosine. As in most biological systems, a certain possibility for error exists and correct pairing is not always an absolute certainty. Occasional mistakes may occur during replication. Adenine may, instead of pairing with thymine, pair with cytosine; guanine, instead of pairing with cytosine, may pair with thymine. In the same way, thymine may pair with guanine or cytosine with adenine. Adenine and guanine are purines; thymine and cytosine are pyrimidines (Fig. 3). In this type of erroneous pairing, a pyrimidine is exchanged for a pyrimidine and a purine for a purine. This type of mutation is called a base-pair transition (Fig. 4). Another type of base-pair substitution exists in which a purine is exchanged for a pyrimidine or *vice versa*. This type of mutation is called a base-pair transversion (Fig. 4).

An exchange of one base pair for another can result in the exchange of one amino acid for another, or stated in another way, the triple base-pair code in the DNA is read but differs from the original code. These mutations are called missense mutations. The effect of a missense mutation on the function of the polypeptides depends on its site in the DNA and the type of amino acid which is exchanged. Some amino acids are charged molecules and, as a result of a base-pair substitution, a neutral amino acid may be exchanged for another neutral, positive, or negative charged amino acid or any other combination. The last two exchanges result in an overall change in the charge of the protein. Thus, this mutant protein will travel with a different velocity in an electric field and can therefore be recognized by electrophoresis. In an average protein, approximately one third of all possible base-pair substitutions lead to changes in the overall charge. Since changes in the charge are technically easy to recognize, many polymorphisms have been found by screening various populations of humans and other living systems. From these studies it can be concluded that many charge differences do not lead to differences in enzyme activity. A missense mutation can result in a polypeptide with an impaired function (leaky mutations), or it can lead to totally nonfunctional proteins.

The terminator codon in mRNA is UAA (called ochre). Normally no tRNA will pair with this codon. Therefore, during the synthesis of proteins, this codon is not read and the elongation of the protein stops at this point. Another codon, UAG (called amber), has a similar effect. Since these codons are located at the end of the gene, they terminate the translation of the mRNA message which specifies the amino acid composition of a polypeptide chain. Tryptophan has the RNA codon UGG; a base-pair transition of the second G to an A will result in a new codon which is UAG, a termination codon. Several of the codons for amino acids can be changed via base-pair substitutions, and these changes lead to either a UAG or a UAA codon which results in prema-

<u>PURINES</u>

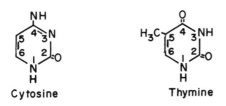

Adenine   Guanine

<u>PYRIMIDINES</u>

Cytosine   Thymine

Normal base pairing patterns

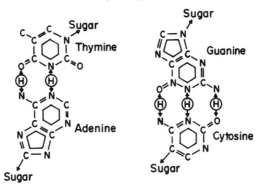

Fig. 3. Base constituents of DNA and pairing patterns
of these bases.

ture termination of polypeptide synthesis. Such changes have been given the
appropriate name of nonsense mutations.

D. Frameshift Mutations

Changes in the DNA code can also occur by a loss or addition of one or
several base pairs, e.g., if the sequence in DNA is GTA-GTA-GTA-GTA-
GTA, we would read GTA five times (GTA is the genetic DNA code for
histidine) and consequently there would be five histidine codons in a row
transcribed to the mRNA. If we omit the second T of this message, it would
then read: GTA-GAG-TAG-TAG-TA? and would be transcribed into the

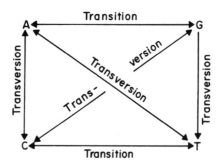

Fig. 4. Possible exchanges of bases in base-pair substitutions.

mRNA codons for histidine (CAU), histidine (CUC), isoleucine (AUC), isoleucine (AUC), etc. Most of the DNA code which follows a frameshift mutation will more than likely be translated into the wrong amino acids. If this erroneous translation of the message runs through a stretch of the polypeptide essential for its function, we are likely to end up with a nonfunctional protein. For instance, if the GTA sequence in the DNA mentioned above had continued and included the triplet pair AAT-TTT (GTA-GTA-GTA-GTA-GTA-AAT-TTT), the loss of the second T would then result in GTA-GAG-TAG-TAG-TAA-ATT-TT?. ATT would subsequently be transcribed to a nonsense codon (UAA) in the mRNA and the translation would therefore be terminated. The opposite can also occur, i.e., a frameshift mutation can nullify a natural terminator codon. Since the mRNA in higher organisms contains bases beyond the terminator codon, the polypeptide being formed will be elongated until a codon for a terminator occurs. This, more than likely, is what has happened in the production of the abnormal human hemoglobin called hemoglobin Wayne. In this condition, the last three amino acids of the β-hemoglobin have been changed and an additional five amino acids have been added.

On either side of a mammalian gene there are regions of DNA which are not translated. For example, if the natural terminator triplet (ATT) in DNA is mutated by a base-pair substitution to the triplet of phenylalanine which is TTT, then the terminator codon will be read as a phenylalanine, and this amino acid would eventually be added to the polypeptide chain. The translation of the mRNA would continue until another nonsense or terminator codon is encountered. This type of mutation is probably the cause of hemoglobin Constant Spring in which normal hemoglobin has five additional amino acids attached to the end of it (Clegg *et al.*, 1971). Additional discussion on the production of abnormal hemoglobins is found in the next section.

Of the available 64 triple codon combinations, 61 are used to code for the 20 essential amino acids. A high degree of redundancy therefore exists for the coding of certain amino acids. For instance, there are six different codons for leucine. Two of these are UUA and UUG. A base-pair transition of the last letter in the UUA codon from A to G will still provide a leucine codon at the original base. This type of mutation will therefore have no influence on the function of the polypeptide. Of the 649 possible codon changes by base-pair substitutions, 134 (25%) will not result in any change in the amino acid sequence (Kimura, 1968).

E. Small Deletions

The physical removal of one or a few genes is called a small deletion. Small deletions are usually not recognizable by the light microscope except in situations involving the polytene chromosomes of *Drosophila*. The effect produced by the deletion of one gene is often similar to that in the class of point mutations in which no functional protein is synthesized. Various test systems have been developed for simultaneous detection of point mutations and small deletions, e.g., specific locus mutation systems. A very efficient way of studying specific locus mutations in mice has been developed by Russell (1951). This assay system detects mutations by using easily recognized morphological characters which are determined by known genetic alterations. In the actual experiment, the animal carrying the wild-type genome (+/+) is treated and then mated to a strain which is homozygous for the recessive mutant gene in that particular locus, e.g., a coat color (a/a). Most of the progeny from this cross will be heterozygous wild-type (+/a). If a mutation had occurred in the wild-type allele, the mutant offspring would be homozygous for the mutant coat color (a*/a) (a* denotes the induced mutation). This animal will then contain one mutant allele from the homozygous mutant parent and one newly induced mutant allele from the treated parent. The new mutation induced in the treated wild-type parent can either be a point mutation or a small deletion in the region of the wild-type a gene. Even if the genetic alteration induced by a mutagenic agent has a recessive lethal effect due to the deletion of neighboring genes as well as the wild-type a locus, the first generation progeny would still be viable and express the mutant phenotype. For example, after exposures to acute X-rays, up to 78% of the newly induced mutations were found to be deletions of not only the wild-type locus but also neighboring genes (Russell and Russell, 1959).

A similar specific locus system has been developed in the fungus *Neurospora crassa* by de Serres (de Serres and Osterbind, 1962; de Serres and Malling, 1971). This well-defined assay system is based on two closely linked genes which control two intermediate steps in the purine biosynthesis pathway [*adenine-3A, (ad-3A); adenine-3B, (ad-3B)*]. A mutation occurring in either of these two genes leads to a requirement for adenine and the accumulation of a purple pigment in the mutant cells. The mutants recovered are heterozygous for these two genes. In this system both point mutations and chromosome deletions can be detected. Results obtained with this system can, however, differ from those obtained in the system just described for mice. For instance, the X-ray doses which gave up to 78% deletions in mice gave only 1% deletions in *Neurospora*. A possible explanation for this different spectrum of genetic alterations may be due to the fact that the mammalian nucleus contains far more repetitive DNA than the nucleus of fungi. It is known that whenever chromosomal DNA has been broken at two places, these breaks can either be repaired back to their normal condition or the region between the breaks can be excised and lost which results in a deletion. It is very likely that, because of the greater amount of repetitive DNA in the mammalian nucleus, the latter process has a higher probability of occurring in the mouse than in *Neurospora* which has less repetitive DNA in its nucleus.

Very interesting small deletions have been discovered in man. It is known that three different hemoglobins have an amino acid compostion closely related to the β chain. They are the two fetal hemoglobin chains G_γ and A_γ and the δ chains. The genes coding for these three hemoglobin chains and the one coding for the β chains are closely linked. The similarity of the amino acid sequences in these four genes suggests that they have arisen by tandem duplications. Individuals carrying the mutant gene from hemoglobin Lepore produce a composite hemoglobin chain in which the first part consists of the δ chain and the latter part the β chain (Badr *et al.*, 1973). In comparison, a normal individual produces a separated δ and β chain. Similarily, hemoglobin Kenya is a product of a γ–β fusion gene (Kendall *et al.*, 1973). Both hemoglobin Lepore and Kenya can be explained on the basis of an uneven crossing-over event which may have been facilitated by the repetitive sequences in duplicated genes.

F. Types of Genes

In a complex organism like a mammal, an enzyme may vary in structure, amount, and occurrence. These variables are either particularly or totally under genetic control. Paigen and Ganschow (1965) have divided the genes determining enzyme functions into four classes:

Structural Genes. These determine the primary amino acid sequence of the polypeptide chain(s) which make up the protein molecule. This group also includes any additional elements that may influence its secondary, tertiary, or quaternary structure.

Regulatory Genes. These specify the nature of the control systems to which the protein or its synthetic apparatus responds as part of its normal function in the cell. This concerns the transient and reversible changes in protein concentration that occur in response to changes in cellular environment or cellular functions.

Architectural Genes. These are concerned with the agents responsible for the integration of a protein into the structure of cells. The function of a protein may be profoundly affected by the site at which it exists, and indeed the function of an enzyme may become irrelevant anywhere except at its proper location.

Temporal Genes. These control the time and place where the three primary gene classes are activated and provide the fundamental program for the sequence of events in differentiation.

A mutation in any of these genes coding for or regulating a particular enzyme will produce a different phenotypical effect. Both dominant and recessive mutations exist in the human population. A mutation in a structural gene is usually recessive. In man, dominant mutations are more common than recessive, which may be due to the many genes involved in the regulation of a single structural gene.

G. Chromosome Abnormalities

Point mutations and small deletions are usually not recognizable under the light microscope, as we have mentioned earlier. The genome can, however, undergo more dramatic changes. A piece of a chromosome can be detached and lost (deletion) or duplicated and inserted into the genome. A piece of a chromosome can also be exchanged with another chromosome (translocation) or turned around within a chromosome (inversion). A whole chromosome may be deleted (resulting in monosomy) or duplicated (trisomy). A change in the whole set of chromosomes can result in either haploidy or polyploidy (see chapter by Hsu and Hirschhorn, Vol. 2).

IV. GENERAL EFFECTS OF MUTATIONS

A. Types of Effects

1. Dominant Lethals

The induction of dominant lethal mutations has been extensively studied in laboratory animals, and today the assay system for dominant lethality is one of the most widely used tests for mutagenicity in whole animals. Dominant lethal mutations are those that cause the death of the heterozygous zygote; hence their genetic nature cannot be tested by breeding experiments, but only inferred from indirect evidence. The male germ cells show a dramatic degree of stage sensitivity which differs for the various mutagenic agents (Ehling *et al.,* 1968). Dominant lethals can express themselves soon after fertilization and result in preimplantation loss or they can be expressed after implantation and are observed as black moles on the uterus wall. Chromosome analysis and observations of micronuclei at early stages indicate that many dominant lethals are caused by chromosome abnormalities.

2. Point Mutations

Many point mutations have little or no effect on the individual and are expressed as normal polymorphism within a population. McKusick (1975) in his encyclopedic treatment of Mendelian traits in man has listed 1142 identifiable abnormal traits which have known modes of inheritance, and 583 of these are listed as being dominant. An additional 1194 traits were also listed as being "in limbo" which means that their mode of inheritance has not been completely ascertained. Even though many point mutations have little or no effect on the carrier, some result in diseases.

B. Consequences to Society

1. The Effect of an Increased Mutation Rate

Most of the data for specific locus mutations in mammals have been derived from experiments in which ionizing radiation was used as the mutagen. Some data are available for chemical mutagens although it has been shown that chemicals can induce specific locus mutations (Ehling, 1974).

The induced mutation rate per locus per rem (roentgen equivalent man) in the mouse is 0.25×10^{-7}; if we assume that the spontaneous rate is 0.5×10^{-6}, then one rem will give an increase in the mutation rate of 1:20. This ratio will, of course, vary with the estimation of the spontaneous rate. The National Research Council of the National Academy of Sciences has taken 5 R (roentgens) as the 30-year limit for the population average in the Radiation Protection Guides. In the BEIR* report prepared by the National Research Council, it is calculated (based on the assumption that a doubling dose is equal to 20 rem) that the effect of 5 rem per generation on a population of one million will, in the first generation, add 1000 new disease cases to the 60,000 already present and will add another 7500 when at equilibrium. It should be pointed out that there are not enough data at the present time to even try to estimate the increase in mutation rates induced by other environmental agents which have mutagenic activity.

---

*The BEIR report—Advisory Committee on the Biological Effects of Ionizing Radiation, The Effects on Populations of Exposure to Low Levels of Ionizing Radiations (National Academy of Sciences–National Research Council, Washington, D.C., 1972).

2. Comparison of the Mutation Spectrum between Ionizing Radiation and Chemicals

The general mutation spectrum varies between the types of ionizing radiation and chemicals. The best data available for comparing the variations in the mutation spectrum between chemicals and radiation have been noted by de Serres and his colleagues in a series of papers published on this subject (see de Serres and Malling, 1971). In these publications they studied the mutation spectrum induced by radiation and chemicals in the fungus, *Neurospora crassa*. In general, radiation was found to be effective in breaking chromosomes and was extremely nucleotoxic. Most alarming, however, was the ability of some chemicals to induce high frequencies of point mutations at concentrations which yielded low lethality. As we have seen earlier, selection against rare recessive point mutations is ineffective. Because of the low lethality of these compounds and the fact that an increase in the frequency of point mutations has to be substantial before they are noticed, a population could be exposed to agents which may induce point mutations a long time before their consequences are recognized.

V. SOMATIC MUTATIONS

It is postulated that mutations in somatic cells occur in the same way as mutations in the germ cells with the overriding difference being, of course, the absence of meiosis and conjugation before the mutation can be expressed. Work with tissue culture has shown that many different types of mutations can be induced in mammalian cells "in culture," e.g., point mutations, chromosome aberrations, etc. Chromosome aberrations have also been observed in mammalian cells *in vivo*, but no system has yet been developed to detect point mutations *in vivo*. Mutations in somatic cells *in vivo* have formed the basis for several hypotheses to explain aging and the occurrence of cancer. The role of somatic mutations in the aging process is still uncertain, but several different results have indicated that somatic mutations may lead to an increase in the probability of a cell to undergo neoplastic transformation. Current research is showing us that there is a good correlation between carcinogenesis and mutagenesis (de Serres, 1975). We can therefore assume with some certainty that there are mutant genes exhibiting either recessive or dominant Mendelian inheritance in the human population that increase the probability for an individual to develop tumors (Fraumeni, 1973). Mutations in somatic cells are either dominant or codominant in character, and in order for a recessive mutation to express itself in a diploid cell, it has to undergo mitotic recombination.

There is considerable research work under way directed toward the development of an assay system for the detection of somatic mutations *in vivo* (see chapter by Malling, Vol. 3). Such a method is the only feasible way by which we can hope to monitor the exposure of the human population to mutagenic agents. Two of the methods now under development look very promising. One of these methods is based on the detection of mutant hemoglobins in red blood cells and the other is based on the detection of mutant variants in sperm by histochemical techniques. With such systems available, we should be able to detect early exposure to carcinogens many years prior to the occurrence of cancer. This capability will make it possible to minimize further exposure to the causative agent either by limiting its use or eliminating it from the environment altogether.

VI. INFLUENCE OF REPAIR ON MUTAGENESIS

Prokaryotes and eukaryotes possess an array of enzymes which can repair spontaneous and induced damage in DNA. After UV irradiation, cyclobutane dimers are formed between neighboring pyrimidine bases in the same DNA strand (see Harm *et al.*, 1971; Sutherland *et al.*, 1973). This damage is repaired by photoreactivation enzymes which open up these rings. Other types of DNA damage are repaired by a set of enzymes which excise the strand of damaged DNA together with approximately 100 nucleotides. This process, which is called excision repair, leaves part of the DNA single stranded. New bases are then incorporated into this area by means of unscheduled DNA synthesis. (For a recent review, see Lieberman, 1975.) After irradiation with X-rays, a third repair system seems to be activated during which only a few bases are excised. Many mutants have been isolated in which various steps of the repair process have been blocked. In man, cells from xeroderma pigmentosum patients are unable, in the great majority of cases, to repair UV-type lesions.

Some of these repair processes may be error-prone and result in mutations, with the possible exception of the photoreactivating enzyme repair systems which represent, for the most part, a less error-prone process. Mutants which are blocked in one repair process may utilize another process to repair the damage and thereby change from a relatively error-free process to one that is more error-prone. Mutants induced as a result of malfunctioning repair processes have a higher sensitivity than the normal wild type to the mutagenic action of some chemicals but not of others (Malling, 1971). This difference in sensitivity is due to their inability to repair specific types of damage to the DNA molecule. Repairless mutants have been used extensively in the prescreening of compounds for mutagenic activity in short-term testing programs. The spectrum of genetic alterations is also different between the repairless mutants and the normal wild types (de Serres, 1974). So even though there is a gain in sensitivity, there is a loss in the quantitative relevance of the data. The work with repairless mutants indicates the possibility that fractions of the human population may be much more sensitive to chemical mutagens than the average, and techniques should therefore be found to identify such individuals.

VII. MUTAGENIC AGENTS

The extensive list of chemical agents shown in Table 2 was selected from the Environmental Mutagen Information Center's Agent Registry. These compounds have been classified according to their primary effect on macromolecules, particularly DNA. With few exceptions, all of the agents shown have either been reported mutagenic in one or more assay systems or reported to have induced chromosome aberrations in one or more systems.

We urge the reader to keep in mind that the reaction mechanism reported between DNA and the chemicals listed in this table are those most often quoted in the literature. These reactions may or may not be the primary mechanism responsible for the induction of mutations; they may not even be directly involved at all. For example, the alkylation of guanine at the N-7 position, long thought to be the primary reaction mechanism leading to the induction of mutations, has been challenged and emphasis is now being placed on the alkylation of this base at the O-6 position as the responsible mechanism leading to the mutagenic event.

To further complicate our understanding of the mutagenic process, there is an array of factors which may influence and/or determine whether a chemical will induce a mutation. These factors have been assembled by Freese (1974) in a recent publication and are shown in Table 1. These factors must be

Table 1. Mechanisms Controlling the Frequency of Genetic Alterations Induced by External Agents (Mutagens)[a]

Absorption, penetration, transport (control entry of cell membranes, organs, blood, gonads, etc.)
Activation, inactivation (chemical or enzymic)
Excretion
Nuclear membrane, chromosome condensation (protect chromosomal DNA)
Specificity of nucleotide kinases and DNA replicase (avoids incorporation of wrong nucleotides into DNA but accepts some analogs of bases and nucleotides)
Excision of wrong bases and repair
Repair of single-stranded lesions (by copying of complementary strand and sealing of gap)
Repair of double-stranded lesions (by "stickiness" and joining of broken ends)
Recombination
pH, temperature
Removal of mutated cells (by phagocytosis, coating with antibodies, etc.)

---

[a]From Freese (1974), courtesy of Charles C Thomas, Publisher, Springfield, Illinois.

considered whenever one is evaluating the mutagenic potential of any chemical agent.

From the information obtained during the preparation of Table 2, we found that there were seven major categories which could be used when classifying chemicals according to their action on DNA. The following is a list of these categories with relevant information concerning each. There are several other minor categories but due to limited space we will not comment on them.

Alkylation. Presently, the largest class of mutagenic agents in the EMIC agent registry is alkylating agents. There are various sites on the DNA molecule which react with alkylating agents as evidenced in Table 2. Sites on the guanine moiety of the DNA, however, seem to be more reactive than those on the other bases, with the N-7 position being generally the most frequently alkylated site and the O-6 position reportedly the most important alkylated site for mutagenicity. It is therefore not surprising to find that the GC base pair of DNA is the more frequent mutated site than the AT pair. Alkylation of guanine may result in either its depurination and eventual removal from the DNA or cause it to pair erroneously thereby inducing mainly transition mutations, i.e., G:C→A:T. Alkylating agents which have more than one active alkyl group (polyfunctional) may also cross-link DNA and inhibit replication. It has also been suggested that mutations may arise during the repair of alkylated DNA where normal base pairs in close proximity to alkylated bases may be excised along with the damaged material. This action then could cause mutations during the repair process following excision repair. Errors could also occur in the repair of strand or chain breaks.

Arylation. Agents classed under this mechanism of action are made up primarily of antibiotics and polycyclic aromatic hydrocarbons. These compounds react with DNA either by means of covalent bonding with existing reactive functional groups or with reactive groups which may result from metabolic processes such as the epoxide moiety formed in some polycyclic aromatic hydrocarbons. Since many of these compounds are polycyclic, they have also been found to intercalate with DNA in the same manner as acridines (Green, 1970). From the available information, it appears that guanine is the base with which these agents most frequently form complexes, and as a result of this action, frameshift mutations may occur.

Intercalation. Compounds in this class form complexes with DNA after wedging themselves (intercalating) into the DNA helix. This action usually causes a distortion of the normal DNA structure brought about by an extension and unwinding of the deoxyribose–phosphate backbone. Such distortions can induce errors in replication by causing the insertion or deletion of a base pair (frameshifts). The structural changes caused by these compounds activate repair enzymes and mutations may arise as mistakes in the repair process. Acridine compounds constitute the largest number of agents in this group.

Also, intercalating compounds which have active side chains, e.g., alkylating groups, are double trouble since they have the ability to both alkylate and intercalate with DNA.

Base Analog Incorporation. The induction of point mutations with base analogs using nonmammalian assay systems is well documented. These agents interrupt the normal function of DNA by modifying the fidelity of replication. The presence of these pseudobases in DNA causes base-pairing errors which can result in transition mutations such as G:C→A:T or A:T→G:C. Purine base analogs may replace the normal DNA purine bases, and pyrimidine analogs may replace the normal DNA pyrimidine bases.

Metaphase Poisons. The agents included in this group have been classed as a result of their ability to mimic the action of the classic metaphase poison colchicine, hence the usage of the term C-mitotic agent. These compounds interfere with spindle formation and consequently chromosome segregation by forming complexes with protein components of the microtubules comprising the spindle fibers. These agents are capable of inducing chromosome structural changes as well as changes in ploidy.

Deamination. Two of the compounds grouped under this heading (nitrous acid and hydroxylamine) are classic mutagens. Their mechanisms of action are probably the best documented cases that can presently be found in the literature. All of the agents under this classification seem to have one feature in common: they all preferentially attack cytosine followed by guanine and to a lesser extent adenine. Thymine is not reactive with these agents since it does not have an amino group. Mutations induced as a result of deamination are of the transition type G:C→A:T or A:T→G:C.

Enzyme Inhibition. Agents classed in this group are those which either inhibit enzymes involved with the biosynthesis of normal purines or pyrimidines or inhibit the function of repair enzymes. These mechanisms are probably not directly responsible for the induction of mutations but rather create a situation which makes it more likely that mutations occur by other mechanisms. For example, interference of the normal synthesis of thymine creates a situation of thymine starvation which can lead to replication errors; in this case, noncomplementary insertions have been suggested as the probable mechanism of mutagenesis. Additionally, the inhibition of repair enzymes by agents in this class may either potentiate the activity of another agent or cause spontaneous errors to persist.

The extensive summary of experimental results shown in Table 2 is obtained from a vast literature which is, for the most part, not readily assembled by the general readership. This list, even though extensive, is only a superficial cataloguing of agents which may or may not prove harmful to human genetic material.

In closing this section, we would like to emphasize that there is no doubt that future investigations will clarify much of our current thinking on how mutations occur, and one must remember this when reviewing the material presented in this section and the rest of the chapter.

Table 2. Mutagenic Action of Chemical Agents

Agent (Chemical Abstracts Service Registry number)	Classification according to primary or most often reported effects on macromolecules[a]	Remarks on mode of action and metabolism[a]	Subchromosomal lesions produced in one or more systems[a]	Chromosome aberrations produced in one or more systems[a]	Dominant lethals (mammals only)[a]
Apholate (52-46-0)	Alkylating agent	c	+	+	c
1,3-Bis(2-chloroethyl)-1-nitrosourea (154-93-8)	Alkylating agent	c	+	+	c
Butylmethanesulfonate (1912-32-9)	Alkylating agent	Major alkylation product: N7-alkylguanine / Minor alkylation product(s)[c]	?	+	c
Captan (133-06-2)	Alkylating agent	Major alkylation product: N7-alkylguanine / Minor alkylation product(s)[c]	+	+	+
Chlorambucil (305-03-3)	Alkylating agent	Major alkylation product: N7-alkylguanine / Minor alkylation product(s)[c]	c	+	c
Cyclophosphamide (50-18-0)	Alkylating agent	Major alkylation product: N7-alkylguanine / Minor alkylation product(s)[c]	+	+	+
Diazomethane (334-88-3)	Alkylating agent	Major alkylation product: N7-alkylguanine / Minor alkylation product(s)[c]	+	c	c
Diepoxybutane (1464-53-5)	Alkylating agent	Major alkylation product: N7-alkylguanine / Minor alkylation product(s): N3-alkyladenine	+	+	?
Diethylnitrosamine (55-18-5)	Alkylating agent	Major alkylation product: N7-alkylguanine	+	?	0

Chemical	Classification	Alkylation products			
(continued)		Minor alkylation product(s): N3-alkyladenine. Presumably the active alkylating derivative is formed during oxidative dealkylation	+	+	+
Diethylsulfate (64-67-5)	Alkylating agent	Major alkylation product: N7-alkylguanine. Minor alkylation product(s): N3-alkyladenine, 06-alkylguanine	0	+	+
Dimethylnitrosamine (62-75-9)	Alkylating agent	Major alkylation product: N7-alkylguanine. Minor alkylation product(s): N3-alkyladenine, 06-alkylguanine. Presumably the active alkylating derivative is formed during oxidative dealkylation	?	+	+
Dimethylsulfate (77-78-1)	Alkylating agent	Major alkylation product: N7-alkylguanine. Minor alkylation product(s)[c]	0	+	+
Ethionine[b] (13073-35-3)	Alkylating agent; amino acid analog	[c]	[c]	+	+

[a] Experimentation from both *in vivo* and *in vitro* studies were considered.
[b] Indicates chemicals which have more than one CAS Registry number representing salt forms.
[c] No information available in the EMIC data base.
[d] Questionable results, ?; negative results reported, 0; positive results reported, +.

(cont'd)

Table 2 (cont'd)

Agent (Chemical Abstracts Service Registry number)	Classification according to primary or most often reported effects on macromolecules[a]	Remarks on mode of action and metabolism[a]	Subchromosomal lesions produced in one or more systems[a]	Chromosome aberrations produced in one or more systems[a]	Dominant lethals (mammals only)[a]
Ethylene dibromide (106-93-4)	Alkylating agent	Major alkylation product: N7-alkylguanine Minor alkylation product(s)[c]	+	[c]	0
Ethylene oxide (75-21-8)	Alkylating agent	Major alkylating product: N7-alkylguanine Minor alkylating product(s): N3-alkyladenine	+	+	?
Ethylenimine (151-56-4)	Alkylating agent	Major alkylation product: N7-alkylguanine Minor alkylation product(s)[c]	+	+	+
Ethyl methanesulfonate (62-50-0)	Alkylating agent	Major alkylation product: N7-alkylguanine Minor alkylation product(s): N3-alkyladenine, 06-alkylguanine	+	+	+
Formaldehyde (50-00-0)	Alkylating agent	?[a]	+	+	0
ICR-191 (17070-44-9)	Alkylating agent	?[a]	+	[c]	[c]
Isopropyl methanesulfonate (926-06-7)	Alkylating agent	Major alkylation product: N7-alkylguanine Minor alkylation product(s): N3-alkyladenine, 06-alkylguanine	+	+	+

Agent	Type	Alkylation products			
Metepa (57-39-6)	Alkylating agent	[c]	+	+	+
Methylazoxymethanol (590-96-5)	Alkylating agent	Major alkylation product: N7-alkylguanine / Minor alkylation product(s): N3-alkyladenine	c	+	+
Methyl methanesulfonate (66-27-3)	Alkylating agent	Major alkylation product: N7-alkylguanine / Minor alkylation product(s): N3-alkyladenine	+	+	+
N-Methyl-N'-nitro-N-nitrosoguanidine (70-25-7)	Alkylating agent	Major alkylation product: N7-alkylguanine / Minor alkylation product(s): N3-alkyladenine, O6-alkylguanine	0	+	+
N-Methyl-N-nitroso-acetamide (7417-67-6)	Alkylating agent	Major alkylation product: N7-alkylguanine / Minor alkylation product(s)[c]	c	c	+
Mitomycin C (50-07-7)	Alkylating agent	Major alkylation product: N7-alkylguanine / Causes cross-linking and prevents replication after metabolic activation	+	+	+
Mustard gas (505-60-2)	Alkylating agent	Major alkylation product: N7-alkylguanine / Minor alkylation product(s): N3-alkyladenine / Induces cross-links	+	+	+
Myleran (55-98-1)	Alkylating agent	Major alkylation product: N7-alkylguanine / Minor alkylation product(s)[c]	+	+	+

(cont'd)

Table 2 (cont'd)

Agent (Chemical Abstracts Service Registry number)	Classification according to primary or most often reported effects on macromolecules	Remarks on mode of action and metabolism[a]	Subchromosomal lesions produced in one or more systems[a]	Chromosome aberrations produced in one or more systems[a]	Dominant lethals (mammals only)[a]
Natulan (366-70-1)	Alkylating agent	[c]	0	+	?
Nitrogen mustard (51-75-2)	Alkylating agent	Major alkylation product: N7-alkylguanine Minor alkylation product(s)[c] Causes cross-linking	+	+	?
N-Nitrosoethylurea (759-73-9)	Alkylating agent	Major alkylation product: N7-alkylguanine Minor alkylation product(s): N3-alkyladenine, 06-alkylguanine	+	+	+
N-Nitrosomethylurea (684-93-5)	Alkylating agent	Major alkylation product: N7-alkylguanine Minor alkylation product(s): N3-alkyladenine, 06-alkylguanine	+	+	+
N-Nitrosomethylurethane (615-53-2)	Alkylating agent	Major alkylation product: N7-alkylguanine Minor alkylation product(s)[c]	+	+	c
N-Nitrosomorpholine (59-89-2)	Allkylating agent	Major alkylation product: N7-alkylguanine Minor alkylation product(s)[c] Requires metabolic activation	+	+	0
N-Nitrosoethylurethane (614-95-9)	Alkylating agent	Major alkylation product: N7-alkylguanine Minor alkylation product(s): N3-alkyladenine	+	+	c

N-Nitroso-2-imidazolidone (3844-63-1)	Alkylating agent	Major alkylation product: N7-alkylguanine Minor alkylation product(s): N3-alkyladenine	+	+	c
Propane sultone (1120-71-4)	Alkylating agent	Major alkylation product: N7-alkylguanine Minor alkylation product(s)c	+	+	c
β-Propiolactone (57-57-8)	Alkylating agent	Major alkylation product: N7-alkylguanine Minor alkylation product(s): N3-alkylguanine	+	+	0
Propylene oxide (75-56-9)	Alkylating agent	Major alkylation product: N7-alkylguanine Minor alkylation product(s): N3-alkyladenine	+	c	c
Propyl methanesulfonate (1912-31-8)	Alkylating agent	Major alkylation product: N7-alkylguanine Minor alkylation product(s):c	+	+	+
Sarcolysine[b] (148-82-3)	Alkylating agent	Specific sites of DNA alkylation are still under investigation; studies to date have indicated that the compound cross-links DNA and causes strand breakage due to weakening of the phosphoester bonds	+	?	0
Streptozotocin (11006-80-7)	Alkylating agent	Major alkylation product: N7-alkylguanine Minor alkylation product(s)c	+	+	0
TEPA (545-55-1)	Alkylating agent	c	+	+	+

Table 2 (cont'd)

Agent (Chemical Abstracts Service Registry number)	Classification according to primary or most often reported effects on macromolecules[a]	Remarks on mode of action and metabolism[a]	Subchromosomal lesions produced in one or more systems[a]	Chromosome aberrations produced in one or more systems[a]	Dominant lethals (mammals only)[a]
Thio-TEPA (52-24-4)	Alkylating agent	[c]	+	+	+
Trenimon (68-76-8)	Alkylating agent	Alkylates the purine moiety of DNA, forms apurine sites, induces strand breakage and cross-links	+	+	+
Triethylenemelamine (51-18-3)	Alkylating agent	Major alkylation product: N7-alkylguanine Minor alkylation product(s): N3-alkyladenine Causes cross-links	+	+	+
Trimethyl phosphate (512-56-1)	Alkylating agent	Major alkylation product: N7-alkylguanine Minor alkylation product(s): N3-alkylguanine	+	+	+
Urethane (51-79-6)	Alkylating agent	Alkylates adenine *in vitro* forming N3-ethyladenine; deaminates adenine *in vivo*	?	+	0
Vinyl chloride (75-01-4)	Alkylating agent	Documentation of the specific alkylation site for this compound not yet available; preliminary work indicates metabolic activation may be required before it becomes biologically active	+	+	[c]

Compound	Type	Mechanism			
N-Acetoxy-2-acetyl-aminofluorene (6098-44-8)	Arylating agent	Binds covalently to the C-8 position of guanine; causes cross-links	+	c	c
2-Acetylaminofluorene (53-96-3)	Arylating agent	Metabolic active products bind covalently to the C-8 position of guanine causing displacement from normal coplanar position; induces cross-links	+	?	0
Aflatoxin B$_1$ (1162-65-8)	Arylating agent	Requires metabolic activation to a reactive product which binds to DNA, the nature of the binding is not understood	+	+	?
1,2-Benzanthracene (56-55-3)	Arylating agent	Metabolically activated epoxide form may possibly intercalate and interact covalently with DNA (via the epoxide grouping) causing frameshift mutations	+/0	c	c
Benzo(a)pyrene (50-32-8)	Arylating agent	Arylation of DNA occurs via covalent bonding with base pairs; metabolically activated epoxide form may possibly intercalate and interact covalently with DNA (via the epoxide grouping) causing frame-shift mutations	+/0	?	?

(cont'd)

Table 2 (cont'd)

Agent (Chemical Abstracts Service Registry number)	Classification according to primary or most often reported effects on macromolecules	Remarks on mode of action and metabolism[a]	Subchromosomal lesions produced in one or more systems[a]	Chromosome aberrations produced in one or more systems[a]	Dominant lethals (mammals only)[a]
7-Bromoethylbenz(a)-anthracene (24961-39-5)	Arylating agent	Reacts covalently at the C-8 position of guanine in DNA, molecule linked in the major grove parallel with the helix and perpendicular with the bases; minor binding positions at the N-4 position of cytosine and N-6 position of adenine has been suggested.	+	+	c
Butter yellow[b] (60-11-7)	Arylating agent	Arylation of DNA occurs via covalent binding to DNA base pairs; requires metabolic activation	+	?	0
Chlorpromazine[b] (50-53-3)	Arylating agent	Binding to nucleic acids mediated by a photo-product formed by irradiation, mechanisms of this reaction still under investigation	c	+	0
Chromomycin A_3 (1409-52-5)	Arylating agent	Complexes preferentially with native DNA and reacts specifically with guanine; reaction is reportedly at the $2\text{-}NH_2$ group located in the minor grove of the DNA helix	c	+	c

1,2,5,6-Dibenzanthracene (53-70-3)	Arylating agent	Metabolically activated epoxide form may possibly intercalate and interact covalently with DNA (via the epoxide grouping) causing frameshift mutations	+	+	c
9,10-Dimethyl-1-2,benzanthracene (57-97-6)	Arylating agent	Arylation of DNA occurs primarily via covalent bonding with constituent bases, particularly guanine and to a lesser extent to adenine; metabolically activated epoxide form may possibly intercalate and interact covalently with DNA (via the epoxide grouping) causing frameshift mutations	+	+	c
N-Hydroxy-2-acetyl-aminofluorine (53-95-2)	Arylating agent	Binds covalently to the C-8 position of guanine	+	c	?
3-Methylcholanthrene (56-49-5)	Arylating agent	Metabolically activated epoxide form may possibly intercalate and interact covalently with DNA (via the epoxide grouping) causing frameshift mutations	+	+	0
3'-Methyl-4-dimethyl-amino azobenzene[b] (55-80-1)	Arylating agent	Binding to DNA occurs primarily at the C-8 position of guanine	+	?	c
Nogalamycin (1404-15-5)	Arylating agent	Reacts with the adenine and/or thymine moieties of DNA	c	+	c

(cont'd)

Table 2 (cont'd)

Agent (Chemical Abstracts Service Registry number)	Classification according to primary or most often reported effects on macromolecules[a]	Remarks on mode of action and metabolism[a]	Subchromosomal lesions produced in one or more systems[a]	Chromosome aberrations produced in one or more systems[a]	Dominant lethals (mammals only)[a]
Streptonigrin (3930-19-6)	Arylating agent	Selectively binds to DNA, specific details not available; inhibits incorporation of adenine	+	+	c
Acridine orange[b] (494-38-2)	Intercalating agent	Intercalation occurs when chemical forms complexes between adjacent nucleotide pair layers by extension and unwinding of the deoxyribose–phosphate backbone; causes frameshift mutations	+	+	0
Acriflavin (86-40-8)	Intercalating agent	Intercalation occurs between adjacent nucleotide-pair layers by extension and unwinding of the deoxyribose–phosphate backbone; in ensuing replication, this causes insertion or deletion of a base pair	+	+	c
Actinomycin D (50-76-0)	Intercalating agent	Intercalations occur in DNA primarily at sites rich in guanine and cytosine pairs	+	+	+

Compound	Mode of action	Description			
9-Aminoacridine (90-45-9)	Intercalating agent	Intercalation occurs between adjacent nucleotide-pair layers by extension and unwinding of the deoxyribose–phosphate backbone, in ensuing replication, this causes insertion or deletion of a base pair	+	+	0
Atabrine (83-89-6)	Intercalating agent; repair enzyme inhibitor	Interferes with repair process; intercalates with DNA in the same manner as acridine	+	+	c
Daunomycin[b] (20830-81-3)	Intercalating agent	Intercalation occurs between adjacent base pairs of helical DNA	c	+	c
Ethidium bromide (1239-45-8)	Intercalating agent	Intercalation occurs when a part of the ethidium ion sandwiches between adjacent base pairs causing an extension of the DNA polymer; compound will complex with both DNA and RNA	+	+	c
Hycanthone (3105-97-3)	Intercalating agent	Intercalation occurs between adjacent base pairs similar to aminoacridines	+	+	?
4-Hydroxyaminoquinoline-N-oxide (4637-56-3)	Intercalating agent	Intercalates with DNA in the same manner as 4-NQO, both purines and pyrimidines form 1:1 charge-transfer complexes with this compound; induces frameshift mutations	+	+	c

(cont'd)

Table 2 (cont'd)

Agent (Chemical Abstracts Service Registry number)	Classification according to primary or most often reported effects on macromolecules[a]	Remarks on mode of action and metabolism[a]	Subchromosomal lesions produced in one or more systems[a]	Chromosome aberrations produced in one or more systems[a]	Dominant lethals (mammals only)[a]
ICR-170 (146-59-8)	Intercalating agent; alkylating agent	Compound intercalates into DNA in much the same manner as other acridines; also forms covalent bonds with DNA via its nitrogen mustard side chain.	+	+	+/0
Lucanthone (479-50-5)	Intercalating agent	Intercalation occurs between adjacent base pairs similar to aminoacridines	+	+	c
LSD (50-37-3)	Intercalating agent; ?[a]	Proposed intercalation occurs with DNA in much the same manner as ethidium bromide causing confirmation changes; also a reaction causing the disassociation of histones from chromosomal DNA has been suggested	+	+	+
Methylene blue (61-73-4)	Intercalating agent	Binding to DNA is mediated by light; details on binding mechanism are not available	+	+	c
4-Nitroquinoline-1-oxide (56-57-5)	Intercalating agent	Intercalates with DNA; forms charge-transfer complexes with adenine and guanine	+	+	0

Compound (CAS)	Class	Mechanism of action			
Proflavin (92-62-6)	Intercalating agent	Intercalation occurs between adjacent nucleotide-pair layers by extension and unwinding of deoxy–ribose–phosphate backbone; causes errors in base pairing	+	+	c
Psoralen (66-97-7)	Intercalating agent	Binds to DNA by a photo-chemical reaction which mediates cross-linkage of the double strands via binding with pyrimidine bases	+	+	c
2,3,7,8-Tetrachloro-dibenzo-*p*-dioxin (1746-01-6)	Intercalating agent	Interacts with DNA in same manner as acridine compounds causing frame-shift mutations	+	c	0
2-Aminopurine (452-06-2)	Base analog	Incorporated into both DNA and RNA, may pair like adenine or guanine	+	c	c
5-Bromodeoxyuridine (59-14-3)	Base analog	Incorporated into replicating DNA in place of thymidine causing base-pairing errors, mainly AT to GC	+	+	0
5-Bromouracil (51-20-7)	Base analog	Incorporated into DNA in place of thymine	+	c	c
2,6-Diaminopurine (1904-98-9)	Base analog	Incorporated into nucleic acids causing transition mutations to occur	+	+	c

(cont'd)

Table 2 (cont'd)

Agent (Chemical Abstracts Service Registry number)	Classification according to primary or most often reported effects on macromolecules[a]	Remarks on mode of action and metabolism[a]	Subchromosomal lesions produced in one or more systems[a]	Chromosome aberrations produced in one or more systems[a]	Dominant lethals (mammals only)[d]
8-Ethoxycaffeine (577-66-2)	Base analog	?[d]	+	+	c
5-Fluorouracil (51-21-8)	Base analog	Incorporated into RNA in place of uracil; mutagenic for many RNA viruses	+	c	0
Maleic hydrazide (123-33-1)	Base analog	Reacts mainly with heterochromatic areas of chromosomes	0	+	0
6-Mercaptopurine (50-44-2)	Base analog	?[d]	+	+	+
Theobromine (83-67-0)	Base analog	c	+	+	0
Theophylline (58-55-9)	Base analog	c	+	+	0
Colcemid (477-30-5)	C-mitotic agent	Action similar to that of colchicine	c	+	c
Colchicine (64-86-8)	Protein-complexing agent	Interferes with spindle formation and chromosome segregation by binding to a protein component of microtubules of the mitotic apparatus, characteristic action called C-mitotic effects	c	+	?
Hydroquinole (123-31-9)	C-mitotic agent	Interacts with spindle fibers producing colchicine-like effects	c	+	c

Agent	Class	Description			
Vinblastine (865-21-4)	C-mitotic agent	Action similar to that of colchicine	c	c	c
Vinblastine sulfate (143-67-9)	C-mitotic agent	Action similar to that of colchicine	0	c	+
Vincristine (57-22-7)	C-mitotic agent	Action similar to that of colchicine	c	c	+
Hydrazine (302-01-2)	Deaminating agent	Reacts mainly with cytosine and guanine in DNA causing GC to AT transitions	0	+	+
Hydroxylamine (7803-49-8)	Deaminating agent	Deaminates the pyrimidine bases cytosine and guanine	+	+	+
O-Methylhydroxylamine (67-62-9)	Deaminating agent; methylating agent	Deaminates cytosine	c	+	+
Nitrous acid (7782-77-6)	Deaminating agent; nitrosifying agent	Nitrosification results in deamination of bases guanine, cytosine, and adenine; also cross-links DNA	0	+	+
Sodium bisulfite (7631-90-5)	Deaminating agent	Deaminates cytosine to form uracil	?	c	+
Azaserine (115-02-6)	Enzyme inhibitor	Irreversibly inactivates certain enzymes concerned in purine biosynthesis; it is not clear whether this is the origin of its mutagenic action.	c	+	+

(cont'd)

Table 2 (cont'd)

Agent (Chemical Abstracts Service Registry number)	Classification according to primary or most often reported effects on macromolecules[a]	Remarks on mode of action and metabolism[a]	Subchromosomal lesions produced in one or more systems[a]	Chromosome aberrations produced in one or more systems[a]	Dominant lethals (mammals only)[a]
Caffeine (58-08-2)	Repair enzyme inhibitor; base analog	Mutagenic action is dependent on DNA synthesis and presence of oxygen; effects are offset by guanosine and adenosine; potentiates activity of other mutagens by inhibiting repair enzymes which could be due to competition with the substrate for the active site on the enzyme or competition with the enzyme for the substrate	+	+	0
5-Fluorodeoxyuridine (50-91-9)	Enzyme inhibitor; base analog	Inhibits thymidylate synthetase; is not incorporated into DNA; produces thymine starvation	c	+	c
Hydroxyurea (127-07-1)	Enzyme inhibitor; oxidating agent	Blocks conversion of nucleotides to deoxynucleotides; inhibits scheduled DNA synthesis	c	+	0
Aminopterin (54-62-6)	Folic acid antagonist	Inhibits thymidylate synthetase; produces thymine starvation	c	+	0
Chloramphenicol (56-75-7)	Folic acid antagonist	Potentiates the mutagenic activity of other agents; blocks synthesis of proteins	c	0	0

Methotrexate (59-05-2)	Folic acid antagonist	?[a]	0	+	0
Cytosine arabinoside (147-94-4)	Nucleoside analog	Inhibits DNA polymerase	0	+	c
5-Iododeoxyuridine (54-42-2)	Nucleoside analog	Incorporated into nucleic acids during replication causing base-pairing errors	0	+	0
Adenine (73-24-5)	Natural purine constituent of nucleic acids	?[a]	+	+	c
EDTA (60-00-4)	Chelating agent	Removal of essential metal ions from nucleic acids has been suggested as a possible mechanism for explaining chromosome-breaking activity	0	+	c
Hydrogen peroxide (7722-84-1)	Oxidizing agent	Inactivates both types of nucleic acids but preferentially RNA via oxidation of pyrimidines at the N-1 position; catalase destroys activity	+	+	c
Spermine[b] (71-44-3)	Antimutagen	Complexes with DNA by a specific interaction with double helix, molecule bridges narrow grove between DNA strands via electrostatic interactions between spermine's protonated imino and amino groups and ionized phosphate groups of nucleotide pairs; acts as an antimutagen.	0	c	0

(cont'd)

Table 2 (cont'd)

Agent (Chemical Abstracts Service Registry number)	Classification according to primary or most often reported effects on macromolecules	Remarks on mode of action and metabolism[a]	Subchromosomal lesions produced in one or more systems[a]	Chromosome aberrations produced in one or more systems[a]	Dominant lethals (mammals only)[a]
Thymidine (50-89-5)	Natural nucleoside	Interferes with pool of nucleotides	+	+	c
AF-2, 2-(2-furyl)-3-(5-nitro-2-furyl)-acrylamide (3688-53-7)	?[a]	Biological activity seems to require presence of a nitro group at the C-5 position on one of the furan rings and a side chain at the C-2 position on this same ring structure; nature of interaction with DNA has yet to be determined.	+	+	0
Allylisothiocyanate (57-06-7)	?[a]	c	+	+	0
Benzene (71-43-2)	?[a]	?[a]	c	+	c
Coumarin (91-64-5)	?[a]	Compound will react with thymine or uracil after irradiation with visible light	c	+	c

Chemical					
DDT (50-29-3)	?[d]	?[d]	0	+	?
Hempa (680-31-9)	?[d]	c	+	+	+
Lead acetate[b] (301-04-2)	?[d]	Damage to chromosomes may be the result of activating lysozymes such as DNase or Pb may interrupt protein or ATP synthesis needed for repair; potentiates mutagenic action of other agents	c	+	0
Mercury chloride[b] (7487-94-7)	?[d]	Hg ions bind to the purines of DNA	c	+	c
Sodium azide (26628-22-8)	c	Unknown, but reaction mechanism seems to induce predominately base-pair substitutions.	+	0	c
2,4,5-T (93-76-5)	c	c	c	+	0

VIII. POTENTIAL MUTAGENS IN THE ENVIRONMENT

There have been several review articles written in the last five years concerning the potential genetic dangers to man from environmental chemicals (Malling, 1970; Barthelmess, 1970; Neel, 1971; de Serres, 1973). Most of these reviews have used the following classifications for the chemicals they covered: pesticides, food additives, drugs, industrial compounds, and pollutants. We have expanded this list in much the same manner as Fishbein (1972) and Fishbein and Flamm (1972a, b) in several of their extensive reviews. These include additional classifications for feed chemicals, naturally occurring chemicals, biologicals (viruses), and laboratory chemicals. It should be emphasized that several chemicals have multiple uses and therefore appear in more than one of the categories shown. Only those compounds which have been referenced in the EMIC file 10 or more times were considered for entry into this classification scheme. The only exception to this was made for entries placed in the category called biologicals. In this case the papers for viruses were considered individually. The chemicals found in the other categories are listed according to the availability of published material. This order indicates the level of current testing* which these compounds have undergone in order to understand their potential mutagenicity. Therefore the presence of any agent, chemical, or biological in this section is not an indication of mutagenicity. Such information on the majority of these agents can be found in Table 2, or literature pertaining to this question may be obtained by writing to the Environmental Mutagen Information Center.

A. Agricultural Chemicals

Agricultural chemicals, mainly pesticides, have been singled out for particular concern to man's health because they are widely used and usually in large amounts (Table 3). There are many representative chemicals which could be listed under this classification but not many of these have been tested for mutagenic effects. Most of the work reported in the references was done primarily in plant and microbial assay systems. Very few studies have been done with mammals with the exception of those compounds proposed for use as chemosterilants.

B. Industrial Compounds

Industrial development in all countries of the world has expanded rapidly since World War II in order to meet the needs and demands of

---

*Based on selections made from 17,500 scientific papers having publication dates covering a 9-year period (1968–1976).

Table 3. Agricultural Chemicals

Compound	Use or proposed use	References available from the EMIC data base
Triethylenemelamine	Chemosterilant	180
Formaldehyde	Insecticide	101
Cycloheximide	Fungicide	95
Maleic hydrazide	Plant-growth regulator	80
TEPA	Chemosterilant	65
DDT	Insecticide	55
Ethylene oxide	Fungicide	43
Metepa	Chemosterilant	39
2,4-D	Herbicide	39
Apholate	Chemosterilant	37
Captan	Fungicide	30
Dichlorvos	Insecticide	27
Hempa	Chemosterilant	26
Aminopterin	Rodenticide	26
Carbaryl	Insecticide	22
2,4,5-T	Herbicide	21
Dieldrin	Insecticide	15
Ethylene dibromide	Fungicide	12
Endrin	Insecticide	7

growing populations. Opportunities for using chemical formulations to meet these needs and demands have grown along with industrial expansion. Therefore, workers in industrial environments have been confronted with higher exposure risks to the chemicals resulting from this technological expansion than the general population. Characteristically, occupational exposure in industrial environments involves higher concentrations and frequently an extended contact which may occur via different exposure routes (e.g., inhalation, skin contact). These factors have made the review of possible health hazards from industrial chemicals of immediate concern.

The list of compounds in Table 4 occur frequently (10 or more times) in the EMIC data base and have some industrial application. A number of these are well-known mutagens in many assay systems including mammals.

C. Food and Feed Additives

This category of chemicals has a more general distribution among the population than those listed in all other categories. A few years ago when the U.S. Food and Drug Administration undertook a reexamination of the substances used in food, mutagenicity was among the items considered. Likewise, many investigators interested in the problem of environmental mutagenesis also considered food additives as a class of compounds which needed immediate evaluation. Papers resulting from these investigations have not been

Table 4. Industrial Compounds

Compound	Use	References available from the EMIC data base
Ethylenimine	Alkylating agent	352
Hydroxylamine	Antioxidant in soap manufacturing; reducing agent in photography and tanning and textile industries	308
Diethylsulfate	Alkylating agent	297
Triethylenemelamine	Alkylating agent	180
Dimethylnitrosamine	Manufacture of dimethylhydrazine	169
Dimethylsulfate	Alkylating agent; manufacture of perfume and dyes	137
Formaldehyde	Textile manufacturing; organic synthesis	101
Urethane	Solvent (molten form)	101
β-Propiolactone	Intermediate in organic synthesis; vapor sterilant	69
Diepoxybutane	Cross-linking agent for textile fibers	62
Hydrogen peroxide	Bleaching agent and oxidant in pharmaceutical industry	62
Ethylene oxide	Textile and food fumigant	43
Metepa	Crease and flameproofing of textiles	39
Benzene	Solvent	34
Cyclohexylamine	Organic synthesis of pesticides and plasticizers	32
Captan	Bacteriostat in soap	30
Hydrazine	Reducing agent in rocket fuel	27
Sodium azide	Preparation of pure sodium and hydrazoic acid	26
Trimethylphosphate	Gasoline additive; paint solvent; methylating agent	22
Mercury chloride	Preservative	20
Diazomethane	Methylating agent	19
Sodium bisulfite	Bleaching agent for wool	18
Lead acetate	Dyeing and printing textiles	17

published, which accounts for the surprisingly small number of chemicals listed in this particular category (Table 5). Lack of publication of results from studies promoted by these interests is due either to their submission to the government for review and evaluation as special reports or because they turned up negative data and negative findings are not generally submitted for

Table 5. Food and Feed Additives

Compound	Use	References available from the EMIC data base
Nitrous acid	Formed from the food preservative sodium nitrite	385
Caffeine	Food additive	360
EDTA	Food additive	45
Ethylene oxide	Food sterilant	43
Sodium cyclamate	Artificial sweetener	23
Sodium bisulfite	Food preservative	18
Hydroquinone	Proposed food antioxidant	15

publication. Only those compounds which are added to food or animal feed during processing are included in Table 5.

D. Naturally Occurring Substances

This section lists compounds which may occur naturally in the environment (Table 6). Selection was based primarily on the same outline Miller (1973) employed in his review of naturally occurring substances that can induce tumors.

There are many compounds thought not to be naturally occurring, but evidence from new research has indicated that they may be. The most notable example of this is the evidence that the carcinogenic pollutants benzo(a)pyrene and 1,2-benzanthracene and other polynuclear aromatic hydrocarbons may be normal products of plant biosynthesis. These compounds are not shown in this list but appear in the list of pollutants (Table 8). Compounds are listed in their EMIC frequency order along with information about how or where they may originate or may be found.

E. Drugs

Only compounds reported to have some medicinal use in humans or animals are included in Table 7. Of the compounds most frequently cited in the EMIC file, the majority are found in this category. It is not surprising that more entries are in this classification than the rest since most of these compounds, e.g., antibiotics and antineoplastic agents, are used in medicine be-

Table 6. Naturally Occurring Substances

Compound	Description	References available from the EMIC data base
Caffeine	Plant alkaloid found in tea, coffee, mate leaves, guarana paste, and cola nuts	360
Colchicine	Plant alkaloid found in numerous species of colchicum as well as other *Liliaceae*	356
Actinomycin D	Antibiotic substance belonging to the actino-mycin complex produced by several *streptomyces* species, such as *Streptomyces parvulus*	338
Mitomycin C	Antitumor antibiotic produced by *Streptomyces caespitosus*	251
LSD	Metabolic by-product formed by cultures of *Claviceps paspali*	139
Daunomycin	Antibiotic of the rhodomycin group isolated from *Streptomyces peucetius*	82
Puromycin	Antibiotic isolated from cultures of *Streptomyces alboniger*	71
Aflatoxin B_1	Metabolite of *Aspergillus flavus.*	57
Vinblastine	Antitumor alkaloid produced from periwinkle (*Vinca rosea*)	54
Leucomycin	Antibiotic produced by *Streptomyces kitasatoensis*	52
Bleomycin	Antitumor antibiotic produced by a *Streptomyces* sp.	50
Vincristine	Alkaloid isolated from *Vinca rosea*	41
Theophylline	Plant alkaloid found in tea leaves	39
Rifampicin	Antibiotic isolated from *Streptomyces mediterranei*	29
Streptonigrin	Antitumor antibiotic produced by *Streptomyces flocculus*	26

Table 6 (cont'd)

Compound	Description	References available from the EMIC data base
Adriamycin	Antitumor antibiotic produced by *Streptomyces peucetius var. caesius*	24
Streptozoticin	Antibiotic produced from *Streptomyces achromogenes*	24
Spermine	Nitrogenous base found in animal sperm	20
Griseofulvin	Antibiotic produced by *Streptomyces griseus*	20
Theobromine	Plant alkaloid found in the cacao bean, cola nuts, and tea	19
Erythromycin	Antibiotic produced by *Streptomyces erythreus*	17
Coumarin	Plant alkaloid found in tonka beans, lavender oil, woodruff (*Asperula* sp) and sweet clover (*Melilotus*)	16
Allylisothiocyanate	Isolated from black mustard (*Brassica nigra*) seeds	10

cause of their abilities to interfere with cellular division and/or suppress cellular metabolism.

F. Pollutants

The chemicals in Table 8 were prepared from a list of compounds termed "toxic pollutants" by the Environmental Protection Agency and from a list compiled by the Energy Research and Development Administration of compounds formed during the processing and/or combustion of fossil fuels. Even though these compounds were compiled by U.S. agencies, all could be called pollutants of air and water of practically all countries.

G. Biologicals (Viruses)

This category contains information only on viruses (Table 9). The selection of viruses was made because of the potential these environmental agents have as inducers of genetic damage and as agents which may potentiate damage induced by other agents. Only viral strains for which information is available in the EMIC data base are included in this category.

Table 7. Drugs

Compound	Use	References available from the EMIC data base
Caffeine	Stimulant	360
Colchicine	Treatment of gout in humans and mammary tumors in female dogs	356
Actinomycin D	Antibiotic	338
Mitomycin C	Antineoplastic agent	251
Nitrogen mustard	Antineoplastic agent	212
Triethylenemelamine	Antineoplastic agent	180
Cyclophosphamide	Treatment of glomerulonephritis; antineoplastic agent	180
Ethidium bromide	Antiparasitic agent (veterinary use only)	156
Hydroxyurea	Antineoplastic agent	146
LSD	Psychological studies	139
Acriflavine	Antiseptic	124
Myleran	Antineoplastic agent	110
Urethane	Antineoplastic agent (veterinary anesthetic)	101
Trenimon	Antineoplastic agent	83
Daunomycin	Antibiotic; antineoplastic agent	82
Colcemid	Antineoplastic agent	81
Methotrexate	Antineoplastic agent	73
Puromycin	Antibiotic; antineoplastic agent	71
β-Propiolactone	Sterilizing vaccines, grafts, plasma	69
TEPA	Antineoplastic agent	65
Hydrogen peroxide	Topical antiseptic	62
5-Fluorodeoxyuridine	Antiviral agent; antineoplastic agent	61
Nalidixic acid	Antibiotic (veterinary)	55
Vinblastine	Antineoplastic agent	54
Bleomycin	Antineoplastic agent; antibiotic	50
Hycanthone	Antischistosomal agent	49
6-Mercaptopurine	Antineoplastic agent	48
Cytosine arabinoside	Antineoplastic agent	47
Ethylene oxide	Sterilizing surgical instruments	43
Streptomycin	Antibiotic; antineoplastic agent	42
Chlorambucil	Antineoplastic agent	41
Vincristine	Antineoplastic agent	41
Theophylline	Muscle relaxant, diuretic, myocardial stimulant	39
Chloroquine	Antimalarial agent	34
Rifampicin	Antibiotic	29
Methylene blue	Diagnostic dye, cyanide antidote (veterinary use only)	28

Table 7 (cont'd)

Compound	Use	References available from the EMIC data base
Dichlorvos	Veterinary antiparasitic agent	27
Sodium fluoride	Anticavity agent	26
Sodium azide	Antihypertensive agent	26
Streptozotocin	Antibiotic; antineoplastic agent	24
Adriamycin	Antibiotic; antileukemic agent	24
Hempa	Antineoplastic agent	24
Chlorpromazine	Sedative; antiemetic	23
Carbaryl	Veterinary antiparasitic agent	22
Mercury chloride	Disinfectant	21
Lucanthone	Antischistosomal agent	21
Griseofulvin	Antibiotic	20
Sarcolysine	Antineoplastic agent	20
Phenylbutazone	Antirheumatoid agent	19
Procarbazine (Natulan)	Antineoplastic agent	19
Urea	Diuretic; antiseptic (veterinary use only)	18
Iododeoxyuridine	Antiviral agent	17
8-Methoxypsoralen	Vitiligo sun protectant	17
Lead acetate	Astringent (veterinary use only)	17
Erythromycin	Antibiotic	17
γ-Lindane	Scabicide; pediculocide	15
Atabrine	Antimalarial agent	15
Dieldrin	Antiparasitic agent (veterinary use only)	15
9-Aminoacridine	Antiseptic	15
Phenobarbital	Sedative; anticonvulsant	13
Trimethypsoralen	Photoprotective agent	10
Mustard oil	Counterirritant	10

H. Laboratory Chemicals

All of the categories used to classify agents, with the notable exception of the entry for laboratory chemicals, are relatively self-explanatory. Laboratory chemicals, which the reader may at first suspect of being a catch-all classification for agents not fitting into other categories, is a valid and pertinent category. It was selected for a specific purpose. The chemicals listed under this heading in Table 10 are those which have extensive use in government, industrial, and academic laboratories because of their abilities to adversely affect hereditary material. We believe their use is extensive enough to warrant recognition as potential hazards to laboratory workers for the same reasons as mentioned in Section VIII. B for industrial workers.

Table 8. Pollutants

Chemical	References available from the EMIC data base
Ethylenimine	352
Dimethylnitrosamine	169
Dimethylsulfate	137
Formaldehyde	101
Dimethylbenzanthracene(s)	89
Benzo(a)pyrene	88
Methylcholanthrene(s)	75
TEPA	65
DDT	55
Hg from mercury-containing compounds	50
Pb from lead-containing compounds	44
2,4-Dichlorophenoxyacetic acid (2,4-D)	39
Cd from cadmium-containing compounds	38
Benzene	34
Dibenzanthracene(s)	32
Benzanthracene(s)	28
Dichlorvos	27
2,4,5-Trichlorophenoxyacetic acid (2,4,5-T)	21
Ozone	18
Acrolein	15

IX. CONCLUSION

We have, in this chapter, attempted to give a general up-to-date picture of our state of knowledge on the mutation process and the chemical agents capable of inducing it.

This endeavor, even though broad in scope, is not comprehensive since that would involve a review of all modern genetics. It is not, therefore, possible to obtain a complete understanding of chemical mutagenesis from the material presented in this chapter. We do, however, hope that we have stimulated your interest and provided you with an awareness of how mutations may occur, the types of agents which may induce mutations, and the possible consequences of mutations to human health.

The material selected for this article is current and came from a unique mutagenesis library and agent registry belonging to the Environmental Mutagen Information Center (EMIC). At this writing, this library and agent registry contain 17,500 documents and testing results on 7000 agents. These EMIC information resources provide the best means available for reviewing the voluminous literature of this research field. In our screening of this vast amount of information, several things became evident. To begin with, an extensive literature had already accumulated on many chemical compounds

Table 9. Biologicals (Viruses)

Virus	Nucleic acid type	References available from the EMIC data base
Rous sarcoma virus	RNA	29
Simian virus 40	DNA	25
Measles virus	RNA	24
Sendai virus	RNA	21
Adenovirus type 12	DNA	19
Polyoma virus	DNA	19
Herpes simplex virus	DNA	17
Influenza virus	RNA	10
Epstein–Barr virus	DNA	9
Newcastle disease virus	RNA	8
Mumps virus	RNA	8
Murine sarcoma virus	RNA	8
Polio virus	RNA	8
Rubella virus	RNA	7
Vaccinia virus	DNA	6
Poliomyelitis virus	RNA	6
Adenovirus	DNA	5
Cytomegalovirus	DNA	5
Mengovirus	RNA	4
Adenovirus type 2	DNA	4
Phage lambda	DNA	4
Phage mu	DNA	3
Frog virus 3	DNA	3
Herpes-type virus	DNA	3
Adenovirus type 18	DNA	3
Adenovirus type 4	DNA	3
Barley stripe mosaic virus	?	3
Friend virus	RNA	3
Sindbis virus	RNA	3
Coxsackie A13 virus	RNA	2
Herpes zoster virus	DNA	2
Nuclear polyhedrosis virus	DNA	2
Phage T4	DNA	2
Poliovirus type 1	RNA	2
Rauscher leukemia virus	RNA	2
Adenovirus type 3	DNA	1
Avian myeloblastosis virus	RNA	1
Big bud virus	?	1
Ectromelia virus	DNA	1
Equine herpes 3 virus	DNA	1
Herpes virus saimiri	DNA	1
Honk Kong influenza virus	RNA	1
Influenza virus type A2	RNA	1
Influenza virus 3	RNA	1
Leukemia virus	RNA	1
Phage mu-1	DNA	1
Phage SP-10	DNA	1
Murine cytomegalovirus	DNA	1
Reovirus	RNA	1
Shope fibroma virus	DNA	1
Simian adenovirus	DNA	1
Teschen disease virus	RNA	1
Vesicular stomatitis virus	RNA	1

Table 10. Laboratory Chemicals

Compound	References available from the EMIC data base
N-Methyl-N'-nitro-N-nitrosoguanidine	1193
Ethyl methanesulfonate	1126
N-Nitrosomethylurea	471
Nitrous acid	385
Methyl methanesulfonate	383
Colchicine	356
Ethylenimine	352
Actinomycin D	338
Hydroxylamine	308
Diethylsulfate	297
N-Nitrosoethylurea	259
Mitomycin C	251
Nitrogen mustard	235
Nitroso-N-methylurethane	183
Triethylenemelamine	180
Cyclophosphamide	180
9,10-Dimethyl-1,2-benzanthracene	179
Dimethylnitrosamine	169
Ethidium bromide	156
5-Bromodeoxyuridine	147
Proflavin	147
Hydroxyurea	146
4-Nitroquinoline-1-oxide	130
Acriflavin	124
2-Aminopurine	112
Myleran	110
Acridine mustard (ICR-170)	102
Urethane	101
Formaldehyde	101
5-Fluorouracil	98
Benzo(a)pyrene	88
Trenimon	83
Colcemid	81
3-Methylcholanthrene	75
Puromycin	71
Isopropyl methanesulfonate	67
TEPA	65

which have an array of environmental distributions and usages. We believe, as a growing number of others do, that many of these chemicals, as well as other yet untested agents, pose a potential threat to man's genetic health. We therefore urge the critical evaluation of the material already published on these compounds and the testing of new compounds before being sanctioned for general use by the population. We would also encourage research directed toward elucidating the basic mechanisms of chemically induced mutations. Hopefully our future research efforts will be directed toward anticipated problems rather than a crisis response to unravel the causes of a catastrophe.

ACKNOWLEDGMENTS

We would like to acknowledge the assistance of all the EMIC staff members, M. D. Shelby, E. S. Von Halle, I. R. Miller, M. M. Brown, R. O. Beauchamp, Jr., M. T. Sheppard, R. E. Brooksbank, Jr., and E. T. Owens. The capable typing assistance of Wilma J. Barnard is also gratefully acknowledged.

REFERENCES

Badr, F. M., Lorlan, P. A., and Lehmann, H., 1973, Haemoglobin P-nilotic containing a β–δ chain, *Nature (London) New Biol.* **242:**107–110.

Barthelmess, A., 1970, Mutagenic substances in the human environment, *in: Chemical Mutagenesis in Mammals and Man* (F. Vogel and G. Röhrborn, eds.), pp. 69–147, Springer-Verlag, New York.

Borst, P., 1972, Mitochondrial nucleic acid, *Annu. Rev. Biochem.* **41:**333–376.

Bunn, C. L., Mitchell, C. H., Lukins, H. B., and Linnane, A. W., 1970, Biogenesis of mitochondria, XVIII. A new class of cytoplasmically determined antiobiotic resistant mutants in *Saccharomyces cerevisiae, Proc. Natl. Acad. Sci. U.S.A.* **67:**1233–1240.

Cairns, J., 1966, Autoradiography of HeLa cell DNA, *J. Mol. Biol.* **15:**372–373.

Clegg, J. B., Weatherall, D. J., and Milner, P. F., 1971, Haemoglobin Constant Spring—a chain termination mutant?, *Nature (London)* **234:**337–340.

Crippa, M., Meza, I., and Dina, D., 1973, Sequence arrangement in mRNA: Presence of poly (A) and identification of a repetitive fragment at the 5' end, *Cold Spring Harbor Symp. Quant. Biol.* **38:**933–942.

Crow, J. F., and Kimura, M., 1970, *An Introduction to Population Genetics Theory, Harper and Row,* New York.

Darnell, J. E., Jelinek, W. R., and Malloy, G. R., 1973, Biogenesis of mRNA: Implications for genetic regulation in mammalian cells, *Science* **181:**1215–1221.

de Serres, F. J., 1973, Genetic test systems to evaluate the mutagenic effects of environmental chemicals, *Pharmacol. Future Man, Proc. 5th Int. Congr. Pharmacol. (1972)* **2:**171–181.

de Serres, F. J., 1974, Mutagenic specificity of chemical carcinogens in micro-organisms, *in: Chemical Carcinogenesis Assays* (R. Montesano, L. Tomatis, and W. Davis, eds.), IARC Scientific Publication No. 10, International Agency for Research on Cancer, Lyon, France.

de Serres, F. J., 1975, The correlation between carcinogenic and mutagenic activity in short-term tests for mutation-induction and DNA repair, *Mutat. Res.* **31:**203–204.

de Serres, F. J., and Malling, H. V., 1971, Measurement of recessive lethal damage over the entire genome and at two specific loci in the ad-3 region of *Neurospora crassa* with a two component heterokaryon, *in: Environmental Chemical Mutagens; Principles and Methods for Their Detection,* Vol. 2 (A. Hollaender, ed.), pp. 311–342, Plenum Press, New York.

de Serres, F. J., and Osterbind, R. S., 1962, Estimation of the relative frequencies of X-ray induced viable and recessive lethal mutations in the ad-3 region of *Neurospora crassa, Genetics* **47:**793–796.

DuPraw, E. J., 1970, *DNA and Chromosomes,* Holt, Reinhart and Winston, New York.

DuPraw, E. J., 1973, Quantiative constraints in the arrangement of human DNA, *Cold Spring Harbor Symp. Quant. Biol.* **38:**87–98.

Ehling, U. H., 1974, Dose-response relationship of specific locus mutations in mice, *Arch. Toxicol.* **32:**19–25.

Ehling, U. H., Cumming, R. B., and Malling, H. V., 1968, Induction of dominant lethal mutations by alkylating agents in male mice, *Mutat. Res.* **5:**417–428.

Fishbein, L., 1972, Pesticidal, industrial, food additive, and drug mutagens, *in: Mutagenic Effects of Environmental Contaminants* (H. E. Sutton and M. I. Harris, eds.), pp. 129–170, Academic Press, New York.

Fishbein, L., and Flamm, W. G., 1972a, Potential environmental chemical hazards. Part 1: Drugs, *Sci. Total Environ.* **1:**15–30.

Fishbein, L., and Flamm, W. G., 1972b, Potential environmental chemical hazards. Part 2: Feed additives and pesticides, *Sci. Total Environ.* **1:**31–64.

Fraumeni, J. F., Jr., 1973, Genetic determinants of cancer, *in: Host Environment Interactions in the Etiology of Cancer in Man* (R. Doll and I. Vodopija, eds.), pp. 49–55, IARC Scientific Publication No. 7, International Agency for Research on Cancer, Lyon, France.

Freese, E., 1974, Genetic effects of mutagens and of agents present in the human environment, *in: Molecular and Environmental Aspects of Mutagenesis* (L. Prakash, F. Sherman, M. W. Miller, C. W. Lawrence and H. W. Taker, eds.), pp. 5–13, Charles C. Thomas, Springfield, Illinois.

Goldring, E. S., Grossman, L. I., Krupnick, D., Cryer, D. R., and Marmur, J., 1970, The petite mutation in yeast, loss of mitochondrial deoxyribonucleic acid during induction of petites with ethidium bromide, *J. Mol. Biol.* **52:**323–335.

Green, B., 1970, Influence of pH and metal ions on the fluorescence of polycyclic hydrocarbons in aqueous DNA solutions, *Eur. J. Biochem.* **14:**567–574.

Harm, W., Rupert, C. S., and Harm, H., 1971, The study of photoenzymatic repair of UV lesions in DNA by flash photolysis, *in: Photophysiology,* Vol. VI (A. C. Giese, ed.), pp. 279–324, Academic Press, New York.

Kendall, A. G., Ojwang, P. J., Schroeder, W. H., and Huisman, T. H. J., 1973, Hemoglobin Kenya, the product of a δ–β fusion gene: Studies of the family, *Am. J. Hum. Genet.* **25:**548–563.

Kimura, M., 1968, Genetic variability maintained in a finite population due to mutational production of neutral and nearly neutral isoalleles. *Genet. Res.* **11:**247–269.

Kimura, M., and Ohla, T., 1971, Protein polymorphism as a phase of molecular evolution, *Nature (London)* **229:**467–469.

Lieberman, M. W., 1976, Approaches to the analysis of fidelity of DNA repair in mammalian cells, *in: International Review of Cytology,* Vol. 45 (G. H. Bourne, J. F. Danielli, and K. W. Jeon, eds.), pp. 1–23, Academic Press, New York.

Malling, H. V., 1970, Chemical mutagens as a possible genetic hazard in human populations, *Am. Ind. Hyg. Assoc. J.* **31:**657–666.

Malling, H. V., 1971, Dimethylnitrosamine: Formation of mutagenic compounds by interaction with mouse liver microsomes, *Mutat. Res.* **13:**425–429.

McKusick, V. A., 1975, Mendelian inheritance in man. *Catalogs of Autosomal Dominant, Autosomal Recessive Phenotypes and X-linked Phenotypes,* 4th ed., Johns Hopkins University Press, Baltimore, Maryland.

Miller, J. A., 1973, Naturally occurring substances that can induce tumors, *in: Toxicants Occurring Naturally in Foods,* pp. 508–549, National Academy of Sciences, Washington, D.C.

Neel, J. V., 1971, The detection of increased mutation rates in human populations, *Perspect. Biol. Med.* **14:**522–537.

Paigen, K., and Ganschow, R., 1965, Genetic factors in enzyme realization, *Brookhaven Symp. Biol.* **18:**99–115.

Perlman, P. S., and Mahler, H. R., 1971, Molecular consequences of ethidium bromide mutagenesis, *Nature (London) New Biol.* **231:**12–16.

Russell, W. L., 1951, X-ray induced mutations in mice, *Cold Spring Harbor Symp. Quant. Biol.* **16:**327–366.

Russell, W. L., and Russell, L. B., 1959, The genetics and phenotypic characteristic of radiation-induced mutations in mice, *Radiat. Res. Suppl.* **1:**296–305.

Smith, C. A., Jordan, J. M., Vinograd, J., 1971, In vivo effects of intercalating drugs on the superhelix density of mitochondrial DNA isolated from human and mouse cells in culture, *J. Mol. Biol.* **59:**255–272.

Sparrow, A. H., Price, H. J., and Underbrink, A. G., 1972, A survey of DNA content per cell and per chromosome of prokaryotic and eukaryotic organisms: Some evolutionary considerations, *Brookhaven Symp. Biol.* **23:**451–494.

Sutherland, B. M., Chamberlin, M. J., and Sutherland, J. C., 1973, Deoxyribonucleic acid photoreactivation enzyme from *Escherichia coli, J. Biol. Chem.* **248:**4200–4205.

Wassom, J. S., 1973, The literature of chemical mutagenesis, *in: Chemical Mutagens—Principles and Methods for their Detection,* Vol. 3 (A. Hollaender, ed.), pp. 271–287, Plenum Press, New York.

Watson, J. D., 1965, *Molecular Biology of the Gene,* W. A. Benjamin, New York.

General Bibliography

Anonymous, 1972, "Ionizing Radiation: Levels and Effects," A report of the U.N. Scientific Committee on the effects of atomic radiation to the General Assembly, Vol. II: Effects, United Nations, New York.

Anonymous, 1973, "The Testing of Chemicals for Carcinogenicity, Mutagenicity and Teratogenicity," Ministry of Health and Welfare, Canada.

Arcos, J. C., and Argus, M. F., eds., 1974, *Chemical Induction of Cancer—Structural Bases and Biological Mechanisms,* Vol. IIA, Academic Press, New York.

Arcos, J. C., and Argus, M. F., eds., 1974, *Chemical Induction of Cancer—Structural Bases and Biological Mechanisms,* Vol. IIB, Academic Press, New York.

Balis, M. E., 1968, Antagonists and nucleic acids, *in: Frontiers of Biology,* Vol. 10 (A. Neuberger and E. L. Tatum, eds.), pp. 1–293, North Holland Publishing, Amsterdam.

Becker, F. F., ed., 1975, *Cancer, A Comprehensive Treatise, Etiology: Chemical and Physical Carcinogenesis,* Vol. I., Plenum Press, New York.

Busch, H., ed., 1974, *The Molecular Biology of Cancer,* Academic Press, New York.

Christensen, H. E., Luginbyhl, T. T., and Carrol, B. S., eds., 1974, *The Toxic Substances List,* National Institute for Occupational Safety and Health, Rockville, Maryland.

Corcoran, J. W., and Hahn, F. E., eds., 1975, *Antibiotics, Mechanism of Action and Antitumor Agents,* Vol. III, Springer-Verlag, New York.

Coutinho, E. M., and Fuchs, F., eds., 1974, *Physiology and Genetics of Reproduction,* Vol. 4B, Plenum Press, New York.

Drake, J. W., 1970, *The Molecular Basis of Mutation,* Holden-Day, San Francisco.

Epstein, S. S., ed., 1971, *Drugs of Abuse: Their Genetic and Other Chronic Nonpsychiatric Hazards,* MIT Press, Cambridge, Massachusetts.

Fishbein, L., Flamm, W. G., and Falk, H. L., 1970, *Chemical Mutagens: Environmental Effects on Biological Systems,* Academic Press, New York.

Freese, E., 1963, Molecular mechanism of mutations, *in: Molecular Genetics* (J. H. Taylor, ed.), pp. 207–269, Academic Press, New York.

Freese, E., 1969, Hereditary DNA alterations, *Angew. Chem., Int. Ed. Engl.* **8:**12–20.

Freese, E., 1971, Structure-activity considerations in potential mutagenicity, *in: Drugs of Abuse: Their Genetic and Other Chronic Nonpsychiatric Hazards,* (S. S. Epstein, ed.), pp. 120–124, MIT Press, Cambridge, Massachusetts.

Gottlieb, D., and Shaw, D., eds., 1967, *Antibiotics, Mechanism of Action,* Vol. 1, Springer-Verlag, New York.

Hahn, F. E., ed., 1971, *Progress in Molecular and Subcellular Biology,* Vol. 2, Research Symposium on Complexes and their Modes of Action, held at Walter Reed Army Institute of Research, Washington, D. C., Springer-Verlag, New York.

Hayes, W., 1968, *The Genetics of Bacteria and Their Viruses,* John Wiley, New York.

Hollaender, A., ed., 1971, *Chemical Mutagens: Principles and Methods for Their Detection,* Vols. I and II, Plenum Press, New York.

Hollaender, A., ed., 1973, *Chemical Mutagens: Principles and Methods for Their Detection,* Vol. III, Plenum Press, New York.

Hook, E. B., Janerich, D. T., and Porter, I. H., eds., 1971, *Monitoring, Birth Defects and Environment, the Problem of Surveillance,* Academic Press, New York.

Kihlman, B. A., 1966, *Actions of Chemicals on Dividing Cells,* Prentice-Hall, Englewood Cliffs, New Jersey.

Loomis. T. A., ed., 1972, *Toxicological Problems, Pharmacology and the Future of Man,* Proceedings of the Fifth International Congress on Pharmacology, Vol. 2, S. Karger, Basel.

Loveless, A., 1966, *Genetic and Allied Effects of Alkylating Agents,* Pennsylvania State Univ. Press, University Park, Pennsylvania.

Nakahara, W. T., Sugimura, T., and Odashima, S., eds., 1972, *Topics in Chemical Carcinogenesis,* University Park Press, Baltimore, Maryland.

Prakash, L., Sherman, F., Miller, M. W., Lawrence, C. W., and Faber, H. W., eds., 1974, *Molecular and Environmental Aspects of Mutagenesis,* Charles C. Thomas, Springfield, Illinois.

Röhrborn, G., 1970, Biochemical mechanisms of mutation, *in: Chemical Mutagenesis in Mammals and Man* (F. Vogel and G. Röhrborn, eds.), pp. 1–15, Springer-Verlag, Berlin.

Sarma, D. S. R., Fajalakshmi, S., and Farber, E., eds., 1975, Chemical carcinogenesis: Interaction of carcinogens with nucleic acid, *in: Cancer 1—A Comprehensive Treatise Etiology: Chemical and Physical Carcinogenesis* (F. F. Becker, ed.), pp. 235–288, Plenum Press, New York.

Schull, W. J., ed., 1962, *Mutations: Second Conference on Genetics,* Univ. of Michigan Press, Ann Arbor, Michigan.

Singer, B., 1975, The chemical effects of nucleic acid alkylation and their relation to mutagenesis and carcinogenesis, *in: Progress in Nucleic Acid Research and Molecular Biology,* Vol. 15 (W. E. Cohn, ed.), pp. 219–284, Academic Press, New York.

Sutton, H. E., and Harris, M. I., eds., 1972, *Mutagenic Effects of Environmental Contaminants,* Academic Press, New York.

Ts'o, P. O., and DiPaolo, J. A., eds., 1972, Chemical carcinogenesis, *in: The Biochemistry of Disease,* Vol. 4, Part A, Marcel Dekker, New York.

Ts'o, P. O., and DiPaolo, J. A., eds., 1972, Chemical carcinogenesis, *in: The Biochemistry of Disease,* Vol. 4, Part B, Marcel Dekker, New York.

Vogel, F., and Rathenberg, R., 1975, Spontaneous mutation in man, *in: Advances in Human Genetics,* Vol. 5 (H. Harris and K. Hirschhorn, eds.), pp. 223–318, Plenum Press, New York.

Vogel, F., and Röhrborn, G., eds., 1970, *Chemical Mutagenesis in Mammals and Man,* Springer-Verlag, Berlin.

Volkenshtein, M. V., 1970, *Molecules and Life: An Introduction to Molecular Biology,* Plenum Press, New York.

Wagner, R. P., and Mitchell, H. K., 1955, *Genetics and Metabolism,* John Wiley, New York.

Wilson, J. G., 1973, *Environment and Birth Defects,* Academic Press, New York.

Wolstenholme, G. E. W., and O'Connor, M., eds., 1969, *Symposium on Mutation as Cellular Process,* J. and A. Churchill, London.

Radiations and Other Physical Agents   5

ROBERT L. BRENT

I. INTRODUCTION

The purpose of this chapter is to describe the effects of exposing mammalian embryos to various forms of ionizing and nonionizing radiation. Although some clinical applications will be alluded to, the primary purpose will be to present the quantitative information derived from animal experiments. A detailed consideration of the clinical implications and clinical interpretation of the hazards of ionizing and nonionizing radiation are discussed in other monographs (Brent and Gorson, 1972; Brent and Harris, 1976).

II. IONIZING RADIATION

A. General Considerations

In considering the embryonic effects of ionizing radiation, one must take into account the differences in the dose distribution arising from radioactive isotopes administrered during pregnancy as compared to X-rays or gamma rays delivered from external sources. The distribution of absorbed energy is relatively uniform in embryos exposed to externally administered high-energy electromagnetic radiation. Radioactive isotopes, however, have a variable distribution in the embryo following administration to the mother, de-

ROBERT L. BRENT · Jefferson Medical College, Thomas Jefferson University, Philadelphia, Pennsylvania. Work supported by Grants E(11-1)3268 and NIH-HD630.

pending on (1) the stage of gestation, (2) whether the radioactive material crosses the placenta, and (3) the biochemical affinities of the material for certain tissues. The absorbed dose distribution also depends on the type of radiation emitted (alpha particles, beta particles, and/or, gamma rays, and on their energies). Thus, the calculation or measurement of absorbed dose and the evaluation of the relative hazards are considerably more complex for radiations absorbed internally from administered radioactive materials than for radiation delivered from external X-ray or gamma ray sources (Brent, 1971). For the purposes of this chapter the terms "rads" or "roentgens" (R) will be used to reflect the exposure referred to in the article being quoted. For practical purposes "rads" refers to the absorbed dose in ergs per gram of tissue and "roentgens" refers to the exposure to an equivalent dose standardized by the number of ionizations produced in air. In most instances the rad and roentgen vary by only 10–15%.

B. Effects of Externally Administered Radiation on the Embryo

The classic effects resulting from irradiation on the mammalian embryo are: (1) intrauterine and/or extrauterine growth retardation; (2) embryonic, fetal, or neonatal death; and (3) congenital malformations. In multitocous animals, all three effects may be seen in one litter of irradiated animals. For the single pregnancy, such as in the human, an abortus may also be malformed as well as growth retarded. These three effects of irradiation are referred to as the "classic triad" of radiation embryologic syndromes. The probability of finding one or more of these effects in an embryo exposed to radiation depends on a number of factors, including: (1) the absorbed dose, (2) the dose rate, and (3) the stage of gestation at which the exposure occurred.

1. Teratogenesis

Most investigators report a dearth of malformations when mouse and rat embryos are irradiated in the preimplantation period, regardless of dose (Brent, 1963, 1970, 1972a,b, 1973a,b; Brent and Bolden, 1967a,b, 1968; Brent and Gorson, 1972; Harvey and Chang, 1962; Rugh *et al.,* 1969; Russell, 1954, 1965). On the other hand, the preimplanted stage of the embryo is the most sensitive to lethal effects of radiation. Thus, 150 rads absorbed on the first day of gestation in the rat has a probability of killing approximately 70% of the embryos exposed, whereas it will kill only 50% of the embryos at the stage at which the embryo is most susceptible to teratogenesis. Nature seems to have evolved a protective mechanism so that radiation (and other teratogenic agents) will either kill the preimplanted embryo or permit it to develop normally, at least morphologically.

There are several possible exceptions to this observation. It has been reported (Russell and Montgomery, 1966; Russell and Saylors, 1963) that exposure of 100 R (roentgens) increases the incidence of XO aneuploidy in a strain of mice that had a spontaneous incidence of this anomaly of 1%. Rugh (1962, 1963, 1965, 1969; Rugh and Grupp, 1959, 1960, 1961) reported that 15–25 R increases the incidence of exencephaly in preimplanted embryos. The latter study was in a strain of mice exhibiting a 1% spontaneous incidence of exencephaly (Rugh, 1969). In our laboratory, we have irradiated mice and rats with 10–150 R during the preimplantation period and have never observed any difference between the controls and the irradiated animals in the incidence of exencephaly or, for that matter, of congenital malformations when using the technique of complete fetal dissection (Brent, 1970, 1967).

In the rat and mouse, radiation has its greatest effectiveness in producing congenital malformations during the organogenetic period (mouse 6½–12th day; rat 9th–13th day), which is approximately equivalent to the 14th–50th day of the human pregnancy (Fig. 1). Gross malformations can be produced in low frequency with 25 R at a very narrow period in rat gestation, namely, the 10th day, which would be equivalent to about the 15th–21st day in the human. With 50 R administered during these stages, one would expect growth retardation; gross, major malformations; and an increased probability of embryonic death. On the other hand, at later stages of the organogenetic period, 50 R would produce no gross congenital malformations and the only evidence of radiation effect would be in the nature of cellular deletions or tissue hypoplasia. Russell (1957) studied the effect of 25 R at a most sensitive period of organogenesis and found changes in the variability of the number of ribs, although there were no gross malformations during the period of maximal sensitivity. Furthermore, this effect can be produced for only a short period during organogenesis, and the minimal dose for gross malformation induction is higher during the remainder of the organogenetic and fetal periods.

In an extensive and complicated study Jacobsen (1965) reports changes in the response of irradiated embryos to low doses of radiation at different times of the year. He utilized exposures of 5 R and greater, but the poor correlation of the incidence of malformations with dose and season leave the results open to question. There is a good possibility that the radiation exposure at the 5-R level did not account for the changing incidence of malformations reported by Jacobsen.

During the fetal stages (mouse 12th–18th day; rat 13th–22nd day), which would be equivalent to the last two trimesters of human pregnancy, defects in development can be produced with doses of 150–400 R, but these are primarily limited to the central nervous system and other organs, which continue to differentiate throughout gestation. Thus, cerebral hypoplasia, microcephaly, cerebellar hypoplasia, and testicular atrophy can be produced by these high doses at specific times during the fetal stages. After the administration of 50 R, minimal tissue hypoplasia or cellular hypoplasia can be identified by some investigators, but others report complete functional and tissue normalcy following these lower radiation doses.

Over the past 20 years it has become increasingly apparent that the fetus is sensitive to the effects of radiation and that, although grotesque abnormalities of major organ systems do not occur, permanent alterations can be produced in many organs and systems. Hicks (1958), Brizzee *et al.* (1961a,b, 1967), Cowen and Geller (1960), Altman *et al.* (1968a,b), Dekaban (1968, 1969) and others have been interested in the selective ability of radiation to affect specific parts of the central nervous system. The specificity of the radiation effect is related to the stage of gestation at which radiation is administered. The severity of the effect is related to the magnitude of the dose of radiation.

Hicks and D'Amato (1963, 1966; Hicks *et al.*, 1959) compared the gross size and structure of a normal adult rat brain with those of animals that received 200 R on the 13th or 17th day of gestation or on the first day after delivery. They determined that the degree of cortical atrophy depends on the

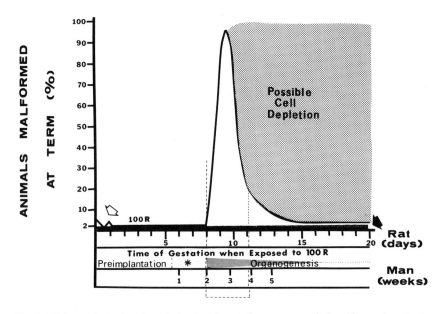

Fig. 1. This graph depicts the relative incidence of gross congenital malformations in the rat following an exposure of 100 R. The arrow on the right side of the ordinate points to the 2% control incidence of congenital malformations seen in this species. The incidence of malformations before the 9th day is the same as in the controls. The light arrow on the left side of the ordinate points to a slight increase in malformations on the 1st day. This was inserted because of the work of Russell in a particular strain of mice (see text). A very large increase in malformations occurs during the early organogenetic period. Note that this corresponds to the 3d and 4th weeks of human pregnancy. This high incidence of gross malformations falls off rapidly as organogenesis diminishes. Note that organogenesis to some degree (CNS) continues to term. Also note that although gross malformations may not be produced during the late fetal stages with 100 R, an irreversible loss of cells occurs. The significance of these cell losses is under study. The asterisk (*) is placed at the stage of implantation to indicate that although malformations are not readily produced at this stage, growth retardation can be induced.

embryonic stage at the time of radiation exposure. A cross section of the most severely affected animals revealed that although there was a marked diminution in the size of the brain, the gross overall organization was normal. In fact, they originally reported that many of these adult animals with abnormal brains exhibited normal motor and behavioral activities. Microscopic examination of the irradiated brains exhibited irregular and bizarre neuron formation. Hicks and D'Amato (1963) reported that exposure of the fetus to 20 R produces discernible but subtle microscopic changes in the adult cortex, although these animals demonstrate no overt neurologic anomalies. Brizzee and Brannon (1972) report no tissue damage in rat embryos irradiated on the 13th day with 25 R, but there was a slight decrease in the cortical mantle. Fractionation of larger doses markedly reduce their effectiveness, indicating that the recovery from radiation damage in the embryo was quite rapid. It is important to extend these studies to determine the extent to which some of the histologic changes are partially or wholly repaired by the time the animal reaches the adult stage in development.

Some quantitative studies performed by Brizzee and coworkers (1967) demonstrate that the reduction in depth of the cortical plate is related to the exposure and exposure rate. Fractionation of the dose drastically reduces the effectiveness of the irradiation.

Other organ systems that have been affected specifically during the fetal stages are the thyroid and the diaphragm on the 12th and 13th day of gestation in the rat (Brent, 1960a,b). Testicular atrophy can be produced by irradiation of the fetus at two periods of gestation (Brent, 1960a; Murphree and Pace, 1960). Testicular atrophy occurs in the rat if the 9-day-old embryo is exposed to 100 R (Brent, 1960a). As little as 50 R administered to 18-day-old rat embryos results in a reduction in the size of the adult testes. The fact that testicular atrophy can be induced by radiation exposure at two widely separated stages of embryonic and fetal development is a remarkable phenomenon and may indicate that, although the result appears similar, the mechanisms at each stage are quite different.

These observations indicate that irradiation of the fetus can alter the structure of adult organs and tissues. In many instances, it is difficult to evaluate the significance of these alterations, since they consist of cell and tissue depletions and, in some instances, organ hypoplasia.

It is difficult to determine the dose of radiation that will not produce any effect in the embryo, since one may consider effects that are present in only one species, or that are completely reversible, or are present but have no deleterious effect on the organism. Thus, it is difficult to speak in generalities about the safe level below which there will be no effect on the mammalian embryo. One must specify the particular effect in question, the species, and even the dose rate of the particular dose (Tables 1 and 2). Table 3 lists a number of studies dealing with low levels of radiation with approximately 10 rads or less of acute or continuous radiation exposure. It would appear from the data presented in Tables 1 and 3 that there may be some pathologic effects in the embryo with exposure of 10 R or less but these effects are

Table 1. A Compilation of the Effects of 10 Rads or Less Acute Radiation at Various Stages of Gestation in Rat and Mouse[a]

	Stage of gestation (days)				
	Preimplantation	Implantation	Early organogenesis	Late organogenesis	Fetal stages
Mouse	0–4½	4½–6½	6½–8½	8½–12	12–18
Rat	0–5½	5½–8	8–10	10–13	13–22
Corresponding human gestation period	0–9	9–14	15–28	28–50	50–280
Lethality	+[b,c]	−	−	−	−
Growth retardation at term	−	−	−	−	−
Growth retardation as adult	−	−	−	−	−
Gross malformations (aplasia, hyperplasia, absence or overgrowth of organs or tissues)	±[d]	−	−	−	−
Cell depletions, minimal but measurable tissue hypoplasia	−	−	−	−	−
Sterility	−	−	−	−	−
Significant increase in germ cell mutations[e]	±	±	±	±	±
Cytogenetic abnormalities	−	−	−	−	−
Neuropathology	−	−	−	−	−
Tumor induction[e,f]	−	−	±	±	±
Behavior disorders[g]	−	−	−	−	−
Reduction in life-span[e]	−	−	−	−	−

[a] Dose fractionation or protection effectively reduces the biologic result of all the pathologic effects reported in this table.

[b] (−) no observed effect; (±) questionable but reported or suggested effect; (+) demonstrated effect; (++) readily apparent effect; (+++) occurs in high incidence.

[c] At this stage the murine embryo is most sensitive to the lethal effects of irradiation. With 10 R in the mouse, Rugh reports a slight decrease in litter size in the mouse.

[d] Rugh reports exencephalia with 15 and 25 R in a strain of mice with a 1% incidence of exencephalia.

[e] The potential for mutation induction exists in the embryonic germ cells or their precursors. Several long term studies indicate that considerably greater doses in mice and rats do not affect longevity, tumor incidence, incidence of congenital malformations, litter size, growth rate, fertility.

[f] Stewart and others have reported that 2 R increases the incidence of malignancy by 50% in the offspring. See text for discussion.

[g] Piontkovskii reports behavioral changes in the rat after 1-R daily irradiation. This work has not been reproduced. See text for discussion.

not gross malformations or growth retardation. Brown (1964) reported that 2 R/day of continuous radiation exposure did not alter any aspect of rat development or growth except for a reduction of litter size after continuous radiation exposure for 11 generations. Havlena *et al.* (1964) irradiated the 36-hr rat embryo with as much as 100 R and reported no difference between the control and surviving irradiated animals with regard to learning and locomotor function as adults. Furthermore, they observed no congenital malformations in the surviving offspring.

Meyer *et al.* (1968) examined the offspring of women who themselves had been exposed to X-ray pelvimetry while *in utero*. There were no positive findings. Segall *et al.* (1964) studied the incidence of malformations in high-dose and low-dose radiation areas but found no evidence that malformation rates were related to background radiation. Tabuchi (1964; Tabuchi *et al.*, 1967) and Kinlen and Acheson (1968) reported diagnostic radiation exposure did not contribute to fetal disorders observed at term. Vorisek (1965) irradiated pregnant rats with 2.5 R/day and observed an increase in postnatal mortality in irradiated rats. The total exposure dose of continuous radiation was 50 R. This observation does not agree with the results of Brown (1964).

Thus, the evidence strongly supports the conclusion that major congenital malformations are highly unlikely to occur in the human after doses of less than 20 rads absorbed from the 14th day of human pregnancy until term. By major malformations, we mean the absence or gross distortion of a tissue, organ, or tissue component, or the identification of easily observable tissue hypoplasia, such as microcephaly, microphthalmia, testicular atrophy, or cerebral hypoplasia. This, of course, does not mean that doses below 20 rads do not produce some deleterious effects that may be subtle and difficult to detect and difficult to document. It is for this reason that the National Council on Radiation Protection and Measurements (NCRP) has established the maximum permissible dose to the fetus from occupational exposure of the expectant mother well below the known teratogenic dose (NCRP, 1971) (Table 4).

2. Growth Retardation

At sufficiently high doses, intrauterine growth retardation can result from radiation exposure delivered any time during gestation *after the time of implantation* (Fig. 2). Hence, the implanted embryo can suffer growth retardation from radiation exposure even at the time of gestation when it is not susceptible to the malforming effects of radiation. The resistance to growth retardation exhibited by the preimplanted embryo at term persists into adulthood. Thus, the irradiated preimplanted embryo that survives until term grows normally in the prepartum and postpartum periods. Animal experiments indicate that all embryos exposed to 100 R or more after implantation will exhibit some degree of growth retardation. This growth retardation can per-

Table 2. A Compilation of the Effects of 100 Rads Acute Radiation on Embryonic Development at Various Stages of Gestation in Rat and Mouse[a]

	Stage of gestation (days)				
	Preimplantation	Implantation	Early organogenesis	Late organogenesis	Fetal stages
Mouse	0–4½	4½–6½	6½–8½	8½–12	12–18
Rat	0–5½	5½–8	8–10	10–13	13–22
Corresponding human gestation period	0–9	9–14	15–28	28–50	50–280
Lethality	+++[b,e]	+	++	±	—
Growth retardation at term	—	+	+++	++	+
Growth retardation as adult	—	+	++	+++	++
Gross malformations (aplasia, hypoplasia, absence or overgrowth of organs or tissues)	—	—	+++	±[d]	—[d]
Cell depletions, minimal but measurable tissue hypoplasia	—	—	±	++	+
Sterility	—	—	±	—	++[c]
Significant increase in germ cell mutations[f]	±	±	±	±	±
Cytogenetic abnormalities[g]	±	—	+++	+	+
Cataracts	—	—	±	++	+
Neuropathology	—	—	+++	++	+
Tumor induction[h]	—	—	±	±	±
Behavior disorders[i]	—	—	+	+	±
Reduction in lifespan[j]	—	—	—	—	—

[a]Dose fractionation or protraction effectively reduces the biologic result of all the pathologic effects reported in this table.

[b](−) no observed effect; (±) questionable but reported or suggested effect; (+) demonstrated effect; (++) readily apparent effect; (+++) occurs in high incidence.

[c]Russell reported that 200 **R** increased the incidence of XO aneuploidy in 2–5% of offspring in mice with a spontaneous incidence of 1%. One hundred **R** kills substantial numbers of mouse and rat embryos at this stage but the survivors appear and develop normally.

[d]One hundred **R** produces changes in the irradiated fetus which are subtle and necessitate detailed examination and comparison with comparable controls.

[e]The male gonad in the rat can be made extremely hypoplastic by irradiation in the fetal stages with 150 **R**. In the mouse the newborn female is most sensitive to the sterilizing effects of radiation. Much of this research on other animals cannot be applied to the human (see text for discussion).

[f]The potential for mutation induction exists in embryonic germ cells or their precursors. The relative sensitivity of the embryonic germ cells when compred to adult germ cells is not known. Several long term studies in animals do not indicate any exceptional differences.

[g]Footnote 1 refers to the aneuploidy produced in a strain of mice with a 1% incidence of spontaneous XO aneuploidy. Bloom has reported a much higher percentage of chromosome breaks in human embryos receiving 100–200 **R** *in utero* than in adults receiving the same dose of irradiation. As yet there have been no diseases associated with this increase in frequency of chromosome breaks.

[h]Animal experiments and the data from Hiroshima and Nagasaki do not support the concept that *in utero* irradiation is much more tumorigenic than extrauterine irradiation. On the other hand Stewart and colleagues and many others report that irradiation from pelvimetry (2 **R**) increases the incidence of leukemia and other tumors.

[i]A statistically significant increase in percentage of mental retardation occurs with this dose of radiation. On the other hand normal intelligence has been found in children receiving much higher doses *in utero*.

[j]Animal experiments indicate that survivors of *in utero* irradiation have a life-span which is longer than groups of animals given the same dose of radiation during their extrauterine life, and the same life expectancy as unirradiated controls.

Table 3. Reported Effects of Low Exposures of Radiation on the Embryo When Administered Throughout Pregnancy

	Organism	Source	Approximate exposure rate per minute (R)	Exposure per day	Exposure per pregnancy	Comments	Effects
Russell et al. (1959)	Mouse	^{137}Cs	0.0086	12.4 R	171 R	1st–18th day	None except shortened breeding period in female
Ronnback[a]	Mouse	^{137}Cs		8.4 R	170 R	During gestation and in some instances 20 days postpartum	None
Vorisek (1965)	Rat	^{60}Co	0.0017	2.5 R	—	Daily during pregnancy	None
Stadler and Gowan[a]	Mouse		0.0015	2.2 R		Continuous through 10 generations	None
Coppenger and Brown (1965)	Rat	^{60}Co	0.0014	2.0 R	—	Continuous through 11 generations	None
Konerman[a]	Mice	^{60}Co	0.007 0.014	10 R 20 R 20 R	180 R 360 R	1st–18th day 1st–18th day Day 6–13 only	None Growth retardation None
Wesley[a]	Humans	Background		0.3 mR	0.1 R	Variation in background radiation	Increase congenital malformation
Gentry	Humans	Background		0.3 mR	0.1 R	Variation in background radiation	Increase congenital malformation

Grahn[a]	Humans	Background		0.3 mR	0.1 R	Variation in background radiation	Background radiation level not a factor in incidence of congenital malformations
Segall et al. (1964)	Humans	Background		0.3 mR	0.1 R	Background radiation	Background radiation level not a factor in incidence of congenital malformations
Kriegel and Langendorf[a]	Mouse	X-ray		2.5 R 5 R 10 R 20 R		Acute dose given daily	No influence No influence No influence Malformations, resorptions, growth retardations
Piontkovskiy (1958, 1961)	Rat	X-ray		1 R	20 R	Acute dose given daily	Functional changes in behavior and motor activity (questionable)
Laskey et al. (1973)	Rat	HTO	Continuous 0.01–10 μCi HTO ml of body water	0.003–3 rad/day	0.066–66	Tritiated water continuous exposure	Male testes reduced 10 μCi; no growth or reproductive impairment in F_1; F_2 generation had reduced weight

[a]Data from Blot and Miller (1973).

Table 4. Estimation of the Acute LD$_{50}$ Dose and Minimal Malforming Doses of Radiation for the Human Embryo Based on Compilation of Mouse, Rat, and Human Data

Age	Approximate minimal lethal dose (R)	Approximate LD$_{50}$ (R)	Minimum dose for non-recuperable growth retardation in adult (R)[a]	Minimum dose for recognizable gross malformation (R)	Minimum dose for induction of genetic, carcinogenic, and minimal cell depletion phenomena
Day 1	10	70–100	[b]	[c]	Unknown
Day 14	25	140	25	—	Unknown
Day 18	50	150	25–50	25	Unknown
Day 28	>50	220	50	25	Unknown
Day 50	>100	260	50	50	Unknown
Late Fetus to term		300–400	50	>50	Unknown

[a] Estimates for maximum dose effective in reducing body weight. Specific organs or measurements may be more or less sensitive.
[b] Surviving embryos are not growth retarded even after high doses of radiation.
[c] Malformation incidence is extremely low even after high doses of radiation.

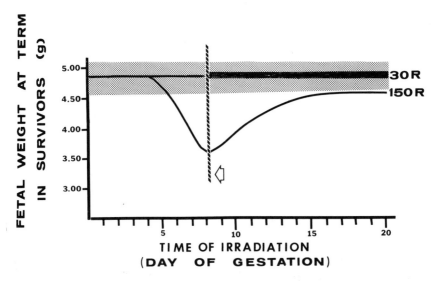

Fig. 2.   The term fetal weight in the Wistar colony of rats in our laboratory averages 4.8 g on the 22nd day when delivered by cesarean section that morning. Note that 150 R causes growth retardation just after implantation but before the stage is sensitive to teratogenesis (arrow pointing to vertical line). From the time of implantation until term, 150 R produced some degree of growth retardation. Note that 30 R does not appear to produce significant deviations in growth, although it may be that the experimental groups are too small to show differences in weight at this low dose.

sist into adulthood (Rugh *et al.,* 1964b). Although embryos exposed to radiation in the early organogenetic period exhibit the most severe intrauterine growth retardation at term, radiation exposure later in pregnancy results in the largest degree of permanent growth retardation (Figs. 2 and 3) (Tables 5 and 6). It appears that the younger embryo can repair more completely some of the radiation-induced deficits. The threshold dose for radiation-induced growth retardation is below 100 R and depends on the stage of gestation as well as the dose rate. Animal embryos irradiated at any stage of gestation with less than 25 R have not exhibited growth retardation as adults (Tables 1, 3–6). When growth retardation does occur, experimental evidence indicates that the effect is due primarily to the direct interaction of the radiation with the fetus rather than to indirect effects produced by the irradiation of the pregnant animal (Brent, 1960a,b; Brent and McLaughlin, 1960).

Growth retardation is closely associated with teratogenesis. In groups of irradiated embryos that exhibit an observable deficit, such as major congenital anomalies or tissue deficiencies, there also is a concomitant reduction in over-all growth as evidenced by a reduction in weight and various physical measurements at term. Furthermore, it appears that growth retardation may be a more sensitive index of radiation damage than is teratogenesis; such may be the case in humans also. The *in utero* victims exposed to the atomic bombs in Japan had more manifestations of growth retardation than major congenital

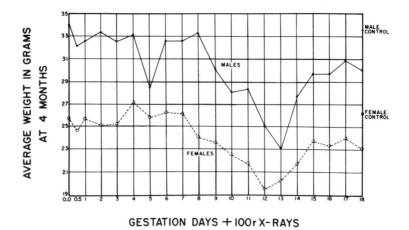

Fig. 3. Weight of control and of irradiated male and female mice at 4 months of age following an exposure of 100 R at gestational stages from conception to the 18th day. Note that permanent growth retardation is not produced until after implantation and is produced at every stage thereafter. (From Rugh *et al.*, 1964b, with permission from the *American Journal of Anatomy*.)

malformations, despite the fact that many of the survivors had received doses of 50–200 rads (Miller, 1956, 1970; Wood *et al.*, 1967a–c; Yamazaki *et al.*, 1954).

3. Lethality

Embryonic death is a component of the triad of radiation effects observed in irradiated mammalian embryos. The LD[50] of the embryos varies throughout gestation, being very high just after fertilization and gradually reaching adult levels near term (Fig. 4, Table 4). The mortality curve has several peaks, possibly due to the variable radiosensitivity of a synchronously dividing zygote during the first few days of pregnancy (Fig. 4). Although the preimplanted embryo has a lower LD_{50} than the embryo during organogenesis, it apparently can recover from the effects of radiation, if implantation is delayed (Hahn and Feingold, 1973). Furthermore, the increased survival following delayed implantation has been attributed to an inhibition of mitosis with an associated prolongation of the time spent in the S phase of the cell cycle. After implantation, the maximal sensitivity of the embryo to the lethal effects of radiation occurs at approximately the same time that the embryo is maximally susceptible to teratogenesis (Figs. 1 and 4).

The period of major variation in the susceptibility of the mouse and rat embryo to the lethal effects of radiation corresponds to the first 21 days of human pregnancy, which is a time when it is very difficult to be precise about the exact stage of gestation. As mentioned before, nature seems to have built

Table 5. Mean Birth Weight in Grams (± SD) of Irradiated and Control Mice[a]

Intra-uterine exposure (R)	N[b]	Mean birth weight, grams								
		0.5[c]	N	7	N	8	N	12	N	16
0	31	1.58 ± 0.24	22	1.57 ± 0.18	19	1.62 ± 0.15	24	1.63 ± 0.15	18	1.65 ± 0.20
30	23	1.63 ± 0.19	18	1.68 ± 0.16	21	1.58 ± 0.23	19	1.60 ± 0.20	19	1.64 ± 0.26
60	21	1.75 ± 0.20	19	1.61 ± 0.13	22	1.67 ± 0.22	24	1.60 ± 0.22	21	1.60 ± 0.16
90	—		21	1.57 ± 0.17	22	1.61 ± 0.17	17	1.49 ± 0.16[d]	19	1.53 ± 0.14[d]

[a]The only groups with significantly reduced weights at birth were the 12- and 16-day-old embryos irradiated with 90 R.
[b]N = number of litters.
[c]Embryonic age at the time of irradiation (days).
[d]Significantly smaller than control term weights ($P = 0.01$).

**Table 6. Weight in Grams of Adult Mice
(150 Days of Age) Irradiated *in Utero*[a]**

In utero exposure (R)	Day of gestation when irradiated			
	7	8	12	16
0	32.58[a]	29.38	30.99	30.04
30	29.78	30.56	30.38	28.43
60	29.28	27.80	24.28	28.47
90	28.46	27.58	23.55	28.12

[a]At least 70 animals in each group. The dotted line indicates a significant difference from the controls at the $P < 0.05$ level of significance.

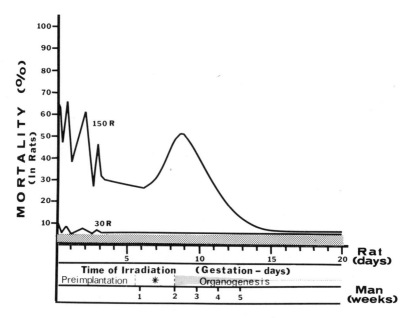

Fig. 4. The lethal effect of radiation in rats is greater on the 1st day. It appears that the LD_{50} shifts at different times of the day during early gestation, possibly because the cells are dividing synchronously, therefore, the zygote will vary in its radiosensitivity, depending on the stage of the cell cycle. Note that the embryo becomes somewhat resistant in the implantation stage and then becomes sensitive to the lethal effect of radiation during early organogenesis. A 30-R dose apparently does increase the resorption rate when radiation takes place on the 1st day. The superimposition of this lethality curve for rats onto the human embryonic development timetable may not be appropriate and is for comparison purposes only, but it allows one to estimate at least roughly the stages at which the human embryo might be most readily killed by high doses of radiation.

in a protective mechanism that decreases the chances of an abnormal embryo, since the range in dose from no effect to the dose that kills all the embryos is quite small during the period of maximum susceptibility to teratogenesis. That probably is a reason why there was such a high incidence of normal children and abortions following the atomic bomb explosions in Japan.

Although most experimental studies concern themselves with mortality at term following intrauterine radiation, there are a number of studies that have been concerned with postpartum mortality (Brent, 1968; Brent and Bolden, 1961; Rugh and Wohlfromm, 1965, Russell and Russell, 1954). Human mortality following intrauterine radiation will be discussed under the category of tumor induction (Section II. C.4). It is apparent that postnatal mortality studies bring into play other variables difficult to control, such as mothering, the litter effect, litter size, etc.

C. Pathologic Findings in Adult Organisms Following Embryonic and Fetal Irradiation

The majority of investigations have been concerned with the pathologic state of the irradiated embryo before or at the time of parturition. In recent years, increasing interest has developed in pathologic alterations of adult organisms that have been irradiated *in utero*. The subject matter covered in this section will deal with pathologic changes that may not be apparent at birth or would not even become manifest until later in life. Thus, the incidence and type of gross malformations which can be recorded in animals at term will not be discussed further except where necessary to support a particular point.

1. Growth Retardation

There is ample evidence to indicate that growth retardation in newborn animals is a common and predictable consequence of *in utero* irradiation. Furthermore, the presence, absence, and degree of retardation are dependent on the dose of radiation and the stage of gestation (Brent, 1960a, 1968; Levy *et al.*, 1953; Pobisch, 1960; Rugh, 1953, 1959; Russell, 1954; Russell and Russell, 1954; Skreb *et al.*, 1963). The irradiated preimplanted and undifferentiated mammalian embryo, when it survives to term, is not growth retarted by even 150 R (Brent and Bolden, 1967a,b, 1968; Russell, 1954). However, the early differentiating embryo is retarded in growth at term, as is the late differentiating embryo (Fig. 2). As the fetus proceeds through gestation, the 150-R exposure is less effective in producing stunting at term. This may be due to a combination of factors in the rat and mouse increasing radioresistance in later gestation and increasing proximity of the irradiation time to parturition. There is evidence to indicate that the differentiating human embryo may be severely growth retarded at term by high doses of radiation

(Dekaban, 1968; Murphy, 1929; Murphy and Berens, 1952). There are cases in the literature reporting both the presence and absence of severe growth retardation when a human fetus has been irradiated late in gestation (Dekaban, 1968; Glass, 1944; Murphy, 1929; Murphy and Berens, 1952; Ronderos, 1961). There are not enough documented survivors of embryos irradiated during the first week of gestation to determine whether the human fetus can be growth retarded at term if irradiated during preimplantation. Furthermore, if the dose of radiation exceeds 200 R at this time, there is a considerable probability that the zygote will not survive (Brent and Bolden, 1968; O'Brien, 1956; Russell, 1954; Wilson, 1954; Wilson et al., 1953a,b).

The effect of intrauterine irradiation on the postpartum growth of mice was examined by Rugh et al. (1964a,b, 1966). They found that 100 R would produce growth retardation in 4-month-old mice if administered on any day from the 8th to the 18th day of gestation (Fig. 3). Irradiation on the 12th and 13th days of gestation resulted in the most severe growth retardation at four months of age. This growth retardation was manifested not only by a decrease in body weight but also by a decrease in measurements of the long bones and skull. It is interesting that the most severe degree of in utero growth retardation is produced several days earlier, but apparently the younger, "more sensitive" embryo has a greater capacity to approach its adult size than does the 12- and 13-day irradiated mouse embryo (Skreb et al., 1963; Wilson, 1954; Wilson et al., 1953a,b). Rugh and his colleagues demonstrated that the surviving embryos irradiated during preimplantation, which did not manifest growth retardation at term, also showed normal growth in the postpartum period.

Although Rugh's data give us an excellent profile of the growth-retarding effects of radiation in the mouse, they do not indicate what the threshold dose is. In the mouse, data from our laboratory indicate that the lowest dose that will produce growth retardation manifest at term is less than 60 R. These data are summarized in Table 5 and indicate that, at term, only the 12-day-old and 16-day-old embryos were retarded at birth after receiving 90 R. Yet, Rugh has shown that the 7-day-old embryos exhibited growth retardation as adults following 100 R. In this instance, it appears that extrauterine growth was a more sensitive index of radiation damage than intrauterine growth. This conclusion is borne out by our own data, which indicate that at some stages in gestation the threshold dose for in utero growth retardation in adults may be between 30 R and 60 R (Table 6). The maximum impact of radiation on postpartum growth was during the early fetal period of mouse gestation.

There are many other studies dealing with the effects of a single exposure to X-rays on the postpartum growth of rodents (Levy et al., 1953; Murphree and Pace, 1960; Pobisch, 1960; Russell, 1954; Skreb et al., 1963) and goats. Murphree and Pace (1960) irradiated rats during the late fetal stages with exposures ranging from 110 to 220 R. Rats irradiated in the 20th day with 150 R were small as adults. All surviving rats irradiated in the late fetal stages with doses of greater than 110 R exhibited growth retardation.

In studies dealing with the indirect effect of radiation on embryogenesis, Brent and McLaughlin (1960) demonstrated that 400 R could be administered to the pregnant mother or to the placenta (Brent, 1960b; Brent and McLaughlin, 1960) without producing growth retardation of the fetus when the fetus or embryo was shielded. If the maternal organisms received 1000 or 1400 R while shielding the embryos, the fetuses showed growth retardation, but in these instances the mother lost considerable weight due to radiation sickness.

The effects of chronic radiation throughout gestation have been studied by a number of investigators (Table 3). A total dose of 200 R administered throughout mouse gestation produced no effects on postpartum growth (Russell et al., 1959). Doses of 2 R/day given continuously for 11 generations, including the gestation periods, resulted in a decrease in litter size, although the birth and weaning weights were normal (Brown, 1964). Doses of 50 R/day from conception to term had a marked effect on intrauterine growth, as would be expected (Coppenger and Brown, 1965). Since two exposures of 50 R during the early fetal stages of rat gestation should produce growth retardation, these investigators also administered chronic irradiation (20 hr/day) to pregnant animals at the rate of 5, 10, and 20 R/day. They were interested primarily in sterility, litter size, mortality, and malformations, but their data indicate that the rat embryo can tolerate 5 and 10 R/day of chronic radiation exposure without detectable growth retardation. Laskey et al. (1973) administered tritiated water continuously from the moment of conception through the F_2 generation. The absorbed dose ranged from 3 to 3000 mrads/day. There was no impairment of growth or reproductive capacity, although the males in the highest exposure group had diminshed testicular size (Table 3).

There is considerable evidence to indicate that radiation-induced growth retardation can occur in mammals after exposure in the neonatal and infant periods. Billings et al. (1966) irradiated 4-day-old rats with 0, 5, 25, or 125 R and noted that growth retardation was observed only in the rats that received 125 R. Petrosyan and Pereslegin (1962) irradiated newborn rats with 50, 100, 200, 300, or 500 R. They noted marked growth retardation of the surviving animals given 300 or 500 R. There are numerous reports in the literature to indicate that the growth of the human neonate or infant can be retarded by sufficiently large amounts of radiation (Conard and Hicking, 1965; Greulick et al., 1953; Miller, 1968; Murphy and Berens, 1952; Reynold, 1954; Sutow et al., 1965). Following the exposure of children in Nagasaki, Hiroshima, and the Marshall Islands to the atomic bomb radiation or fallout, there were many studies indicating that not only infants and children were susceptible to the growth-retarding effects of radiation but also that the younger children were more susceptible to these growth-retarding effects. Although there was definite growth retardation present in boys exposed to fallout radiation in the Marshall Islands, it is difficult to be certain if this was due to external radiation plus thyroid irradiation, since an overall dose of 1000 rads to the thyroid by radioactive iodine was estimated to have been absorbed by some of the people (Conard and Hicking, 1965; Sutow et al.,

1965). There is little doubt that doses of 100 rads or greater contributed to growth retardation in children in both Hiroshima and Nagasaki.

The extrauterine growth of irradiated human embryos has been studied carefully by a number of investigators stationed in Hiroshima and Nagasaki after the atomic bomb explosions (Burrow *et al.*, 1964, 1965; Miller, 1956, 1970; Plummer, 1952; Tabuchi, 1964; Wood *et al.*, 1967a–c; Yamazaki *et al.*, 1954). There was a significant reduction in head circumference, height, and weight. The head-circumference reduction was most significant in children who were believed to have received doses greater than 50 rads. The presence of three factors were associated with the greatest reduction in head circumference: (1) symptoms of radiation sickness in the mother, (1) distance of less than 1500 meters from hypocenter, and (3) less than 15 weeks of gestation. It also was of interest to note that there was no "catch-up" growth, since the smallness in head size was maintained through adolescence. The measurable growth retardation in the group exposed *in utero* was not as severe as the growth retardation observed in children irradiated between one and two years of age (Toyooka *et al.*, 1963). This finding agreed with the results of some animal experiments, but then the environment of the infant following the atomic bomb detonation might have been a contributing factor in growth retardation.

Finally, there are scores of individual case reports in the literature describing the clinical status of children who were exposed to therapeutic doses of radiation while *in utero* (Cohen *et al.*, 1973; Dekaban, 1968; Glass, 1944; Murphy, 1929; Ronderos, 1961, Scharer *et al.*, 1968; Stahl, 1963). Many of these case reports are incomplete because either the exact time of gestation during which the radiation was given or the data on the growth of the child are inadequate. Dekaban collated the results from 26 cases and reported them to be similar to the cases in Hiroshima and Nagasaki. The most common finding following high doses of *in utero* radiation is mild to severe growth retardation, and the most likely time for this to be produced is during the 3rd to 20th week of gestation (Fig. 5).

2. Sterility

Laboratory investigations have been directed along two avenues. Some investigators have been interested in the functional capacity of an organism to sire or deliver offspring. In many instances, investigators have attempted to predict the presence or absence of sterility on the basis of histologic interpretation, but it will be obvious that this cannot be done.

The response of the embryonic male and female gonad to irradiation will be discussed separately.

a. Male. Radiation exposure of the male rat embryo and fetus has yielded both interesting and unpredictable results. Exposure of the 9-day-old rat embryo to 100 R results in a low frequency of severe testicular hypoplasia or

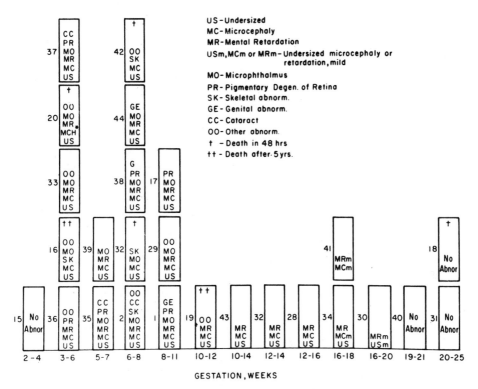

Fig. 5. The spectrum of malformations in 26 human infants irradiated *in utero*. Note that all infants receiving therapeutic doses of radiation were growth retarded if they were irradiated between 3 and 20 weeks of gestation. Note that in no instance is there an associated malformation in structures other than the nervous system without either microcephaly or mental retardation. (From Dekaban, 1968, with permission from the *Journal of Nuclear Medicine.*)

aplasia. Only about 3% of the animals will be so affected, and these animals are sterile (Brent, 1960a). The mechanism of the testicular aplasia is not known. Testicular hypoplasia or aplasia is not produced in the rat during the late organogenetic or early fetal period but appears again following irradiation on the 17th through the 22nd day of rat gestation (Erickson *et al.*, 1963; Murphree and Pace, 1960). As little as 50 R diminished testicular size from 3.5 to 2.5 g, whereas 150 R resulted in testicles that weighed only 0.5 g. Although maximum reduction in testicular weight occurred in animals irradiated from the 19th through 22nd day, minimal mating activity occurred in animals irradiated from the 18th through the 20th day. Male fetuses that received 150 R on the 18th, 19th, or 20th day were sterile as adults and made no attempt to mate. Brown (1964) exposed pregnant rats to various doses of daily radiation for 20 hr each day. The rats received 2, 5, 10, 20, or 50 R/day throughout their intrauterine life. All male animals receiving 10, 20, or 50 R/day were sterile. A second study performed by Brown limited the daily irradiation from the 15th day of pregnancy to the 23rd postpartum day to 1, 2, 5, 10, 20, and

40 R. The three highest dosage groups of male offspring were sterile from total doses of 300, 600, and 1200 R fractionated irradiation. Ershoff and Brat (1960) also noted significant testicular injury following 300 R of acute radiation exposure on the 18th day (17.92 R/min) and less damage from 300 R (0.03 R/min) administered from the 13th to the 20th day.

Radiation exposure of neonatal male rats also resulted in permanent sterility, provided that the exposure of acute radiation was 300 R or greater (Petrosyan and Pereslegin, 1962). If the young rat is irradiated on the 18th or 17th day postnatally with 50, 150, 300, or 500 R (100 R/min), testicular damage can be produced in only 3% of the seminiferous tubules of 18-day-old animals given 300 R (Leonard *et al.*, 1964). It appears, therefore, that, following parturition, the rat testes rapidly lose the radiosensitivity demonstrated during the last few days of pregnancy.

Erickson *et al.* (1963) exposed pregnant pigs at various stages of gestation to 400 R. Organization of the sex cords does not begin until the 50th day of gestation in the male pig fetus, yet 12% of male fetal pigs irradiated with 400 R prior to this time were sterile as adults. After the 50th gestational day, the incidence of sterility rose sharply (animals irradiated at 50 days had a 50% incidence of sterility). Sterility rose to 70% and 75% after irradiation of 55- and 75-day-old male pig fetuses, respectively. Pregnant goats were exposed on the 13th, 40th, and 50th days of gestation to 300 R. The male offspring were not tested for fertility, but there was a 32% reduction in sperm count, 26% less sperm activity, and an increase of 133% in abnormal spern (O'Brien *et al.*, 1966).

Radiation-induced sterility in the female mouse has been studied more than male sterility. Rugh and Wohlfromm (1964a,b) exposed 10.5–12.5-day-old mice to 100 R but did not observe sterility in the males. Although there was some reduction in the number of matings, the litters sired were normal in size. Petrosyan and Pereslegin (1962) reported that 300 R administered to neonatal male rats sterilized them, but Rugh and Wohlfromm (1964a,b) found that 400 R to the neonatal male mouse did not permanently sterilize. In fact, they administered 100 R on each day of gestation or 400 R to sexually maturing mice and concluded that these doses "are not sufficient to seriously affect the reproductivity of the male over their normal reproductive period." The specific stage of maximal sensitivity for sterilization occurs during the latter part of mouse gestation (16th day) (Rugh and Jackson, 1958). Laskey *et al.* (1973) observed that male rats exposed to up to 3 rads/day from tritiated water from the time of conception were fertile as adults, although they exhibited small testes.

b. Female. The effects of radiation on the developing ovary are more fascinating and even less predictable than those reported for the male embryo and neonate. Beaumont (1962) exposed pregnant rats at either 13.5, 15.5, 17.5, or 19.5 days postcoitum to 100 R and then determined the population of oocytes in the female offspring at 100 days of age. Animals irradiated at the 15.5-day stage with 100 R had the greatest depopulation of oocytes, namely, 241 rather than the 5025 in control rats.

These findings help explain the results of Ershoff and Brat (1960). They reported that 300 R protracted radiation exposure from the 13th to the 20th day was more effective than one acute dose of 300 R on the 18th day. This is understandable, since the 18th day of gestation is not the day when the ovary is the most sensitive to radiation. The course of protracted radiation exposure straddled the sensitive period for the ovary in the rat (15.5 days) as demonstrated by Beaumont (1962). Although sterility was not tested, there were histologic changes in the gonad.

In pregnant Wistar rats exposed to tritiated water and tritiated thymidine continuously through pregnancy, total postnatal oocyte counts were reduced to 50% by 1450 μCi of tritiated water (Haas *et al.*, 1973). Total aplasia of the ovaries was produced by 5800 mCi. Tritiated thymidine was 10 times more effective than tritiated water in its effect on the developing ovary. Brown (1964) compared the sterilizing effects of low-dose, protracted radiation exposure in the male and female rat. After 11 generations of irradiation, the rats exposed to 2 R/day (20 hrs/day) exhibited a decrease in litter size. The male rat, overall, was more sensitive than the female predominantly due to the highly radiosensitive period in the testes during the last few days of rat gestation (Langendorff and Neumann, 1972).

Irradiation of the ovaries has resulted in some interesting discoveries. If large doses of radiation (200–400 rad) are administered to one ovary 3–4 days before conception, the unirradiated ovary will ovulate an increase number of eggs to make up for the reduction in number of eggs ovulated from the irradiated ovary (Hahn and Feingold, 1973). If one ovary is irradiated with 4500 rad and the other ovary removed on the fourth day of pregnancy, the size of the implantation sites and the number of live embryos will be reduced. This effect can be reversed by administration of estradiol-17-beta and progesterone (Gibbons and Chang, 1973).

The mouse presents a different picture. First of all, the female mouse can be sterilized by a dose of radiation below 100 rads and is, therefore, much more readily sterilized than the females of most other species and the male mouse. Second, the mouse ovary has its greatest sensitivity in the neonatal period rather than *in utero*. Rugh and Wohlfromm (1964a,b), tested the fertility of mice exposed to 100 R from the time of conception to the age of two months postpartum. Exposure to 100 R on any day of mouse gestation from fertilization to the 18th day did not result in sterile females, although the litter size of adult females that were irradiated on the 13th day of gestation was smaller (Rugh and Wohlfromm, 1966). An increase in ovarian cyst formation was noted in adult mice irradiated on the 12th and 13th days of gestation. Only 44% of the mice irradiated as newborns were fertile. One-, two-, and three-week-old mice receiving 100 R were sterile, whereas those receiving the same exposure at 1 and 2 months had 16% and 14% fertility, respectively. Sterility was not produced by 30 R given to the newborn mouse or 1-, 2-, or 3-week-old mice. In fact, 30 R to the newborn mouse did not decrease fertility, but the fertility of 1-, 2-, or 3-week-old mice was decreased (Rugh and Wohlfromm, 1964b).

The explanation for these findings was demonstrated by Peters (1961), Peters and Levy (1964), Russell and Oakberg (1963), and others. Mice receiving 20 R on the day of birth had 50% of their small oocytes destroyed. On the other hand, the 1-, 2-, and 3-week-old mice that received 20 R had 95%, 98%, and 100% of their small oocytes destroyed, respectively. The exact mechanism of oocyte killing is not understood, but the cells die shortly after radiation exposure, resulting in partial destruction of the pool supplying future mature ova. Although 50% of the small oocytes are killed in the neonatal mouse, there are 85% of the normal number of oocytes in the ovary at maturity. On the other hand, a 3-week-old mouse that receives 20 R will have only 1% of the normal number of oocytes at maturity. The change in oocyte sensitivity is believed to be due to the synchrony of mouse oocyte development and the transformation of oocytes of the newborn in the early diplotene stage to the highly sensitive late diplotene and pachytene stages in the 1–3-week-old mice.

During this sensitive period, the oocyte also is very sensitive to changes in dose rate. Female mice, 10-days-old, exposed to 25 R at a dose rate of 0.009 R/min had 10 times the number of surviving oocytes than at 285 R/min. This indicates that there are repair processes in the oocyte (Rugh and Wohlfromm, 1965). Russell *et al.* (1959) exposed pregnant mice to 200 R protracted over the entire period of gestation and noted a slight reduction in adult female fertility (Table 3).

Pregnant goats exposed to 300 R on the 13th day of gestation had a normal number of oocytes as adults (O'Brien *et al.*, 1966). Fractionation of the 300 R exposure, with exposures at 30, 40, and 50 days, resulted in a decrease in the number of follicles in the ovaries of the surviving goats to 59% of the controls.

Apparently, the rhesus monkey is more resistant to the sterilizing effects of radiation, since the minimal dose to sterilize the monkey fetus was approximately 2000 R. Although the primary oocytes disappeared within two days following 140–300 R in the mouse and rat at the primordial stage, the monkey needed 700 R to produce the same results (Baker and Beaumont, 1967; Baker, 1970). The radiation resistance of the monkey oocytes may be due to the asynchronous development of the germ cells.

There are few data on the effect of radiation *in utero* on the fertility of the offspring in the human. The large number of survivors in Hiroshima and Nagasaki exposed *in utero* have not lived long enough to have complete fertility histories. In 2345 women who had been exposed to irradiation at Hiroshima and Nagasaki there was no observable reduction in fertility in the exposed population even when the dose was greater than 100 R (Blot and Sawada, 1972). This population included children and adult women, and none were *in utero* at the time of exposure.

Mondorf and Faber (1968) studied the fertility of 180 females who had been examined fluoroscopically for intussusception during their infancy. They found no difference between the number of children born and the age distribution of births in the controls and the exposed girls. The gonadal exposure was estimated to be 1–5 R. Tabuchi (1964) and Tabuchi *et al.* (1967) reviewed the data of the Atomic Bomb Casualty Commission in Japan and found that menarche was delayed in girls exposed to high doses of radiation while *in utero* (1–3 months).

It is evident that it is quite difficult to generalize about the effects of radiation *in utero* or during the neonatal period on subsequent sterility. There are marked sex and species differences in the response to radiation (Baker, 1970). However, from the data on both humans and animals we have accumulated to date, we can tentatively conclude that acute doses below 25 rads absorbed by the fetus during gestation will not result in sterility in the human male or female.

3. Mutagenic and Other Genetic Effects

The most controversial aspect of applied radiation biology deals with the effects of radiation on genetic material and the clinical consequences of genetic alteration in both a quantitative and qualitative sense (Brent, 1972a,b). One has to be careful to define the area of concern which might involve:

1. The effect of ovarian and testicular radiation some time before conception resulting in point mutations and/or resulting in cytogenetic changes (Reisman *et al.*, 1967; Sato, 1966).
2. The effect of embryonic irradiation on embryonic development secondary to the consequence of genetic alteration in rapidly proliferating embryonic tissue.
3. The effect of embryonic radiation on developing gonadal tissue. Are the embryonic gonads similarly affected by the mutagenic effects of radiation?

The last question is the easiest to answer because there are so few data available. There are few experiments which were designed to examine the alteration in the mutation rate of the F_1 and succeeding generations following *in utero* radiation (Frolen, 1970). Furthermore, the mammalian embryo cannot survive high doses of radiation *in utero;* therefore there is a limit to the amount of radiation that can be utilized in an *in utero* genetic experiment. It is possible to give the adult testes much larger doses of radiation as part of any genetic experiment. Finally, since the embryonic reproductive tissue is very sensitive to radiation in other respects, infertility can readily be produced by irradiating the mouse and rat during gestation and, therefore, these animals would be of no use in studying succeeding generations. There are several experiments that have been reported from which we can indirectly infer that the embryonic gonad does not have a drastically increased susceptibility to the mutagenic effects of radiation. Coppenger and Brown (1965, 1967) and Stadler and Gowen (1964) (Table 3) followed rat and mouse populations for 10 and 11 generations with continuous exposure at various levels of radiation.

There was nothing in their experimental results that would indicate that the embryonic gonad was more susceptible to the mutagenic effects of radiation. But it must be remembered that an answer to the above-mentioned specific question was not provided in these long-term experiments. Selby (1973a,b) attempted to answer this question by studying X-ray-induced specific mutation rates in young and newborn male mice. In the nine experimental groups dealing with the young newborn mouse, none had mutation rates significantly higher than the adults. Furthermore, the newborn mouse had mutation rates significantly lower than that of the adult. Selby (1973a,b) also refers to the work of Carter, who found that the fetal male gonad, when irradiated, manifested a lower mutation rate than when the adult testis was irradiated. In fact the radiation of the fetal testes and newborn testes resulted in comparable but lower mutation rates. Embryonic gonadal radiation using fast neutrons resulted in the following mutation rates: 42×10^{-8}/locus/rad for male and 58×10^{-8}/locus/rad for female embryonic germ cells (Searle and Phillips, 1971). It is concluded that the intrinsic radiosensitivity of male and female genetic material is similar in mice and that in the male the early embryonic stages are not much less sensitive than spermatogonia in the adult. It seems likely that the immature dictyate oocyte is the only germ-cell stage in the female from which mutations cannot occur. These investigators also studied the incidence of translocations in adult males irradiated as embryos. The overall translocation frequency was 1.2% which is lower than would be expected with irradiation of adult spermatogonia (Searle et al., 1971). The field of translocation induction in irradiated adult animals will not be covered in this manuscript, except to point out that hereditary translocations can be produced by adult gonadal irradiation, that the rate of induction is dose dependent, that few of these cells are involved in viable offspring, that fractionation and protraction markedly reduce the incidence of translocations, and that as little as 5 R can increase the incidence of translocations (Ford et al., 1969; Leonard and Deknudt, 1969; Lyon et al., 1970; Muramatsu et al., 1973).

If the testis is given large doses of radiation (1000 R), a predictable series of events occurs in the mouse (Rugh and Skaredoff, 1971). Nine days after the radiation the males will tend to become sterile, although there will be no effect on the incidence of mating attempts or positive vaginal plugs in receptive females. Within the first few days after radiation, there will be no alteration in the fertilizing ability of the irradiated, mature spermatozoa. There will be a linear drop in implantation averages as postirradiation days accumulate suggesting impending sterility and preimplantation mortality (Rugh and Skaredoff, 1971; Schroder and Hug, 1971). In the offspring, no bizarre anomalies appeared and the incidence of malformations was the same as in the control population, suggesting that radiation-damaged sperm resulted in early resorptions rather than the survival of anomalous offspring. Postimplantation losses consist of resorbed embryos and early and late deaths. The late deaths seem to be nongenetic while the early deaths, which were dose dependent, represent genetically induced postimplantation losses (Schroder and Hug, 1971). In experiments with ^{90}Sr administered before conception, there was a reduction in litter size with certain exposure levels, but this did not appear in subsequent generations from the offspring of surviving embryos (Frolen, 1970). An important aspect of these experiments was the fact that X-rayed fathers were also crossed with other strains of mice and the reduction was always less when heterozygous offspring were involved. Since heterozygosity is a protection against X-ray anomaly production, these data may indicate that offspring of the human species might have less response than those of genetically pure species (Rugh and Skaredoff, 1971).

Other studies indicate that spermatogonial depletion following X irradiation depends on the age of the adult animal (Erickson and Martin, 1973). If the stage of spermatogenesis is specifically analyzed for the induction of dominant lethals, decreasing mutagenic radiosensitivity can be ranked as follows: early spermatids > spermatocytes II > spermatocytes I > spermatozoa (Schroder and Hug, 1971). Not only may there be a variability in radioresistance among spermatogonia at various stages of development but also among spermatogonia of the same apparent stage. It has been suggested that selection of radioresistant cell lines of spermatogonia is one possible explanation for the reduced effect of repeated small doses of radiation on the frequency of translocations. Searle *et al.* tested this hypothesis (1971) and failed to find any increased radioresistance to translocation induction after a large initial dose of radiation.

The studies dealing with the effect of radiation on mutation rates have made important contributions to our understanding of genetics. They have also raised the anxiety level of the population since the quantitative aspects, the importance of recovery and repair, and the selection against damaged germ cells are frequently not emphasized.

As an example, Spalding and Brooks (1972) irradiated 46 generations of male mice when they were 26 days old. A comparison of the F_6 with the F_{46} data showed no degradation in reproductive fitness during 40 generations of X-ray exposure. In a large series of patients exposed to the atomic bomb as adults, no increased incidence of congenital malformations in the F_1 generation has been reported.

Since we do know that high, acute doses of radiation can cause mutations, cytogenetic aberrations, and carcinogenesis, what is the impact of fractionation and protraction on these effects? The protective standards established from recommendations by the NCRP are based on a linear extrapolation from high dose to low dose. In reality, there are many experiments that contradict this linear relationship, including references already mentioned in this section (Lyon *et al.*, 1970). Russell demonstrated that as one lowers the dose rate, the mutagenic effect falls off exponentially. For example, with a total dose of 600 R, the low-dose-rate group had one eighth the induced mutation rate of the high-dose-rate group. Other investigators have demonstrated the DNA has very sophisticated mechanisms for repairing damage. Okada was able to demonstrate that 1 rad caused approximately five chromosome breaks/cell rather than the usual reported incidence of mutations of 10^{-7}/locus/rad. These data indicate that practically every single-stranded chromosome break is repaired and that only a very small fraction of radiation-induced chromosome breaks are not repaired. These experiments do raise the issue as to whether all point mutations are caused by improperly repaired cytogenetic abnormalities.

The following examples will give some quantitative information on what impact radiation might have on the incidence of genetic disease, and they further emphasize that radiation delivered during pregnancy has a much greater impact on the developing embryo than on the F_1 generation.

Radiation of the testes can result in abnormal spermatozoa as evidenced by pycnotic nuclei and cytogenetic abnormalities. The cytogenetic abnormalities disappear with time, so that several months after radiation the incidence of abnormal sperm and spermatozoa may almost be back to the pre-radiation level. Of course, if a large enough dose of radiation is absorbed, permanent sterility may be produced. During the several weeks after radiation exposure, there is an increased probability of a male fathering offspring with translocations. As an example, if male animals absorb 20 rads/day and are mated at random, one might expect 17–20 translocations per 10,000 offspring, which is slightly higher than one might expect from unirradiated offspring. In effect 100–200 rads radiation to a population of males would produce imperceptible changes in the F_1 population, whereas 100–200 rads delivered during early organogenesis would be devastating, resulting in 50% lethality and 50–70% malformations.

In the female the story is similar in that preconception radiation is far less deleterious (with regard to the frequency of malformations) than is postconception radiation. There is one complication in that there is more species variation in oogenesis than in spermatogenesis. The most sensitive period of oogenesis to radiation in the mouse is during the first three weeks after birth. As little as 30 R can destroy 95–99% of the future ova. There appears to be no comparable period in the rhesus monkey or human, although each oocyte goes through this same sensitive phase but not all within a short period of time. This, of course, relates to oogonia depletion and not mutation incidence.

Finally, the little bit of data that we have about the genetic effects of irradiating the embryo on its future offspring indicate that, if anything, the embryo has a similar response to the mutagenic effect of embryonic gonadal irradiation as the adult and, in certain instances, the induced mutation rate is even lower than that of the adult. These data differ markedly from the information dealing with the induction of malignancy in the adult and embryo, since it has been suggested that the embryo is much more sensitive to the tumorigenic effects of radiation.

4. Induction of Malignancy

The embryo and fetus have been considered more "sensitive" to the effects of radiation than the adult. This conclusion resulted from observations made on the lethal effects of radiation. The increased sensitivity is due primarily to the destruction of cells or the disorganization of complicated developmental processes at critical periods of development. The embryo does have the ability, however, to repair cellular radiation damage to some degree and at least to replace cells that are destroyed. The question then arises as to whether the embryo or neonate that survives radiation and appears normal is susceptible to long-term effects of radiation, such as shortening of life-span or tumor induction. How does the sensitivity of the embryo to these long-term effects compare with the sensitivity of the adult organism (Tables 1 and 2)?

Stewart and her colleagues (Stewart, 1972, 1973; Stewart and Kneale, 1970; Stewart *et al.*, 1956, 1958) suggested that the human embryo was more sensitive to the leukemogenic effects of radiation, and more recently they have concluded that other tumors also occur more frequently in persons exposed *in utero* during diagnostic radiologic procedures (primarily pelvimetry). The present estimate is that 1–2 rads of *in utero* radiation exposure increases the chance of leukemia developing in the offspring by a factor of 1.5–2.0 over the natural incidence. This increase is considerably greater than the increase resulting from 2 rads delivered to adults. In fact, 2 rads delivered to an adult population probably would not make a perceptible change in the incidence of leukemia even for very large population groups (Lewis, 1957). Lilienfeld (1966) reviewed the epidemiologic considerations with respect to leukemogenesis. The results achieved by Lilienfeld (1966), McMahon and Hutchinson (1964), Graham *et al.* (1966), Polhemus and Koch (1969), Yamazaki *et al.* (1954), Ager *et al.* (1965), and Ford and Patterson (1959) support the thesis that diagnostic radiation absorbed *in utero* increases the risk of leukemia. Six of nine studies reported in Lilienfeld's paper indicate an increase in leukemia risk of 1.3–1.8 following diagnostic radiation exposure *in utero*. Lilienfeld states: "When one considers the variety of control groups used and the sampling variability, the results are remarkably consistent in showing an excess frequency of leukemia among children of radiation-exposed pregnant mothers." Diamond *et al.* (1973) have extended the studies of Lilienfeld and have corroborated their earlier finding of a higher incidence of leukemia (threefold) in children exposed to diagnostic radiation *in utero*. They also report that this effect did not occur in the black population.

There are a number of interesting associations in these data that should be pointed out. In Stewart's studies there was a higher incidence of abortion history in the mothers receiving pelvimetry, and also the children in the pelvimetry group had a higher incidence of upper respiratory infections. Others have reported that infants from families with a strong family history of allergy are also more susceptible to radiation-induced leukemia when exposed to diagnostic radiation *in utero* (Natarajan and Bross, 1973). The problem with these data is that in some instances patients with allergic history and no preconception radiation had a higher frequency of leukemia than some groups that had received irradiation *in utero*. Tabuchi *et al.* (1967) reported no increase in leukemia following diagnostic radiologic procedures. In some of the studies that did not report an increased risk of leukemia, the number of patients was small. Of the 86 humans exposed *in utero* at Nagasaki, none developed leukemia (Burrow *et al.*, 1965). These individuals received considerably higher doses of radiation than those patients in the previous studies. Kato (1971) studied 1300 individuals exposed to the A bomb while *in utero* and observed no increased incidence in malignancy in the first 24 hr of follow-up, although there was an increase mortality in the first year of life and after 10 years of age.

It is of interest that Graham *et al.* (1966) reported an increased risk of leukemia that was identical whether a mother received radiation from diagnostic procedures shortly before or after conception. Hoshino *et al.* (1965) reported no increase in leukemia in a study of 17,000 children of parents receiving radiation before conception. The question arises as to what extent the same biases that contribute to the increased risk of leukemia in the cases of radiation exposure before conception also affect the *in utero* radiation cases. Graham *et al.* (1966) pointed out that children of mothers with a history of abortion or stillbirth also had children with a higher risk of leukemia. Neutel and Buck (1971) found that childhood malignancy occurred more often in offspring from mothers who smoked. Fasal *et al.* (1971) reported that infants who were heavier at birth were more likely to get leukemia. It appears that whenever one looks for association of an event with the occurrence of leukemia, it is readily found.

It has been thought that the work of Bloom *et al.* (1968) might explain the increased leukemia risk of patients exposed to radiation *in utero*. They examined 38 survivors of radiation *in utero* who received more than 100 rads and found a much higher incidence of chromosome rearrangement in the irradiated group. Kucerova (1970) reported a higher incidence of cytogenetic abnormalities in the offspring of a woman who received nearly 4000 rads for carcinoma of the cervix during the last trimester. Since many instances of leukemia are associated with either diseases transmitted genetically or diseases associated with chromosome abnormalities, the work of Bloom *et al.* seemed important. Fraumeni and Miller (1967) reviewed the diseases associated with leukemia. These are Down's syndrome, Klinefelter's syndrome, D trisomy, Bloom's syndrome, Fanconi's syndrome, the congenital agammaglobulinemias, ataxia telangiectasia, and the Wiskott–Aldrich syndrome. The problem with this line of investigation is that several laboratories have now reported that human fetuses or young infants exposed to diagnostic radiologic procedures did not have an increased incidence of cells with chromosome abnormalities (MacIntyre *et al.*, 1965; Reisman *et al.*, 1967; Sato, 1966). Thus, it appears that we cannot support a cytogenetic hypothesis to explain the increased risk of leukemia from pelvimetry. On the other hand, we do not have enough data to reject this hypothesis.

At present, it is not clear whether radiation exposure in either pre-or postconception cases is a causative or associative factor in the increased incidence of leukemia. Miller (1970) and others (Brent and Gorson, 1972; Jablon, 1973; Wright, 1973) present a dissenting vote to the conclusions of Stewart and all the reports to support her hypothesis. Miller (1970) says

> Minimal doses of X-ray are equally oncogenic whether exposure occurred before conception or during pregnancy, whether the neoplasm studied was leukemia or

any other major cancer of childhood, and whether the study was based on interviews which may be biased, or on hospital records. Taken in aggregate, the similarity of results, in the absence of a dose response effect or of supporting data from
animal experimentation, raises a question about the biologic plausibility of a causal
relationship.

Furthermore, Miller points out that siblings of leukemic children have an
incidence of leukemia of 1 in 720 per 10 years which is greater than the 1 in
2000 risk of leukemia following pelvimetry exposure and the 1 in 3000 probability of leukemia in the general population of children followed for 10 years.
Stewart and Kneale's (1970) recent publication on this subject reinforces the
contention that radiation may not be the etiologic factor responsible for the
induction of malignancy, since the *unirradiated* siblings of the irradiated patient population with a higher incidence of leukemia also had a higher incidence of malignancy. The incidence was greater than in control siblings and
control patients. This observation certainly would indicate that genetic or
other environmental factors may be of more importance than radiation in
tumorigenesis.

Rugh *et al.* (1966) irradiated mice with 100 R on each day of gestation and
observed the incidence of tumors in the offspring. There was no statistical
increase in the incidence of tumors in adult animals from irradiation *in utero*
on any day. Brent and Bolden (1961, 1968) exposed pregnant mice to 30, 60,
and 90 R after ½, 7½, 8½, 12½, and 16½ days of gestation. They also did not
observe an increase in the incidence of tumors.

At present, a number of investigators believe that *in utero* exposure to
even small amounts of radiation increases the risk of leukemia and other
malignancies, whereas other investigators seriously question such conclusions.
Until the mechanism is understood, there always will be doubt concerning the
role of *in utero* radiation in leukemia induction. Certainly, the interesting field
of transplacental carcinogenesis reinforces the importance of the question of
in utero radiation leukemogenesis (Miller, 1971; Rice, 1973). The mechanisms
involved in these phenomena may be important in understanding both the
causes and prevention of malignancy.

It appears that infants do have an increased susceptibility to the
leukemogenic effects of radiation. Irradiated children under the age of 10 in
Hiroshima and Nagasaki had six times greater risk of leukemia than the
control population. Kitabatake (1966a,b) reported 1197 leukemic and 581
lymphoma patients and the association of these diseases with diagnostic and
therapeutic doses of radiation. Only about 20% of the patients in this group
were children, and there was no association between previous diagnostic irradiation and an increased risk of leukemia. There was a definite increase in
the risk of leukemia following radiation therapy. This also was true for lymphoma, but the risk was not as great. Simpson *et al.* (1955) reported an increased incidence of leukemia in children following radiation of the thymus in
infancy. From these data there is no way to determine whether the infant has
greater susceptibility to leukemia than the adult, since there is no disease or
radiation therapy protocol in the adult that can be compared with the treat-

ment of children with an enlarged thymus. There are not many experiments that can be used for comparison. Billings *et al.* (1966) reported the incidence of tumors in 4-day-old rats following 0, 5, 25, and 125 R. None of the groups had an increase in the incidence of tumors despite the fact that the 125-R group had a 10% reduction in life-span. On the other hand, Toth (1968) reviewed experiments dealing with chemical carcinogenesis in the newborn. He reported that in some instances the newborn had a greater susceptibility to chemical carcinogens compared with adults, although other carcinogens were more effective when administered to adult animals.

There is ample evidence to support the hypothesis that therapeutic doses of radiation of 250–1000 R can induce carcinoma of the thyroid (Albright and Allday, 1967; Burke *et al.*, 1967; Duffy and Fitzgerald, 1950; Hagler *et al.*, 1966; Nichols *et al.*, 1965; Polhemus and Koch, 1969; Simpson *et al.*, 1955; Toyooka *et al.*, 1963a,b). In most instances, the carcinoma followed radiation therapy to an enlarged thymus. Other forms of therapy were radiation to the tonsil and lymph nodes (Hagler *et al.*, 1966), radioactive iodine therapy (Burke *et al.*, 1967), and radiation therapy for the treatment of adolescent acne (Albright and Allday, 1967). It has been estimated that the threshold dose for inducing carcinoma of the thyroid in childhood is 180 R, and it has been reported that diagnostic radiation exposures do not increase the incidence of carcinoma of they thyroid (Silverman, 1966) presumably because of the much lower doses involved. High doses of radiation to 6 and 7-week-old rats resulted in an increase in thyroid adenomas and a slight increase in carcinoma of the thyroid. No laboratory has determined simultaneously the susceptibility of infant and adult rates to radiation-induced thyroid cancer.

In conclusion, it appears that high doses of radiation can produce malignancy in children. There may even be an increased susceptibility to radiation-induced tumorigenesis in children. The increased incidence of cancer in children exposed *in utero* to diagnostic radiation has to be clarified in view of the fact that higher doses of radiation to animal embryos and the children exposed *in utero* at Hiroshima and Nagasaki have not resulted in a significant increase in the incidence of malignancy (Kato, 1971; Wood *et al.*, 1967a–c). One cannot overstress either the importance of the multiplicity of factors involved or the difficulties in their identification and control. Even laboratory experiments dealing with tumor production are difficult to interpret. For example, Ross and Brass (1965) reported that the incidence of spontaneous tumors varied with the diet and weight of the animals. Heavier animals on high-protein diets had a higher incidence of tumors than did the lighter rats on low-protein diets. Hence, there are many unanswered questions concerning the relationship between leukemia and malignancy and *in utero* radiation exposure.

5. Neuropathologic, Neurologic, and Behavior Abnormalities

There are a number of publications that review, in part, this subject and should be noted as excellent source material (Cowen and Geller, 1960;

Furchtgott, 1963; Hicks, 1958; Hicks and D'Amato, 1966; Murphy, 1929; Rugh and Grupp, 1961).

Probably the most important information available to investigators interested in the effects of radiation on the brain and behavior can be derived from reports dealing with irradiated human embryos and fetuses. The reports suggest that children exposed *in utero* to doses of radiation above 100 R and during the 3rd–12th week have an increased probability of being microcephalic, mentally retarded, or both (Blot and Miller, 1973; Burrow *et al.*, 1964; Kucerova, 1970; Miller, 1956, 1968, 1970; Wood *et al.*, 1967a–c; Yamazaki *et al.*, 1954) (Fig. 5). Blot and Miller (1973) found that the head size of some children irradiated *in utero* in Hiroshima and Nagasaki diminished even if the dosage was in the 10–20-rad range, which included some neutron radiation, although the mentality of this low-dosage group was normal. It is possible that other environmental factors contributed to the increased incidence of small head size following the atomic bomb explosions. There are case reports of pregnant women being treated for carcinoma of the cervix who bore microcephalic infants that were mentally retarded. Dekaban (1968) reviewed 26 cases in the literature and contributed a case report of his own, namely, a woman in the third month of pregnancy who was treated for rheumatoid spondylitis. The fetus received 900 R. The child was severely microcephalic, having hypoplasia of the brain and dilated ventricles. The next most common pathologic finding was microphthalmia. It was present in the majority of infants irradiated from the 2nd to the 20th week with high doses of radiation. Mental retardation was the third most common finding in this group of patients. There are scores of case reports in the literature (Blot and Miller, 1973; Geets, 1967; Glass, 1944), but they cannot all be discussed individually. Several representative cases demonstrate the variability in response of the human fetus. Stettner (1921) reported that an infant born to a mother who received radiation therapy for a uterine myoma during the latter part of the first trimester had not only microcephaly and other malformations but also general disturbances of coordination. Scharer *et al.* (1968) reported an infant with microcephaly and severe retardation of motor development whose mother was treated with radiation therapy. Another example is of interest because the child is reported to be normal and the mother, who was treated for carcinoma of the cervix during the sixth and seventh month of pregnancy, was cured and is living 11 years after the birth of the child (Ronderos, 1961). She was a 40-year-old woman treated with X-rays and radium over a 2-week-period. The 6½-month fetus was exposed to 754 R from X-rays and 134 R from radium. The child is now 11 years of age, has grown normally, and appears mentally normal. It is likely that exposure at this level resulted in some neuronal depletion. Driscoll *et al.* (1963) described the neuropathologic finding in two fetuses that received 400 and 500 R, one at 16 weeks and the other at 22 weeks of gestation. Therapeutic abortion was performed following the radiation therapy, and there was neuronal depletion in the cerebral cortex and the cerebellum. The histology was not inconsistent with the results of irradiating other mammalian fetuses later in gestation. The survival without neurologic sequelae following exposure of a 6-month fetus to 988 R protracted radiation suggests that repair processes occur in the developing brain. The fact that this radiation was protracted may be very important in explaining the results in this particular case (Brizzee *et al.*, 1967; Ronderos, 1961).

The exposure of pregnant animals to high doses of radiation from the initiation of differentiation to parturition results in a high incidence of central nervous system malformations. The exposure used by most investigators has ranged between 50 and 400 R, with the lower doses being utilized during differentiation and the higher doses during the late fetal periods. Hicks (1958), Hicks and D'Amato (1966), and Cowen and Geller (1960) have worked out in detail the types of malformations that are produced in the offspring of pregnant rats at various stages of gestation and various doses of radiation. Their contribution was the concept that the central nervous system in the rat maintains its radiation sensitivity throughout gestation. Many investigators were aware that the mammalian fetal brain could be severely malformed by exposure to 50–200 R during the early organogenetic period (Roizin *et al.*, 1962; Rugh and Grupp, 1959; Russell, 1954; Wilson *et al.*, 1953b).

The timetable that has been developed by Hicks and D'Amato and Cowen and Geller can be summarized as follows: Animals irradiated on the 15th–16th day of gestation with 200 R acute exposure had severe forebrain defects. Those animals that survived the radiation and grew into adulthood exhibited microcephaly, heterotopia, corpus callosum agenesis, and interhemispheric third ventricles. Absence of the corpus callosum was not observed in animals irradiated after the 17th day. By the 21st day of gestation, radiation effects on the cortex were still demonstrable but were less severe than after radiation exposure on the 16th or 17th day of gestation. Cerebellar malformations appeared after irradiation later in gestation. Maximal depression of cerebellar growth occurred with exposure after day 18, and cerebellar hypoplasia could be observed following irradiation of rats in the immediate postpartum period. The histologic pictures associated with these malformations in adult animals are well described by Hicks and D'Amato (1966). It can be stated that besides the gross morphologic alterations there was evidence of neuron depletion and "scrambling" of neuron fibers. The area and extent of neuron depletion depended on the dose and stage of gestation at which radiation was administered. Kameyama *et al.* (1972) Kameyama and Hoshino (1972) have studied the postnatal development of mice irradiated *in utero* with 200 R. Radiation-induced microcephaly remained the same or progressed into hydrocephaly depending on the day of gestation the embryos were irradiated. The development of the hydrocephalus may be related to vascular disorders of the cortex. Although radiation-induced hydrocephaly has been rarely reported in the human following *in utero* irradiation, it is commonly produced by radiating the pregnant rat and mouse (Kameyama *et al.*, 1972; Kameyama and Hoshino, 1972; Senyszyn and Rugh, 1969). The threshold dose for producing recognizable neurologic, neuroanatomic, and neurophysiologic changes following *in utero* irradiation is somewhat controversial and has been advocated to be as high as 40–50 R or as low as 1 R (Yamazaki, 1966).

Dekaban (1969) has made some generalizations concerning the mouse that are similar to findings for the rat, based on studies involving the administration of 200 R to pregnant mice at various stages of gestation. He states:

> Diencephalon, olfactory system and a part of the limbic system had low sensitivity in the range of irradiation used; the hippocampus and basal ganglia were moderately sensitive structures and severe damage was caused by irradiation between the 10th and 13th gestational days, while milder abnormalities occurred after exposure between the 14th and 16th gestational days.

There appears to be little disagreement concerning the deleterious effect of high doses of radiation, but there is some difference of opinion concerning the ability of the irradiated embryo to repair some of the damage that is produced by high or low doses of radiation. Hicks and D'Amato (1963, 1966) reported subtle changes in the histologic development of the brain following exposures to 10, 20, 30, 40, or 50 R at various periods of rat gestation. The brains were studied from a few hours after exposure until some animals were 5 months old. One problem with the study is that some of their experimental and control groups consisted of one animal each, although usually two animals were examined in each stage. These authors reported that exposures of 10–40 R regularly initiated changes that led to abnormalities of individual neurons and their organization in the cerebral cortex. They considered the threshold exposure for these phenomena to be 10 R just after birth and below 10 R in late fetal life (Hicks and D'Amato, 1963):

> There was persistent retardation of differentiation of all layers of the cortex well into postnatal life after irradiation on the 16th or 18th day (rat gestation). In addition to cytologic and architectural abnormalities, a notable outcome of irradiation of these fetal rats was a paucity of neurons in all the cortical layers, with a compensatory increase of neuropil so that there was little diminution in cortical volume. The paucity of neurons was attributed to an injury to the source of neurons, the primitive proliferative–migratory cell system. The injury was sublethal and persistent and must have been transmitted through a number of cell generations because whole series of cell migrations to all layers of cortex, extending over a week in the 16-day irradiated animal, was altered. The nature of the injury which seemed to make the proliferative–migratory cells 'forget' how many neurons a cortex should have but not what volume the cortex should occupy is unknown. It is suggested that in producing persistent or delayed changes, radiation may have more specific effects than is often supposed.

While Hicks and D'Amato (1963) have been impressed with the sequelae of high and low doses of radiation to the developing central nervous system, Altman *et al.* (1968a,b) and Brizzee *et al.* (1961a,b; 1967) have been interested in the ability of the fetal and newborn rat to repair radiation damage to the central nervous system. Brizzee and coworkers found that fractionation or protraction of the total dose of radiation could ameliorate markedly the reduction in cortical thickness produced by radiating the fetal brain. If the dose rate is low enough or the intervals between exposure long enough, the repair processes of the cells of the developing nervous system can function quite well. Brizzee *et al.* (1961a,b) were impressed with these repair processes even in the groups exposed to high doses at high dose rates. They said: "Although the total number of cells in the cortex is drastically reduced by *in utero* irradiation at the dose and rate levels employed, the surviving cells tend strongly to follow normal growth patterns and establish their normal spatial integrity." Altman *et al.* (1968b) exposed 3-day-old rats to 200 R and found that marked depletion of the granular layer was evident by 4 hr. Yet, those animals that were allowed to reach adulthood had a cerebellum that was functionally and histologically normal.

The studies by Hicks, Altman, Brizzee and their coworkers were all performed with different doses of radiation and at different stages of gestation and the neonatal period. These experimental differences alone may explain why the residual changes in the adult organism were different. In any event, there are two important matters that need further clarification: (1) the lowest doses of radiation that cause irreparable damage to the central nervous system at the various periods of gestation, and (2) the rate and mechanisms of repair of the developing nervous system irradiated at various stages of gestation.

There are several other interesting reports concerning the nervous system in adult organisms that have been irradiated *in utero* or during the neonatal period. Pregnant rats exposed to 100 R on the 14th day of gestation exhibited a delay in neuromuscular development, as measured by response to electroshock (Vernadakis *et al.*, (1966). Exposure of the newborn rat to 300 R accelerated the maturation of the lactate dehydrogenase isoenzyme patterns of the developing brain (Bonavita *et al.*, 1965). In irradiated 2-day-old rats the total protein content of the brain and the activity of a number of enzymes was depressed (DeVellis *et al.*, 1967). Rugh (1963) exposed rats on the 9th day of pregnancy to 100 R. Those animals that survived the effects of radiation until 24 weeks of age had eletroencephalographic (EEG) patterns that were the same as those of the controls. Mice that had severe brain abnormalities had abnormal EEG patterns from birth and were destined to die prematurely. Berry and Eayrs (1970) studied evoked cortical potentials in mature rats irradiated on the 19th–21st day of gestation. Adult animals that had been irradiated on the 21st day of gestation exhibited increased conduction velocities, lowered thresholds, and raised amplititude, unlike those irradiated on the 19th day of gestation. Many of the altered responses were not produced by irradiating adult animals.

Furchtgott (1963) mentioned in his review that there had been a marked increase in the number of publications dealing with the alterations in behavior of adult animals that had been irradiated *in utero*. Most investigations have demonstrated that the effect of irradiating a pregnant mammal was due primarily to the direct effect of the radiation absorbed by the embryo (Brent, 1960a,b; Brent and Bolden, 1967a,b; Brent and McLaughlin, 1960; Neifakh, 1957). In the field of behavior, the question arose as to whether an indirect effect might be more plausible, and so a number of investigations have been carried out to determine whether irradiation of only the mother could contribute to behavioral changes in the offspring. Brizzee *et al.* (1961a,b) demonstrated that exposure of the mother to 100, 200, or 300 R did not affect the ability of the offspring to function in a water maze. Meier and Foshee (1962) used a Lashley III maze to test animals that had been irradiated *in utero* or that came from litters whose mothers had been irradiated. They were convinced that irradiation of the mother was more detrimental than irradiation of the fetus. The problem with these experiments was that the shielding and radiation techniques were very primitive. Levinson (1952, 1959, 1962) suggested that more investigations are needed to resolve the question.

Animals that have been exposed to enough radiation *in utero* manifest certain obvious clinical manifestations as adults. Sikov *et al.* (1960) exposed rats on the 10th day of pregnancy to 20 or 100 R and on the 15th day to 50 or 185 R. The offspring of the 185-R, 15-day group exhibited diminution of the grasp reflex at birth, progressive hindlimb ataxia, loss of ability to hop, nystagmus, and circling. Some females had myoclonic jerks in response to sensory stimuli. Some males had priapism at three months of age. Pobisch (1960) observed ataxia and trembling in the offspring of irradiated pregnant rabbits. Ataxia occurred in adult rats exposed during the newborn period to 300 R (Petrosyan and Pereslegin, 1962; Ershoff *et al.*, 1962).

The *in utero* irradiated offspring of pregnant rats exposed early and late in gestation to 100 and 600 R had difficulty in learning (Brizzee *et al.*, 1961a,b; Levinson, 1952). Tests for locomotor function demonstrated diminished function. Werboff *et al.* (1961) irradiated pregnant rats on the 5th, 10th, 15th, and 20th days of gestation, and the group demonstrating the greatest reduction in motor activity as adults was the group exposed to 100 R on the 15th day. It was surprising that the offspring of irradiated 5-day pregnant animals also that had depressed motor activity, since differentiation had not begun at the time of the radiation exposure. Walker and Furchtgott (1970) studied the effects of prenatal X irradiation on the acquistion, extinction, and discrimination of a classically conditioned response. They concluded that prenatal irradiation seemed to influence primarily the speed of initial learning. Lipton (1966) exposed pregnant animals on the 17th day of pregnancy or newborn rats 1 or 3 days old to 200 R. All three groups had locomotor deficits. Lipton also correlated these neurologic findings with neuropathologic studies on these animals. Although the animals irradiated on the 17th day of gestation had cytoarchitectural anomalies of the cerebrum, the cerebellar architecture was not as disarranged as in the animals irradiated at 1 and 3 days of age. When 16-day pregnant rats were exposed to 100 or 200 R, the offspring that had received 200 R had increased locomotor activity, which decreased as the animals grew older (Furchtgott *et al.*, 1968; Wechkin *et al.*, 1961). They interpreted these results to mean that the irradiated rats exhibited fearfulness when exposed to novel situations, but less so with increased experience.

It has been reported that the offspring of pregnant rats exposed to 150 R between the 17th and 22nd days had a decrease in mating activity (Hupp *et al.*, 1960). Administration of 200 R to rats on the 16th day of pregnancy caused depression of olfactory sensation in the adult progeny (Furchtgott and Kimbrell, 1967). Rat fetuses receiving 200 R had a diminished ability to adjust to food deprivation cycles (Tacker and Furchtgott, 1963). The critical flicker frequency was increased in adult rats that had been irradiated on the 16th day of gestation (Sharp, 1968). These animals were similar to controls when tested for the acquisition, retention, and extinction of a conditioned emotional response (Sharp, 1965).

The susceptibility to audiogenic seizures in irradiated and control offspring was tested by several groups. Adult rats that had been irradiated on the 5th and 10th days of gestation had a decreased susceptibility to audiogenic seizures whereas adult rats irradiated on the 15th and 20th days of gestation had an increased susceptibility to audiogenic seizures (Werboff *et al.,* 1961). Furthermore, the relationships were independent of dosage. Cooke *et al.* (1964) exposed pregnant rats to 2 R/day for the first 10 days of gestation for a total exposure of 20 R. They found that only the female offspring had an increased susceptibility to audiogenic seizures, and suggested that audiogenic seizures may serve as an indicator of low-level radiation damage. Geller (1970, 1973) irradiated pregnant rats during the first third of pregnancy with 250 R and was unable to demonstrate an increased susceptibility to audiogenic seizures in the adult offspring. Thus these three experiments give conflicting results regarding audiogenic seizures.

Brown (1964) reported that 11 generations of rats had received 2 R/day chronic ^{60}Co gamma exposure without manifesting behavioral effects as measured by response to gravity, balance, reaction time, and endurance.

Stahl (1963) has reviewed the work of scientists in the U.S.S.R. in this field and summarizes the many experiments utilizing 200 R at various fetal stages.

> When the Pavlovian terminology is converted into more conventional modern concepts, one sees that it compares fairly closely with parameters such as emotionality and anxiety level in test animals and it is these same parameters which have often shown deviations in U.S. work on antenatal effects. The published Soviet results are, therefore, very similar in their implications to the large amount of U.S. work on the same subject.

The only work from the Soviet Union that is difficult to accept is the work of Piontkovskii (1958, 1961). He exposed pregnant rats to 1 R/day for the 20 days of rat gestation. The irradiated offspring exhibited an increase in early motor activity and adapted to test chambers more slowly. The results of this experiment, like those of Cooke *et al.* (1964), are difficult to explain, considering the low dose rates that were used. Hicks and D'Amato (1966) reviewed their own work, which combines histopathologic knowledge of the irradiated offspring and behavioral testing. They reported that adult rats exposed to 200 R on the 13th, 15th, or 19th day of gestation were able to discriminate visual patterns as well as the controls. Rats with small cerebral hemispheres and scrambled cerebral cortices (200 R, 17 days) learned to press a bar in response to signals as well as controls rats, but did not have the motivation of the control rats when time was limited.

Ader and Deitchman (1972) reported that it was very important to consider the maternal organism in the postnatal development of animals that had been irradiated *in utero*. Behavioral changes and growth reduction were observed in nonirradiated offspring whose mothers had been irradiated. Furthermore, the development of animals that had been irradiated *in utero* differed, depending on whether they were raised by irradiated or nonirradiated mothers. As long as there is little attempt to correlate pathologic neuroanatomic findings with behavioral or functional changes, it is unlikely that this line of investigation will reveal any startling new concepts. Hicks refers to the combination of neuropathology and functional studies as functional teratology. It is the combination of these two disciplines that will aid in the understanding of the development and adult function of the brain.

6. Eye Defects

A very complete study of cataract production in the mouse following *in utero* was done by Rugh *et al.* (1964a). Although 10 R, at any stage of gestation, did not increase the incidence of cataracts, 100 R, at most stages, did. The peak incidence of 60% followed the administration of 100 R to ova before the first cleavage. The lack of stage specificity of this effect led Rugh *et al.* to infer that response was not a direct effect on the embryo. Human embryos and children exposed to much higher doses of irradiation had a lower incidence of cataracts than reported in the experiments with mice. No radiation cataracts occurred in the *in utero* survivors of atomic bomb radiation exposure in Hiroshima and Nagasaki (Miller, 1968). Among 26 cases human embryos that had been exposed to therapeutic doses of radiation, only two surviving infants had cataracts (Dekaban, (1968). Furthermore, one child irradiated during the first two weeks of gestation exhibited no malformations.

Cataracts have been produced in rats that were exposed in the newborn period to 300 R (Petrosyan and Pereslegin, 1962) and, in some instances, microphthalmia was produced. Eye anomalies can be produced in over 90% of rat embryos exposed to 100 R on the 10th day of gestation. These include anopthalmia, microphthalmia, and coloboma (Wilson *et al.*, 1953a). On the 11th day similar malformations are produced by 100 R but in a lower incidence (Wilson, 1954). No malformations of the eye were produced when rat embryos were irradiated early on the 9th day of gestation (Wilson *et al.*, 1953a). The lowest dose that can produce any eye malformations at the most sensitive stage (10th day) of gestation is approximately 25 R, although subtle changes may be produced by lower exposures. Strange and Murphree (1972) exposed rat embryos to 100 R on the 11th day of gestation and noted an interesting exposure–rate relationship. The incidence of malformations increased as the exposure increased from 1 to 10 R/min, but then the incidence fell at exposure rates of 25 and 47 R/min. Rugh and Skaredoff (1969) studied the retina of monkeys that had been irradiated *in utero* with 200–300 R. This dose of radiation left the retina deficient in cellular requisites. Even rosette formation persisted into adult life. Greim *et al.* (1967) reported no increase in the incidence of eye anomalies in human infants exposed to radiation *in utero* from diagnostic X-ray examinations. Jacobsen and Mellemgaard (1968) reported 11 cases of eye abnormalities in 184 patients either exposed *in utero* or born to parents who were exposed to radiation from diagnostic examinations prior to conception. There were seven different eye abnormalities but no cataracts. It seems unlikely that the radiation was responsible for these eye malformations, but additional studies are required to obtain more reliable data and to clarify the etiologic factors. As pointed out by Miller (1970) and reiterated many times by us, *one should be suspicious of the validity of conclusions regarding any agent that appears to produce the same spectrum and incidences of malformation or disease, regardless of whether the agent is administered before or after conception.*

7. Longevity

At present, there is no body of data dealing with *in utero* irradiation of human embryos that can be used to determine whether the human embryo and fetus are more sensitive to the nonspecific life-shortening effects of radiation than is the adolescent or adult human being. A preceding section of this monograph did present information which indicated that the human embryo was more sensitive to the carcinogenic effects of radiation. As an example, metabolic alterations could be very important in determining the long-term health of the adult organisms that were irradiated *in utero* (Nair, 1969; Rugh *et al.*, 1966; Rugh and Wohlfromm, 1965; Sikov *et al.*, 1972). There is no evidence from animal studies that the irradiated animal embryo is more sensitive to the tumorigenic effects of radiation. Tumor induction is only one contributing factor to shortening of life and, therefore, the controversy over whether the embryo has an increased susceptibility to radiation-induced tumors will answer only part of the question dealing with longevity. Besides tumor induction, somatic mutations, induced premature aging of connective tissue, and deleterious effects on the immunologic system or renal system (Freedman and Keehn, 1966; Scharer *et al.*, 1968) might shorten life-span. There has been minimal work dealing with the effect of *in utero* radiation on the immunologic and hematopoietic status of the adult organism (Svigris, 1958; Swasdikul and Block, 1972).

The planning and execution of animal experiments concerned with the nonspecific life-shortening effects of *in utero* irradiation should observe the following precautions:

1. The dose of radiation should not be teratogenic.
2. The litter size should not be altered by the *in utero* irradiation.
3. A minimal number of procedures should be performed on animals in a longevity study.

There are few longevity studies dealing with the irradiated embryo that have considered these pitfalls. Brent and Bolden (1961) reported that 100 R administered to 9-day-old rat embryos shortened their adult life considerably, but many of these adult rats had obvious congenital malformations. A study just completed in our laboratory indicates that irradiation of embryonic mice at five periods of gestation does not reduce life expectancy in the surviving embryos when the embryos were exposed to 30, 60, or 90 R (Brent, 1968, 1970; Brent and Bolden, 1961). It is obvious that more experiments are needed regarding nonspecific life-shortening effects.

D. Background and Fallout Radiation

The natural background radiation varies widely throughout the world from about 80 mrads/year to well over 1000 mrads/year in some monazite areas of India and Brazil. In the United States, all but perhaps a tiny percentage receive a background radiation dose between 100 and 400 mrads/year. The average for the United States population probably is around 125 mrads/year. Of this amount, about 40 mrads on the average are from cosmic rays, 60 mrads are from terrestrial sources (radionuclides in the air and ground), and about 25 mrads are from radionuclides (^{40}K, ^{14}C, radium isotopes, and decay products) deposited within the body. It is assumed that the fetus also is exposed to a background radiation rate of about the same level. Residual fallout radiation (at present) adds less than 4% of the background dose rate, and contributions from other man-made sources (excluding medical sources of radiation) such as nuclear power reactors and consumer products is much less than 1%.

It appears, therefore, that the fetus receives a dose of about 90 mrads throughout the gestation period (Table 7). Although the background radiation may contribute a small part (less than 1%) to the spontaneous mutation rate in the prenatal as well as the postnatal period in humans, both the magnitude and rate are so low that background radiation absorbed by the embryo during gestation is thought not to be a factor in the normal incidence of congenital malformations, intrauterine or extrauterine growth retardation, or embryonic death.

E. Radiation Effects of Radionuclides Administered During Pregnancy

The effect of various radioisotopes administered to the pregnant mammal (Table 8) has been less well studied than the effects of externally administered radiation. Furthermore, one cannot generalize the effects of administered radionuclides because, depending on the chemical form, type, and energy of the emitted radiation: (1) they may have specific target organs, (2) they may or may not cross the placenta, (3) the distribution of radiation may

Table 7. Absorbed Dose from Naturally Occurring Background Radiation During Pregnancy[a]

	Millirads to soft tissue per nine months	
	Mother	Embryo
^{40}K	14–18	10–14
^{14}C	0.5–1.3	0.5–1.3
External terrestrial	36	36
Cosmic rays	37	37
Total	90	<86

[a]The dose to the embryo is less because the embryo has a higher water content and a higher extracellular volume than the adult. Ossification does not occur in the early stages of pregnancy, therefore radium and strontium localization would occur only in the latter portion of gestation. Neither makes a significant contribution.

be quite nonrandom, (4) the metabolism of radioactive elements or compounds may vary greatly from individual to individual because of individual biologic variations or because of the disease state of the individual, and (5) the change in dose rate with time usually is a multiexponential function and often difficult to evaluate (Cloutier *et al.*, 1973). Radioisotopes administered to the mother may also effect the newborn if they are administered shortly before birth, since many are excreted in the breast milk (Berke *et al.*, 1973; Wyburn, 1972).

In any event, before one can estimate the potential hazard of administering a radioactive nuclide or compound to a pregnant woman, one must determine with some accuracy: (1) total dose to the fetus or a particular fetal tissue, (2) the dose rate and how it varies with time, and (3) the stages of gestation during which the radiation is delivered.

1. Radioactive Iodine

Until recently (Sikov *et al.*, 1972), the radioactive isotopes of iodine were the radionuclides used most commonly in nuclear medicine. The two most important ones are ^{131}I and ^{125}I. Technetium 99m, a much less hazardous radioisotope, is rapidly becoming the most extensively used radionuclide and is of greatest value in "imaging" studies. If the dose of radioactive iodine is high enough, it can cause ablation of the thyroid and even inhibition of growth of the underlying trachea (Sikov *et al.*, 1972). Women who have received very high doses of ^{131}I (up to 175 mCi) before conception have not had an increased incidence of malformations in their offspring, nor have their children had an increased incidence of cytogenetic breaks in their peripheral lymphocytes (Einhorn *et al.*, 1972).

Table 8. Average Absorbed Dose to Mother and Fetal Thyroid for Various Diagnostic Tests Utilizing Radioactive Isotopes[a]

| Radionuclide | Purpose of procedure | Whole body dose | | | Fetal thyroid dose |
| | | Mother | | Fetus | |
		millirads/μCi administered	millirads per typical procedure	millirem to fetus	millirads per typical procedure
[131I]NaI	Thyroid scan	0.45	27.0	15.0	5000
[131I]RISA	Plasma volume	1.70	17.0	10.0	5000
[131I]Oleic acid	Lipid absorption	0.65	32.5	17.0	5000
[131I]Rose bengal	Liver function, scan	0.36	36.0	19.0	5000
[131I]Hippuran	Kidney function, scan	0.04	0.6	0.3	100
[99mTc]Pertechnetate	Brain scan	0.01	100.0	—	—
[99mTc]Sulfur colloid	Liver scan	0.015	30.0	—	—
[131I]RISA	Placentography	—	15.0	7.0	4900
Technetium-99m	Placentography	—	9.0	2.0	10
51Cr (RBC)	Placentography	—	4–12	3–4	3–4

[a]Data from Tabuchi (1964).

The fetal thyroid readily takes up iodine by the 10th week and will continue to do so for the remainder of gestation. Furthermore, fetal thyroid avidity for iodide is greater than maternal avidity (Sternberg, 1970). Although inorganic iodide readily crosses the placenta, iodine attached to proteins, hormones, and even radioactive rose bengal is less likely to cross the placenta. However, a significant amount of iodide usually is released from the labeled compounds and becomes available to the fetus. *There probably is no form of radioactively labeled iodide compound that does not release some iodine to the circulation after administration.*

Pathologic effects, including thyroid destruction, have been reported in the fetus when therapeutic (ablative) doses of ^{131}I have been administered to pregnant women (Sternberg, 1970). Tracer doses of radioactive iodine have not been reported to produce a deleterious effect on the fetus. There remains, nevertheless, a concern over the possibility of inducing thyroid cancer in susceptible individuals by prenatal exposure to even small amounts of radioactive iodine. It is prudent, therefore, to avoid administering radioactive iodine during pregnancy, unless its use is essential for the medical care of the mother and there is no adequate substitute therapy; if necessary, radioactive iodine is best administered during the first 5–6 weeks of human pregnancy, when the fetal thyroid is not yet developed. Even in this circumstance, the total dose to the embryo should be estimated and considered (see Table 8). Note that the whole-body dose to the embryo from ^{131}I is relatively insignificant (less than 1% of that to the fetal thyroid.)

2. Other Isotopes

Inorganic radioactive postassium, sodium, phosphorus, cesium, and strontium cross the placenta readily. Experiments with radioactive phosphorus and strontium indicate that if the dose is large enough, embryonic pathology and death can be induced (Frolen, 1970; Sikov and Noonan, 1958). Tritium has a markedly different effect on the developing embryo, depending on the compound that is labeled (Haas *et al.*, 1973; Laskey *et al.*, 1973). Tritiated thymidine had 10 times the effect of tritiated water in producing intrauterine aplasia of the ovaries.

Technetium 99m is a radioactive isomer that has become, in recent years, the most important radionuclide for diagnostic imaging procedures. Its usefulness depends on its almost optimal gamma-ray energy (140 KeV), its short half-life (6 hrs), and the fact that it emits no beta rays. The absence of beta radiation and the short half-life account for the much lower dose of radiation to the patient as compared to the same amount of ^{131}I. ^{99m}Tc pertechnetate is trapped to some extent by the throid gland, but it is not incorporated into thyroid hormones. Furthermore, ^{99m}Tc is so rapidly excreted that within 24–48 hr after the administration of a diagnostic dose of ^{99m}Tc to a nursing mother, the levels of ^{99m}Tc would be safe for infant consumption while the comparable test utilizing ^{131}I might necessitate two weeks before safe levels were reached in the mother's breast milk (Berke *et al.*, 1973; Wyburn, 1972).

If a radioactive material is administered to a pregnant woman, the radiation dose to embryo or fetus or particular fetal organ can be calculated or estimated by a medical physicist trained in nuclear medicine if the relevant parameters are known (amount and chemical form of the isotope and its radiation characteristics, site of administration, diagnostic or therapeutic problem, stage of gestation, etc.) (Cloutier *et al.*, 1973). Then the potential hazard or risk can be judged before recommendations are made regarding whether the procedure should be done or whether a therapeutic abortion or other action should be considered if the radioactive material has already been administered.

Although the calculation of the embryonic or fetal dose should be individualized, published estimates of the approximate fetal and maternal exposure are available. Table 8 summarizes the approximate dose to the patient for various isotope procedures. In many instances, the embryo will receive much less than its mother because the chemical form of the radioactive substance prevents its transfer to the embryo from the mother. Even in procedures in which the fetal thyroid absorbs significant doses (> 5 rads), the fetal whole-body exposure is quite low. In fact, ^{99m}Tc placentography localizes the placenta with approximately 1% of the dose that would be required for a radiologic procedure.

Although pregnant women quite frequently are exposed to radiation during diagnostic X-ray procedures, exposure of such women to radioactive isotopes from nuclear medicine procedures is relatively rare. There are several reasons: (1) diagnostic X-ray procedures are performed more frequently, (2) physicians are more aware of the need to question patients about pregnancy before nuclear medicine procedures are done, (3) therapeutic doses of radionuclides are utilized primarily in patients with maligancies, that is, in an older population, in whom pregnancy is much less likely, and (4) vague symptoms associated with pregnancy may simulate intraabdominal disease and warrant diagnostic X-ray procedures. In any event, anxiety and medicolegal problems produced by the administration of radionuclides during pregnancy are insignificant compared to those arising from the exposure of pregnant women to diagnostic X-ray procedures (Brent, 1967).

When radioisotopes are to be utilized in a woman of childbearing age, the following procedure is recommended: (1) record the date of the last menstrual period, (2) if the menstrual period is past due, determine whether it is possible that the woman could be pregnant, (3) if pregnancy is a possibility, determine the stage of gestation and estimate the dose to the fetus or fetal target organ if the nuclear medicine procedure is to be done, (4) discuss this information with the referring physician and delay the procedure, if possible, until the possibility of pregnancy has been ruled out, and (5) if the patient is pregnant or if the procedure must be done before the possibility of pregnancy is ruled out, discuss the importance of the examination and the relative risks involved to both fetus and mother with the referring physician and with the patient and/or responsible member of the family. Record this information and the summary of the discussion on the patient's chart.

As a general principle, it is prudent to avoid carrying out *in vivo* nuclear medicine procedures on women beyond the second week of pregnancy. Fortunately, most nuclear medicine procedures can be postponed or other diagnostic procedures not involving radiation exposure to the fetus can be substituted. The question of possible pregnancy should be looked into very carefully, of course, in the case of proposed therapeutic application of radioactive material. In the case of diagnostic nuclear medicine studies, however, if the patient's menstrual period has not been delayed, the risk is very small, since it is highly unlikely that an embryo, if present, would have proceeded to the stage of early organogenesis before the examination is carried out.

If, after the radioactive material has been administered to the patient, it is discovered that the patient was pregnant at the time of the study, and if the dose of radiation is large enough to represent a significant hazard to the embryo or fetus (greater than 10 rad), the physician and patient (or family) must decide on the relative merits of terminating the pregnancy.

F. Ionizing Radiation Summary

Much information is now available regarding the effects of ionizing radiation on the embryo and fetus. Although most of the more reliable data are derived from animal experiments, particularly the mouse and rat, many of the conclusions are applicable to other mammals, including the human, at least in a qualitative sense. Some of the conclusions are summarized in this section.

The classic effects of radiation on the mammalian embryo are: (1) intrauterine and extrauterine growth retardation, (2) embryonic, fetal, or neonatal death, and (3) gross congenital malformations. The structure most readily and consistently affected by radiation is the central nervous system. If the acute, *in utero* dose is below 30 R, these classic effects of radiation are never observed *together* in experimental animals nor, in all likelihood, in the human.

The *absorbed* dose, the dose *rate,* and the *stage of gestation* are important in determining the effect of irradiating a mammalian embryo. Many embryopathologic effects are reduced significantly by decreasing the dose rate to allow recovery processes to function. Several experiments have demonstrated that some mammalian species can be exposed continuously to 2–5 rads/day throughout pregnancy without perceptibly affecting the offspring.

The preimplanted embryo, having an LD_{50} below 100 R at some stages, is sensitive to the lethal effects of radiation but is not readily malformed by radiation. On the contrary, embryos in the preimplantation stage of embryonic development that survive radiation exposure are developmentally normal and exhibit normal intrauterine and extrauterine growth. Possible experimental exceptions to this observation were discussed earlier (Fig. 1 and Table 1).

Teratogenesis. The peak incidence of gross malformations occurs when the fetus is irradiated during the early organogenetic period, although cellular, tissue, and organ hypoplasia can be produced by radiation throughout organogenesis, fetal, and neonatal periods, if the dose is high enough (Fig. 1 and Table 1).

Growth Retardation. Growth retardation at term can occur as a result of radiation exposure given at any stage of gestation after implantation. Thus, although the growth of the surviving embryo exposed before implantation is not retarded, the implanted embryo is more susceptible to growth retardation than it is to the malforming effects of radiation.

Lethality. The embryonic LD_{50} dose of radiation is *lowest* during the early preimplantation period, rises during implantation, and falls again during organogenesis. From this stage on, the LD_{50} dose gradually approaches the adult LD_{50} dose in the late fetal stages (Fig. 4 and Table 4).

The neonatal death rate following irradiation is highest in the surviving embryos irradiated during the early organogenetic period.

There is no stage of gestation during which an exposure of 50 R is not associated with a significant probability of an observable embryopathologic effect: increased incidence of death during the preimplanted period, malformations during the early organogenetic stage, and cell deletions and tissue hypoplasia during the fetal stages (Figs. 1 and 2).

Manifestation of in Utero Irradiation in the Adult. Growth retardation produced by irradiation *in utero* is not only seen at term but also persists into the postpartum period and into adulthood. The resistance to growth retardation exhibited by the irradiated preimplanted embryo at term persists into adulthood. Growth retardation is caused by radiation reaching the fetus and not by radiation to other parts of the pregnant mammal. Animal experiments indicate that all embryos exposed to 100 R or more after implantation will exhibit some degree of growth retardation as adults. Although the embryos irradiated in the early organogenetic period exhibit the most intrauterine growth retardation at term, the irradiated early fetus exhibits the largest degree of permanent growth retardation. This may be due to the fact that the younger embryo can repair more completely some of the radiation-induced deficits. The threshold dose for radiation-induced growth retardation is well below 100 R, but the level depends on the stage of gestation and the dose rate or dose fractionation.

Fertility. There are marked differences between the intrauterine and extrauterine sensitivity of the male and female to radiation-induced sterility. In the mouse, the male is more sensitive to the effects of irradiation *in utero* and the female is more sensitive in the neonatal period. There also are species differences in the minimal sterilizing dose and the optimal stage for sterilization of females. This is primarily due to the nature of the development of ova. In those species in which the ova develop synchronously and have the bulk of the developing oocytes at the same sensitive stage of development, low doses of radiation can eliminate a very high percentage of the future ova. Sterilization following irradiation depends on many factors—sex of species, the stage of development at the time of irradiation, dose, and dose rate. It appears that the fetal primate is more resistant to radiation sterilization than either the rat or mouse.

Mutation. Although there is very little experimental data, one must assume that embryonic germ cells are susceptible to the mutagenic effects of radiation throughout gestation. There is evidence that cytogenetic abnormalities persist following *in utero* irradiation. As yet, such abnormalities have not been associated with clinical disease, and there is some controversy as to whether low doses (< 10 R) or low dose rates increase the incidence of these cytogenetic defects. There is no data to support the contention that point mutations occur at a higher frequency in adult gonads that were irradiated *in utero* and, in fact, there are some indications that the fetal or embryonic gonads are less susceptible to the mutagenic effects of radiation.

Tumor Induction. There is no doubt that high doses of radiation administered to man or experimental animals can be carcinogenic. The question discussed in this monograph, however, is whether the embryo is more susceptible to the leukemogenic or tumorigenic effects of radiation than is the adult organism. Many studies have reported an apparent increase in leukemia and other neoplasia in human offspring who had received as little as 2 rads *in utero*. However, there are many epidemiologic and etiologic questions still unanswered, and the leukemogenic and carcinogenic effects of low radiation doses and dose rates remain largely in doubt at the time of this writing.

CNS and Behavior. Large doses of radiation used in radiation therapy can produce microcephaly and mental retardation in human offspring irradiated *in utero*. Reproducible central nervous system defects of various types depending on stage of gestation can be produced by radiation throughout the fetal period in experimental animals.

Finding and recognizing radiation-induced deleterious effects in offspring irradiated *in utero* becomes increasingly difficult with decreasing doses (less than 50 R) because of their low probability of occurrence and the high natural incidence of defects. From a clinical point of view, an absorbed dose of *10 rads* to the fetus at any time during gestation can be considered as a *practical* threshold for the induction of congenital defects; below this the probability of producing adverse effects is nil. It appears that the fetal nervous system has significant repair processes for radiation damages. There is some evidence that *in utero* irradiation can also alter animal behavior. However, there has been little attempt to correlate the neuropathologic with the behavioral changes.

Eye. Radiation-induced cataracts and other eye defects have been reported following *in utero* irradiation, but there are inconsistencies between the human and animal data. Irradiation of 100 rad would produce a very high incidence of eye malformations if the radiation exposure occurred during the early portion of the 4th week of human gestation.

Life-span. One of the best indices of the long-term effects of radiation is the reduction in life-span of the animals irradiated *in utero.* There are no such studies with irradiated humans, and the only adequately performed studies indicate that the surviving mammalian embryo is less susceptible to the life-shortening effect of radiation than is the irradiated adult organism.

Fallout Radiation. It appears highly improbable that fallout radiation (at the present level) or variations in natural background radiation significantly affect the incidence of congenital malformations, intrauterine or extrauterine growth retardation, or embryonic death.

Clinical Application. Problems arising from the irradiation of pregnant or fertile women during clinical procedures and recommendations for dealing with them are discussed at length in this monograph, including the specific definition of an elective radiographic procedure that can be used in clinical situations. The wisdom of limiting elective radiologic examinations of the lower abdomen and pelvis to the first part of the menstrual cycle, as recommended by the ICRP and the NCRP, is questioned, since available data are insufficient to guide us in making a choice between the hazards of irradiating the ovum just before conception and the hazards of irradiating the zygote just after fertilization.

Much of the unwarranted anxiety over the effects of small doses of radiation to the fetus could be allayed and many unnecessary lawsuits could be avoided if the radiologist, radiation therapist, or practitioner in nuclear medicine, as the case may be, would investigate the possibility that the patient is pregnant before instituting the *elective* radiologic procedure involving radiation to the pelvis or lower abdomen. If pregnancy is a possibility, the various available courses of action and their relative advantages and disadvantages should be discussed with the patient and/or family and the referring physician, and a written summary of such a discussion and resulting decisions should be entered into the patient's record. If on the other hand, the patient presents symptoms or signs related to the abdomen that need immediate attention that fall into the category of a nonelective radiologic procedure necessitating less than 10 rads, the procedure should be performed. In this instance, whether the woman is in the first or second half of the menstrual cycle should be ignored. Even if we develop the ability to diagnose pregnancy from the moment of conception, it will still be impossible to ignore undiagnosed abdominal or pelvic complaints or signs in a woman of childbearing age.

III. MICROWAVE RADIATION

A. General Consideration

The biologic effects of microwave radiation have been studied by many investigators, and there are several recent comprehensive review articles dealing with this topic (Sher, 1970; Milroy and Michaelson, 1971; Michaelson, 1969, 1971a,b). There is controversy as to whether microwave radiation produces its biologic effects purely by means of hyperthermia or whether there are nonthermal effects of microwave radiation. McRee (1971) and Carpenter (1962) studied the minimal exposure and temperature necessary for the induction of cataracts in rabbits at frequencies of 2.45, 8.236, and 10.05 GHz. Although their studies are carefully performed and carefully analyzed, they neither prove nor disprove a nonthermal biologic effect of microwave radiation; however, the results do indicate the importance of hyperthermia in cataract production. Michaelson discusses the biologic and physical aspects of microwave radiation in an excellent review article (Michaelson, 1969). He points out that "the energy value of one quantum of microwave radiation $(0.024-4.0 \times 10^{-6}$ eV) is much too low to produce the type of excitation necessary for ionization, no matter how many quanta are absorbed." Furthermore, chemical activations involving orbital transitions of electrons are rarely encountered with radiation of wavelengths longer than ultraviolet radiation. Therefore, he indicates that the major effect of microwave exposure is thermal.

B. Hyperthermia

A pertinent question is whether exposure to microwave radiation produces its biologic effect in the mammalian embryo primarily or totally by its thermal effect. In order to answer this question one could compare the effects of elevating the embryonic temperature by several methods. There is an extensive literature dealing with the effect of hyperthermia on embryonic development. In most instances hyperthermia was produced by raising the environmental temperature until the homeostatic mechanisms of the maternal organism were no longer able to maintain normal body temperature (see Chapter 12).

There have been a number of experiments dealing with the effect of heat stress on the preimplanted zygote and very young embryo (Shah, 1956; Chang, 1957; Alliston et al., 1965; Ulberg and Burfening, 1967; Elliot et al., 1968; Thwaites, 1970; Burfening et al., 1969; Howarth, 1969; Fernandez-Cano, 1958). Elevation of temperature in the range that permits survival of the adult mammal resulted in a decrease in implantation and viable offspring, regardless of whether the heat stress was administered in vivo or in vitro. In prolonged heat stress to pregnant ewes during the first several weeks of gestation, Thwaites (1970) attempted to reverse the embryonic wastage by treating the pregnant ewes with thyroxine, progesterone or cortisol. But none of these hormones reversed the heat-induced embryonic wastage, although adrenal insufficiency, when added to heat stress, increased embryonic wastage (Thwaites, 1970). Fernandez-Cano (1958) demonstrated that environmental temperature elevation in the pregnant rat was more deleterious during the preimplantation period than during the embryonic or fetal stages. Pregnant rats maintained at 104°F for 5 hr on specific days of pregnancy had the following resorption rates: 64% (1–2 days); 31% (3–4 days); 16% (6–7 days); 12% (10–12 days); 2.4% (controls). Chang (1957), using a bacterial pyrogen, markedly increased the resorption rate in pregnant rabbits by raising their temperature during the preimplantation period.

An elevated temperature can decrease the embryo's ability to withstand other environmental hazards. The deleterious effects of experimental interruption of the blood supply to the pregnant uterus is increased at higher temperatures (Brent and Franklin, 1960; George *et al.*, 1967). As an example, an animal whose uterine blood supply is interrupted for 2½ hr, but whose uterus is maintained at 4°C, will deliver normal offspring. On the other hand, 30 min of vascular interruption of a pregnant uterus maintained at 40°C will result in complete resorption (George *et al.*, 1967). Thus, the elevation of the embryonic temperature by raising the ambient temperature, by heating the uterus, or by injecting pyrogens into the mother, has resulted in embryonic wastage during the implantation and organogenetic stages.

These data, when compared to the effects of embryonic hyperthermia produced by microwave radiation, do indicate that the same degree of hyperthermia from microwave radiation is more embryotoxic. This would indicate that microwave radiation has a significant nonthermal effect, but a rebuttal to this thesis is presented further along in the discussion.

C. Microwave and Shortwave Radiation

Microwave radiation and shortwave radiation have been utilized in the past to increase the temperature of pelvic organs for therapeutic purposes, such as the treatment of gonorrhea, pelvic inflammatory disease, endometriosis, carcinoma of the uterus, and pelvic peritonitis (Gellhorn, 1928; Schumacher, 1936). Temperature elevations of 1–2°C are readily attained and can be maintained for 2–3 hr without harmful effects. The use of a vaginal electrode permitted Gellhorn (1928) to obtain pelvic temperatures of 115°F in order to "kill harmful bacteria." Although the author was concerned about the harmful effects of this level of radiation, he did not allude to specific complications—nor was there any follow-up of the patients. In rabbits, the ovaries have been exposed to microwave radiation before conception, and the offspring from these animals were not only normal at birth but grew normally (Chizzolini, 1957). The authors' primary consideration was whether the radiation exposure "increases the estrual manifestations and favors pregnancy." They concluded that microwave exposure did not alter the histologic appearance of the ovary and resulted in a "constant and striking luteinizing effect—after radartherapy" (Chizzolini, 1957). Superovulation has been reported following shortwave radiation in the mouse (Ludwig and Ries-Bern, 1942).

Shortwave radiation (10 mHz) was administered to rabbits from the 29th day of life through several matings and pregnancies (Boak *et al.*, 1932). The total exposure ranged between 30 and 75 hr, which raised the animals' temperature to between 41 and 42°C. There was no interference with mating, fertilization, or development of the young *in utero.* The exposed rabbits did have smaller litter sizes, but the difference was not significant.

Several generations of rats were exposed to a weak field of shortwave radiation (75–100 mHz). Although there were some differences between the controls and experimental animals, the reproductive capacity of the exposed animals was not markedly affected (Harmsen, 1954).

Pobzhitkov *et al.* (1961) used a "super-high frequency" low-intensity field (0.344 mW/cm^2) and exposed mice during and before pregnancy. The exposure lasted for 30 min. Exposure prior to pregnancy did not affect the embryos in a subsequent pregnancy, but exposure during pregnancy resulted in stillbirths, postpartum growth retardation, and increased postpartum deaths.

Dietzel and Kern (1970) and Hofmann and Dietzel (1966) reported that the exposure of pregnant rats to 27 mHz resulted in an increased resorption rate. They estimated the temperature of the irradiated abdominal contents by placing a thermistor probe in the rectum. Irradiation was directed into the pelvis through the closed abdomen of the pregnant rat, which was restrained. Hyperthermia up to 42.7°C, following 10-min exposures, resulted in increased resorptions to irradiated preimplanted and postimplanted embryos. Elevation of the uterine temperature to 42°C for 10 min on the first and second day of pregnancy resulted in 65% of the zygotes failing to come to term. A similar exposure on the 7–8th day resulted in 28% resorptions. These investigators concluded that "shortwave diathermy is absolutely contradicted during pregnancy" (Hofmann and Dietzel, 1966).

On the other hand Rubin and Erdman (1959) reported four case histories of women treated with microwave radiation (2450-mHz, 100-W machine) for chronic pelvic inflammatory disease who were pregnant or became pregnant during the course of the therapy. Three women delivered normal infants, and the fourth, who received eight treatments during the first 59 days of pregnancy, aborted on day 67. This woman delivered a normal baby following a subsequent pregnancy during which she received microwave therapy. Rubin concluded "microwaves did not interfere with ovulation and conception." The discrepancy in the conclusions drawn by Dietzel and Kern (1970) and Rubin and Erdman (1959) may be explained by the wavelength of the radiation that was used and the size of the organism. Rubin was working with humans and using 2450-mHz radiation. The 2450-mHz radiation is not able to penetrate human tissue and produce a significant thermal effect beyond 3–4 cm, while 27.5-mHz radiation can readily heat mammalian tissues 10–12 cm inside an organism. Therefore, the contradictory conclusions of Rubin and Dietzel may be explained purely on the basis of variation in dosimetric parameters.

In our own laboratory, 2450-mHz microwave radiation was utilized to expose the pregnant rat uterus. It can produce significant heating to a depth of several centimeters in tissue. Therefore, exposure of the rat uterus on the

9th day of gestation should result in uniform temperature elevations throughout the embryonic site. Of course, if an organ were 10 cm in diameter, one would not expect that the inner portion of the organ would be adequately exposed with the 2450-mHz frequency. On the other hand, shortwave radiation with 27.5 mHz can penetrate deep within the human pelvis and, therefore, Hofman and Dietzel's (1966) warning about 27.5-mHz radiation from his studies in the rat may be appropriate. They are not necessarily appropriate for the 2450-mHz microwave generator which is the frequency of most microwave ovens and most diathermy machines.

For short periods of exposure (1–3 min), the internal temperature of the embryonic site and the surface temperature will be the same. On the other hand, exposures of 10–30 min can result in marked discrepancies between the surface and deep temperatures of an embryonic site. This is reflected in the observation that surface temperatures of 40 and 41°C, when maintained for 30 min result in very high resorption rates. Raising a pregnant mammal's temperature to 40°C by altering the environmental temperature may increase the resorption rate, but it will not result in extremely high resorption or abortion rates. Thus, the high resorption rate in the rat produced by microwave radiation can either be due to (1) a nonthermal effect of microwave radiation that is related to the flux density of the exposure and not the elevated temperature, or (2) the fact that the embryo is less able to dissipate heat than the maternal organs. Long exposure may result in much greater temperature elevations within the embryo than on the surface of the embryonic site. Temperature measurements of various organs within the pregnant mammal support the concept that the embryo is at a physiological disadvantage when attempting to dissipate heat. Hart and Faber (1965) reported that the mean fetal temperature was 0.25°C higher than the adjacent maternal organs of control pregnant rabbits. The embryos should be able to dissipate heat by conduction and radiation as well as adjacent maternal organs. However, maternal organs can readily dissipate heat via the skin and respiratory tract, since the maternal surface can cool the blood which has circulated through a deep maternal organ (Mali, 1969). Convection of heat via the blood circulation is an important method of cooling deep organs. On the other hand, the embryo is isolated from maternal circulation and therefore, none of its blood ever passes through the maternal skin and lungs which are the heat-exchange organs. In the case of microwave radiation, the embryo will not be at any disadvantage if the exposure is of short duration, but as the exposure extends to 10–30 min, then the embryonic temperature should rise considerably above the surface temperature of the uterus, which has the benefit of the maternal circulation as a coolant. It has been impractical to place a thermocouple or thermistor probe within the embryo during radiation and obtain meaningful results, since the wire probe can affect the local temperature. Immediately after a 30-min exposure of 41°C, the embryos is 0.5–1°C warmer than the uterus, but this could be due to the fact that the uterine surface cools faster then the embryo or that the embryo was much warmer at the end of a 30-min exposure.

It is not surprising that the embryo is vulnerable to microwave radiation, since it has some properties which are similar to the lens which is widely accepted as the critical organ when discussing microwaves (Michaelson, 1969; McRee, 1971; Carpenter, 1962). Both the lens and the embryo are bathed in a fluid compartment which is not directly in contact with the maternal circulation. The adult organism has minimal awareness of changes in temperature in the embryo and the lens. These two factors would permit marked elevations in temperature following microwave exposure with minimal perception of the change in temperature by the adult organism. It appears that the embryo may be just as susceptible or more susceptible than the lens to the effects of microwave radiation. As the embryo develops, the amniotic fluid volume increases and the thermal isolation of the embryo increases. Thus, the circumstances which prevent the embryo from dissipating absorbed heat energy worsen as the embryo develops.

When the maternal temperature was elevated to 104°F, the embryonic mortality decreased as the exposure was varied from the preimplantation period to the later stages of gestation, (Fernandez-Cano, 1958). In fact, in the later stages of gestation, an elevated environmental temperature has been reported to produce growth retardation in the ewe as its most severe effect (Shelton and Huston, 1968). On the other hand, in the microwave experiments performed in our laboratory, the sensitivity of the 8-, 10-, and 12-day-old embryo to microwave exposure did not diminish. Since elevating the maternal temperature by raising the environmental temperature delivers heat to the embryo primarily by conduction and radiation, one could not expect the embryonic temperature to be equal to or slightly higher than the temperature of the maternal pelvic organs. On the other hand, microwave exposure delivers the heat directly to the embryo, and the larger the embryonic volume the more difficult it will be for the embryo to dissipate the heat. Thus, it is not surprising that the embryolethal effect of microwave radiation persists into stages of gestation when elevated environmental temperatures have a minimal effect.

Because of the susceptibility of the rat embryo and mammalian lens to the injurious effects of microwave, one might conclude that the embryo and lens are equally susceptible to all forms of microwave and shortwave radiation. As with X radiation, the exposure rate (density of radiation per unit time, mW/sec) and the wavelength are of utmost importance in determining the hazards or effectiveness of radiowave, shortwave, and microwave radiation. One cannot talk about the hazards of any type of radiation unless one defines the essential quantitative and qualitative characteristics of the exposure. The clinical applications of this approach are detailed in recent reviews dealing with X radiation (Brent and Gorson, 1972; Brent, 1972).

As an example, an exposure of 2450-mHz radiation to the surface of the body will always result in a higher exposure to the eye than the embryo because of the inability of 2450 mHz to penetrate tissue more than several centimeters. On the other hand, radiation of a lower frequency, such as in shortwave or radiowave radiation, could penetrate completely through a large mammalian organism. In essence, one practical application of these studies is that leakage from microwave ovens (2450-mHz variety) even above the maximum permissible level of 1 or 10 mW/cm² present no hazard to the human mammalian embryo because the surface exposure is attenuated by several orders of magnitude at the depth of the human uterus.

To those familiar with the precise method of determining the exposure or dose of high-energy radiation, such as X-rays or gamma rays, the measurement of the exposure to microwaves is much less exact. It is true that the exposure can be measured in mW/cm² just as X-ray exposure can be measured in rads. But one should note that the rad is a measure of absorption per unit volume, while microwave is measured in surface exposure. Furthermore, small changes in the microwave radiating conditions, such as position, side shielding, organism size, and distance from the wave guide, outlet, or antenna can markedly alter the exposure rate. That is why most biological experiments measure the temperature of the radiated tissue as the method of approximating the exposure. The variability of absorption of microwave and the reflection of microwaves back into the radiating source can modify the output of the microwave generator. These variables are not part of the problems of determining the exposure from X radiation.

Our original goal in studying the embryonic effects of microwave radiation was to determine whether it had any potential in experimental nonsurgical interruption of pregnancy. It appears that this potential exists and that longer wavelengths, with their ability to penetrate deeper into tissue, is the direction to be taken in order to improve on the ability to irradiate the embryo. As the wavelength is lengthened, the ability to utilize an antenna to direct the radiation becomes more difficult. On the other hand, the greater susceptibility of the embryo to the thermal elevating potential of these types of radiation diminishes the need for precisely directed radiation, since the durrounding maternal tissues will not reach the same temperature levels as the embryo and there are means available to lower the maternal temperature to further protect the pelvic maternal organs.

The ability to elevate the temperature of the maternal organs offers other experimental opportunities:

1. Physiological and biochemical studies can be performed on the heat-stressed embryo within the body of a non-heat-stressed mother (Brent and Wallace, 1972; Brent et al., 1971).
2. The effectiveness of therapeutic and embryotoxic drugs can be studied in embryos with elevated temperatures. One of our original hypotheses was that embryotoxic agents might be effective at lower doses when administered to hyperthermic embryos, thus offering the possibility that hyperthermic embryos could be affected by drugs or chemicals at dosages which would be completely innocuous to the maternal organism. Such studies would not only have application in

the field of experimental abortifacients, but also in the field of cancer chemotherapy, since microwave, shortwave, and radiowave radiation offer the potential for specific heating of various parts of the body.

3.   The marked sensitivity of the mammalian embryo to noxious agents, the short duration of the rodent pregnancy, and the obvious indicators of embryonic abnormality all indicate that the embryo may offer an excellent opportunity to determine whether a nonthermal effect of microwave radiation exists.

D. Microwave Radiation Summary

It appears that the embryo, like the mammalian eye, is uniquely sensitive to the deleterious effects of microwave radiation. The hypothesis to explain the sensitivity of these two structures is the relative isolation of the fluid compartments of these organs from the heat-dissipating abilities of the maternal circulation.

This hypothesis emphasizes the importance of the thermal effects of microwave radiation, although there is no conclusive evidence presented to indicate that nonthermal effects do not exist. Although microwave, shortwave, and radiowave radiation offer potential for experimental models of interrupting pregnancy, physiological and biochemical studies of hyperthermic embryos, evaluation of therapeutic and toxic agents in the isolated hyperthermic embryo, and the testing of combined hyperthermia–cancer chemotherapy treatment, there is no indication that 2450-mHz radiation at or significantly above the present maximum permissible exposure should be hazardous to the human embryo, although the hazard may increase with lower frequencies. Because of the difficulty in determining the level of exposure and the variability of deep-tissue exposures of various wavelengths of radiowave, shortwave, and microwave radiation, it will be necessary to study the radiation of various wavelengths, since it would be unsound to apply the conclusions of these experiments to the entire spectrum of low-energy electromagnetic waves. Finally, the embryo may offer a model that will enable one to determine the magnitude and significance of a nonthermal effect, if one exists.

IV. ULTRASONIC RADIATION

The use of ultrasound for diagnostic purposes has considerably increased in the past decade. In the field of obstetrics, ultrasonography has been utilized to determine the size and number of fetuses (Leopold and Asher, 1974; Thompson and Makowski, 1971; Doust, 1973), to determine fetal maturity (Leopold and Asher, 1974; Garrett and Robinson, 1971; Campbell, 1969; Sabbagha et al., 1974; Sabbagha and Turner, 1972), to localize the placenta (Leopold and Asher, 1974; King, 1973; Kobayashi *et al.*, 1970; Kohorn *et al.*, 1969; Winsberg, 1973a,b; Gottesfield *et al.*, 1966), to determine the presence or absence of embryonic malformations (Leopold and Asher, 1974; Winsberg, 1973a,b; Mitchell and Bradley-Watson, 1973; Hellman *et al.*, 1973; Campbell, 1969), to determine the presence of hydatiform mole (Gottesfield *et al.*, 1967; Leopold, 1971), to determine the presence of fetal death (Gottesfield, 1970), to determine the location of intrauterine contraceptive devices (Janssens *et al.*, 1973; Doust, 1973; Winters, 1966; Nemes and Kerenyi, 1971), and to determine the physiologic status of the fetus (Robinson, 1973).

Although the majority of investigators have alluded to the probability that diagnostic ultrasonography is not hazardous to the embryo (Smyth, 1966; Kirsten *et al.*, 1963; Warwick *et al.*, 1970, Leopold and Asher, 1974; Mannor *et al.*, 1972; Dewhurst *et al.*, 1972; Curzen, 1972), there are some reports that embryopathology has been produced with a wide range of intensities of ultrasonic radiation (Shoji *et al.*, 1975). Shoji and coworkers (1972) reported embryonic malformations in the mouse following exposure to protracted periods of ultrasonic radiation from a Doppler fetal detector. Experience in our own laboratory with diagnostic ultrasound equipment has been that embryonic development is not altered by this level of exposure (Brent, 1971).

As with all forms of potential physical hazards there are exposures, doses, or quantities of the hazard which produce pathological results. Furthermore, as the exposure is reduced, so is the hazard. As with most physical or chemical agents there exist exposure levels with questionable risk and prescribed benefits. Diagnostic radiation fits this prototype quite well. In our society we expect that the risks of any commercially used agent should be very small, but in most instances it is very difficult to prove no effect when utilizing very low exposures of a particular physical agent, drug, or chemical.

With specific regard to ultrasound, one might ask: Can ultrasonic radiation produce tissue or cellular damage at any level of exposure? Of course the answer is yes. Anyone who has used an ultrasonic probe for the emulsification of solutions or the disruption of cells knows that the ultrasonic generator of several hundred watts can disrupt cells and tissues. Cell disruption and cleaning ultrasonic generators are designed for continuous operation, while diagnostic ultrasonic equipment employs microsecond pulses of ultrasound with 1–2 sec pauses. It is difficult enough to extrapolate the effects of high-amplitude, continuous exposure to lower levels. It is even more difficult when one realizes that the high-amplitude equipment produces continuous ultrasound while the diagnostic equipment is pulsed.

Ultrasonic perturbation produces a number of physical effects in tissues, if the amplitude of the radiation is large enough. Thermal effects have been reported to produce tissue destruction (Harvey, 1930; Basauri and Lele, 1962; Lele, 1963). The extent of tissue destruction is related directly to the intensity or amplitude of the ultrasonic radiation. The effect of irradiation is somewhat dependent on the tissue temperature at the time of radiation initiation. As an example, the extent of a destructive lesion is reduced in size if the temperature of the tissue is lowered before the onset of radiation (Basauri and Lele, 1962); furthermore, these investigators believe that a substantial portion of ultrasonic tissue damage is the result of thermal damage. Cavitation and acoustic streaming are other phenomena that can affect tissue structure and viability (Coakley *et al.*, 1971; Neppiras, 1969; Eller and Flynn, 1969). The occurrence and size of cavitation in tissues or body fluids will depend on the intensity, frequency, and pulse duration of the ultrasonic radiation (Rooney, 1972). There appears to be some evidence that the mechanical and thermal effects of ultrasonic radiation act synergistically at least with ultrasonic energies of 0.5–100 W. The unanswered question concerns the magnitude and significance of these deleterious effects from ultrasonic irradiations in the low milliwatt range. In otherwords, what is the threshold for significant thermal and cavitation effects, and what is the relation to frequency, amplitude, and pulsing of ultrasonic radiation?

The biologic effects of ultrasonic radiation on the embryo can be evaluated by monitoring embryonic growth, embryonic death, and the incidence of congenital malformations in exposed embryos. Although there are few experiments dealing with the embryo, there are numerous studies dealing with the effects of ultrasonic radiation on mammalian chromosomes. The latter studies are closely related to embryonic development because cytogenetic damage could result in abnormalities of the gametes and, therefore, the developing embryo as well as interfere with cellular kinetics in the exposed developing embryo. Although there are several studies which report an increase in chromosome damage, the vast majority of cytogenetic studies have reported that diagnostic ultrasonic radiation does not increase the incidence of chromosome breakage (Lele and Pierce, 1972; Abdulla *et al.*, 1972; Buckton and Baker, 1972; Coakley, 1971; Boyd *et al.*, 1971, Lucas *et al.*, 1972; Bobrow *et al.*, 1971; Abdulla *et al.*, 1972; Ikeuchi *et al.*, 1973; Rott and Soldner, 1973; Watts *et al.*, 1972; Watts and Stewart, 1972). Another report indicated that amniotic cells obtained by amniocentesis following diagnostic ultrasonography did not grow as well as cells not previously exposed to ultrasound. This report has proven to be unsubstantiated (Robinson, 1973). Genetic studies have been negative. There are reports of no observable mutations in the offspring of mice exposed to microwave radiation. Macromolecular degradation has been reported with ultrasonic radiation of 20-kHz frequency and from a 500-W source. Thus, if the energy is great enough, it is possible to denature proteins and degrade small molecules (O'Shea and Bradbury, 1972). The quantitative aspects of microwave radiation are therefore extremely important in considering its effect on the developing embryo (Nyborg, 1968). Thus, although all the reports indicate that diagnostic ultrasound in the milliwatt range is not harmful to the mammalian embryo, Sikov (1973) has proposed high-dosage ultrasonic radiation as a potential tool for inducing abortion.

It is obvious that a great deal more information on the interaction of ultrasonic radiation and the mammalian embryo must be obtained before we can be certain about the quantitative and qualitative aspects of the effect of ultrasonic irradiation on embryonic development.

REFERENCES

Ionizing Radiation

Ader, R., and Deitchman, R., 1972, Prenatal materal X-irradiation: Maternal and offspring effects, *J. Comp. Physiol. Psychol.* **78:**202–207.

Ager, E., Schuman, L., Wallace, H., Rosenfield, A., and Gullen, W., 1965, An epidemiological study of childhood leukemia, *J. Chron. Dis* **18:**113–132.

Albright, E. C., and Allday, R. W., 1967, Thyroid carcinoma after radiation therapy for adolescent acne vulgaris, *J. Am. Med. Assoc.* **199:**280–281.

Altman, J., Anderson, W., and Wright, K., 1968a, Gross morphological consequences of irradiation of the cerebellum in infant rats with repeated doses of low-level X-ray, *Exp. Neurol.* **21:**69.

Altman, J., Anderson, W., and Wright, K., 1968b, Reconstitution of the external granular layer of the cerebellar cortex in infant rats after low-level X-irradiation, *Anat. Rec.* **163:**453–471.

Baker, T. G., and Beaumont, H. M., 1967, Radiosensitivity of oogonia and oocytes in the foetal and neonatal monkey, *Nature* **214:**981.

Baker, T. G., 1970, Comparative aspects of the effects of radiation during oogenesis, *Mutat. Res.* **11:**9–22.

Beaumont, H., 1962, Effect of irradiation during foetal life on the subsequent structure and secretory activity of the gonads. *J. Endocrinol.* **24:**325–339.

Berke, R. A., Hoops, E. C., Kereiakes, J. C., and Saenger, E. L., 1973, Radiation dose to breastfeeding child after mother has ⁹⁹ᵐ-MAA lung scan, *J. Nucl. Med.* **14:**51–52.

Berry, M., and Eayrs, J. T., 1970, Evoked cortical potentials in prenatally irradiated rats. *Exp. Neurol.* **28:**411–426.

Billings, M., Yamazaki, J., Bennett, L., and Lamson, B., 1966, Late effects of low dose whole body X-irradiation of four-day-old rats, *Pediatrics* **38:**1047–1056.

Bloom, A. D., Neriishi, S., and Archer, P. G., 1968, Cytogenetics of the *in utero* exposed of Hiroshima and Nagasaki. *Lancet* **2:**10.

Blot, W. J., and Miller, R. W., 1973, Mental retardation in utero exposure to the atomic bombs of Hiroshima and Nagasaki, *Radiology* **106:**617–619.

Blot, W. J., and Sawada, H., 1972, Fertility among female survivors of the atomic bombs of Hiroshima and Nagasaki. *Am. J. Hum. Genet.* **24:**613–622.

Bonavita, V., Amore, G., Avellone, S., and Guarneri, R., 1965, Lactate dehydrogenase isoenzymes in the nervous tissue. V. The effect of X-rays on the enzyme of the developing and adult rat brain, *J. Neurochem.* **12:**37–43.

Brent, R. L., 1960a, The effect of irradiation on the mammalian fetus, *Clin. Obstet. Gynecol.* **3:**928–950.

Brent, R. L., 1960b, The indirect effect of irradiation on embryonic development. II. Irradiation on the placenta, *Am. J. Dis. Child.* **100:**103–108.

Brent, R. L., 1963, The modification of the teratogenic and lethal effects of irradiation to the mammalian fetus, *in: Effects of Ionizing Radiation on the Reproductive System* (W. D. Carlson and F. X. Gassner, eds.), pp. 451–462, Pergamon Press, New York.

Brent, R. L., 1967, Medicolegal aspects of teratology, *J. Pediatr.* **71:** 288–298.

Brent, R. L., 1968, The long term effects of embryonic and fetal irradiation, *Pediatr. Res.* **2:**291–

Brent, R. L., 1970, Effects of radiation on the foetus, newborn and child, *in: Late Effects of Radiation* (R. M. Fry, D. Grahn, M. L. Griem, and J. H. Rust, eds.), pp. 23–60, Taylor & Francis, London.

Brent, R. L., 1971, The response of the 9½-day-old-rat embryo to variations in dose rate of 150 R X-irradiation, *Radiat. Res.* **45:**127–136.

Brent, R. L., 1972a, Irradiation in pregnancy, *in: Davis' Gynecology and Obstetrics,* Vol. 2 (J. J. Sciarra, ed.), pp. 1–32, Harper & Row, New York.

Brent, R. L., 1972b, Protecting the public from teratogenic and mutagenic hazards, *J. Clin. Pharmacol.* **12:**61–70.

Brent, R. L., 1973a, Radiation exposure during first trimester: When is abortion indicated?, Questions and Answers, *J. Am. Med. Assoc.* **224:**536.

Brent, R. L., 1973b, Radiation and teratogenesis in pregnancy, in: *Pathobiology of Development,* pp. 76–96, Wilkins & Wilkins, Baltimore.

Brent, R. L., and Bolden, B. T., 1961, The long term effects of low dosage embryonic irradiation, *Radiat. Res.* **14:**453–454.

Brent, R. L., and Bolden, B. T., 1967a, The indirect effect of irradiation on embryonic development. III. The contribution of ovarian irradiation, uterine irradiation, oviduct irradiation and zygote irradiation to fetal mortality and growth retardation in the rat, *Radiat. Res.* **30:**759–773.

Brent, R. L., and Bolden, B. T., 1967b, The indirect effect of irradiation on embryonic development. IV. The lethal effects of maternal irradiation on the first day of gestation in the rate, *Proc. Soc. Exp. Biol. Med.* **125:**709–712.

Brent, R. L., and Bolden, B. T., 1968, Indirect effect of X-irradiation on embryonic development. V. Utilizaton of high doses of maternal irradiation on the first day of gestation. *Radiat. Res.* **36:**563–570.

Brent, R. L., and Gorson, R. O., 1972, Radiation exposure in pregnancy, *in: Current Problems in Radiology* (R. D. Moseley, Jr., D. H. Baker, R. O. Gorson, A. Lalli, H. B. Latourette, and J. Quinn, III, eds.), Vol. 2, pp. 1–48, Medical Publishers, Chicago, Illinois.

Brent, R. L., and Harris, M., 1976, *The Prevention of Embryonic, Fetal, and Perinatal Disease,* U.S. Government Publication (1976).

Brent, R. L., and McLaughlin, M. M., 1960, The indirect effect of irradiation on embryonic development. I. Irradiation of the mother while shielding the embryonic site, *Am. J. Dis. Child.* **100:**94–102.

Brizzee, K. R., and Brannon, R. B., 1972, Cell recovery in foetal brain after ionizing radiation, *Int. J. Radiat. Biol.* **21:**375.

Brizzee, K. R., Jacobs, L., and Kharetchko, X., 1961a, Effects of total body X-irradiation *in utero* on early postnatal changes in neuron volumetric relationships and packing density in cerebral cortex, *Radiat. Res.* **14:**96–103.

Brizzee, K. R., Jacobs, L., Kharetchko, X., and Sharp, J., 1961b, Quantitative histologic and behavioral studies on effects of fetal X-irradiation in developing cerebral cortex of white rats, *in: Response of the Nervous System to Ionizing Radiation* (T. J. Haley and R. S. Snider, eds.), Academic Press, New York.

Brizzee, K. R., Jacobs, L. A., and Bench, C. J., 1967, Histologic effect of total body X-irradiation in various dose fractionation patterns on fetal cerebral hemisphere, *Radiat. Res.* **31:**415–429.

Brown, S., 1964, Effects of continuous low intensity radiation on successive generations of the albino rat, *Genetics* **50:**1101–1113.

Burke, G., Levinson, M. J., and Zitman, I. H., 1967, Thyroid carcinoma ten years after sodium iodine I^{131} treatment, *J. Am. Med. Assoc.* **199:**247–251.

Burrow, G., Hamilton, H., Hrubec, Z., Amamoto, K., Matsunaga, F., and Brill, A., 1964, Study of adolescents exposed in utero to the atomic bomb, Nagasaki, Japan. I. General aspects: Clinical laboratory data, *Yale J. Biol. Med.* **36:**430–444.

Burrow, G., Hamilton, H., and Hrubec, Z., 1965, Study of adolescents exposed *in utero* to the atomic bomb, Nagasaki, Japan, *J. Am. Med. Assoc.* **192:**357–364.

Cloutier, R. J., Smith, S. A., and Watson, E. E., 1973, Dose to the fetus from radionuclides in the bladder, *Health Phys.* **25**:147–161.

Cohen, Y., Tatcher, M., and Robinson, E., 1973, Radiotherapy in pregnancy. *Radiol. Clin. Biol.* **42**:34–39.

Conard, R., and Hicking, A., 1965, Medical findings in Marshallese people exposed to fallout radiation, *J. Am. Med. Assoc.* **192**:457–459.

Cooke, J., Brown, S., and Krise, G., 1964, Prenatal chronic gamma irradiation and audiogenic seizures in rats, *Exp. Neurol.* **9**:243–248.

Coppenger, C. J., and Brown, S. O., 1965, Postnatal manifestations in albino rats continuously irradiated during prenatal development, *Texas Rep. Biol. Med.* **23**:45–55.

Coppenger, C. J., and Brown, S. O., 1967, The gross manifestations of continuous gamma irradiation on the prenatal rat, *Radiat. Res.* **31**:230–242.

Cowen, D., and Geller, L. M., 1960, Long term pathological effects of prenatal X-irradiation on the central nervous system of the rat. *J. Neuropathol. Exp. Neurol.* **19**:488–527.

D'Amato, C. J., and Hicks, S. P., 1965, Effects of low levels of ionizing radiation on the developing cerebral cortex of the rat. *Neurology* **15**:1104–1116.

Dekaban, A., 1968, Abnormalities in children exposed to X-radiation injury to the human fetus, Part 1., *J. Nucl. Med.* **9**:471–477.

Dekaban, A., 1969, Effects of X-radiation on mouse fetus during gestation: Emphasis on distribution of cerebral lesions, Part 2, *J. Nucl. Med.* **10**:67–77.

DeVellis, J., Schjeide, A., and Clemente, C., 1967, Protein synthesis and enzymic patterns in the developing brain following X-irradiation of newborn rats. *J. Neurochem.* **14**:499–511.

Diamond, E. L., Schmerler, H., and Lilienfeld, A. M., 1973, The relationship of intra-uterine radiation to subsequent mortality and development of leukemia in children, *Am. J. Epidemiol.* **97**:283–313.

Driscoll, S., Hick, S., Copenhaver, E., and Easterday, C., 1963, Acute radiation injury in two human fetuses. *Arch. Pathol.* **76**:113–119.

Duffy, B. J., Jr., and Fitzgerald, P. J., 1950, Thyroid cancer in childhood and adolescence: Report 28 cases, *Cancer, N.Y.* **3**:1018.

Einhorn, J., Hulten, M., Lindsten, J., Wicklund, H., and Zetterqvist, P., 1972, Clinical and cytogenetic investigation in children of parents treated with radioiodine, *Acta Radiol.* **2**:193–208.

Erickson, B., Murphree, R., and Andrews, J., 1963, Effects of prenatal gamma irradiation on the germ cells of the male pig, *Radiat. Res.* **20**:640–648.

Erickson, B. H., and Martin, P. G., 1973, Influence of age on the response of rat stem spermatogonia to X-radiation, *Biol. Reprod.* **8**:607–612.

Ershoff, B., and Brat, V., 1960, Comparative effects of prenatal gamma radiation and X-irradiation on the reproductive system of the rat, *Am. J. Physiol.* **198**:1119–1122.

Ershoff, B., Steers, C., and Kruger, L., 1962, Effects of radioprotective agents on foot deformities and gait defects in the prenatally X-irradiated rat, *Proc. Soc. Exp. Biol. Med.* **111**:391–394, 1962.

Fasal, E., Jackson, E. W., and Klauber, M. R., 1971, Birth characteristics and leukemia in childhood, *J. Natl. Cancer Inst.* **47**:501.

Ford, C. E., Searle, A. G., Evans, E. P., and West, B. J., 1969, Differential transmission of translocations induced in spermatogonia of mice by irradiation, *Cytogenetics* **8**:447–470.

Ford, D., and Patterson, T., 1959, Fetal exposure to diagnostic X-rays and leukemia and other malignant diseases in childhood, *J. Natl. Cancer Inst.* **22**:1093–1104.

Fraumeni, J. F., and Miller, R. W., 1967, Epidemiology of human leukemia: Recent observations, *J. Natl. Cancer Inst.* **38**:593–605.

Freedman, L., and Keehn, R., 1966, Urinary findings of children who were in utero during the atomic bombings of Hiroshima and Nagasaki, *Yale J. Biol. Med.* **39**:196–206.

Frolen, H., 1970, Genetic effects of ^{90}Sr on various stages of spermatogenesis in mice, *Acta Radiol.* **9**:596–608.

Furchtgott, E., 1963, Behavioral effects of ionizing radiations, *Psychol. Bull.* **60**:157–200.

Furchtgott, E., and Kimbrell, G. McA, 1967, Olfactory discrimination in prenatally X-irradiated rats. *Radiat. Res.* **30:**217–220.

Furchtgott, E., Tacker, R., and Draper, D., 1968, Open-field behavior and heart rate in prenatally X-irradiated rats, *Teratology* **1:**201–206.

Geets, W., 1967, Influence des radiations ionisantes sur le development du systeme nerveux, *Acta Genet. Med. Gemell.* **16:**275–309.

Geller, L. M., 1970, Audiogenic seizure susceptibility of rats X/irradiated in utero during first one-third of pregnancy, *Exp. Neurol.* **29:**268–280.

Geller, L. M., 1973, Audiogenic seizure susceptibility of rats X-irradiated in utero late in pregnancy, *Exp. Neurol.* **38:**135–143.

Gibbons, A. F. E., and Chang, M. C., 1973, The effects of x-irradiation of the rat ovary on implantation and embryonic development, *Biol. Reprod.* **9:**343–349.

Gillanders, L. A., 1973, Radiography of potentially pregnant females, *Br. Med. J.* **1:**291–292.

Glass, S. J., 1944, Dwarfism associated with microcephalic idiocy and renal rickets, *J. Clin. Endocrinol. Metab.* **4:**47.

Graham, S., Levin, M. L., Lilienfield, A. M., Schuman, L. M., Gibson, R., Dowd, J. E., and Hempleman, L., 1966, Preconception, intrauterine and postnatal irradiation as related to leukemia, *Natl. Cancer Inst. Monogr.* **19:**347–371.

Greim, M. D., Meier, P., and Dobben, G. D., 1967, Analysis of the morbidity and mortality of children irradiated in fetal life, *Radiology* **88:**347–349.

Greulick, W. W., Crismon, C. S., Turner, M. A., Greulich, M. A., and Okumoto, Y., 1953, The physical growth and development of children who survived the atomic bombing of Hiroshima and Nagasaki, *J. Pediatr.* **43:**121.

Haas, R. J., Schreml, W., Fliedner, T. M., and Calvo, W., 1973, The effect of triatiated water on the development of the rat oocyte after maternal infusion during pregnancy, *Int. J. Radiat. Biol.* **23:**603–609.

Hagler, S., Rosenblum, P., and Rosenblum, A., 1966, Carcinoma of the thyroid in children and young adults. Iatrogenic relation to previous irradiation, *Pediatr.* **38:**77–81.

Hahn, E. W., and Feingold, S. M., 1973, Abscopal delay of embryonic development after prefertilization X-irradiation. *Radiat. Res.* **53:**267–272.

Harvey, E. B., and Chang, M. C., 1962, Effects of radiocobalt irradiation of pregnant hamsters on the development of embryos. *J. Cell. Comp. Physiol.* **59:**293–305.

Havlena, J. M., Werboff, J., and Sikov, M. R., 1964, X-irradiation of the 36-hour rat embryo: Neonatal mortality and postnatal activity, earning and seizure susceptibility, *Radiat. Res.* **22:**193.

Hicks, S. P., 1953, Effects of ionizing radiation on the adult and embryonic nervous system, *in: Metabolic and Toxic Diseases of the Nervous System,* Proceedings of the Association for Research in Nervous and Mental Disease, Vol. 32 (H. H. Merritt and C. C. Hare, eds.), pp. 439–462, Williams and Wilkins, Baltimore.

Hicks, S. P., 1958, Radiation as an experimental tool in mammalian developmental neurology, *Physiol. Rev.* **38:**337–356.

Hicks, S. P., and D'Amato, C. J., 1963, Low dose radiation of developing brain, *Science* **141:**903–905.

Hicks, S. P., and D'Amato, C. J., 1966, Effects of ionizing radiation on mammalian development, *in: Advances in Teratology* (D. H. M. Woollam, ed.), pp. 196–243, Logos Press, London.

Hicks, S. P., D'Amato, C. J., and Lowe, M. L., 1959, The development of the mammalian nervous system. I. Malformations of the brain, especially the cerebral cortex, induced in rats by radiation, *J. Comp. Neurol.* **113:**435–469.

Hoshino, T., Itoga, T., and Kato, H., 1965, Leukemia in the offspring of parents exposed to the atomic bomb at Hiroshima and Nagasaki, presented at Japanese Association of Hematology, March 28–30.

Hupp, E., Pace, H., Furchtgott, E., and Murphree, R., 1960, Effect of fetal irradiation on mating activity in male rate. *Psychol. Rep.* **7:**289–294.

Jablon, S., 1973, Comments, *Health Phys.* **24:**257–258.

Jacobsen, L., 1965, A retrospective study of the possible teratogenic effects of diagnostic pelvic X-irradiation. *Proc. XIth Int. Congr. Radiol.*, Excerpta Medica International Congress, Series No. 105, pp. 1372–1375.

Jacobsen, L., and Mellemgaard, L., 1968, Anomalies of the eyes in descendants of women irradiated with small X-ray doses during age of fertility, *Acta Ophthal.* **46:**352–354.

Kameyama, Y., and Hoshino, K., 1972, Postnatal manifestation of hydrocephalus in mice caused by prenatal X-irradiation, *Congenital Anomolies* **12:**1–9.

Kameyama, Y., Hayashi, Y., and Hoshino, K., 1972, Abnormal vascularity in the brain mantle with X-ray induced microcephaly in mice, *Congenital Anomolies* **12:**147–156.

Kato, H., 1971, Mortality in children exposed to the A-Bombs while in utero. *Am. J. Epidemiol.* **93:**435.

Kinlen, L. J., and Acheson, E. D., 1968, Diagnostic irradiation, congenital malformations and spontaneous abortion, *Br. J. Radiol.* **41:**648–654.

Kitabatake, T., 1966a, Retrospective survey on medical irradiation and leukemogenesis in Japan, *Tokoku J. Exp. Med.* **90:**25–34.

Kitabatake, T., 1966b, Relationship between medical irradiation and development of malignant lymphoma: A retrospective survey. *Nippon Acta Radiol.* **26:**891–893.

Kucerova, M., 1970, Long-term cytogenetic and clinical control of a child following intrauterine irradiation, *Acta Radiol.* **9:**353–361.

Langendorff, H. U. M., and Neumann, G. K., 1972, Die wirkung einer fraktionierten rontgenbestrahlung auf die fertilitat von in utero bestrahlten mausen. *Strahlentherapie* **144:**324–337.

Laskey, J. W., Parrish, J. L., and Cahill, D. F., 1973, Some effects of lifetime parental exposure to low levels of tritium on the F_2 generation, *Radiat. Res.* **56:**171–179.

Leonard, A., and Deknudt, Gh., 1969, Dose-response relationship for translocations induced by X-irradiation in spermatogonia of mice, *Radiat. Res.* **40:**276–284.

Leonard, A., Imbaud, F., and Maisin, J., 1964, Testicular injury in rats irradiated during infancy, *Br. J. Radiol.* **37:**764–768.

Levinson, B., 1952, Effects of neonatal irradiation on learning, *J. Comp. Physiol. Psychol.* **45:**140–145.

Levinson, B., 1959, Effects of neonatal irradiation on learning in rats, *J. Comp. Physiol. Psychol.* **52:**53–55.

Levinson, B., 1962, Comment on Meier's "Prenatal anoxia and irradiation: Maternal-fetal relations," *Psychol. Rep.* **10:**173–174.

Levy, B., Rugh, R., Lunin, L., Chilton, N., and Moss, M., 1953, The effect of a single subacute X-ray exposure to the fetus on skeletal growth: A quantitative study, *J. Morphol.* **93:**561–571.

Lewis, E. B., 1957, Leukemia and ionizing radiation, *Science* **125:**865–972.

Lilienfeld, A. M., 1966, Epidemiological studies of the leukemogenic effects of radiation, *Yale J. Biol. Med.* **39:**143–164.

Lipton, M., 1966, Locomotor behavior and neuromorphologic anomalies in prenatally and postnatally irradiated rats, *Radiat. Res.* **28:**822–829.

Lyon, M. F., Morris, T., Glenister, P., and O'Grady, E., 1970, Induction of translocations in mouse spermatogonia by X-ray doses divided into many small fractions, *Mutat. Res.* **9:**219–223.

MacIntyre, M., Stenchever, M., Wolf, B., and Hempel, J., 1965, Effect of maternal antepartum exposure to X-rays on leukocyte chromosomes of newborn infants, *Obstet. Gyneco.* **25:**650–656.

McMahon, B., and Hutchinson, G. B., 1964, Prenatal X-ray and childhood: A review, *Acta Unio Int. Contra Cancrum* **20:**1172–1174.

Meier, G. W., and Foshee, D. P., 1962, Indirect foetal irradiation effects in the development of behavior, *in: Effects of Ionizing Radiation on the Nervous System*, Proceedings of the Symposium of the International Atomic Energy Agency, pp. 245–259, Vienna.

Meyer, M., Diamond, E., and Merz, T., 1968, Sex ratio of children born to mothers who had been exposed to X-rays in utero, *Johns Hopkins Med. J.* **123:**123–127.

Miller, R. W., 1956, Delayed effects occurring within the first decade after exposure of young individuals to the Hiroshima atomic bomb, *Pediatrics* **18:**1–18.

Miller, R. W., 1968, Effects of ionizing radiation from the atomic bomb on Japanese children, *Pediatrics (Suppl.)* **41:**257–264.

Miller, R. W., 1970, Epidemiological conclusions from radiation toxicity studies, *in: Late Effects of Radiation* (R. J. M. Fry, D. Grahn, M. L. Griem, and J. H. Rust, eds.), Taylor & Francis, London.

Miller, R. W., 1971, Transplacental chemical carcinogenesis in man, *J. Natl. Cancer Inst.* **47:**1169.

Mondorf, L., and Faber, M., 1968, The influence of radiation on human fertility, *J. Reprod. Fertil.* **15:**165–169.

Muramatsu, S., Nakamura, W., and Eto, H., 1973, Relative biological effectiveness of x-rays and fast neutrons in inducing translocations in mouse spermatogonia, *Mutat. Res.* **19:**343–347.

Murphree, R., and Pace, H., 1960, The effects of prenatal radiation on postnatal development in rats. *Radiat. Res.* **12:**495–504.

Murphy, D. P., 1929, The outcome of 625 pregnancies in women subjected to pelvic radium or roentgen irradiation. *Am. J. Obstet. Gynecol.* **18:**179–187.

Murphy, W. T., and Berens, D. L., 1952, Late sequelae following cancericidal radiation in children, a report of 3 cases. *Radiology* **58:**35–42.

NCRP press conference, January 26, 1971, Washington, D. C. (news release).

Nair, V., 1969, An ontogenetic study of the effects of exposure to X-irradiation on the pharmacology of barbiturates, *Chicago Med. Sch. Q.* **28:**9–25.

Natarajan, N., and Bross, I. D. J., 1973, Preconception radiation and leukemia, *J. Med.* **4:**276–281.

Neifakh, A. A., 1957, Role of the maternal organism in the irradiation illness of fetal mice, *Dokl. Akad. Nauk. Biol. Sci. Sect.* **116:**821–824.

Neutel, C. I., and Buck, C., 1971, Effect of smoking during pregnancy on the risk of cancer in children, *J. Natl. Cancer Inst.* **47:**59.

Nichols, C., Lindsay, S., Sheline, G., and Chaikoff, I., 1965, Induction of neoplasms in rat thyroid glands by X-irradiation of a single lobe, *Arch. Pathol.* **80:**177–183.

O'Brien, C. A., Hupp, E. W., Sorensen, A. M., and Brown, S. O., 1966, Effects of prenatal gamma radiation on the reproductive physiology of the Spanish, *Am. J. Vet. Res.* **27:**711–721.

O'Brien, J. P., 1956, Vertebrate radiobiology: Embryology, *Annu. Rev. Nucl. Sci.* **6:**423–453.

Peters, H., 1961, Radiation sensitivity of oocytes at different stages of development in the immature mouse, *Radiat. Res.* **15:**582–593.

Peters, H., and Levy, E., 1964, Effect of irradiation in infancy on the mouse ovary. A quantitative study of oocyte sensitivity. *J. Reprod. Fertil.* **7:**37–45.

Petrosyan, S., and Pereslegin, I., 1962, Acute radiation sickness of newborn rats and its remote consequences, *Med. Radiol.* **5:**38–45 (in Russian).

Piontkovskii, I. A., 1958, Certain properties of the higher nervous activity in adult animals irradiated prenatally by ionizing radiations; the problem of the effect of ionizing irradiation on offspring. *Byull. Eksp. Biol. Med.* **46:**77–80. (in Russian).

Piontkovskii, I. A., 1961, Some peculiarities of higher nervous activity in adult animals subjected to radiation in utero. Part III, *Byull. Eksp. Biol. Med.* **51:**27–31 (in Russian).

Plummer, G., 1952, Anomalies occurring in children exposed in utero to the atomic bomb in Hiroshima, *Pediatrics* **10:**687–692.

Pobisch, R., 1960, Die Einwirkung der Röntgenstrahlen auf den Kaninchenembryo mit besonderer Berücksichtigung der postnatalen Entwicklung, *Radiol. Austriaca* **11:**19–82.

Polhemus, D., and Koch, R., 1969, Leukemia and medical irradiation, *Pediatrics* **23:**453–461.

Reisman, L. E., Jacobson, A., Davis, L., Kasahara, S., and Kelly, S., 1967, Effects of diagnostic X-rays on chromosomes in infants. A preliminary report, *Radiology* **89:**75–80.

Reynold, E. L., 1954, Growth and development of Hiroshima children exposed to the atomic bomb. Three-year study (1951–1953). Atomic Bomb Casualty Commission, *Technical Report* 20–59, Hiroshima.

Rice, J. M., 1973, An overview of transplacental chemical carcinogenesis, *Teratology* **8:**113–196.

Roizin, L., Rugh, R., and Kaufman, M. A., 1962, Neuropathologic investigations of the X-irradiated embryo rat brain, *J. Neuropathol. Exp. Neurol.* **21:**219–243.

Ronderos, A., 1961, Fetal tolerance to radiation. *Radiology* **76:**454–456.

Ross, M. H., and Bras, G., 1965, Tumor incidence patterns and nutrition in the rat, *J. Nutr.* **87:**245–260.

Rugh, R., 1953, Vertebrate radiobiology: Embryology, *Annu. Rev. Nucl. Sci.* **3:**271–302.

Rugh, R., 1959, Vertebrate radiobiology: Embryology, *Annu. Rev. Nucl. Sci.* **9:**493–522.

Rugh, R., 1962, Neurological sequelae to low level X-irradiation of the developing embryo, *in: Effects of Ionizing Radiation on the Nervous System* (T. J. Haley and R. S. Snider, eds.), pp. 202–224, International Atomic Energy Agency, Vienna.

Rugh, R., 1963, Ionizing radiations and the mammalian embryo, *Acta Radiol.* **1:**101–113.

Rugh, R., 1965, Effect of ionizing radiations, including radioisotopes, on the placenta and embryo, *Birth Defects, Orig. Art. Ser.* **1:**64–73.

Rugh, R., 1969, Normal incidence of brain hernia in the mouse, *Science* **163:**407.

Rugh, R., and Grupp, E., 1959, The radiosensitivity of the embryonic nervous system, *Bull. Sloane Hosp. Women* **5:**49–52.

Rugh, R., and Grupp, E., 1960, Fractionated X-irradiation of the mammalian embryo and congenital anomalies, *Am. J. Roentgenol.* **84:**125–144.

Rugh, R., and Grupp, E., 1961, Neuropathological effects of low level X-irradiation of the mammalian embryo, *Mil. Med.* **126:**647–664.

Rugh, R., and Jackson, S., 1958, Effect of fetal X-irradiation upon the subsequent fertility of the offspring, *J. Exp. Zool.* **138:**209–221.

Rugh, R., and Skaredoff, L., 1969, X-rays and the monkey fetal retina, *Invest. Ophthalmol.* **8:**31–40.

Rugh, R., and Skaredoff, L., 1971, The immediate and delayed effects of 1000 R X-rays on the rodent testes, *Fertil. Steril.* **22:**73.

Rugh, R., and Wohlfromm, M., 1964a, X-irradiation sterilization of the premature female mouse, *Atompraxis* **10:**511–518.

Rugh, R., and Wohlfromm, M., 1964b, Can X-irradiation prior to sexual maturity affect the fetility of the male mammal (mouse)? *Atompraxis* **10:**33–42.

Rugh, R., and Wohlfromm, M., 1965, Prenatal X-irradiation and postnatal mortality, *Radiat. Res.* **26:**493–506.

Rugh, R., and Wohlfromm, M., 1966, Resistance of the prenatal female mouse to X-ray sterilization, *Fertil. Steril.* **17:**396–410.

Rugh, R., Duhamel, L., Chandler, A., and Varma, A., 1964a, Cataract development after embryonic and fetal X-irradiation, *Radiat. Res.* **22:**519–534.

Rugh, R., Duhamel, L., Osborne, A. W., and Varma, A., 1964b, Persistent stunting following X-irradiation of the fetus, *Am. J. Anat.* **115:**185–197.

Rugh, R., Duhamel, L., and Skaredoff, L., 1966, Relation of embryonic and fetal X-irradiation to life time average weights and tumor incidence in mice, *Proc. Soc. Biol. Med.* **121:**714–718.

Rugh, R., Wohlfromm, M., and Varma, A., 1969, Low dose X-ray effects on the preclevage mammalian zygote. *Radiat. Res.* **37:**401–414.

Russell, L. B., 1954, The effects of radiation on mammalian prenatal development, *in: Radiation Biology, Part II* (A. Hollaender, ed.), McGraw-Hill, New York.

Russell, L. B., 1957, Effects of low doses of X-rays on embryonic development in the mouse. *Proc. Soc. Exp. Biol. Med.* **95:**174–178.

Russell, L. B., 1965, *in: Ciba Foundation Symposium on Preimplanted Stages of Pregnancy* (G. E. W. Wolstenholme, ed.), p. 217, J. & A. Churchill, London.

Russell, L. B. and Montgomery C. S., 1966, Radiation-sensitivity differences within cell division cycles during mouse cleavage, *Int. J. Radiat. Biol.* **10:**151–164.

Russell, W. L., and Oakberg, E. F., 1963, The cellular basis and aetiology of the late effects of irradiation on fertility in female mice, *in: Cellular Basis and Aetiology of Late Somatic, Effects of Ionizing Radiation* (R. J. C. Harris, ed.), pp. 224–232, Academic Press, New York.

Russell, L. B., and Russell, W. L., 1954, An analysis of the changing radiation response of the developing mouse embryo, *J. Cell. Comp. Physiol.* **43:**103–149.

Russell, L. B., and Saylors, C. L., 1963, The relative sensitivity of various germ-cell stages of the mouse to radiation-induced non-dysjunction, chromosome losses and deficiencies, *in: Sobels: Repair from Genetic Radiation,* pp 313–342, Pergamon Press, New York.

Russell, L. B., Badgett, S. K., and Saylors, C. L., 1959, Comparison of the effects of acute, continuous and fractionated irradiation during embryonic development, *in: A Special Supplement to International Journal of Radiation Biology* (A. A. Buzzati-Traverso, ed.), pp. 343–359, Immediate and Low Level Effects of Ionizing Radiation Conference held in Venice, Taylor & Francis, London.

Sato, H., 1966, Chromosomes of irradiated embryos, *Lancet* **1:**551.

Scharer, K., Muhlethaler, J., Stettler, M., and Bosch, H., 1968, Chronic radiation nephritis after exposure in utero, *Helv. Paediatr. Acta* **23:**489–508.

Schroder, J. H., and Hug, O., 1971, Dominante lethal Mutationen in der Nachkommenschaft bestrahlter mannlicher Mause. I. Untersuchung der Dosiswirkungsbeziehung und des Unterschiedes zwischen ganz und Teilkorperbestrahlung bei meiotischen und postmeiotischen Keimzellenstadien, *Mutat. Res.* **11:**215–245.

Searle, A. G., and Phillips, R. J. S., 1971, The mutagenic effectiveness of fast neutrons in male and female mice, *Mutat. Res.* **11:**97–105.

Searle, A. G., Evans, E. P., and Beechey, C. V., 1971, Evidence against a cytogenetically radioresistant spermatogonial population in male mice, *Mutat. Res.* **12:**219–220.

Segall, A., MacMahon, B., and Hannigan, M., 1964, Congenital malformations and background radiation in northern New England, *J. Chron. Dis.* **17:**915–932.

Selby, P. B., 1973a, X-ray-induced specific-locus mutation rate in newborn male mice, *Mutat. Res.* **18:**63–75.

Selby, P. B., 1973b, X-ray-induced specific-locus mutation rates in young male mice, *Mutat. Res.* **18:**77–88.

Senyszyn, J. J., and Rugh, R., 1969, Hydrocephaly following fetal X-irradiation, *Radiology* **93:**626–634.

Sharp, J., 1965, The effects of prenatal X-irradiation on acquisition, retention, and extinction of a conditioned emotional response, *Radiat. Res.* **24:**154–157.

Sharp, J., 1968, Critical flicker frequency in albino rats following prenatal X-irradiation, *Radiat. Res.* **33:**22–29.

Sikov, M. R., and Noonan, T. R., 1958, Anomalous development induced in the embryonic rat by the maternal administration of radiophosphorus, *Am. J. Anat.* **103:**137–162.

Sikov, M., Meyer, J., Resta, C., and Lofstrom, J., 1960, Neurological disorders produced by prenatal X-irradiation of the rat. *Radiat. Res.* **12:**472(abstract 151).

Sikov, M. R., Mahlum, D. D., and Howard, E. B., 1972, Effect of age on the morphologic response of the rat thyroid to irradiation by iodine ^{131}I, *Radiat. Res.* **49:**233–244.

Silverman, F. N., 1966, Thyroid carcinoma and X-irradiation, Thereapeutic radiation versus diagnostic roentgenology, *Pediatrics* **38:**943–945.

Simpson, C. L., Hemplemann, L., and Fuller, L., 1955, Neoplasia in children treated with X-ray in infancy for thymic enlargement, *Radiology* **64:**840–845.

Skreb, N., Bijelic, N., and Lukovic, G., 1963, Weight of rat embryos after X-ray irradiation, *Experientia* **19:**1–4.

Spalding, J. F., and Brooks, M. R., 1972, Comparative litter and reproduction characteristics of mouse populations with X-ray exposure, including 45 generations of male progenitors, *Proc. Soc. Exp. Biol. Med.* **141:**445–447.

Stadler, J., and Gowen, J. W., 1964, Observations on the effects of continuous irradiation over 10 generations on reproductivities of different strains of mice, *in: Proceedings of an International Symposium on the Effects of Ionizing Radiation on Reproductive Systems* (W. D. Carlson and F. X. Gassner, eds.), Pergamon Press, New York.

Stahl, W., 1963, Recent Soviet work on antenatal effects of radiation and a discussion of future implications of general work in this field, *in: Proceedings of the Symposium on Prenatal Behavior,*

Washington, D. C., October 5–7, Division of Biology and Medicine, U. S. Atomic Energy Commission, Conference Report 631008, p. 37.

Sternberg, J., 1970, Irradiation and radiocontamination during pregnancy, *Am. J. Obstet. Gynecol.* **108:**490–514.

Stettner, E., 1921, Ein weiterer Fall einer Schadingung einer menschlichen Frucht durch Röntgen Bestrahlung, *Jahrb. Kinderheikd. Phys. Erzieh.* **95:**43–51.

Stewart, A., 1973, The carcinogenic effects of low level radiation. A re-appraisal of epidemiologists methods and observations, *Health Phys.* **24:**223.

Stewart, A. M., 1972, Mycloid leukaemia and cot deaths, *Br. Med. J.* **187:**423.

Stewart, A., and Kneale, G. W., 1970, Radiation dose effects in relation to obstetric X-rays and childhood cancers, *Lancet* **1:**1185–1188.

Stewart, A., Webb, D., Giles, D., and Hewitt, D., 1956, Malignant disease in childhood and diagnostic irradiation in utero, *Lancet* **2:**447.

Stewart, A., Webb, D., and Hewitt, D., 1958, A survey of childhood malignancies, *Br. Med. J.* **1:**1495–1508.

Strange, J. R., and Murphree, R. L., 1972, Exposure-rate response in the prenatally irradiated rat: Effects of 100 R on day 11 of gestation to the developing eye, *Radiat. Res.* **51:**674–684.

Sutow, W., Conrad, R., and Griffith, K., 1965, Growth status of children exposed to fallout radiation on Marshall Islands, *Pediatrics* **36:**721–731.

Svigris, A., 1958, Sensitivity of hemopoietic organs of pregnant animals and their fetuses to ionizing radiations, *Akush. Ginek.* **34:**11–17 (in Russian).

Swasdikul, D., and Block, M., 1972, Effect of radiation upon the "embryonic" thymus, *Radiat. Res.* **50:**73–84.

Tabuchi, A., 1964, Fetal disorders due to ionizing radiation, *Hiroshima J. Med. Sci.* **13:**125–173.

Tabuchi, A., Nakagawa, S., Hirai, T., Sato, H., Hori, I., Matsuda, M., Yano, K., Shimada, K., and Nakao, Y., 1967, Fetal hazards due to X-ray diagnosis during pregnancy, *Hiroshima J. Med. Sci.* **16:**49–66.

Tacker, R., and Furchtgott, E., 1963, Adjustment to food deprivation cycles as a function of age and prenatal X-irradiation, *J. Genet. Psychol.* **102:**257–260.

Toth, B., 1968, A critical review of experiments in chemical carcinogenesis using newborn animals, *Cancer Res.* **28:**727–738.

Toyooka, E. T., Pifer, J. W., and Hempelmann, L. H., 1963a, Neoplasms in children treated with X-rays for thymic enlargement. III. Clinical description of cases., *J. Natl. Cancer Inst.* **31:**1379–1405.

Toyooka, E. T., Pifer, J. W., Crump, S. L., Dutton, A. M., and Hempelmann, L. H., 1963b, Neoplasms in children treated with X-rays for thymic enlargement. II. Tumor incidence as a function of radiation factors, *J. Natl. Cancer Inst.* **31:**1357–1377.

Vernadakis, A., Curry, J. H., Maletta, G. J., Irvine, G., and Timiras, P. A., 1966, Convulsive responses in prenatally irradiated rats, *Exp. Neurol.* **16:**57–64.

Vorisek, P., 1965, Einfluss der kontinuierlichen intrauterinen Bestrahlung auf die perinatale Mortalität der Frucht, *Strahlentherapie* **127:**112–120.

Walker, S., and Furchtgott, E., 1970, Effects of prenatal X-irradiation on the acquisition, extinction and discrimination of a classically conditioned response, *Radiat. Res.* **42:**120–128.

Wechkin, S., Elder, R., Jr., and Furchtgott, E., 1961, Motor performance in the rat as a function of age and prenatal X-irradiation. *J. Comp. Physiol. Psychol.* **54:**658–659.

Werboff, J., Broeder, J., Havlena, J., and Sikov, M., 1961, Effects of prenatal X-ray irradiation on audiogenic seizures in the rat, *Exp. Neurol.* **4:**189–196.

Wilson, J. G., 1954, Differentiation and reaction of rat embryos to radiation, *J. Cell. Comp. Physiol.* **43:**11–37.

Wilson, J. G., Brent, R. L., and Jordan, H. C., 1953a, Differentiation as a determinant of the reaction of rat embryos to X-irradiation, *Proc. Soc. Exp. Biol. Med.* **82:**67–70; also *U. S. A. E. C. D. U. R.*-243.

Wilson, J. G., Jordan, H. C., and Brent, R. L., 1953b, Effects of irradiation on embryonic development. II. X-rays on the ninth day of gestation in the rat, *Am. J. Anat.* **92:**153–188.

Wood, J. W., Johnson, K. G., and Omori, Y., 1967a, In utero exposure to the Hiroshima atomic bomb. An evaluation of head size and mental retardation: Twenty years later, *Pediatrics* **39:**385–392.

Wood, J., Johnson, K., Omori, Y., Kawamoto, S. and Keehn, R., 1967b, Mental retardation in children exposed in utero to the atomic bombs in Hiroshima and Nagasaki, *Am. J. Public Health* **57:**1381–1390.

Wood, J., Keehn, R., Kawamoto, S. and Johnson, K., 1967c, The growth and development of children exposed in utero to the atomic bombs in Hiroshima and Nagasaki, *Am. J. Public Health* **57:**1374–1380.

Wright, F. W., 1973, Diagnostic radiology and the fetus, *Br. Med. J.* **3:**693–694.

Wyburn, J. R., 1972, Human breast milk excretion on radionuclides following administration of radiopharmaceuticals, *J. Nucl. Med.* **14:**115–117.

Yamazaki, J. N., 1966, A review of the literature on the radiation dosage required to cause manifest central nervous system disturbances from in utero and postnatal exposure, *Pediatrics* **37:**877–903.

Yamazaki, J., Wright, S., and Wright, P., 1954, Outcome of pregnancy in women exposed to the atomic bomb in Nagasaki, *Am. J. Dis. Child.* **87:**448–463.

Microwave Radiation

Alliston, C. W., Howarth, B., Jr., and Ulberg, L. C., 1965, Embryonic mortality following culture *in vitro* of one and two-cell rabbit eggs at elevated temperatures, *J. Reprod. Fertil.* **9:**337–341.

Boak, R. A., Carpenter, C. M., and Warren, S. L., 1932, Studies on the physiological effects of fever temperatures. II. The effect of repeated shortwave (30 meter) fever on growth and fertility of rabbits, *J. Exp. Med.* **56:**725–739.

Brent, R. L., 1972, Irradiation in pregnancy, *Davis' Gynecology and Obstetrics*, Vol. 2 (J. J. Sciarra, ed.), pp. 1–32, Harper & Row, New York.

Brent, R. L., and Franklin, J. B., 1960, Uterine vascular clamping new procedure for the study of congenital malformations, *Science* **132:**89–91.

Brent, R. L., and Wallace, J. D., 1972, The utilization of microwave radiation in developmental biology research, *Pediatr. Res.* **6:**431.

Brent, R. L., Franklin, J. B., and Wallace, J. D., 1971, The interruption of pregnancy using microwave radiation, *Teratology* **4:**484.

Brent, R. L., and Gorson, R. O., 1972, Radiation exposure in pregnancy, *in: Current Problems in Radiology* (Robert D. Moseley, Jr., David H. Baker, Robert O. Gorson, Anthony Lalli, Howard B. Latourette, and James L. Quinn, III, eds.), Vol. 2, pp. 1–48, Year Book Medical Publishers, Chicago, Illinois.

Burfening, P. J., Alliston, C. W., and Ulberg, L. C., 1969, Gross morphology and predictability for survival of 4-day rabbit embryos following heat-stress during the first cleavage division, *J. Exp. Zool.* **170:**55–60.

Carpenter, R. L., 1962, An experimental study of the biological effects of microwave radiation to the eye, *Rome Air Development Center Rep. TDR. 62-131.*

Chang, M. C., 1957, Effect of pyrogen on embryonic degeneration in the rabbit, *Fed. Proc.* **16:**21.

Chizzolini, M., 1957, Liverprotection effect of trigonelline on fatty liver from carbon tetrachloride: experimental research, *Attual. Ostet. Ginecol.* **3:**1221–1229.

Dietzel, F., and Kern, W., 1970, Fehlgeburt nach Kurzwellenbehandlung—tierexperimentelle Untersuchungen, *Arch. Gynaekol.* **209:**237–255.

Elliot, D. S., Burfening, P. J., and Ulberg, L. C., 1968, Subsequent development during incubation of fertilized mouse ova stressed by high ambient temperatures, *J. Exp. Zool.* **169:**481–486.

Fernandez-Cano, L., 1958, Effect of increase or decrease of body temperature and hypoxia on pregnancy in the rat, *Fertil Steril.* **9:**455–459.

Gellhorn, G., 1928, Diathermy in gynecology, *J. Am. Med. Assoc.* **90:**1005–1008.

George, E. F., Franklin, J. B., and Brent, R. L., 1967, Altered embryonic effects of uterine vascular clamping in the pregnant rat by uterine temperature control, *Proc. Soc. Exp. Biol. Med.* **124:**257–260.

Harmsen, H., 1954, Uber die biologische Wirksamkeit von ultrakurz Wellen niederer Feldstarke auf Ratten, *Arch. Hyg.* **738:**278–297.

Hart, F. N., and Faber, J. J., 1965, Fetal and maternal temperatures in rabbits, *J. App. Physiol.* **20:**734–741.

Hofmann, D., and Dietzel, F., 1966, Aborte und Missbildungen nach Kurzwellendurehflutung in der Schwangerschaft, *Geburtschilfe Frauenheilkd.* **26:**378–390.

Howarth, B., Jr., 1969, Embryonic survival in adrenalectomized rabbits following exposure to elevated ambient temperature and constant humidity, *J. Anim. Sci.* **28:**80–83.

Labhsetwar, A. P., 1972, New antifertility agent—an orally active prostaglandin-ICI 74, 205, *Nature,* **238:**400–401.

Ludwig, F., and Ries-Bern, J. V., 1942, Die Beeinflussung der embryonalen Entwicklung durch Kurzwellen, *Arch. Gynaekol.* **173:**323–324.

Mali, J. W. H., 1969, Some physiological aspects of the temperature of the body surface, *Bibl. Radiol.* **5:**8–21.

McRee, D. I., 1971, Threshold for lenticular damage in the rabbit eye due to single exposure to CW microwave radiation: An analysis of the experimental information at a frequency of 2.45 GHz, *Health Phys.*. **21:**763–769.

Michaelson, S. L., 1969, Biological effects of microwave exposure, *in: Biological Effects in Health Implication of Microwave Radiation,* Symposium Proceedings, Richmond, Virginia, Sept. 17–19 (S. F. Cleary, ed.), pp. 35–58, U. S. Public Health Service, Rockville, Maryland.

Michaelson, S. M., 1971a, Biomedical aspects of microwave exposure, *Am. Ind. Hyg. Assoc. J.* **32:**338–345.

Michaelson, S. M., 1971b, Soviet views on the biological effect of microwaves—an analysis, *Health Phys.* **21:**108–111.

Milroy, W. C., and Michaelson, S. M., 1971, Biological effects of microwave radiation, *Health Phys.* **20:**567–575.

Pincus, G., 1966, Control of conception by hormonal steroids, *Science* **153:**493–500.

Pobzhitkov, V. A., Tyagin, N. V., and Grebeschechnikova, A. M., 1961, The influence of a super-high frequency pulsed electromagnetic field on conception and the course of pregnancy in white mice, *Byull. Eksp. Biol. Med.* **51:**615–618.

Rubin, A., and Erdman, W. J., II, 1959, Microwave exposure of the human female pelvis during early pregnancy and prior to conception, *Case Rep. Am. J. Phys. Med.* **38:**219–220.

Schumacher, P. H., 1936, Technik und Erfolg des Ureterenkatheterismus bei einfacher Harnstauung und bei Nierenbeckenenentzundung in der Schwangerschaft, *Zentralblatt. Gynaekol.* **60:**435–441.

Shah, M. K., 1956, Reciprocal egg transplantation to study embryo-uterine relationship in heat-induced failure of pregnancy in rabbits, *Nature* **117:**1134–1135.

Shelton, M., and Huston, J. E., 1968, Effects of high temperature stress during gestation on certain aspects of reproduction in the ewe, *J. Anim. Sci.* **27:**153–158.

Sher, L. D., 1970, Symposium on Biological Effects and Health Implications of Microwave Radiation, Richmond, Virginia, Sept. 17–19, 1969, *Med. Res. Eng.* **9:**12–13.

Thwaites, C. J., 1970, Embryo mortality in the heat of the corpus luteum, thyroid and adrenal glands, *J. Reprod. Fertil.* **21:**95–107.

Ulberg, U. L., and Burfening, P. J., 1967, Embryo death resulting from adverse environment on spermatozoa or ova, *J. Anim. Sci.* **26:**571–577.

Weber, J., Zak, K., Malee, I., and Hontela, S., 1969, Einfluss der Kurzwellen—Diathermie auf die intrauterine Temperatur, *Zentralblatt Gynaekol.,* **60:**1923–1924.

Ultrasonic Radiation

Abdulla, U., Talbert, D., Lucas, M., and Mullarkey, M., 1972, Effect on ultrasound on chromosomes of lymphocyte, *Br. Med. J.* **3:**797–799.

Basauri, L., and Lele, P. P., 1962, A simple method for production of trackless focal lesions with focused ultrasound: Statistical evaluation of the effects of irradiation on the central nervous system of the cat, *J. Physiol.* **160:**513–534.

Bobrow, M., Bleaney, B., Blackwell, N., and Unrau, A. E., 1971, Absence of any observed effect of ultrasonic irradiation on human chromosomes, *J. Obstet. Gynaecol. Br. Commonw.* **78:**730–736.

Boyd, E., Abdulla, U., Donald, I., Fleming, J. E. E., Hall, A. J., and Ferguson-Smith, M. A., 1971, Chromosome breakage and ultrasound, *Br. Med. J.* **2:**501–502.

Brent, R. L., 1971, unpublished results.

Buckton, K. E., and Baker, N. V., 1972, An investigation into possible chromosome damaging effects of ultrasound on human blood cells, *Br. J. Radio.* **45:**340–342.

Campbell, S., 1969, The prediction of fetal maturity by ultrasonic measurement of the biparietal diameter, *J. Obstet. Gynaecol. Br. Commonw.* **76:**603–609.

Coakley, W. T., 1971, Acoustic detection of single cavitation events in focused fields in water at 1 MHz, *J. Acoust. Soc. Am.* **49:**792–801.

Coakley, W. T., Hughes, D. E., Slade, J. S., and Laurence, K. M., 1971, Chromosome aberrations after exposure to ultrasound, *Br. Med. J.* **1:**109–110.

Čurzen, P., 1972, The safety of diagnostic ultrasound, *Practitioner,* **209:**822.

Dewhurst, C. J., Beazley, J. M., and Campbell, S., 1972, Assessment of fetal maturity and dysmaturity, *Am. J. Obstet. Gynecol.* **113:**141.

Doust, B. D., 1973, Role of ultrasound in obstetrics and gynecology, *Hosp. Pract.* **8:**143–152.

Eller, A., and Flynn, H. G., 1969, Generation of subharmonics of order one-half by bubbles in a sound field, *J. Acoust. Soc. Am.* **46:**722–727.

Garrett, W., and Robinson, D., 1971, Assessment of fetal size and growth rate by ultrasonic echoscopy, *Obstet. Gynecol.* **38:**525.

Gottesfield, K., 1970, The ultrasonic diagnosis of intrauterine fetal death, *Am. J. Obstet. Gynecol.* **108:**623.

Gottesfield, K., Thompson, H., Holmes, J., *et al.,* 1966, Ultrasonic placentography—a new method for placental localization, *Am. J. Obstet. Gynecol.* **96:**538.

Gottesfield, K., Taylor, E., Thompson, H., *et al.,* 1967, Diagnosis of hydatiform mole by ultrasound, *Obstet. Gynecol.* **30:**163.

Harvey, E. N., 1930, Biological aspects of ultrasonic waves, a general survey, *Biol. Bull.* **59:**306–325.

Hellman, L. M., Kobayashi, M., and Cromb, E., 1973, Ultrasonic diagnosis of embryonic malformations, *Am. J. Obstet. Gynecol.* **115(5):**615–623.

Ikeuchi, T., Sasaki, M., Oshimura, M., Azumi, T., Tsuji, K., and Shimizu, T., 1973, Ultrasound and embryonic chromosomes, *Br. Med. J.* **1:**112.

Janssens, D., Vrijens, M., Thiery, M., and Van Kets, H., 1973, Ultrasonic detection, localization and identification of intrauterine contraceptive devices, *Contraception* **8(5):**485–495.

King, D., 1973, Placental ultrasonography, *J. Clin. Ultrasound* **1:**21.

Kirsten, E. V., Zinsser, H. H., and Reid, J. M., 1963, The effect of one Mc ultrasound on the genetics of mice, *IEEE Trans. Ultrasonics Engineering* UE-10, p. 112.

Kobayashi, M., Hellman, L., and Fillisti, L., 1970, Placental localization by ultrasound, *Am. J. Obstet. Gynecol.* **106:**279.

Kohorn, E., Secker-Walker, R., Morrison, J., *et al.,* 1969, Placental localization, *Am. J. Obstet. Gynecol.* **103:**868.

Lele, P. P., 1963, Effects of focused ultrasonic radiation on peripheral nerve, with observations on local heating, *Exp. Neurol.* **8:**47–83.

Lele, P. P., and Pierce, A. D., 1972, The thermal hypothesis of the Mechanism of ultrasonic focal destruction in organized tissues, *in: Interactions of Ultrasound and Biological Tissues—Workshop Proceedings* (J. M. Reid and R. R. Sikov, eds.), U.S. D. H. E. W. Publications (FDA) 73–8008 BRH/DBE, 73–1, Washington, D.C.

Leopold, G. R., 1971, Diagnostic ultrasound in the detection of molar pregnancy, *Radiology* **98:**171.

Leopold, G. R., and Asher, W. M., 1974, Ultrasound in obstetrics and gynecology, *Radiol. Clin. North Am.* **12**(1):127.

Lucas, M., Mullarkey, M., and Abdulla, U., 1972, Study of chromosomes in the newborn after ultrasonic fetal heart monitoring in labour, *Br. Med. J.* **3**:795–796.

Mannor, S. N., Serr, D. M., Tamari, I., Meshorer, A., and Frei, E. H., 1972, The safety of ultrasound in fetal monitoring, *Amer. J. Obstet. Gynecol.* **113**:653–661.

Michell, R. C., and Bradley-Watson, P. J., 1973, The detection of fetal meningocoele by ultrasound B scan, *J. Obstet. Gynaecol. Br. Commonw.* **80**:1100–1101.

Nemes, G., and Kerenyi, T. D., 1971, Ultrasonic localization of the IUCD, *J. Obstet. Gynecol.* **109**:1219–1220.

Neppiras, E. A., 1969, Subharmonic and other low-frequency emissions from bubbles in sound-irradiated liquids, *J. Acoust. Soc. Am.* **49**:792–801.

Nyborg, W. L., 1968, Mechanisms for non-thermal effects of sound, *J. Acoust. Soc. Am.* **44**:1302–1309.

O'Shea, J. M., and Bradbury, J. H., 1972, The effect of ultrasonic irradiation on proteins, *Aust. J. Biol. Sci.* **26**:583–590.

Robinson, A., 1973, Thymoxamine ineffective against spasticity, *Lancet* **2**(7844):1504.

Robinson, H. P., 1973, Fetal heart rates as determined by sonar in early pregnancy, *J. Obstet. Gynaecol. Brit. Commonw.*, **80**:805–809.

Robinson, T. C., ard Lele, P. P., 1972, An analysis of lesion development in the brain and in plastics by high-intensity focused ultrasound at low-megahertz frequencies, *J. Acoust. Soc. Am.* **51**:1333–1351.

Rooney, J. A., 1972, Shear as a mechanism for sonically induced biological effects, *J. Acoust. Soc. Am.* **52**:1718–1724.

Rott, H. D., and Soldner, P., 1973, The effect of ultrasound on human chromosomes *in vitro*, *Humangenetik* **20**:103–112.

Sabbagha, R. E., and Turner, H. J., 1972, Methodology of B-scan sonar cephalometry with electronic calipers and correlation with fetal birth weight. *Obstet. Gynecol.* **40**:74–81.

Sabbagha, R. E., Turner, H. J., Rockette, H., Mazer, J., and Orgill, J., 1974, Sonar BPD and fetal age: Definition of the relationship, *Obstet. Gynecol.* **43**(1):7–14.

Shoji, R., Murakami, U., and Shimizu, T., 1975, Influence of low-intensity ultrasonic irradiation on prenatal development of two inbred mouse strains, *Teratology* **12**:227–231.

Sikov, M. R., 1973, Ultrasound: Its potential use for the termination of pregnancy, *Contraception* **8**(5):429–438.

Smyth, M. G., 1966, Animal toxicity studies with ultrasound at diagnostic power levels, in: *Diagnostic Ultrasound* (C C. Grossman, H. J. Holmes, C. Joyner, and E. W. Purnel, eds.), pp. 296–299, Plenum Press, New York.

Thompson, H., and Makowski, E., 1971, Estimation of birth weight and gestational age, *Obstet. Gynecol.* **37**:44.

Warwick, R. R., Pond, J. B., Woodward, B., and Connolly, C. C., 1970, Hazards of diagnostic ultrasonography—A study with mice, *IEEE Trans. Sonics Ultrasonics* SU-17, pp. 158–164.

Watts, P. L., and Stewart, C. R., 1972, The effect of fetal heart monitoring by ultrasound on maternal and fetal chromosomes, *J. Obstet. Gynaecol. Brit. Commonw.* **79**:715–716.

Watts, P. L., Hall, A., and Fleming J., 1972, Ultrasound and chromosome damage, *Br. J. Radiol.* **45**:335.

Winsberg, F., 1973a, Echocardiography of the fetal and newborn heart, *Invest. Radiol.* **7**:52.

Winsberg, F., 1973b, Echographic changes with placental ageing, *J. Clin. Ultrasound* **1**:52.

Winters, H. S., 1966, Ultrasound detection of intrauterine contraceptive devices, *Am. J. Obstet. Gynecol.* **95**:880–882.

Infectious Diseases 6

JEROME E. KURENT and JOHN L. SEVER

I. INTRODUCTION

The first association between intrauterine viral infection and human congenital malformations was described more than 30 years ago (Gregg, 1941). Since then, additional intrauterine and neonatal infections of humans and experimental animals have proved to be etiologically related to a variety of malformations and lesions.

This chapter will review the currently available information related to teratologic effects caused by microorganisms. It will be apparent that viruses are etiologically important for most of the teratology of infectious disease to be described. The past few years have witnessed an upsurge of scientific information related to viruses and their adverse effects on developing and newborn organisms. One very significant consequence of the virus–host interaction during early stages of development is the production of malformations which may resemble classic defects of embryogenesis. These defects are in many respects similar to defects classically ascribed to genetic, vascular, or toxic etiologies, but perhaps for lack of a more obvious cause. It may be of the utmost significance that most of the experimental models of malformations induced by viruses have analogies in human disease. Most of the latter, however, are of unknown etiology. The elucidation of experimental models may contribute to the future understanding of naturally occurring human malformations as well as to teratologic mechanisms in general.

Viruses may cause malformations and lesions by a variety of different mechanisms. If the target cells are those of an immature, rapidly dividing population, a defect resembling classic defects of embryogenesis may result,

JEROME E. KURENT and JOHN L. SEVER · Infectious Diseases Branch, National Institute of Neurological and Communicative Disorders and Stroke, National Institutes of Health, Bethesda, Maryland.

such as cerebellar hypoplasia. If relatively mature stabilized cell populations are destroyed, cavitary changes characteristic of cystic malformations, such as porencephaly, may develop.

Mechanisms of viral-induced cellular damage may include acute cell death, persistent noncytolytic infection, immune-mediated cell damage, and mitotic inhibition alone or in combination with each other. Other mechanisms may also exist.

The majority of malformations and lesions due to viruses thus far described involve the central nervous system (CNS). As a result, in the following discussion major emphasis will be placed on CNS disease following viral infection of the developing nervous system. This will include consideration of naturally occurring and experimental models, both of which are listed in Table 1. Finally, it will be seen that the principles governing physical and chemical teratogenesis also apply to the teratology of infectious disease.

II. INFECTIOUS AGENTS TERATOGENIC FOR HUMANS AND EXPERIMENTAL ANIMALS

A. Infections of the Human Fetus and Neonate

1. Rubella Virus

The classical model for intrauterine infection of the human fetus is rubella virus. Following the first recognized association of maternal rubella infection with infant cataracts (Gregg, 1941), the expanded rubella syndrome was described. Microcephaly, microphthalmia, encephalitis, mental retardation, blindness, deafness, hepatitis, valvular heart lesions, myocarditis with myocardial necrosis, and lesions of the long bones were seen (Korones et al., 1965; Swan et al., 1943; Cooper et al., 1969). Hepatosplenomegaly, pneumonitis, thrombocytopenia, and petechiae may also occur as consequences of intrauterine rubella infection. Fetal disease may thus be manifested as malformations and lesions of numerous organ systems.

One of the most striking epidemiological features of intrauterine rubella virus infection is the gestational age-dependency, which determines both the frequency of fetal disease and the specific malformations or lesions that follow. Maternal infection during the first month of pregnancy is associated with abnormal infants in 50% of cases. When infection occurs during the second month 22% are affected, and 6–8% of infants are malformed when maternal infection happens during the third through fifth months of gestation (Dekaban et al., 1958, Michaels and Mellin, 1960; Lundstrom, 1962). Figure 1 (Dekaban et al., 1958) demonstrates that the majority of recognized defects caused by rubella virus follow infection during the first trimester.

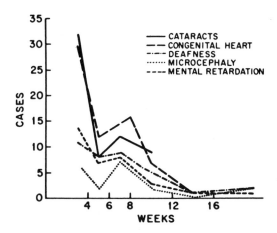

Fig. 1. Frequency of abnormalities in 108 offspring related to time of maternal rubella. (From Dekaban *et al.*, 1958, reproduced with permission.)

During congenital rubella infection, the placenta is infected in virtually 100% of cases. Although placental infection precedes fetal infection, it does not necessarily lead to fetal infection in every instance (Monif *et al.*, 1965). However, if rubella virus crosses the placenta and the fetus is reached, virus may become widely disseminated throughout most of the major organ systems or be limited to relatively few organs.

Following birth of the infant infected with rubella, virus may be recoverable from the nasopharynx, urine, and lymphocytes for months to years later. In one instance, rubella virus was isolated from a cataractous lens at 3 years of age (Menser *et al.*, 1967). The virus may be maintained as a persistent infection, and often in the presence of high-titered neutralizing antibody. Although the mechanisms of viral persistence are poorly understood, it has been suggested that severe defects of the host lymphoid system may be important in this regard. Thus immunologic defects may be manifested in a high incidence of dysgammaglobulinemia (Soothill *et al.*, 1966; Hancock *et al.*, 1968), delayed development of antibody-producing potential (Kenrick *et al.*, 1965; Florman *et al.*, 1973), defective lymphocyte function (Montgomery *et al.*, 1967; Olson *et al.*, 1967), and the failure to maintain lasting immunity to rubella virus (Doege and Kim, 1967; Menser *et al.*, 1967).

Most recently, rubella virus has been associated with chronic progressive panencephalitis that is clinically very similar to subacute sclerosing panencephalitis caused by measles virus (Townsend *et al.*, 1975; Weil *et al.*, 1975). This association was observed in children who had well-documented congenital rubella many years prior to the development of chronic degenerative CNS disease. Rubella virus was isolated from the brain of one of the affected children (Weil *et al.*, 1975).

Table 1. Teratology of Infectious Diseases: Selected Models of CNS Disease

Virus genus	Defect	Host
1. Arenavirus Lymphocytic choriomeningitis	Cerebellar hypoplasia, ataxia	Newborn mouse, rat
2. Herpes virus		

Cytomegalovirus	Microcephaly, hydrocephalus,
	    intracerebral calcifications,
	    seizures, mental retardation
		Human fetus and
		newborn

Herpes virus	Microcephaly, intracerebral
	    calcifications, mental
	    retardation, seizures
		Human fetus and
		and newborn

3. Myxovirus
    Influenza
	Defects of neural tube
	    flexure and closure;
	    micrencephaly
		Chick embryo

	Hydrocephalus	Newborn monkey, hamster
	Hydrocephalus	Fetal rhesus monkey
4. Orbivirus		
Bluetongue virus	Hydranencephaly, porencephaly	Fetal lamb, neonatal mouse, newborn hamster
5. Paramyxovirus		
Mumps virus	Hydrocephalus	Newborn hamster

Newcastle disease virus	Defects in neural
	    tube closure and flexure;
	    micrencephaly
		Chick embryo

Parainfluenza virus type 2	Hydrocephalus	Newborn hamster
6. Parvovirus		
Feline panleukopenia virus	Cerebellar hypoplasia	Fetal and newborn kitten, ferret
Rat virus	Cerebellar hypoplasia	Rat, hamster, cat, ferret
Minute virus	Cerebellar hypoplasia	Neonatal mouse
7. Reovirus Reovirus type 1	Hydrocephalus	Newborn hamster, mouse, rat, ferret

8. Togavirus
    Bovine diarrhea–
    mucosal disease virus
	Cerebellar hypoplasia,
	    hydrocephalus, blindness
		Fetal calf, fetal lamb

Hog cholera virus	Cerebellar hypoplasia,
	    hypomyelinogenesis,
	    congenital termors,
	    microcephaly
		Fetal pig

Rubella virus
    Venezuelan equine
    encephalitis virus
	Microcephaly, microphthalmia,
	    blindness and deafness,
	    mental retardation, other
	    multisystem disease
	    involving cardiovascular
	    and other systems
		Human fetus

Venezuelan equine
    encephalitis virus
	Cerebral cysts, microcephaly,
	    cataracts
		Fetal rhesus
		monkey

Table 1 (cont'd)

Virus genus	Defect	Host
	Hydrocephalus, micro-cephaly, microphthalmia, cerebral cysts	Human fetus
9. Nonviral		
Toxoplasma gondii	Hydrocephalus, micro-cephaly, periventricular cerebral calcifications, hydranencephaly, seizures, mental retardation.	Human fetus
Treponema pallidum	Hydrocephalus, seizures, mental retardation, tooth and bony defects	Human fetus

2. Cytomegalovirus

Cytomegalovirus (CMV) is a DNA-containing virus of the herpes group. It is responsible for cytomegalic inclusion disease, named for the pathognomonic inclusion-bearing cells seen in urine of children infected during intrauterine life. This virus is ubiquitous and approximately 3% of women demonstrate evidence of infection during pregnancy, as shown by viral serological study (Sever and White, 1968) and virus isolation from urine of pregnant women (Hildebrandt *et al.*, 1967). Unlike rubella infection, it is not clear during which stage of intrauterine life the fetus is most susceptible to infection with CMV. This is largely due to the difficulties in ascertaining maternal infection, since most are not clinically apparent.

The major CNS manifestations of congenital CMV are microcephaly, hydrocephaly, encephalitis with cerebral calcifications, microphthalmia, blindness, seizures, and mental retardation. Pneumonitis, thrombocytopenia and anemia, petechiae, and kernicterus are also noted (Medearis, 1975; Hanshaw, 1966a). It has been suggested that CMV may be responsible for a large portion of unexplained cases of microcephaly (Hanshaw, 1966b).

To assess the overall public health problem represented by CMV is difficult. This is illustrated by the fact that minimal CNS disease may not be detectable until months to years later. Recent evidence suggests that approximately 0.5–1.5% of newborns delivered yearly in the U.S. are infected with CMV (Hanshaw, 1971; Starr *et al.*, 1970; Kumar *et al.*, 1973). Approximately 10% of infected infants may later demonstrate sequelae such as low I.Q. and high-tone hearing loss (Reynolds *et al.*, 1974). Thus at least 1 in 1000 live births (for a total of 3000–5000 newborn per year) in the U.S. are damaged by CMV.

CMV may be isolated from peripheral lymphocytes and urine of congenitally infected children from months to years later. The chronicity of congenital CMV infection is consistent with a deficiency of the host's immune response to CMV. However, direct evidence in support of this is not available.

3. Herpes Virus

Herpes simplex virus, a DNA virus, occurs as two distinct serological strains. Type 1 is associated with oral lesions, e.g., "cold sores" or "fever blisters," while type 2 is associated with genital infection (Nahmias and Dowdle, 1968). In the adult, herpes virus has a predilection for tissues derived from embryonic ectoderm, such as oral mucous membranes and the CNS. Generalized infection of the developing fetus or newborn may include involvement of mesodermal tissues as well.

Infection of the human neonate is often associated with disseminated infection, which may include acute necrotizing encephalitis, hepatosplenomegaly, and disseminated intravascular coagulopathy. The majority of neonatal herpes encephalitis is caused by type 2 herpes, which is acquired following direct contamination of the infant by the infected birth canal (Nahmias *et al.*, 1969).

Intrauterine infection with herpes virus is a rare occurrence and is usually associated with the type 2 strain. Congenital disease may produce a clinical picture similar to that seen with CMV. The intrauterine-infected infant may have microcephaly with cerebral calcifications, chorioretinitis, and mental retardation (South *et al.*, 1969). Less frequently, type 1 herpes virus may also be associated with intrauterine infection. The affected newborn may demonstrate microcephaly, retinal dysplasia, intracerebral calcifications, seizures, and mental retardation (Florman *et al.*, 1973).

4. *Toxoplasma gondii*

Toxoplasma gondii is an ubiquitous intracellular protozoan, and it infects humans, other mammals, and birds. The natural reservoir is probably the rodent, and adult infection is probably most often acquired by ingesting undercooked contaminated meat products (Boughton, 1970; Beverley, 1973). More recently, the role of the domestic cat as a reservoir of infection has been emphasized (Peterson *et al.*, 1972). Clinically apparent infection in the adult is uncommon, and consists of an infectious mononucleosis-like syndrome including lymphadenopathy and a febrile course from which the patient recovers without sequelae.

Intrauterine infection with *T. gondii* may range from an apparently normal infant to one with severe and permanent CNS damage (Desmonts and Couvreur, 1974). Manifestations may include microcephaly with periventricular cerebral calcifications, hydrocephalus, seizures, and mental retardation (Eichenwald, 1960; Hume, 1972). Hydranencephaly has also been reported

(Altschuler, 1973). In the majority of cases, intrauterine infection probably occurs as a result of primary maternal infection. The possible significance of the chronic carrier state in congenital toxoplasmosis has been suggested, however (Beverley, 1959; Sanger and Cole, 1955; Remington *et al.*, 1961). Disease in affected infants is manifested in one of two general forms: one in which defects are limited primarily to the CNS, and the other in which generalized disease is seen (Eichenwald, 1960).

5. Syphilis

Trepenoma pallidum, the spirochete causing syphilis, may infect the fetus via transplacental transmission. Intrauterine infection may involve many different organ systems. Some of the more incapacitating effects involve the CNS. Thus hydrocephalus, seizures, mental retardation, and hemiplegia represent congenital defects resulting from intrauterine infection by *T. pallidum* (Silverstein, 1962; Holder and Knox, 1972). Additionally, deformities of the teeth and skeleton may occur.

The relative resistance of the human fetus to intrauterine infection by *T. pallidum* during the first four or five months of gestation has been attributed to the well-developed Langhan's layer of the chorion present during that time. Alternatively, infection may occur during an earlier period of gestation, but pathology may not develop until the fetus is capable of responding with sufficient inflammatory reaction (Silverstein, 1962, 1964).

B. Experimental Models of Viral Teratology

1. Influenza Virus

Influenza A virus was one of the first viruses reported to cause anomalies of the developing embryo (Hamburger and Habel, 1947). Infection of the chick embryo with influenza A resulted in micrencephaly and failure of neural tube development along with failure of neural tube closure as demonstrated in Figs. 2 and 3 (Johnson *et al.*, 1971). It was demonstrated that influenza virus infection was limited to nonneural tissues (Johnson *et al.*, 1971). Thus immunofluorescent antibody study of the neural tube, neural crest, notocord, and surrounding mesenchymal tissues failed to demonstrate viral antigen. Viral antigen was , however, present in chorion and amnion and later throughout the entire extraneural ectoderm, myocardium, and areas of the gut mucosa. Malformations of infected tissues were not seen, however. It was suggested that failure of synthesis or transport of nutrients required by the developing neural tube may have resulted indirectly from infection of extraneural tissues.

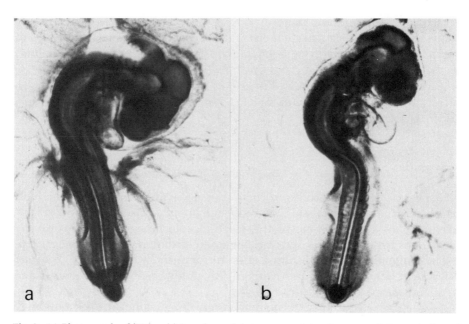

Fig. 2. (a) Photograph of 72-hr chick embryo showing normal development of the cephalic and cervical flexures. Embryo was inoculated with virus-free suspension 24 hr previously. Alum cochineal stain. (b) Photograph of 72-hr chick embryo showing defect of cervical and cephalic flexures and collapse of the metencephalic portion of the primitive brain. Embryo was inoculated with influenza A virus 24 hr previously. Alum cochineal stain. (From Johnson *et al.*, 1971) reproduced with permission, photographs courtesy of Dr. Kenneth P. Johnson.)

Intracerebral inoculation of fetal rhesus monkeys at 100 days gestation with a Hong Kong strain of influenza virus caused gross hydrocephalus which was evident at birth (London *et al.*, 1973). Cross sections of brains from newborn monkeys demonstrated massive ventricular dilatation and cortical atrophy, as demonstrated in Fig. 4. Fetuses inoculated with influenza virus previously neutralized with antiserum to influenza virus failed to develop CNS defects.

The conflicting data relating to a possible role for influenza virus in causing human congenital malformations and childhood neoplasms has been recently reviewed (MacKenzie and Houghton, 1974). The suggested defects have included CNS malformations, circulatory malformations, and cleft lip and reduction deformities. It was concluded that convincing evidence for a positive association does not exist at this time, but the need for further indepth study was stressed.

2. Newcastle Disease Virus

Newcastle disease virus (NDV) caused defects of the developing chick neural tube (Robertson *et al.*, 1955). These are similar to those observed with influenza virus, but the pathogenesis appears to be different. Unlike the influenza-induced anomalies, NDV antigen is demonstrable in cells of the caudal neural tube by the fluorescent antibody technique (Williamson *et al.*, 1965). It is thus presumed that NDV-induced neural tube defects in the chick embryo result from a more direct mechanism than occurs with influenza.

3. Parvoviruses: Feline Panleukopenia Virus, Rat Virus, H-1 Virus, and Minute Virus

The parvoviruses, which include feline panleukopenia virus (FPL), rat virus (RV), H-1 virus, and minute virus, will be discussed as a group. These viruses share similar biological and physicochemical characteristics (Siegl *et al.*, 1971) and produce similar CNS malformations.

Until the demonstration of FPL etiology, cerebellar hypoplasia of domestic cats was considered to be a genetically determined disorder. The transmissibility of this condition was first demonstrated by its production in experimental animals utilizing cerebellar filtrates from diseased kittens (Kilham and Margolis, 1966, 1967). Intracerebrally inoculated kittens and ferrets developed a permanent intractable ataxia within days of inoculation. Grossly, the cerebellum was hypoplastic and was characterized histologically by a highly specific loss of granular cells in the cortex which has been referred to as granuloprival hypoplasia. Figures 5 and 6 demonstrate the pathologic anatomy resulting from FPL infection of the developing cerebellum. Im-

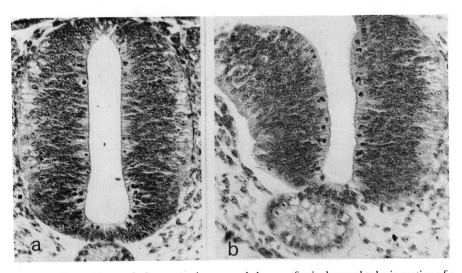

Fig. 3. (a) Photomicrograph demonstrating normal closure of spinal neural tube in section of a control 72-hr chick embryo (hematoxylin and eosin stain). (b) Photomicrograph demonstrating failure of spinal neural tube closure of 72-hr chick embryo, 24 hr after inoculation with influenza A virus (hematoxylin and eosin stain). (From Johnson *et al.*, 1971, reproduced with permission, photographs courtesy of Dr. Kenneth P. Johnson.)

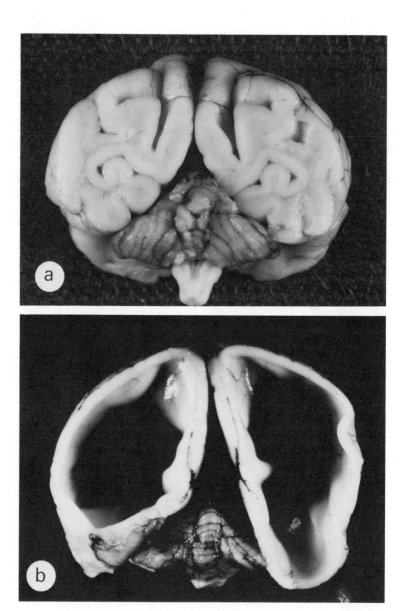

Fig. 4. (a) Newborn rhesus monkey inoculated with influenza A virus plus antiserum at 100 days gestation. Coronal section through cerebral hemispheres demonstrates normal appearance of brain (b) Newborn rhesus monkey inoculated with influenza A virus alone at 100 days gestation. Coronal section of brain at same level as control animal demonstrates gross dilatation of posterior horns of lateral ventricles. (Unpublished photographs courtesy of Dr. William T. London.)

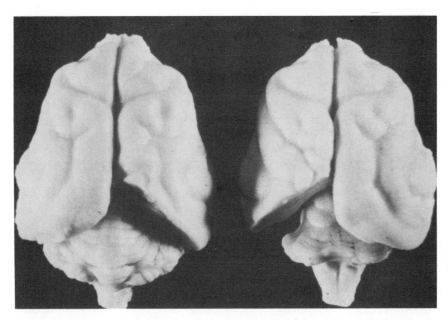

Fig. 5. Brain on left represents the dorsal view of a normal 53-day-old ferret brain. A portion of the posterior cerebrum has been cut away to expose the cerebellum. Brain on right is dorsal view of a 53-day-old ferret brain inoculated on the first postnatal day with feline panleukopenia virus. Striking hypoplasia of the cerebellum is demonstrated beneath the cutaway portion of the cerebrum. Cerebrum appears grossly normal. (From Margolis and Kilham, 1968, reproduced with permission, photograph courtesy of Dr. George Margolis.)

munofluorescent study of infected cerebellum localized the viral antigen to the cells of the granular cortex; the rest of the cerebellar cortex and deep cerebellar nuclei were unaffected. FPL virus was identified as the transmissible agent, and subsequent inoculation of this virus readily produced clinical cerebellar ataxia, with the expected anatomic and histological features (Margolis and Kilham, 1968, 1970).

FPL-induced cerebellar hypoplasia is now known to occur as a transplacental infection (Kilham *et al.*, 1971). Following birth, affected kittens may shed virus in the urine for months even in the presence of passively acquired maternal antibody.

Rat virus (RV), another parvovirus, also caused permanent cerebellar ataxia in the progeny of infected pregnant hamster rats and ferrets (Kilham and Margolis, 1964; Margolis and Kilham, 1968). As with FPL infection, RV-infected animals demonstrated a hypoplastic cerebellum with histological features of cortical granular cell hypoplasia. Intrauterine fetal death and resorption, as well as hepatitis in surviving progeny, may occur.

H-1 virus, which was originally isolated from normal human fetuses and placentas, caused cerebellar hypoplasia in progeny of infected pregnant hamsters (Kilham and Margolis, 1967). In addition, mongoloid-like facies

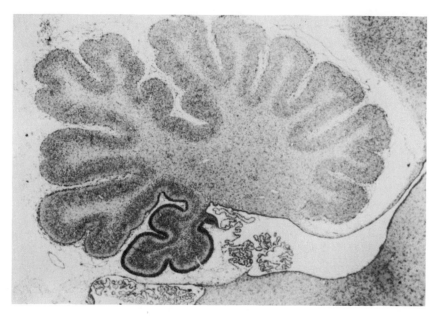

Fig. 6. Cerebellum of a 14-day-old ferret inoculated on the first postnatal day with feline panleukopenia virus. Except for the lingula, the entire external germinal layer has been destroyed. In the lingula the external germinal layer forms a prominent feature, and the granule layer has begun to develop. (From Margolis and Kilham, 1968, reproduced with permission, photograph courtesy of Dr. George Margolis.)

reminiscent of Down's syndrome in humans occurs (Toolan, 1960), although chromosomal abnormalities were not seen. No association, however, has been demonstrated between human disease and H-1 virus.

Minute virus given by the intracerebral route caused cerebellar hypoplasia in newborn mice (Kilham and Margolis, 1970). The defect, however, was not associated with clinically apparent neurological disease.

The parvoviruses, exemplified by FPL, RV, H-1, and minute virus have an important biological property which is strongly associated with their teratogenic potential. Parvoviruses as a class, have a propensity to infect immature dividing cells of the brain which are in a state of increased mitotic activity. These are typified by cells in the external granular layer of the cerebellar cortex.

4. Bluetongue Virus

Bluetongue virus, a double-stranded RNA virus of the orbivirus group, produced gestational age-dependent CNS malformations in sheep (Osburn *et al.*, 1971a,b). The wild virus produces an acute respiratory and gastrointestinal disease in adult sheep, with a mortality rate of approximately 5% (Howell and Verwoerd, 1971). A chick-embryo-adapted vaccine strain of bluetongue virus effectively protected adult sheep from acute disease when given prior to natural exposure. Vaccinated ewes, however, produced "dummy lambs," which demonstrated severe structural brain damage (Young and Cordy, 1964).

Experimental intramuscular inoculation of pregnant ewes with the vaccine strain of bluetongue virus produced severe noninflammatory defects in the brains of progeny which were strongly reminiscent of those observed in human disease (Richards and Cordy, 1967). Hydranencephaly, a condition characterized by replacement of the cerebral hemispheres with fluid-filled membranous sacs, resulted when pregnant ewes were inoculated at 50–54 days gestation. Inoculation at 70–78 days resulted in the production of porencephaly, which designates the presence of multiple cystic malformations within the parenchyma of the cerebral hemispheres. Inoculation at 100 days gestation, however, failed to produce gross CNS defects. Immunofluorescent antibody study localized the highly selective viral infection to the subependymal cells of the telencephalon.

The pathogenesis of bluetongue virus CNS defects has been elucidated by utilizing a nonlethal mouse-adapted strain of bluetongue virus (Narayan and Johnson, 1972). It was shown that an acute encephalitis occurred following intracerebral inoculation of the neonatal mouse and that resolution of the acute infection by host reparative processes eventually resulted in bilateral symmetrical malformations. These defects were of predictable distribution in the cerebral cortex, basal ganglia, hippocampal formation, and other regions. Mice studied serially over a period of several days demonstrated that the acute infection progressed over established migratory paths of the infected cells. Immunofluorescent study of animals 2–4 days after inoculation showed that viral antigen was limited to cells of the subventricular zone. Over the subsequent 4 days, the infection progressed along established migratory paths of infected cells into the basal ganglia, cortex, and other anatomic areas subsequently malformed. The infection was considered to have spread either by cell-to-cell virus transmission or via transport by infected cells actively migrating to their anatomic destinations (Narayan and Johnson, 1972).

The relative immunological immaturity of infected animals has been considered as possibly playing a role in the development of bluetongue virus-induced anomalies (Richards and Cordy, 1967). More recently, however, it was shown that cyclophosphamide and antithymocyte serum treatment did not enhance the severity of lesions in mice (Narayan et al., 1972). The latter finding suggested that the relative stage of immunological development was probably not the major factor responsible for the age-dependent nature of bluetongue virus-induced malformations. In newborn mice cyclophosphamide actually inhibited virus growth, and less severe defects were seen compared to untreated animals. This may have been secondary to an antimetabolitic effect on differentiating CNS cells, effectively reducing the virus-susceptible cell population (Narayan et al., 1972).

5. Mumps Virus

Newborn hamsters inoculated intracerebrally with a nonneurotropic strain of mumps virus developed gross hydrocephalus (Fig. 7) (Johnson *et al.*, 1967). This was a fortuitous observation in that this virus strain was being used as a control in conjunction with a neurotropic strain of mumps virus that caused acute encephalitis. Obstructive hydrocephalus was considered to have occurred secondary to occlusion or stenosis of the aqueduct of Sylvius and other critical CSF pathways. Serial study of newborn hamsters infected with mumps virus subsequently demonstrated that an acute infection of the ependymal cells lining the aqueduct of Sylvius, lateral ventricles, and foramen of Monroe was produced. Following the acute ependymitis, host reparative processes resulted in stenosis or atresia of passages between closely apposed surfaces of certain key CSF bottleneck areas (Johnson and Johnson, 1968). The resultant end-stage defect closely resembled classical defects of embryogenesis.

Immunofluorescent antibody study during the acute ependymitis stage demonstrated the presence of mumps viral antigen in ependymal cells (Fig. 8) (Johnson and Johnson, 1968). Virus reached maximum titers by the fifth day after inoculation, but was not detectable by day nine. Disappearance of virus coincided with the appearance of neutralizing antibody. The clinically silent infection was characterized histologically at first by perivascular inflamma-

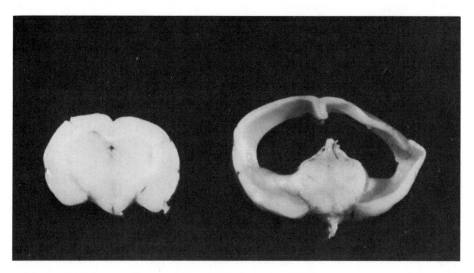

Fig. 7. Coronal sections of hamster brains 4 weeks after inoculation of mumps virus. Brain on left which was inoculated with virus plus antiserum appears normal; brain on right, inoculated with virus alone, shows typical hydrocephalus. (From Johnson *et al.*, 1967, reproduced with permission, photograph courtesy of Dr. Richard T. Johnson.)

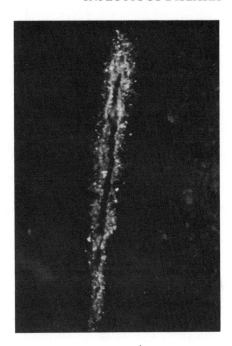

Fig. 8. Horizontal section of hamster brain stem 7 days after intracerebral inoculation of mumps virus. Immunofluorescent antibody technique demonstrates viral antigen limited to ependymal cells lining cerebral aqueduct. (Unpublished photograph courtesy of Dr. Richard T. Johnson.)

tion, with mononuclear cells first appearing on the fourth day after inoculation. The inflammatory response was resolved by the 14th–16th day, and the silent infection gave rise to dilatation of the lateral and third ventricles just after the second week. Severe hydrocephalus was present by the third week, with all infected animals demonstrating total occlusion of the aqueduct of Sylvius (Johnson *et al.,* 1967; Johnson and Johnson, 1968). The end-stage defect was characterized by a lack of inflammatory reaction, or any other histological markers which would have indicated the existence of prior viral infection.

6. Reovirus

Reovirus type 1 when inoculated by the intracerebral route caused obstructive hydrocephalus in newborn hamsters (Kilham and Margolis, 1969; Margolis and Kilham, 1969a). As with mumps virus previously described, a clinically silent infection of ependymal cells led to advanced hydrocephalus by 3–4 weeks postinoculation. Dilatation of the lateral ventricles was the prominent grossly evident pathological change. End-stage hydrocephalic brains demonstrated no evidence of prior inflammation or other hallmarks of viral infection. Thus no glial scars or ependymal rosettes were present at sites of obliterated CSF passageways which might have suggested antecedent acute viral infection. In contrast, the defective brains resembled those seen in developmental malformations, except when calculus or hemosiderin granules were present.

The pathogenesis of reovirus-induced hydrocephalus included the appearance of cytoplasmic inclusions in ependymal cells and choroid plexus epithelial cells at 4–6 days following intracerebral inoculation (Margolis and Kilham, 1969a). A moderate inflammatory exudate, which was present in the ventricles, the parenchymal perivascular spaces bordering the ventricles, and diffusely in the leptomeninges, was observed within 24 hr of appearance of the inclusions. Denudation of ventricular walls due to necrosis and sloughing of ependyma occurred by 10–12 days after inoculation. Reparative processes were advanced by the end of the third week and included astroglial and capillary proliferation at areas of ependymal ulceration. Obliteration of surviving ependyma frequently resulted. Adjacent walls became adherent to one another as a result of host repair mechanisms, with consequent obstruction of CSF pathways. Critical areas included the third ventricle, foramen of Monroe, aqueduct of Sylvius, angles of the fourth ventricle, and the central canal of the spinal cord. The Arnold–Chiari malformation may result as a complication of hydrocephalus in the hamsters (Margolis and Kilham, 1969b).

7. Lymphocytic Choriomeningitis Virus

Lymphocytic choriomeningitis virus (LCM) may be harbored as an asymptomatic chronic persistent infection in adult mice (Hotchin and Cinits, 1958; Oldstone and Dixon, 1969, 1970). The infection may have originally occurred as an intrauterine or newborn infection.

Newborn mice and rats inoculated intracerebrally with LCM developed cerebellar hypoplasia (Cole *et al.*, 1971). Inoculation with LCM at four days of age resulted in cerebellar necrosis secondary to granule cell infection. Infected rats survived the acute infection but developed permanent cerebellar ataxia. If neonatal rats were given intracerebral LCM at one or seven days of age, however, they developed only a transient ataxia of approximately one month's duration. In contrast, newborn mice died of acute choriomeningitis, and this lethal effect superceded any possible clinical effects of cerebellar granule cell necrosis such as that observed in rats.

Extensive granule cell infection appears to be necessary for the development of the cerebellar lesion in rats and mice. It is not known, however, why other areas of the neuroparenchyma, such as the hippocampus and cerebral cortex, did not develop pathologic changes even though they contained viral antigen by immunofluorescence (Cole *et al.*, 1971).

The immunopathogenesis of cerebellar disease in the rat has been demonstrated. Passively administered anti-rat-thymocyte serum given before or after inoculation with LCM prevented the development of clinical ataxia (Cole *et al.,* 1971). Brains of treated animals demonstrated little, if any, cerebellar pathology. Fluorescent antibody staining of the cerebellum, however, indicated that the degree of LCM virus infection was comparable in antithymocyte-treated animals and untreated animals which developed ataxia. This suggested that the host's immune response against cell-associated viral antigen, and not virus infection *per se,* was responsible for cerebellar pathology.

8. Hog Cholera Virus

Immunization of pregnant sows at 20–90 days gestation with hog cholera virus (HCV) vaccine resulted in progeny with severe CNS disease. This included cerebellar hypoplasia, hypomyelinogenesis, congenital tremors, and microcephaly in newborn pigs (Emerson and Delez, 1965; Johnson *et al.,* 1970). Hydrocephalus, small cerebral gyri, and cerebellar agenesis were also noted. HCV infection was not restricted to CNS tissues, but also included defects of the skeletal and urogenital systems.

HCV was present as a persistent noncytolytic infection. The virus has been demonstrated in the cerebellar external granular layer of full-term fetuses by virus isolation and fluorescent antibody staining (Johnson and Byington, 1972). Classic hallmarks of viral infection were absent; there was no evidence of cellular necrosis, inclusions, or inflammation, and serum of infected fetuses did not contain specific antiviral antibody to HCV.

Fetal damage was considered possibly to result from HCV inhibition of cell division and function (Johnson and Byington, 1972). Hypomyelinogenesis and delayed myelination may reflect the pathological effects of HCV on oligodendroglia, the myelin forming elements of the CNS (Emerson and Delez, 1965). CNS damage has also been suggested to result from hypoxia secondary to vasculitis induced by HCV.

9. Venezuelan Equine Encephalitis Virus

The association of CNS pathology in progeny of mothers infected with Venezuelan equine encephalitis (VEE) virus has been reported in humans (Wenger, 1967). Offspring of mothers infected during an epidemic of VEE demonstrated a variety of CNS pathology, including encephalitis, hydranencephaly, hydrocephalus, microcephaly with microphthalmia, and cerebral cysts. A correlation was suggested between the stage of gestation during which maternal VEE infection occurred and the specific nature of CNS pathology in the progeny.

Recently CNS anomalies in newborn rhesus monkeys were produced in our laboratories with a vaccine strain of VEE (Levitt *et al.,* 1973). Fetal

monkeys given intracerebral VEE at 100 days of gestation developed hydrocephalus, cerebral cysts, and dense bilateral cataracts which were evident
at birth. Cystic changes are depicted in Fig. 9.

10. Bovine Viral Diarrhea–Mucosal Disease Virus

Bovine viral diarrhea–mucosal disease virus (BVD-MD) is an agent that
causes a contagious disease in adult cattle characterized by fever, gastroen-
teritis, diarrhea, and abortion (Jensen *et al.*, 1968). Abortion has occurred in
cattle following experimental exposure to BVD-MD virus (Gillespie *et al.*,
1967; Shope, 1968), and abortions have also occurred following natural out-
breaks of the disease in adult animals (Olafson *et al.*, 1946; Swope and
Luedke, 1956). BVD-MD virus has also caused cerebellar hypoplasia and
ocular defects consisting of lens opacities in progeny of infected cattle (Ward,
1971; Khars *et al.*, 1970a,b; Scott *et al.*, 1973). Granuloprival hypoplasia was
associated with experimental inoculation with BVD-MD virus early in gestation (79 days), while more severe focal changes ranging from mild to severe
depletion of the cortical layers to complete destruction of the cortex and folial
white matter occurred following inoculation later in gestation (107–150 days)
(Brown *et al.*, 1973).

III. THE DEVELOPMENTAL BASIS OF VIRAL TERATOLOGY

A. Altered Embryogenesis

1. Apparent Agenesis Following Host Reparative Mechanisms

The most extensively studied models which fulfill criteria of primary
agenesis or atresia of the cerebral aqueduct are mumps (Johnson *et al.*, 1967;
Johnson and Johnson, 1968) and reovirus (Kilham and Margolis, 1969; Margolis and Kilham 1969a) induced hydrocephalus. As described earlier, both
viruses when inoculated into appropriate newborn rodents produced a sub-
clinical ependymitis which subsequently resolved. Tissue repair mechanisms
resulted in stenosis or occlusion of the aqueduct of Sylvius, with subsequent
massive dilatation of the lateral and third ventricles. The obliterated ependymal cells normally lining the surfaces of the aqueduct were replaced by
what appeared to be normal brainstem tissue, in the absence of glial nodules.
The latter, if present, might have served as indirect evidence for an antecedent inflammatory process. The resultant defects, however, fulfilled all the
criteria for primary agenesis or atresia of the cerebral aqueduct. Thus

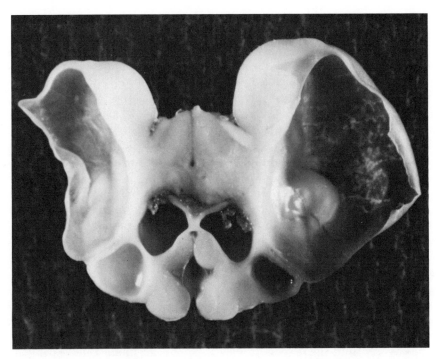

Fig. 9. Three-month-old rhesus monkey inoculated with Venezuelan equine encephalitis virus vaccine at 100 days gestation. Coronal section of brain at level of infundibulum demonstrates cysts present bilaterally. (Unpublished photograph courtesy of Dr. William T. London.)

there was an absence of inflammation and gliosis, and aqueductules and normal tissue between residual ependymal cells were present. Although the histology is consistent with primary agenesis of cerebral aqueducts, it is obvious that these defects were of viral etiology.

Adult mice inoculated with influenza virus likewise developed obstructive hydrocephalus secondary to occlusion of CSF pathways (Johnson and Johnson, 1972). Adult animals, however, demonstrated a glial reaction at area of stenosis. It was suggested that the closer resemblance of the defect in newborn animals to primary agenesis reflected a difference in the immature host response to tissue injury when compared to the adult animal.

2. Hypoplasia Resulting from Interruption of Normal Cell Migration and Differentiation

Immature cells in an active state of differentiation and migration are very susceptible to certain viral infections. Two infections in particular produce defects which strongly resemble degenerative processes. Feline panleukopenia virus (FPL) and other parvoviruses have a marked propensity to infect the external granular layer of the cerebellar cortex. Transplacental infection, as well as intracerebral inoculation of newborn animals, resulted in this highly selective infection of the developing brain. The early infection was characterized by a viral inclusion body phase, followed by more advanced cytopathic changes with subsequent resolution of necrotic cells (Margolis and Kilham, 1968, 1970). The result is a severe degree of cerebellar hypoplasia limited to the external granular layer, with sparing of the deep cerebellar nuclei and associated structures. The general histological appearance was one of a degenerative process, without any hallmarks of previous viral infection or inflammatory process.

Bluetongue virus infection in sheep also demonstrated the affinity of certain infectious agents for immature, actively dividing cells of the developing embryo. The defects which were strikingly gestational age-dependent resulted from a highly specific infection of the subependymal cells of the fetal telencephalon (Osburn *et al.*, 1971a). This cell population, which is adjacent to the ventricles, normally proliferates and migrates outward to form the neurons and glia of the cerebral hemispheres. This germinal zone was very large in the first trimester lamb, and infection caused massive cerebral necrosis and cavitation. Resolution lead to the noninflammatory defect of hydranencephaly, characterized by replacement of both cerebral hemispheres by membranous fluid-filled sacs. Infection of the germinal cells during midgestation resulted in a more focal acute encephalitis. Resolution of the acute infection led to the end-stage defect consisting of fluid-filled porencephalic cysts.

3. Defects of Neural Tube Flexure and Closure Due to Viral Influences on Neural Tube Development

Chick embryos inoculated at 48 hr of age with influenza A virus demonstrated failure of neural tube flexure and closure, and also collapse of the primitive brain (Hamburger and Habel, 1947; Johnson *et al.*, 1970). These defects occurred over a 24-hr period and were usually lethal for affected embryos. The possible analogy of these neural tube defects to those in humans, such as meningomyelocele, has been suggested (Johnson, 1972). Failure of neural tube closure in the chick embryo is not associated with cellular necrosis of the neural tube, notochord, or neural crest. Viral antigen was not demonstrated in affected neural tissues, but was present diffusely in chorionic and amniotic membranes as well as focally in the primitive myocardium and gut (Johnson *et al.*, 1970). It was suggested that the neural tube was probably adversely affected by an indirect mechanism which altered early organogenesis. Since the neural tube was avascular at the stage of development under study, it was considered that amniotic fluid may have a primary role in its early formation. Thus infection of the extraembryonic membranes by influenza A virus might have altered this fluid. An alternative mechanism suggested that nonneural ectoderm may in some indirect manner affect neural tube closure.

It was mentioned earlier that Newcastle disease virus (NDV) likewise produced defects of neural tube closure and flexure. This probably occurred by a more direct mechanism than influenza A, since NDV antigen was detectable in cells of the caudal neural tube by the immunofluorescent antibody technique (Williamson *et al.*, 1965).

B. Postnecrotic and Cavitary Changes Following Acute Encephalitis

1. Cystic Malformations

Bluetongue virus caused a severe acute encephalitis in fetal lambs which reflected a highly selective infection of primordial cells in the subventricular layer of the telencephalon. Bluetongue virus demonstrated an important feature of certain intrauterine and newborn infections, that of an exquisite cellular specificity. The culmination of this infection in the gross cavitary cerebral defects noted was preceded by a well-defined series of events. Examination of the static end-stage gross defects, however, demonstrated no evidence of prior infection. Instead they closely resembled noninflammatory defects of embryogenesis or degenerative changes.

The nature of the bluetongue virus–cell relationship is still incompletely understood. It is not known, for instance, why other immature dividing cell populations, such as the external granular layer of the cerebellum and immature cells of the retina, were not pathologically affected.

IV. POSSIBLE VIRAL-INDUCED TERATOLOGIC MECHANISMS OCCURRING AT THE CELLULAR LEVEL

There are several well-defined mechanisms by which viruses can damage host cells. These may range from acute cell death on through the induction of persistent noncytolytic infection. Host tissues can be injured by other means as well, such as through alterations of normal mitotic rates or via immune-mediated pathogenetic mechanisms. Still other more obscure mechanisms may be involved in viral teratogenesis.

A. Acute Cell Death

Acute cell death is one of the most common ways by which viral-induced malformations arise. Although the specific nature of the anatomic defects may vary according to which cells are affected, the destruction of critical cell populations without their replacement underlies many models under consideration. Frequently, cellular necrosis was accompanied by host reparative mechanisms which indirectly resulted in end-stage defects indistinguishable from classic defects of embryogenesis.

The parvoviruses demonstrated very well the role of acute cell death within discrete CNS cell populations in determining the specific nature of viral malformations. Feline panleukopenia (FPL) and rat virus (RV) infection of the external granular layer of the cerebellar cortex in susceptible hosts caused the development of cerebellar hypoplasia. Following the acute infection and inflammatory response, tissue repair mechanisms replaced the damaged surfaces with tissue components not readily distinguished as of inflammatory or glial origin. The end result was one resembling a degenerative process. It is noteworthy that before the viral etiology of FPL-induced cerebellar hypoplasia was elucidated, the defect was considered to be of genetic or toxic origin.

The acute encephalitis which is ultimately responsible for the development of bluetongue virus encephalopathies further exemplified the importance of acute cellular destruction as a pathogenetic mechanism in viral teratology. With bluetongue virus the highly selective acute infection which resulted in death of the subependymal cells of the forebrain proceeded to the noninflammatory cavitary defects of hydranencephaly and porencephaly. Although the nature of the resultant defects caused by bluetongue virus are clearly different from those caused by the parvoviruses, they both occurred secondary to acute cell death. RV and bluetongue virus adapted to hamsters infect the same cell populations in neonatal hamsters and replicate comparable amounts of infectious virus (Becker *et al.*, 1974). However, strikingly different pathological reactions are evoked, and the sequelae are likewise different, as already described. It was not clear why rat virus did not evoke an inflammatory response while bluetongue virus did. It was suggested, however, that the basis for cellular susceptibility for these two viruses infecting the same cell populations was different. Cellular susceptibility to rat virus may be related to increased mitotic activity, while that to bluetongue virus may be related to a lack of cellular differentiation (Winn *et al.*, 1974).

Obstructive hydrocephalus in susceptible newborn rodents following infection by mumps virus, reovirus, and parinfluenza type 2 also occurred by means of acute cytolytic changes. As described earlier, virtually all of these infections were characterized by an acute ependymitis, with subsequent resolution. Virus was readily demonstrable during the acute stages of infection, but this was no longer possible once the gross defects were evident. They then resembled classic defects of embryogenesis, including agenesis and atresia.

B. Persistent Noncytolytic Infection

Certain virus infections involve a persistent virus–cell interaction in which the parasitized elements are not necessarily destroyed. The effects of such an infection may range from no apparent deleterious effect to one in which hypoplastic organs result. The virus effector mechanism may include inhibition of normal mitotic rates by inhibitory proteins (Plotkin and Vaheri, 1967) or possibly other unknown mechanisms. Impairment of cellular growth potential may be involved in the production of virus-induced malformations in the fetal pig caused by HCV and possibly in human malformations caused by rubella virus.

A variety of multisystem defects may occur in animals infected with HCV. The CNS is often severely involved. Cerebral and cerebellar hypoplasia probably result from a persistent noncytocidal infection with HCV which presumably interferes with the normal development and maturation of CNS tissues. Hypomyelinogenesis, a hallmark of HCV intrauterine infection, may result from persistent infection of the oligodendrocyte. This cell is responsible for the formation and maintenance of myelin within the CNS. Piglets with CNS disease continue to shed virus in the urine for weeks after birth, even in the presence of maternally acquired specific antibody.

Some features of rubella infection in the human fetus suggested that a persistent, noncytocidal infection may be present. Virus may be shed for months after the birth of congenitally affected infants, and virus has been isolated from cataracts 3 years after birth. Direct evidence for the existence of persistent infection of the fetal brain in congenital rubella, however, is lacking. Furthermore, the CNS cellular specificity of rubella virus is unknown. Recently, however, it has been reported that chronic panencephalitis has occurred more than 10 years after congenital rubella virus infection. Rubella virus was recovered from the brain of one of these patients, and antibody to rubella virus was found in the cerebrospinal fluid.

C. Immune-Mediated Cellular Damage

The classical model for immune-mediated CNS damage is that of LCM infection in the rodent. Anti-thymocyte antibody passively administered to virus-susceptible animals prevented development of cerebellar hypoplasia. Clinical disease was also prevented by a variety of immunosuppressive treatments, including X irradiation (Hotchin and Weigard, 1961), corticosteroids (Larsen, 1969; Hotchin and Cinits, 1958), chlorambucil (Larsen, 1969), neonatal thymectomy (East et al., 1964) and antilymphoid serum (Gledhill, 1967; Hirsch et al., 1967). Further support for immune-mediated cerebellar hypoplasia caused by LCM is provided by the demonstration that CNS disease can be induced by passively administered immune splenocytes.

Late disease caused by LCM consisted of chronic glomerulonephritis due to deposition of immune complexes in renal glomeruli (Hotchin and Collins, 1964; Oldstone and Dixon, 1969). Viral antigen plus specific antibody and complement were demonstrated in renal glomeruli and were also found in the peripheral circulation as immune complexes. The precise immunopathogenetic mechanisms responsible for CNS disease, however, remain to be determined. Immune complexes have not been demonstrated within the CNS.

D. Inhibition of Mitosis

Rubella virus may possibly cause organ hypoplasia by an inhibition of normal mitotic rates. Inhibition of cellular proliferation has been demonstrated *in vitro* by cells obtained from congenitally infected children. This may result from the elaboration of a protein which is inhibitory to other cells, such as demonstrated in fibroblast tissue culture cells obtained from children with congenital rubella (Plotkin and Vaheri, 1967).

As mentioned earlier, HCV may exist as a persistent noncytolytic infection. Organ hypoplasia is thought to result at least in part from an inhibition of normal mitotic rates by HCV.

E. Chromosomal Abnormalities

Many different viruses have caused chromosomal abnormalities *in vitro*. Herpes virus (Boiron *et al.*, 1966; Michols, 1966), rubella (Plotkin *et al.*, 1965), and CMV (Hartmann and Brunnemann, 1972) have been associated with alterations of chromosomal structure when observed in cell culture. However, the changes are relatively nonspecific, and their relationship to viral teratogenic potential has not been established. Likewise, *in vivo* chromosomal changes caused by rubella virus (Michols, 1966) have not been correlated with pathological changes in the host.

F. Vasculitis

The exact role of vasculitis in producing malformations of the CNS has not been clearly defined. It has been suggested, however, that vasculitis of the cerebral circulation during intrauterine infection may cause microinfarcts with cerebral calcifications in congenital rubella (Singer *et al.*, 1967). It is possible that a similar mechanism could be at least partly responsible for cerebral calcifications and organ hypoplasia seen in congenital toxoplasmosis, CMV, and herpes virus infections (Catalano and Sever, 1971). Vasculitis may also occur in hog cholera virus infection (Emerson and Delez, 1965) and bluetongue virus infection (Winn *et al.*, 1974).

G. Nonspecific Factors

The significance of factors not related directly to viral infection on the production of malformations and lesions is poorly understood. Constitutional but nonspecific effects of infection, such as fever and possible production of toxic metabolites, may theoretically affect the placental fetal unit. The role of such factors in viral teratogenesis remains to be defined, but cannot be ruled out at this time.

V. PRINCIPLES OF TERATOLOGY AS APPLIED TO INFECTIOUS AGENTS

Wilson (1973) has proposed six general principles of teratology as they relate to the developing embryo. Their application to infectious agents has been previously illustrated (Mims, 1968). Table 2 is a list of these principles and their relevance to teratology caused by infectious agents with particular reference to rubella virus.

A. Susceptibility to Teratogen Depends on Genotype of Conceptus

The ability of a teratogenic agent to cause malformations frequently depends on the genetic constitution of the host. In the broadest sense, this is evident at the species level. Rubella virus teratology is limited to the human embryo, and animal models have not been successfully produced.

B. Susceptibility to Teratogen Depends on Developmental Stage When Exposure Occurs

The varying age-dependent susceptibility of the embryo to teratogenic agents is well known regarding physical and chemical agents. Indeed models of viral-induced teratology demonstrates this principle very well. The highest incidence of congenital rubella anomalies occurs when infection takes place during the first month, and may be responsible for congenital defects in approximately 50% of infants subsequently born. About 22% of infants were congenitally affected following infection during the second month, and 6–8% of infants were malformed following maternal infection during the third month. As with physicochemical teratogens, no teratologic effects were noted when rubella infection occurred prior to blastula formation. This was supported by the observation that only one of 47 infants born following maternal rubella infection which occurred within two weeks of the postmenstrual period had evidence of congenital rubella. The period of greatest susceptibility to congenital rubella is during differentiation and early organogenesis.

C. Teratogenic Agents Cause Malformations According to Specific Pathogenetic Mechanisms

The concept of "agent specificity" has been applied to situations where a particular teratogenic agent has been associated with a recognizable pattern of

Table 2. Principles of Teratology as Applied to Infectious Agents

	Physicochemical agents	Rubella virus
1. Susceptibility to teratogenesis depends on the genotype of conceptus.	Yes	Yes, limited to human fetus
2. Relative susceptibility to teratogenesis is a function of developmental stage when exposure occurs.	Yes	Yes
a. Predifferentiation	No	No
b. Early differentiation	Yes	Yes
c. Advanced organogenesis	No	No
3. Teratogenic agents act in specific ways (mechanisms) on developing cells and tissues to initiate sequence of abnormal developmental events (pathogenesis).	Yes	Probably
4. The access of the teratogen to fetal tissues may be influenced by maternal factors.	Yes	Yes; (a) Maternal specific antiviral immunity confers protection; (b) Placenta may serve as physical barrier to infectious agents.
5. The manifestations of deviant development are death, malformation, growth retardation, and functional deficit.	Yes	Yes
6. Manifestations of deviant development increase in frequency and degree as dosage increases, from the no-effect to the totally lethal level. Corollary: High embryotoxicity associated with minimal maternal toxicity.	Yes	Unknown
	Yes	Yes

defects. This is relevant to the problem of how teratogenic agents initiate abnormal development and how these early changes eventually culminate in specific malformations. Laboratory models of experimental viral teratogenesis have been particularly useful in attempting to solve these problems. Studies of bluetongue virus encephalopathies, for example, provided excellent samples of how early changes induced by teratogens may be related to the pathogenesis of end-stage CNS malformations. By means of serially studied animals, it has been demonstrated that gestational age-dependent CNS anomalies may be caused in a highly predictable fashion by bluetongue virus. The age-dependent defects were initiated by a highly specific infection and subsequent destruction of the subventricular cells of the telencephalon which eventually terminated in the formation of bilateral symmetrical cavitary defects of the cerebral hemispheres. The exquisite cellular specificity of other viral infections of the developing CNS was also demonstrated by mumps virus and reovirus infection of ependymal cells lining cerebral aqueducts. These acute infections resulted in obstructive hydrocephalus. FPL infection of the external granular layer of the cerebellar cortex resulted in cerebellar hypoplasia and intractable ataxia which was a direct consequence of a highly selective infection resulting in acute cellular death. Human viral-induced congenital malformations, however, do not lend themselves to study as readily as experimental models.

The study of virtually all models of viral-induced teratology has demonstrated that a final common pathway for teratogenesis does indeed exist for them, as suggested by Wilson for physical and chemical teratogens (Wilson, 1973). All of the teratologic effects caused by infectious agents appear to occur secondary to cell death or impaired cell growth potential. This in turn may be related to an insufficient number of cells or cell products to carry out normal morphogenesis or to serve normal organ function.

D. Maternal Factors May Influence the Relative Teratogenicity of an Agent

Developing tissues are more sensitive to adverse environmental influences than are mature somatic tissues (Wilson, 1959, 1973). The embryo, however, may be protected from teratogenic agents by at least several different mechanisms which relate primarily to the protective effects afforded by the maternal physiology and anatomy. Toxic and potentially teratogenic chemicals may fail to reach the embryo in teratogenic doses as a result of detoxification, excretion, and other mechanisms provided by the mother. Likewise the placental unit is considered to play a role as a physical barrier to infectious agents of potential danger to the fetus, although viruses are capable of crossing the placenta under certain conditions. The maternal immunologic status may also be an important modulator of fetal infection. Maternal antibody to a specific virus, such as rubella, may be associated with marked protection conferred to the fetus. Indeed, widespread use of live attenuated rubella vaccine has been associated with a marked decrease in congenital rubella since its implementation.

E. The Manifestations of Deviant Development Are Death, Malformations, Growth Retardation, and Functional Deficit

The development of malformations is but one of four manifestations of deviant intrauterine development. These include death, malformation, growth retardation, and functional deficit. During early organogenesis, embryonic death and malformations may be associated with most physical and chemical teratogenic agents (Wilson, 1973). This is also true of intrauterine rubella virus infection.

The pattern observed with congenital rubella is consistent with the fact that both malformations and mortality usually decrease with increasing gestational age with regard to physicochemical teratogens. It is unknown, however, whether intrauterine death reflects severe degrees of malformations incompatible with life or whether death occurs by another mechanism independent of malformation. As intrauterine development progresses, death and malformation attributable to rubella infection diminish. The relative likelihood of growth retardation and functional deficit rises correspondingly, however, with increasing gestational age of rubella infection.

F. Manifestations of Deviant Development Increase in Frequency and Degree as Dosage Increases, from the No-Effect to the Totally Lethal Level

Increasing the dosage of a physical or chemical teratogen may be associated with greater risk to the fetus. The possible dose–response effect of viral teratogens, however, has not been adequately explored. A corollary to the dose–response principle of teratogenic agents concerns the relative toxicity of a teratogen for the mother compared to embryo. Experimental teratology has frequently demonstrated the association of high embryotoxicity with minimal maternal toxicity for many teratogenic agents. This observation has also been borne out with congenital rubella and other congenital infections. Mild to absent maternal illness may be associated with devastating effects on the fetus following rubella infection. Mild maternal infection with bluetongue virus vaccine may be associated with the severe fetal encephalopathies already discussed. Maternal FPL infection may also be associated with mild illness, while severe permanent cerebellar ataxia occurs in the progeny.

VI. DISCUSSION

The association of maternal rubella virus infection with cataracts in infants was first reported by Gregg (1941). This represented the first evidence that infectious agents were capable of causing congenital deformities. The expanded congenital rubella syndrome was subsequently described which included a vast array of multisystem malformations and lesions. Numerous additional associations between infectious agents and malformations and lesions following infection of the developing organism have since been described. Selected examples are listed in Table 1. They have included both naturally occurring and experimental models which have been reviewed in detail (Blattner, 1974; Catalano and Sever, 1971; Fuccillo and Sever, 1973; Johnson and Mims, 1967; Johnson, 1972; Kurent and Sever, 1975; Mims, 1968; Plotkin, 1975; Sever, 1971; Sever and White, 1968). Other infectious agents such as varicella-zoster virus (Siegel, 1973; Srabstein *et al.*, 1974) and mycoplasma may also be capable of causing developmental defects (Bøe *et al.*, 1973; Kohn *et al.*, 1975).

Except for the developmental defects associated with *Toxoplasma gondii* and *Treponema pallidum*, all of the extensively studied infections have been viral in nature. These have included a number of experimental models for viral-induced teratology. Practically all the associations between infectious agents and malformations have involvement of the CNS as a major component. The viruses capable of causing malformations are represented by several major virus groups and are not limited to one specific virus type.

Viruses may cause malformations by several different mechanisms, but each is usually characteristic of a particular agent. The resultant anatomic defect may be indistinguishable from classical defects of embryogenesis, including agenesis, atresia, aplasia, and hypoplasia. Degenerative changes of the cerebral hemispheres may occur, such as hydranencephaly and porencephaly. Malformations of unknown etiology, but similar to those caused by viruses, have been previously attributed to genetic, toxic, and vascular etiologies, particularly as they occur in human. Viral-induced anomalies frequently show no indication of antecedent infection or inflammatory response and thus have resembled developmental defects. Frequently there is no trace of infectious virus or viral antigen at the time end-stage defects were apparent.

The final common pathway by which viral-induced malformations occur appears to be by way of an adverse effect on the cell. A similar scheme probably operates during physicochemical teratogenesis, as well. Viral teratogenesis may occur by a variety of cytopathogenetic mechanisms. These include acute cell death as a direct result of virus infection, persistent non-cytolytic infection, cell death following immunologic damage, and alterations in normal mitotic rates. The specific teratologic roles, if any, of viral-induced chromosomal abnormalities and vasculitis, are not known for certain at this time.

Teratologic principles which are operational in physicochemical teratogenesis also apply to viral teratology. These include the influence of genetic constitution of the host in determining susceptibility to a specific teratogen. Such susceptibility is probably present at least at the species level. Susceptibility to a viral teratogen may vary greatly according to the age of the fetus or the newborn animal. Furthermore, the specific nature of the malformations may be a direct function of the developmental stage during which infection occurred. A specific virus may cause malformations of a predictable type and by pathogenetic mechanisms characteristic for that agent. Maternal factors such as specific maternal immunity may prevent fetal infection by neutralizing viral infectivity.

The consequences of intrauterine infection may include any of four basic features. These are death, malformation, growth retardation, and functional deficit; they appear to reflect the stage of development when infection occurs. Finally, the teratology of infectious diseases parallels that of physicochemical teratogenesis in another regard. The developing fetus may demonstrate severe developmental defects following exposure to a teratogenic infectious agent, yet maternal illnesss or toxicity may be minimal, if present at all.

One of the most intriguing observations resulting from the study of the teratology of infectious diseases regards the possible relationship of experimental animal models to human disease. Virtually all of the experimental models for CNS malformations have analogs in human pathology. The majority of human malformations have been generally ascribed to developmental defects caused by genetic, toxic, or vascular factors. The animal models could have as easily been attributed to similar etiologies were it not for their proven viral origin as demonstrated under laboratory conditions. The gross and microscopic anatomy of the end-stage noninflammatory defects would certainly not enable one to distinguish them on etiologic grounds. This further emphasizes the need for exhaustive efforts to demonstrate infectious etiologies for human CNS malformations.

ACKNOWLEDGMENT

The authors gratefully acknowledge the assistance of Ms. Nelva Reckert in the preparation of this manuscript.

REFERENCES

Altschuler, G., 1973, Toxoplasmosis as a cause of hydranencephaly, *Am. J. Dis. Child.* **125**:251.

Becker, L. E., Narayan, O., and Johnson, R. T., 1974, Comparative studies of viral infections of the developing forebrain. I. Pathogenesis of rat virus and bluetongue vaccine virus infection in neonatal hamsters, *J. Neuropathol. Exp. Nerol.* **33**:519.

Beverley, J. K. A., 1959, Congenital transmission of toxoplasmosis through successive generations of mice, *Nature* **183**:1348.

Beverley, J. K. A., 1973, Toxoplasmosis, *Br. Med. J.* **2**:475.

Blattner, R. J., 1974, The role of viruses in congenital defects, *Am. J. Dis. Child.* **128**:781.

Bøe, Ø., Diderichsen, J., and Matre, R., 1973, Isolation of Mycoplasma hominis from cerebrospinal fluid, *Scand. J. Infect. Dis.* **5**:285.

Boiron, M., Tanzer, J., Thomas, M., and Hanope, A., 1966, Early diffuse chromosome alterations in monkey kidney cells infected *in vitro* with herpes simplex virus, *Nature* **209**:737.

Boughton, C. R., 1970, Toxoplasmosis, *Med. J. Aust.* **2**:418.

Brown, T. T., DeLahunta, A., Scott, F. W., Kahrs, R. F., McEntee, K., and Gillespie, J. H., 1973, Virus induced congenital anomalies of the bovine fetus. II. Histopathology of cerebellar degeneration (hypoplasia) induced by the virus of bovine viral diarrhea-mucosal disease, *Cornell Vet.* **63**:561.

Catalano, L. W., and Sever, J. L., 1971, The role of viruses as causes of congenital defects, *Annu. Rev. Microbiol.* **25**:255.

Cole, G. A., Gilden, D. H., Monjan, A. A., and Nathanson, N., 1971, Lymphocytic choriomeningitis virus: Pathogenesis of acute central nervous system disease, *Fed. Proc.* **30**:1831.

Cooper, L. Z., Ziring, P. R., and Ockerse, A. B., 1969, Rubella: Clinical manifestations and management, *Am. J. Dis. Child.* **118**:18.

Dekaban, A., O'Rourke, J., and Corman, T., 1958, Abnormalities in offspring related to maternal rubella during pregnancy, *Neurology* **8**:387.

Desmonts, G., and Couvreur, J., 1974, Congenital toxoplasmosis, *N. Engl. J. Med.* **290**:1110.

Doege, T. C., and Kim, K. S. W., 1967, Studies of rubella and its prevention with immune globulin, *J. Am. Med. Assoc.* **200**:584.

East, J., Parrott, D. M. V., and Seamer, J., 1964, The ability of mice thymectomized at birth to survive infection with lymphocytic choriomeningitis virus, *Virology* **22**:160.

Eichenwald, H. F., 1960. A study of congenital toxoplasmosis with particular reference on clinical manifestations, sequelae, and therapy, *in: Human Toxoplasmosis* (J. C. Siim, ed.), pp. 41–49, Williams and Wilkins, Baltimore.

Emerson, J. L., and Delez, A. L., 1965, Cerebellar hypoplasia, hypomyelinogenesis, and congenital tremor of pigs associated with prenatal hog cholera vaccination of sows, *J. Am. Vet. Med. Assoc.* **147**:47.

Florman, A. L., Gershon, A. A., Blackett, P. R., and Nahmias, A. J., 1973, Intrauterine infection with herpes simplex virus—resultant congenital malformations, *J. Am. Med. Assoc.* **225**:129.

Fuccillo, D. A., and Sever, J. L., 1973, Viral teratology, *Bacteriol. Rev.* **37**:19.

Gillespie, J. H., Bartholomew, P. R., Thompson, R. G., and McEntee, K., 1967, The isolation of noncytopathic virus diarrhea virus from two aborted bovine fetuses, *Cornell Vet.* **57**:564.

Gledhill, A. W., 1967, Protective effect of anti-lymphocyte serum on murine lymphocyte choriomeningitis, *Nature* **214**:178.

Gregg, N., 1941, Congenital cataracts following German measles in the mother, *Trans. Ophthalmol. Soc. Aust.* **3**:35.

Hamburger, V., and Habel, K., 1947, Teratogenic and lethal effects of influenza A and mumps viruses on early chick embryos, *Proc. Soc. Exp. Biol. Med.* **66**:608.

Hanshaw, J. B., 1966a, Congenital and acquired cytomegalovirus infection, *Pediatr. Clin. North Am.* **13**:279.

Hanshaw, J. B., 1966b, Cytomegalovirus complement fixing antibody in microcephaly, *N. Engl. J. Med.* **275**:476.

Hanshaw, J. E., 1971, Congenital cytomegalovirus infection. A fifteen year perspective, *J. Infect. Dis.* **123**:555.

Hartmann, M., and Brunnemann, H., 1972, Chromosome abberations in cytomegalovirus-infected human diploid cell culture, *Acta Virol.* **16**:176.

Hildebrandt, R. J., Sever, J. L., Margileth, A. M., and Callaghan, C. A., 1967, Cytomegalovirus in the normal pregnant woman, *Am. J. Obstet. Gynecol.* **98**:1125.

Hirsch, J. J., Murphy, F. A., Russe, H. P., and Hicklin, M. D., 1967, Effects of thymocyte serum on lymphocytic choriomeningitis virus infection in mice, *Proc. Soc. Exp. Biol. Med.* **125**:980.

Holder, W. R., and Knox, J. M., 1972, Syphilis in pregnancy, *Med. Clin. North Am.* **58**:1151.

Hotchin, J., and Cinits, P., 1958, Lymphocytic choriomeningitis infection of mice as a model for the study of latent virus infection, *Can. J. Microbiol.* **4**:149.

Hotchin, J., and Collins, D., 1964, Glomerulonephritis and late onset disease of mice following neonatal virus infection, *Nature* **203**:1357.

Hotchin, J., and Weigard, H. 1961, The effect of pre-treatment with x-rays on the pathogenesis of lymphocytic choriomeningitis in mice. I. Host survival, virus multiplication and leukocytosis, *J. Immunol.* **87**:675.

Howell, P. G., Verwoerd, D. W., 1971, Bluetongue virus, *Virol. Monogr.* **9**:35.

Hume, O. S., 1972, Toxoplasmosis and pregnancy, *Am. J. Obstet. Gynecol.* **114**:703.

Jensen, R., Kennedy, P., Ramsey, F., McKercher, D., Collier, J. C., and Fox, F. H., 1968, Report of the ad hoc committee on terminology for the symposium on immunity to the bovine respiratory disease complex, *J. Am. Vet. Med. Assoc.* **152**:940.

Johnson, K. P., and Byington, D. P., 1972, Hog cholera virus: Multiple malformations produced by persistent tolerant infection of fetal swine, *Teratology* **5**:259.

Johnson, K. P., and Johnson, R. T., 1972, Granular ependymitis. Occurrence in myxovirus infected rodents and prevalence in man, *Am. J. Pathol.* **67**:511.

Johnson, K. P., Ferguson, L. C., Byington, D. P., and Redman, D. R., 1970, Multiple fetal malformations due to persistent viral infection. I. Abortion, intrauterine death, and gross abnormalities in fetal swine infected with hog cholera vaccine virus, *Lab. Invest.* **30**:608.

Johnson, K. P., Klasnja, R., and Johnson, R. T., 1971, Neural tube defects of chick embryos: An indirect result of influenza A virus infection, *J. Neuropathol. Exp. Neurol.* **30**:68.

Johnson, R. T., 1972, Effects of viral infection on the developing nervous system, *N. Engl. J. Med.* **287**:599.

Johnson, R. T., and Johnson, K. P., 1968, Hydrocephalus following viral infection: The pathology of aqueductal stenosis developing after experimental mumps virus infection, *J. Neuropathol. Exp. Neurol.* **27**:591.

Johnson, R. T., and Mims, C., 1967, Pathogenesis of viral infections of the nervous system, *N. Engl. J. Med.* **278**:23–30, 84–92.

Johnson, R. T., Johnson, K. P., and Edmonds, C. J., 1967, Virus-induced hydrocephalus: Development of aqueductal stenosis in hamsters after mumps infection, *Science* **157**:1006.

Kahrs, R. F., Scott, F. W., and deLahunta, A., 1970a, Bovine viral diarrhea-mucosal disease, abortion, and congenital cerebellar hypoplasia in a dairy herd, *J. Am. Vet. Med. Assoc.* **156**:851.

Kahrs, R. F., Scott, F. W., and deLahunta, A., 1970b, Congenital cerebellar hypoplasia and ocular defects in calves following bovine viral diarrhea-mucosal disease infection in pregnant cattle, *J. Am. Vet. Med. Assoc.* **156**:1443.

Kenrick, K. E., Slinn, R. F., Dorman, D. C., and Menser, M. A., 1965, Immunoglobulin and rubella virus antibodies in adults with congenital rubella, *Lancet* **1**:548.

Kilham, L., and Margolis, G., 1964, Cerebellar ataxia in hamsters inoculated with rat virus. *Science* **143**:1047.

Kilham, L., and Margolis, G., 1965, Cerebellar disease in cats induced by inoculation of rat virus, *Science* **148**:244.

Kilham, L., and Margolis, G., 1966, Viral etiology of spontaneous ataxia of cats, *Am. J. Pathol.* **48**:991.

Kilham, L., and Margolis, G., 1967, Congenital infections of cats and ferrets by feline panleucopenia virus manifested by cerebellar hypoplasia, *Lab. Invest.*. **17**:465.

Kilham, L., and Margolis, G., 1969, Hydrocephalus in hamsters, ferrets, rats, and mice following inoculations with reovirus type 1. I. Virologic studies, *Lab Invest.* **21**:183.

Kilham, L., and Margolis, G., 1970, Pathogenicity of minute virus of mice (MVM) for rats, mice, and hamsters, *Proc. Soc. Exp. Biol. Med.* **133**:1447.

Kilham, L., Margolis, G., and Colby, E. D., 1971, Cerebellar ataxia and its congenital transmission in cats by feline panleukopenia virus, *J. Am. Vet. Assoc.* **158**:888.

Kohn, D. F., Chou, S. M., and Kirk, B. E., 1975, Mycoplasma induced hydrocephalus in newborn rats. Presented at the 51st Meeting of the American Association of Neuropathologists, New York, New York.

Korones, S. B., Ainger, L. E., Monif, G. R. G., Roane, J., Sever, J. L., and Fuste, R., 1965, Congenital rubella syndrome: New clinical aspects with recovery of virus from affected infants, *J. Pediatr.* **67**:166.

Kumar, M. L., Nankervis, G. A., and Gold, E., 1973, Inapparent congenital cytomegalovirus infection: A follow-up study. *N. Engl. J. Med.* **288**:1370.

Kurent, J. E., and Sever, J. L., 1975, Pathogenesis of intrauterine infections of the brain, *in: Biology of Brain Dysfunction*, Vol. 3 (G. E. Gaull, ed.), pp. 307–341, Plenum Press, New York.

Larsen, J. H., 1969, The effect of immunosuppressive therapy on murine lymphocytic choriomeningitis infection, *Acta Pathol. Microbiol. Scand.* **77**:433.

Levitt, N. H., London, W. T., Kent, S. G., and Sever, J. L., 1973, In utero induction of cataracts and hydrocephalus in rhesus monkeys using Venezuelan equine encephalitis virus. Presented at the Annual Meeting of the American Society for Microbiology, Miami.

London, W. T., Kent, S. G., and Sever, J. L., 1973, Influenza virus—A teratogen in rhesus monkeys. Presented at the Annual Meeting of the Federation of the American Society for Experimental Biology, Atlantic City, New Jersey.

Lundstrom, R., 1962, Rubella during pregnancy: A follow-up study of children born after an epidemic of rubella, 1951, with additional information on prophylaxis and treatment of natural rubella. *Acta Paediatr.* **81**:(Suppl)133:1.

MacKenzie, J. S., and Houghton, M., 1974, Influenza infections during pregnancy: Association with congenital malformations and with subsequent neoplasms in children, and potential hazards of live virus vaccines, *Bacteriol. Rev.* **38**:356.

Margolis, G., and Kilham, L., 1968, In pursuit of an ataxic hamster or virus induced cerebellar hypoplasia, *in: The Central Nervous System, Some Experimental Models of Neurological Diseases, International Academy of Pathology Monograph No. 9*, pp. 157–183, Williams and Wilkins, Baltimore.

Margolis, G., and Kilham, L., 1969a, Hydrocephalus in hamsters, ferrets, rats, and mice following inoculations with reovirus type 1. II. Pathologic studies, *Lab. Invest.* **21**:189.

Margolis, G., and Kilham, L., 1969b, Experimental virus induced hydrocephalus. Relation to pathogenesis of the Arnold-Chiari malformation, *J. Neurosurg.* **31**:1.

Margolis, G., and Kilham, L., 1970, Cerebellar ontogenetic patterns: Key to a virus code, *in: The Cerebellum in Health and Disease* (W. S. Fields and W. D. Willis, Jr., eds.), pp. 353–379, Warren H. Green, St. Louis, Missouri.

Medearis, D. N., 1957, Cytomegalic inclusion disease. An analysis of the clinical features based on the literature and six additional cases, *Pediatrics* **19**:467.

Menser, M. A., Harley, J. D., Hertzberg, R., Dorman, O. C., and Murphy, A. M., 1967, Persistence of virus in lens for three years after prenatal rubella, *Lancet* **2**:387.

Michaels, R. H., and Mellin, G. W., 1960, Prospective experience with maternal rubella and the associated congenital malformations, *Pediatrics* **26**:200.

Michols, W. W., 1966, The role of viruses in the etiology of chromosomal abnormalities, *Ann. Hum. Genet.* **18**:81.

Mims, C. A., 1968, Pathogenesis of viral infections of the fetus, *Prog. Med. Virol.* **10**:194.

Monif, G. R. G., Sever, J. L., Schiff, G. M., and Traub, R. G., 1965, Isolation of rubella from products of conception, *Am. J. Obstet. Gynecol.* **91**:1143.

Monjan, A., Gilden, D. H., Cole, G. A., and Nathanson, N., 1971, Cerebellar hypoplasia in neonatal rats caused by lymphocytic choriomeningitis virus, *Science* **171**:194.

Montgomery, J. R., South, M. A., Rawls, W. E., Melnick, J. L., Olson, G. B., Dent, P. B., and Good, R. A., 1967, Viral inhibition lymphocyte response to phytohemagglutinin, *Science* **157**:1068.

Nahmias, A. J., and Dowdle, W. R., 1968, Antigenic and biologic differences in herpesvirus hominis, *Prog. Med. Virol.* **10**:110.

Nahmias, A. J., Dowdle, W. R., Josey, W. E., Naib, Z. M., Painter, C. M., and Luce, C., 1969, Newborn infection with herpesvirus hominis types 1 and 2, *J. Pediatr.* **75**:1194.

Narayan, O., and Johnson, R. T., 1972, Effects of viral infection on nervous system development. I. Pathogenesis of bluetongue virus infection in mice, *Am. J. Pathol.* **68**:1.

Narayan, O., McFarland, H. F., and Johnson, R. T., 1972, Effects of viral infection on nervous system development. II. Attempts to modify bluetongue virus-induced malformations with cyclophosphamide and antithymocyte serum, *Am. J. Pathol.* **68**:15.

Olafson, P., MacCallum, A. D., and Fox, F. H., 1946, An apparently new transmissible disease of cattle, *Cornell Vet.* **36**:205.

Oldstone, M. B. A., and Dixon, F., 1969, Pathogenesis of chronic disease associated with persistent lymphocytic choriomeningitis viral infection. I. Relationship of antibody production to disease in neonatally infected mice, *J. Exp. Med.* **129**:483.

Oldstone, M. B. A., and Dixon, F., 1970, Pathogenesis of chronic disease associated with persistent lymphocytic choriomeningitis viral infection. II. Relationship of the anti-lymphocytic choriomeningitis immune response to tissue injury in chronic lymphocytic choriomeningitis disease, *J. Exp. Med.* **131:**1.

Olson, G. B., South, M. A., and Good, R. A., 1967, Phytohemagglutinin unresponsiveness of lymphocytes from babies with congenital rubella, *Nature* **214:**695.

Osburn, B. I.; Silverstein, A. M., Prendergast, R. A., Johnson, R. T., and Parshall, C. J., 1971a, Experimental viral-induced congenital encephalopathies. I. Pathology of hydranencephaly and porencephaly caused by bluetongue vaccine virus, *Lab. Invest.* **25:**197.

Osburn, B. I., Johnson, R. T., Silverstein, A. M., Prendergast, R. A., Jochim, M. J., and Levy, S. E., 1971b, Experimental viral-induced congenital encephalopathies. II. The pathogenesis of bluetongue virus infections in fetal lambs, *Lab. Invest.* **25:**206.

Peterson, D. R., Tronca, E., and Bonin, P., 1972, Human toxoplasmosis prevalence and exposure to cats, *Am. J. Epidemiol.* **96:**215.

Plotkin, S. A., 1975, Routes of fetal infection and mechanisms of fetal damage, *Am. J. Dis. Child.* **129:**444.

Plotkin, S. A., and Vaheri, A., 1967, Human fibroblasts infected with rubella virus produce a growth inhibitor, *Science* **156:**659.

Plotkin, S. A., Boue, A., and Boue, J. E., 1965, The *in vitro* growth of rubella virus in human embryo cells, *Am. J. Epidemiol.* **81:**71.

Remington, J. S., Jacobs, L., and Melton, M. L., 1961, Congenital transmission of toxoplasmosis from mother, *J. Infect. Dis.* **108:**163.

Reynolds, D. W., Stagno, S., Stubbs, K. G., Dahle, A. J., Livingston, M. M., Saxon, S. S., and Alford, C. A., 1974, *N. Engl. J. Med.* **290:**291.

Richards, W. P. C., and Cordy, D. R., 1967, Bluetongue virus infections pathologic responses of nervous systems in sheep and mice, *Science* **156:**530.

Robertson, G. G., Williamson, A. P., and Blattner, R. J., 1955, A study of abnormalities in early chick embryos inoculated with newcastle disease virus, *J. Exp. Zool.* **129:**5.

Sanger, V. L., and Cole, C. R., 1955, Toxoplasmosis. VI. Isolation of toxoplasma from milk, placentas, and newborn pigs of asymptomatic carrier sows, *Am. J. Vet. Assoc.* **16:**536.

Scott, F. W., Kahrs, R. F., deLahunta, A., Brown, T. T., McEntee, K., and Gillespie, J. H., 1973, Virus induced congenital anomalies of the bovine fetus. I. Cerebellar degeneration (hypoplasia) ocular lesions and fetal mummification following experimental infection with bovine viral diarrhea mucosal disease virus, *Cornell Vet.* **63:**536.

Sever, J. L., 1971, Virus infections and malformations, *Fed. Proc.* **30:**114.

Sever, J. L., and White, L. R., 1968, Intrauterine viral infections, *Annu. Rev. Med.* **19:**471.

Shope, R. E., 1968, Comments on bovine viral diarrhea-mucosal disease, *J. Am. Vet. Med. Assoc.* **152:**769.

Siegl, G., Hallauer, C., Novak, A., and Kronauer, 1971, Parvoviruses as contaminants of permanent human cell lines. II. Physiochemical properties of the isolated viruses, *Arch. Gesamte Virusforsch.* **35:**91.

Siegel, M., 1973, Congenital malformations following chickenpox, measles, mumps, and hepatitis, *J. Am. Med. Assoc.* **226:**1521.

Silverstein, A. M., 1962, Congenital syphilis and the timing of immunogenesis in the human fetus, *Nature* **194:**196.

Silverstein, A. M., 1964, Ontogeny of the immune response, *Science* **144:**1423.

Singer, D. B., Rudolph, A. J., Rosenberg, H. S., Rawls, W. E., and Boniuk, M., 1967, Pathology of the congenital rubella syndrome, *J. Pediatr.* **71:**665.

Soothill, J. F., Hayes, K., and Dudgeon, J. A., 1966, Immunoglobulin in congenital rubella, *Lancet* **1:**1385.

South, M. A., Tompkins, W. A. F., Morris, R., and Rawls, W. E., 1969, Congenital malformations of the central nervous system associated with genital (type 2) herpes virus, *J. Pediatr.* **75:**13.

Srabstein, J. C., Morris, N., Larke, R. P. B., de Sa, D. J., Castelino, B. B., and Sum, E., 1974, Is there a congenital varicella syndrome? *J. Pediatr.* **84:**239.

Starr, J. G., Bard, R. D., Jr., and Gold, E., 1970, Inapparent congenital cytomegalovirus infection, Clinical and epidemiologic characteristics in early infancy, *N. Engl. J. Med.* **282:**1075.

Swan, C., Tosterin, A. L., Moore, B., Mayo, H., and Black, B. H. B., 1943, Congenital defects in infants following infectious diseases during pregnancy, with special reference to relationship between German measles and cataract, deaf-mutism, heart disease and microcephaly, and to period of pregnancy in which occurrence of rubella is followed by congenital abnormalities, *Med. J. Aust.* **2:**201.

Swope, R. E., and Luedke, A., 1956, A mucosal disease in cattle in Pennsylvania, *J. Am. Vet. Med. Assoc.* **129:**111.

Toolan, H. W., 1960, Experimental production of mongoloid hamsters, *Science* **131:**1446.

Townsend, J. J., Baringer, J. R., Wolinsky, J. S., Malamud, N., Mednick, J. P., Panitch, H. S., Scott, R. A. T., Oshioro, L. S., and Cremer, N. E., 1975, Progressive rubella panencephalitis: Late onset after congenital rubella. *N. Engl. J. Med.* **292:**990.

Ward, G. M., 1971, Bovine cerebellar hypoplasia apparently caused by BVD-MD virus. A case report, *Cornell Vet.* **61:**179.

Weil, M. L., Itabashi, H. H., Cremer, N. E., Oshiro, L. S., Lennette, E. H., and Carnay, L., 1975, Chronic progressive panencephalitis due to rubella virus simulating SSPE, *N. Engl. J. Med.* **292:**994.

Wenger, F., 1967, Necrosis cerebral masiva del feto en casos de encefalitis equina Venezulano, *Invest. Clin.* **21:**13.

Williamson, A. P., Blattner, R. J., and Robertson, G. G., 1965, The relationship of viral antigen to virus-induced defects in chick embryos: Newcastle disease virus, *Dev. Biol.* **12:**498.

Wilson, J. G., 1973, *Environment and Birth Defects,* Academic Press, New York.

Winn, K., Becker, L. E., and Herndon, R. M., 1974, Comparative studies of rat virus and bluetongue vaccine virus infections. II. Electron microscopic studies of rat virus and bluetongue virus vaccine virus infections, *J. Neuropathol. Exp. Neurol.* **33:**530.

Young, S., and Cordy, D. R., 1964, An ovine fetal encephalopathy caused by bluetongue vaccine virus, *J. Neuropathol. Exp. Neurol.* **23:**635.

Nutritional Deficiencies and Excesses

7

LUCILLE S. HURLEY

I. INTRODUCTION

Dietary derangement as an agent of teratogenesis has special significance in the history of teratology. The production of congenital anomalies by deficiences of vitamin A and riboflavin marked the beginning of the experimental stage of this science. Since then, these and other nutritional manipulations have served as valuable tools in the development of models, as well as providing information of intrinsic interest in relation to both basic and clinical questions.

In addition to these considerations, there are other factors which make it important for a teratologist to be aware of the influence of nutrition on prenatal development. Although an investigator may be unaware of it, every teratological episode either in the laboratory or in nature involves nutrition at least to some extent, since all animals must derive certain essential compounds and elements from their environment in order to maintain body processes. Thus, unless a worker in the field of teratology is alert to the possible effects of nutritional deficiencies and excesses on developmental processes, interpretation of data may be distorted.

This chapter attempts to summarize present knowledge on the influence of nutritional factors during the prenatal period on development in mammals. The large literature on nutrition during the neonatal and early postnatal period has not been included.

LUCILLE S. HURLEY · Department of Nutrition, University of California, Davis, California 95616.

II. MAJOR NUTRIENTS

A. General Malnutrition

1. War and Famine

The effects of general malnutrition can be studied in human populations under severe conditions of famine such as have occurred during wars. This subject is well reviewed by Hytten and Leitch (1971). Smith (1947a,b) examined birth statistics from the Netherlands for a period of 8 months in 1944–1945 during World War II when food supplies in that country were extremely limited. Amenorrhea and irregularity of menstrual cycles were common leading to a marked decrease in birth rate. The data were not statistically significant since the sample size was too small. However, for infants conceived during famine, incidence of miscarriages and abortions, stillbirths, neonatal deaths, and malformations were all higher than for prewar births. It is thought that the adverse effects of malnutrition were diminished because the women were in a relatively well-nourished condition prior to the famine period.

More severe starvation occurred during the siege of Leningrad in World War II. Antonov (1947) has described the appalling effects of hunger at this period in combination with bombing, bitter cold, and extreme physical exertion. Although Antonov did not report an increase in congenital malformations, he noted that the condition of the infants at birth was very poor, with a high incidence of stillbirths and premature births. Birth weight was low, the infants were very inert, and neonatal mortality was high.

2. Clinical Studies

Clinical studies of maternal nutrition as related to the condition of the newborn infant have been difficult to conduct and have yielded some conflicting results. However, the consensus now indicates that the diet of pregnant women can affect development of the fetus (Committee Maternal Nutrition, 1970). An early well-controlled clinical study, done in the 1940s, showed a definite correlation between the quality of the mother's diet during pregnancy and the condition of the infant at birth (Burke et al., 1943). Furthermore, birth weights and birth length correlated positively with the amount of protein in the diet (Burke and Stuart, 1948). Similar findings were reported by Jeans et al. (1944) who also found increased incidence of premature births in poorly nourished women.

In more recent clinical investigations, cardiovascular abnormalities (Naeye, 1965) and retarded development of postnatal bone growth (Wilson et al., 1967; Tuncer, 1970) were seen in infants malnourished before birth.

About 10% of infants with fetal growth retardation had physiological problems postnatally that resulted from congenital anomalies (Wilson *et al.,* 1967).

Primrose and Higgins (1971) carried out a clinical study in Montreal in which the dietary history of pregnant women was recorded and nutrition education was given. If the women were poor, nutritional supplements were also provided. The incidence of stillbirths, perinatal mortality, and neonatal mortality was lower in the group with improved nutrition than in Quebec or Canada as a whole.

3. Significance of Low Birth Weight

The studies so far available have not provided convincing proof that general maternal malnutrition during pregnancy in humans increases the incidence of gross congenital malformations. It should be remembered, however, that other parameters of fetal development such as birth weight may provide an indication of the presence of congenital anomalies not necessarily grossly visible at birth. There is, in fact, some evidence that low birth weight is correlated with congenital anomalies (van den Berg·and Yerushalmy, 1966; Scott and Usher, 1966). Thus, fetal malnutrition as evidenced by low birth weight is associated with increased congenital anomalies. Low birth weight is in general an important index of the condition of the newborn. There is a very high correlation between low birth weight and perinatal mortality (Brimblecombe *et al.,* 1968; Bergner and Susser, 1970). Low birth weight is correlated with poverty and certain racial groups, and these two factors are themselves correlated with malnutrition (Bergner and Susser, 1970; Rosa and Turshen, 1970; Naeye *et al.,* 1969, 1971). In human studies, it is seldom possible to differentiate these various factors.

4. Animal Studies

The effects of undernutrition, that is, restriction of total food intake, have been studied extensively in animals (Roeder and Chow, 1972). Influences of prenatal maternal restriction on postnatal development of central nervous system function will be considered below (see Section II. A. 5). Chow (1964; Chow and Lee, 1964) showed that when the food intake of pregnant and lactating rats was restricted to 75% or 50% of the amount normally eaten, the offspring were permanently stunted in growth. Similar effects were seen if the dietary restriction was limited to gestation alone (Blackwell *et al.,* 1969). Under such conditions, no malformations were observed (Berg, 1965), although some investigators postulated that metabolic derangements occurred (Hsueh *et al.,* 1970).

Although no observations have been reported of congenital malformations resulting from restriction of the maternal dietary intake in experimental animals, some permanent anatomical aberrations have been seen. For example, skeletons of young from undernourished female rats were smaller and less mature than controls even after more than three months of *ad libitum* feeding (Dickerson and Hughes, 1972). Similarly, rats subjected to prenatal maternal undernutrition showed perineural damage even when fed *ad libitum* from the neonatal period (Sima and Sourander, 1973).

In recent years, the effects of maternal undernutrition on the fetus have often been studied in terms of cellular growth. (For recent review, see Winick *et al.*, 1973.) Depending in part upon timing during gestation, maternal undernutrition generally leads to a decreased number of cells in the placenta and in various fetal organs, in both animals and man (Brasel and Winick, 1972).

5. Brain Development and Behavior

In recent years considerable attention has been devoted to the effects of prenatal or early postnatal undernutrition on growth and development of the brain and its functional maturation. There is now a large literature on this subject, too extensive to be reviewed here. However, although undernutrition does not appear to cause gross malformations of the brain, there are abnormalities of development resulting from this prenatal insult which will be briefly summarized (see Chase, 1973; Scrimshaw and Gordon, 1968; Barnes, 1971; Latham and Cobos, 1971; Winick, 1969; Zamenhof and van Marthens, 1974).

Winick and Noble (1966) showed that dietary restriction during the period of hyperplastic growth of an organ caused a reduced number of cells, while restricted food intake during hypertrophic growth caused a reduction in cell size. Decreased cell size can be overcome by supplementary feeding, but decreased cell number appears to be permanent. The effects of undernutrition on cell number in the brain are thus largely dependent on the timing of the nutritional restriction relative to development. However, since the brain is composed of more than one cell type, effects of undernutrition on cellular growth may be variable for different cell types. In addition to cellular growth, brain weight and cortical thickness are also decreased (Clark *et al.*, 1973).

Undernutrition during critical periods of brain growth and development affects the chemical composition of the brain as well as its cellular growth. Decreased myelination is accompanied by decreased concentration of cholesterol, cerebrosides, and other components of myelin (Dobbing, 1972; Chase, 1973). Undernutrition in early life can thus have permanent effects on brain size, number of cells, and chemical composition. Whether such effects are caused by prenatal or postnatal undernutrition depends on the timing of brain growth and development with respect to birth for each species.

The meaning for brain function of these anatomical and chemical parameters, which are known to be affected by undernutrition, has been investigated in both experimental animals and human beings. It is now well established that malnutrition during critical periods brings about changes in the behavior of animals. However, since learning in both animals and man results from an interaction of response and stimulus, the malnourished animal whose responsiveness to stimuli is depressed is less able to learn than is the normal one. Enrichment with sensory stimuli in the environment as well as with food may influence the development of learning and overcome some effects of nutritional or sensory deprivation (Barnes, 1971). In this connection, the recent report that prenatal exposure to the Dutch famine of 1944–1945 does not seem to be related to the mental performance of young men at age 19 is of interest (Stein *et al.,* 1972). Thus, although early nutritional deprivation may result in permanent anatomical or biochemical changes, causal relationships to permanent and irreversible alterations in intellectual capacity, especially in humans, remain uncertain at present.

B. Fasting

Fasting during pregnancy has teratogenic effects in some strains of mice, but similar reports have not been published for any other species. Runner and Miller (1956) observed vertebral and costal malformations and exencephaly in 19-day fetuses of mice of strain 129 subjected to total fasts of 24 or 30 hr during the 7th, 8th, 9th, or 10th days. Fasts of 40 hr were incompatible with pregnancy. Small quantities of glucose or amino acids counteracted these effects. The teratogenicity of fasting in mice was confirmed in other strains (Miller, 1962; Nishikawa, 1963).

McClure (1961) investigated the pathogenesis of embryonic mortality caused by fasting and concluded that it involved failure of hypophyseal gonadotrophic function. Effects of fasting on embryonic development were modified by maternal age and weight (Yasuda *et al.,* 1966) and decreased by intermittent fasting before pregnancy (Terada, 1970).

In human subjects, concentrations of plasma glucose, insulin, and blood ketone acid rose more rapidly in pregnant women fasted for 84 hr during the second trimester of pregnancy than in nonpregnant controls (Felig and Lynch, 1970).

C. Protein and Amino Acids

1. Protein Deficiency

When general malnutrition or undernutrition is imposed during pregnancy, protein becomes a limiting nutrient. Thus many similarities are seen between effects of undernutrition and of protein deficiency, and in fact, few studies have attempted to differentiate between caloric deprivation, protein deficiency, and protein–calorie malnutrition. The relation of dietary protein levels to reproduction in the rat was first studied with purified diets by Nelson and Evans (1953). They found a high incidence of resorptions with low protein diets (containing 0, 2.5, and 5% protein), but no congenital malformations were seen. Pregnancy was maintained, and fetuses were carried to term when estrogen and progesterone were injected into the mothers (Nelson and Evans, 1954, 1955b), even though maternal weight loss was very high. Again, no malformations were apparent at term, although fetal weight at term was lower than normal.

Adrenocortical steroid or ACTH administered also maintained pregnancy in protein deficient rats, apparently through mobilization of maternal tissue protein (Berg et al., 1967). A protein-free diet appeared to have less deleterious effects on embryonic mortality in the pig than in the rat (Pond et al., 1968).

The most extensive and well-controlled series of studies on prenatal maternal protein deficiency in rats has been carried out by Zeman and her colleagues. In experiments in which the effects of prenatal deficiency were differentiated from those of postnatal (lactational) deprivation, these workers found decreased birth weight, birth length, liver and kidney weight, and increased heart, brain, and thymus weights in relation to body weight in young of protein deficient dams as compared with controls (Zeman, 1967). Decreased cell number was found in the fetus at term (Zeman and Stanbrough, 1969) and this effect appeared to be irreversible (Zeman, 1970; Allen and Zeman, 1971).

In addition, kidneys of newborn young from protein-restricted rats had fewer and less well-differentiated glomeruli, proportionately more connective tissue, and fewer collecting tubules (Zeman, 1968). Reduced kidney function was found as well as morphological changes in this organ (Hall and Zeman, 1968). Both morphological and functional differences appeared to be irreversible postnatally even with increased feeding (Allen and Zeman, 1973a,b). Other tissues also showed morphological abnormalities, especially intestine and skelton (Shrader and Zeman, 1969; 1973a; Loh et al., 1971). Many of these changes appeared to lead to postnatal developmental defects which may be related to persistent hormonal inadequacies in these offspring (Shrader and Zeman, 1973b; Zeman et al., 1973).

2. Amino Acids

a. Deficiency. Relatively few studies have been made of effects of deficiencies of specific amino acids on prenatal development in mammals. Pike (1951) found cataracts in offspring of rats given a diet deficient in tryptophan. With diets containing amino acid mixtures, the omission of one essential amino acid (except for lysine), resulted in a high number of resorptions (Niiyama et al., 1970, 1973). If methionine, valine, or isoleucine were omitted, birth weight and placental weight were decreased. Zamenhof and his colleagues (1974), studying prenatal brain development in rats, found that omission of tryptophan, lysine, or methionine produced decreases in birth weight and in cerebral weight, cerebral cell number, and protein content. The effects were similar to those of deprivation of total protein. Thus, omission of single essential amino acids may be as harmful as total absence of dietary protein.

Persaud and Kaplan (1970) studied the effects of hypoglycin-A, a leucine analog, on development of rats when given to the pregnant female intraperitoneally. Most fetuses at term had malformations including gastroschisis, encephalocele, syndactyly, and stunting. The possible relationship of this compound (naturally occurring in certain West Indian fruit) to human malformation would be clearer if these investigators had given the analog by mouth rather than by injection and had included a control group with added leucine. Nevertheless, the experiment suggests the importance of leucine for fetal development.

b. Excess. Excessive amounts of single amino acids also have deleterious effects on fetal development. Cohlan and Stone (1961) found that addition of large amounts of lysine or leucine resulted in retarded fetal growth. Excess leucine also has been found to cause resorptions and malformations (Persaud and Tiesj, 1969).

Excessive amounts of phenylalanine are extremely deleterious to the developing embryo and fetus, as has been shown in rats, monkeys, and humans. In rhesus monkeys, hyperphenylalaninemia was induced by dietary means, resulting in offspring with reduced learning behavior (Kerr et al., 1968). Woolley and van der Hoeven (1964) have observed similar effects in newborn mice. Luse et al. (1970) fed pregnant rats a high level of phenylalanine and in addition gave them daily intraperitoneal injections of the amino acid. They found fetal death, resorption, and congenital malformations in the offspring.

In humans, maternal phenylketonuria may result in intrauterine growth retardation, mental retardation, microcephaly, and other congenital malformations, including skeletal anomalies, cardiac defects, esophageal atresia, agenesis of spleen, and lung abnormalities. Spontaneous abortions have occurred as well (Fisch et al., 1969). One phenylketonuric mother, who had three retarded children while on an unrestricted diet, gave birth to a normal infant during a subsequent pregnancy when she took a low phenylalanine diet (Allan and Brown, 1968). An interesting speculation regarding the interaction of maternal nutrition and genetic factors in the mental retardation associated with amino acid disorders has recently been made by Bessman (1972).

D. Carbohydrates

Although carbohydrates are important in the metabolism of the developing embryo, the effect of deficiencies or excesses of carbohydrates as a whole in the diet of pregnant females has not been reported. Excessive amounts of galactose have been shown to be deleterious to the developing mammal *in utero*. Bannon and coworkers (1945) fed pregnant rats a diet containing 25% galactose and found histological abnormalities of the lens in the fetuses. Segal and Bernstein (1963) recognized the clinical implications of this observation and extended the studies. In experiments in which dietary levels of 40% galactose were fed to pregnant rats, they found a high incidence of cataracts in the newborn young. Haworth *et al.* (1969), using similar conditions, found retardation of body growth and brain growth and reduced numbers of brain cells in offspring of galactose-fed females.

These findings in experimental animals are consistent with clinical observations of congenital galactosemia, which leads to mental retardation (Isselbacher, 1959) and lens opacities (Christian and Diamond, 1961). The experimental studies with pregnant rats fed high-galactose diets suggest that the effects of congenital galactosemia in humans may be initiated *in utero* and women known to be heterozygous for the galactosemia gene should take a diet low in lactose during pregnancy (Segal and Bernstein, 1963).

E. Lipids

1. Deficiency

Lipids are extremely important to the development and growth of the mammalian fetus. The lipid composition of the maternal diet can have profound effects on composition and metabolism of the fetus (for review see Roux and Yoshioka, 1970), but no malformations have been reported as a result of dietary lipid deficiencies.

A diet deficient in fat will not support normal pregnancy, lactation, or neonatal survival because of lack of essential fatty acids (EFA), especially linoleic (Burr and Burr, 1930; Evans *et al.*, 1934a,b; Deuel *et al.*, 1954). A low level of EFA in the diet of rats for two generations caused low birth weight in the offpsring and a lower concentration of cerebrosides in their brains than in controls (Alling *et al.*, 1972).

2. Excess

High cholesterol feeding to rabbits during pregnancy resulted in low fetal weight and high fetal mortality (Popjak, 1946; Zilversmit *et al.*, 1972), but the same effect was not seen in rats. On the other hand, a drug known to increase the cholesterol level of the blood (Tuchmann-Duplessis and Mercier-Parot, 1964), as well as a drug known to decrease it (Roux and Aubry, 1966), have been shown to be teratogenic in experimental animals.

III. VITAMINS

A. Fat-Soluble Vitamins

1. Vitamin A

a. Deficiency. The importance of vitamin A for normal prenatal development of mammals has been known since 1933, when Hale (1933) reported that pigs without eyeballs were born to a vitamin A-deficient sow. This was, indeed, the first report of the production of a congenital malformation by environmental means, and it led eventually to the development of experimental teratology (See Chapter 1). Hale found further that offspring of vitamin A-deficient sows, in addition to various eye defects, had other anomalies such as accessory ears, subcutaneous cysts, cleft lip, and misplaced kidneys (Hale 1934, 1935, 1937).

Great progress was made in experimental study of the teratogenic effects of vitamin A deficiency by the work of Warkany and his colleagues in a series of detailed and comprehensive studies in rats. Eye defects consisted of colobomas, eversion, retinal abnormalities, defects of cornea, "open eye," and, most frequently, a fibrous retrolenticular membrane was seen in place of the vitreous body (Warkany and Schraffenberger, 1944, 1946; Warkany and Roth, 1948). Abnormalities of the genitourinary tract also occurred in offspring of vitamin A-deficient rats. These included fused kidneys, absence of male accessory glands, and lack of vaginal development, as well as keratinizing metaplasia in epithelia derived from the embryonic urogenital sinus in fetuses (Wilson and Warkany, 1947, 1948). Cardiovascular abnormalities were also pronounced and consisted of defects in the interventricular septum, the aorticopulmonary septum, and various aortic arch anomalies (Wilson and Warkany, 1950). Diaphragmatic hernia was also observed to result from maternal dietary deficiency of vitamin A (Andersen, 1949; Wilson *et al.*, 1953). Wilson *et al.* (1953) analyzed the syndrome of anomalies produced by vitamin A deficiency and concluded that the malformations resulting from maternal vitamin A deficiency were determined during the period of active organogenesis. Urogenital and eye anomalies in offspring of vitamin A-deficient rats were also studied by Roux and his collaborators (1962a,b), who found further that such defects could result even when the deficiency was so mild that the mother herself appeared to be normal.

In rabbits, relatively mild deficiencies of vitamin A caused ovum infertility and degeneration leading to loss of fertilized ova before implantation (Lamming *et al.*, 1954a). With vitamin A deficiency during late gestation, resorption, abortion, and ocular abnormalities were seen in fetuses (Lamming *et al.*, 1954b) and hydrocephalus in newborn and postnatal young (Millen *et al.*, 1953, 1954; Lamming *et al.*, 1954b; Millen and Woollam, 1956).

Although the teratogenicity of vitamin A deficiency was first discovered in pigs, extensive investigation of its effects in this species was reported only recently. Palludan (1966a), in a detailed monograph, described the congenital anomalies produced by vitamin A in pigs and analyzed especially the morphogenesis of the eye in both normal and abnormal embryos.

The teratogenic effects of vitamin A deficiency have been described extensively in a number of species, but the mechanism of action remains unknown. There is some evidence that the role of vitamin A in sulfation may be involved in effects on myelination (Clausen, 1969).

b. Excess. In the course of a study of hypervitaminosis A in adult rats, Cohlan observed that several pregnant animals produced offspring with congenital malformations. A subsequent experiment showed that gross anomalies of the skull and brain occurred in 54% of the offspring of females given 35,000 I.U. of vitamin A from the 2nd, 3rd, or 4th to the 16th day of gestation (Cohlan, 1953). Since this report, numerous studies have been published describing in detail the effects of various doses of vitamin A at various times in a number of species. Notable among these is an extensive series of studies by Giroud and his collaborators, who confirmed the observation of Cohlan on the teratogenicity of vitamin A (Giroud and Martinet, 1954) and showed further that it produced many types of anomalies depending on the gestational stage when it was administered (Giroud and Martinet, 1955a–c, 1956a–d). Anencephaly and cleft palate appeared to be the most common malformations, but anophthalmia and other eye defects such as cataract and open eye, spina bifida, syndactyly, and face malformations also occurred. Giroud and his collaborators further described at the histological level other anomalies produced by excess vitamin A, including hydronephrosis, hydroureter, encephalocele, and meningocele (Giroud *et al.,* 1958a–c). Dental malformations have also been found (Masi *et al.,* 1966).

Although most studies of hypervitaminosis A teratogenicity have been carried out in rats, excessive amounts of the vitamin have also been shown to produce congenital malformations in other species. The mouse has been studied in some detail (Giroud and Martinet, 1959a,b) and has been shown to respond with abnormal development to relatively small doses (Giroud and Martinet, 1962). In the guinea pig, rabbit (Giroud and Martinet, 1959a,c), hamster (Martin-Padulla and Ferm, 1965; Ferm, 1967), and pig (Palludan, 1966b) hypervitaminosis A also produces congenital malformations.

c. Mechanism of Action. The morphogenesis of vitamin A-induced malformations of the brain was studied both at the anatomical level (Giroud *et al.,* 1959; Morriss, 1972) and at the cellular level using tritiated thymidine (Langman and Welch, 1966, 1967). All of these approaches seemed to indicate that anomalous development was brought about through abnormalities in differential growth rates between various tissues (loss of synchrony), perhaps through interference with DNA synthesis in neuroepithelial cells and a lengthening of the cell cycle, leading to a decreased rate of cell proliferation. The work of Nanda (1971,1974) suggesting that excess vitamin A retards the growth of the palatal processes so that they do not come in contact with each other at the optimal time for palatal closure is consistent with this idea.

Other work suggests that alterations in the metabolism of sulfated mucopolysaccharides, perhaps in mucopolysaccharide biosynthesis, are at the basis of the teratogenic effects of vitamin A (Kochhar and Johnson, 1965; Kochhar, 1968; Solursh and Meier, 1973); it could also be related to the general effect of the vitamin on the membranes of cells and intracellular organelles (Dingle and Lucy, 1965). Takekoshi (1964) has suggested that the thyroid gland is involved in vitamin A-induced teratogenesis.

Most of the work with hypervitaminosis A was done with the ester form of the vitamin, retinyl acetate. However, in 1967 Kochhar (1967) reported that retinoic acid was also teratogenic and indeed produced congenital malformations with much smaller doses than did other vitamin A compounds, being about 40 times more active. Kochhar (1973) made use of the high activity of retinoic acid as a teratogen and its rapid elimination from the body to produce limb abnormalities in mouse embryos and to relate them to specific embryonic stages. This work also suggested that retinoic acid had a disruptive effect on spatial organization of mesenchymal cell condensations, resulting in abnormally shaped cartilage models.

d. Postnatal Effects of Prenatal Hypervitaminosis A. The effect of relatively mild teratogenic doses of vitamin A on learning ability of the offspring was studied in rats. When excess vitamin A was given to rats on the 8th, 9th, and 10th days of pregnancy, few malformations were seen, but maze learning was impaired (Butcher *et al.*, 1972). Treatment with excess vitamin A on days 14 and 15 or on days 17 and 18 of gestation also produced subtle behavioral deficits, but of different types, without obvious neurological impairment (Hutchings *et al.*, 1973; Hutchings and Gaston, 1974).

e. Human. Although knowledge is extensive regarding teratogenic effects of both deficiency and excess of vitamin A in experimental animals, very little research has been done on possible effects in humans. Sarma (1959) reported a case of a woman severely deficient in vitamin A to the point of blindness who gave birth to a premature baby with microcephaly and anophthalmia. Observations made by McLaren (1968) also suggest that intrauterine vitamin A deficiency (or other nutritional deficiences as well) may be responsible for eye abnormalities and impaired vision in children.

Another case history concerned a woman who took excessive amounts of vitamin A throughout pregnancy (25,000 i.u. daily for the first 3 months and 50,000 i.u. daily thereafter) and produced a child with congenital urogenital anomalies (Bernhardt and Dorsey, 1974). Gal and coworkers (1972) found that maternal serum vitamin A levels postpartum were significantly higher in mothers of infants with CNS malformations than in mothers of normal babies. Fetal liver vitamin A levels were consistent with this finding.

It is obvious that insufficient information is available on vitamin A in human pregnancy and its possible relationship to either gross congenital malformations or more subtle postnatal deficits. Fundamental research is needed on placental transfer of vitamin A to the fetus, the relationship of maternal dietary level of the vitamin to neonatal and postnatal condition of the infant, and correlation of maternal, fetal, and neonatal serum levels of the vitamin.

2. Vitamin D

a. Deficiency. Vitamin D is necessary for the absorption and utilization of calcium. Thus, vitamin D deficiency during pregnancy results in abnormal development of the fetal skeleton. If sufficient quantities of vitamin D are available in the diet of the mother, the fetuses will have nearly normal concentrations of calcium and phosphorus even if the maternal diet is suboptimal in these elements (for review of early literature, see Swanson and Iob, 1939). Warkany (1943) found that the offspring of rats given a rachitogenic diet showed skeletal abnormalities at birth which resembled those of rickets. Similar observations were made by Wallis (1938) in cattle.

In humans, vitamin D deficiency in association with low calcium intake leads to the development of fetal rickets (Maxwell, 1934) and may lead to neonatal tetany and enamel hypoplasia of the teeth (Purvis *et al.*, 1973; Rosen *et al.*, 1974). There is also some evidence that poor dentition and high susceptibility to caries may result (Mellanby and Coumoulos, 1944).

Recently, the biochemical role and metabolism of vitamin D have been elucidated. The relationship of the active metabolites of the vitamin to biochemical defects of genetic disorders involving vitamin D dependency have led to new treatments for these diseases (Scriver, 1970, 1974; DeLuca, 1974).

b. Excess. The effect of hypervitaminosis D has been studied in rats and rabbits. It was shown in early work and has since been confirmed that excess vitamin D in the diet of pregnant rats produced low birth weight and low concentrations of calcium and phosphorus in the offspring (Sontag *et al.*, 1936; Potvliege, 1972). In rats, hypercalcemia is not a usual effect of excessive administration of vitamin D (Ornoy *et al.*, 1968). In rat fetuses of mothers receiving large dosages of vitamin D, a decrease in wet weight, ash weight, and in calcium and phosphorus content of the whole body and of fetal bones was observed, with impairment of osteogenesis of the fetal long bones (Ornoy *et al.*, 1969). Abnormal bone development persisted during the postnatal period in such rats, evidenced by thick periosteum, thin compact bone, and multiple fractures as early as the 5th day of life, with impaired fracture healing (Ornoy *et al.*, 1972).

Growth of offspring of rats which received vitamin D in excessive amounts during pregnancy was retarded, and the retardation persisted throughout life, becoming more marked with time. There was apparently some fundamental damage to fetal osteogenic tissues induced by the teratogenic effect of large amounts of vitamin D_2 (Ornoy *et al.*, 1972). However, it should be noted that in these studies no attempt was made to control for the possible effect of maternal hypervitaminosis D on lactation, thus the possible effects of impaired nutrition during the suckling period were not taken into account. In studies with rats, Potvliege (1972) concluded that the pregnant female is less susceptible to toxic effects of vitamin D than the nonpregnant female.

Teratogenic effects of hypervitaminosis D have also been observed in rabbits. In this species, excessive amounts of vitamin D during pregnancy produced hypercalcemia in the mothers, and in the fetuses anomalies of the aorta which resembled the supravalvular aortic stenosis syndrome (SASS) in children (Friedman and Mills, 1969; Friedman and Roberts, 1966). In these rabbit fetuses as well as in children with the syndrome, cranial, facial, and dental anomalies were observed. There was hypoplasia of the mandible, congenital absence of teeth, microdontia, dysgnathia, enamel hypoplasia, and malocclusion. Peculiar facies, premature closure of the cranial bones, strabismus, odd shaped ears, and low birth weight were also noted. Chemical studies revealed that the blood level of the vitamin in both mothers and fetuses was much higher in rabbits treated with high levels of vitamin D than in controls, showing that transplacental passage occurred. Serum calcium levels were also higher in both fetuses and maternal animals.

c. Human. Idiopathic infantile hypercalcemia, a genetic disease in children, appears to be similar in many of its features to the effect of excessive vitamin D in the pregnant rabbit (see above). This disease appears to be related to deranged vitamin D metabolism (Fellers and Schwartz, 1958; Forfar *et al.,* 1959).

In humans there is little information on the possible effects of hypervitaminosis D during pregnancy. Goodenday and Gordan (1971) have reported that women receiving vitamin D treatment for hypoparathyroidism gave birth to children who appeared to be completely normal when examined at ages ranging from 6 weeks to 16 years. None of the children had any of the cardiovascular or craniofacial stigmata associated with infantile hypercalcemia. However, in these hypoparathyroid patients the large amounts of vitamin D did not produce hypercalcemia in the mother, thus in this respect the patients differed from the experimental animals in which fetal anomalies have been produced.

3. Vitamin K

Vitamin K is essential for normal clotting of the blood. Thus the effects of vitamin K deficiency involve hemorrhage and death from bleeding. An early report of what was later recognized to be vitamin K deficiency appeared in the description of sweet clover disease in cattle (Schofield, 1923). Sweet clover disease, as was shown later, causes a deficiency of vitamin K because of ingestion of the vitamin K analog, dicoumarol. When cattle eat this material, they produce calves which die of hemorrhage. "The active well-developed calf born of a cow fed for 13 days on damaged sweet clover hay dies of hemorrage 28 hours after birth" (Roderick, 1929).

In experimental animals, rabbits were shown to have a high rate of abortion when given a vitamin K-deficient diet (Moore et al., 1942). Even when the compound was given for only 2 days, the fetuses died in utero. When given lower levels of dicoumarol, newborn rabbits had a very low level of prothrombin at birth with a tendency to hemorrhage, although the mothers appeared to be normal (Kraus et al., 1949). Quick (1946) showed that pups of dogs given dicoumarol in therapeutic doses in the last week of pregnancy had reduced prothrombin levels at birth in spite of normal prothrombin levels of the mothers.

Since dicoumarol is used in human medicine in the prevention and treatment of clotting diseases (thromoboembolic diseases), considerable concern has been expressed as to whether this drug given to pregnant women has any deleterious effects on the growth and development of the fetus. Patients with thromboembolism or with mitral valve prosthesis are subject to long-continued treatment with dicoumarin anticoagulants. Several cases have been reported in which fetal anomalies were found in infants born to such women. In one, the infant was blind and mentally retarded (DiSaia, 1966). In another, the infant was born with hypoplastic nasal structures (Kerber et al., 1968). In a third report, fetal or neonatal death appeared to result (Palacios-Macedo et al., 1969). Other studies, however, suggested that oral anticoagulants of the coumarin type are not teratogenic or deleterious to the fetus if the therapy is properly managed (Olwin and Koppel, 1969; Fillmore and McDevitt, 1970; Hirsh et al., 1970; Finnerty and Mackay, 1962).

4. Vitamin E

Vitamin E was first discovered in studies relating to the influence of nutrition on reproduction in rats. Animals given a diet known later to be deficient in vitamin E were sterile (Evans and Bishop, 1922). By administering small doses of vitamin E to pregnant rats on a deficient diet, Cheng and her collaborators (1957) were able to maintain pregnancy in the rat and produce teratogenic effects in the offspring. When vitamin E-deficient rats were given 2 mg of DL-α-tocopherol acetate by stomach tube on the 10th day of gestation, the resorption rate was high and 37% of the live fetuses showed one or more malformations at term. A large variety of abnormalities was seen including exencephalus, umbilical hernia, scoliosis, club feet, cleft lip, syndactyly, anencephalus, and kinked tail. The percentage of abnormal young varied according to the composition of the ration and the amount of vitamin E it contained (Cheng et al., 1960). Tocopherol levels of the fetuses did not seem to be correlated with the vitamin E intake of the mothers (Cheng et al., 1961). Gortner and Ekwurtzel (1965) were not able to reproduce the findings of

Cheng concerning teratogenic effects of vitamin E. They found only minor effects on fetal development when vitamin E-deficient diets were given to pregnant rats.

Using a different approach, Kenney and Roderuck (1963) produced a syndrome resembling eclampsia in rats by giving tocopherol to chronically depleted rats at a critical time in fetal development. Gestation was prolonged, and most of the young were stillborn. Maternal mortality at parturition was high, but the dead fetuses did not appear to be grossly abnormal. Fetal deaths and resorption also occur in mice as a result of vitamin E deficiency (Suomalainen, 1950; Lee *et al.*, 1953).

B. Water-Soluble Vitamins

1. Ascorbic Acid

a. Deficiency. Unlike the other vitamins, ascorbic acid is required by only a few animal species, namely, man and other primates, guinea pig, and certain tropical birds. Thus, observations on the effects of ascorbic acid deficiency during pregnancy in mammals have been limited to a few studies with guinea pigs. Mouriquand and coworkers (1935) reported that in guinea pigs given a vitamin C-deficient diet spontaneous abortions resulted, especially between the 26th and 30th days of gestation.

More recently, fetal and uterine tissues in ascorbic acid deficiency have been studied biochemically and histologically (Rivers *et al.*, 1970). Because of the role of ascorbic acid in collagen synthesis, these investigators measured ascorbic acid, proline, and hydroxyproline levels, and concluded that collagen synthesis was impaired in both fetal and uterine tissues from deficient animals. Histologically there were abnormalities in the mucopolysaccharide, collagen, and elastin components. Since the maternal animals were given the deficient diet only between days 20 and 35, the mothers themselves showed no outward signs of vitamin C deficiency.

Since the human disease scurvy has been known for hundreds of years and extensive studies of vitamin C deficiency have been made in humans and in experimental animals, it is surprising that little work has been reported on ascorbic acid deficiency in pregnancy. In the Vanderbilt cooperative study of maternal and infant nutrition, the possible relationships between intake and serum level of vitamin C in pregnant women was studied in relation to the course and outcome of pregnancy. In general, serum levels decreased during pregnancy in these women except in the group which had a high level of ascorbic acid intake. Congenital malformations did not differ between the groups with the highest and lowest serum levels, but increased frequency of premature birth was associated with the lowest intake of vitamin C and the lowest serum concentrations seen in the study (Martin *et al.*, 1957).

b. Excess. Massive doses of vitamin C during pregnancy have been shown to be detrimental in several species. Guinea pigs given high doses of vitamin C showed infertility and fetal mortality although no toxic effects were seen in the pregnant females (Neuweiler, 1951). Similar findings were reported by Mouriquand and Edel (1953) in the same species. These investigators also observed abortions. Samborskaya and Ferdman (1966) reported stillbirths and abortions in guinea pigs, rats, and mice. In women given large doses of ascorbic acid, termination of early pregnancy was reported. However, no control group was used in this study.

2. B-Complex Vitamins

a. Riboflavin

i. Deficiency. Following the reports of Hale on congenital malformations in vitamin A-deficient pigs, riboflavin was the second nutrient deficiency of which was shown to produce anomalies of fetal development. Warkany and Schraffenberger (1943) reported that maternal dietary riboflavin deficiency in rats caused malformations of the newborn young consisting of shortness of the mandible and the long bones, fusion of ribs, abnormal centers of ossification, syndactylism, brachydactylism, and cleft palate. Later work (Warkany and Deuschle, 1955) described dentofacial changes, mainly micrognathia, cleft palate and associated dental anomalies, hydrocephalus, and eye defects (Grainger *et al.*, 1954).

The discovery of teratogenic effects produced by dietary riboflavin deficiency in rats was soon confirmed by Giroud and Boisselot (1947) and Bologna and Piccioni (1956). The latter also showed that if the deficient diet was given to females 60 days before mating, no young were produced, while with 40 days of dietary deficiency before pregnancy, young were carried to term, but about half were abnormal. Teratogenic effects of riboflavin deficiency occurred with extremely mild deficiency states in the mother (Giroud *et al.*, 1952; Potier de Courcy and Terroine, 1968b).

Nelson and her colleagues (1956b) made use of the riboflavin antimetabolite galactoflavin to produce acute transitory riboflavin deficiency in pregnant rats. Skeletal anomalies similar to those reported previously were found and, in addition, soft tissue defects not previously reported in the cardiovascular and urogenital systems, diaphragm, and epidermis. Kalter and Warkany (1957) also used galactoflavin in combination with a riboflavin-deficient diet to produce anomalies in mice and found that the frequencies of types of malformations differed in various inbred strains.

More recently, Shepard *et al.*, (1968) have described histologically and histochemically the development of limb malformations in riboflavin-deficient, galactoflavin-treated embryos. Although the use of antimetabolites to produce vitamin deficiencies has helped to clarify temporal relationships in teratogenic nutritional deficiencies, it should be noted that toxic effects unrelated to the antivitamin action may be involved. In pigs Ensminger *et al.* (1947) found that riboflavin deficiency during pregnancy resulted in young that were dead at birth or died within 48 hr thereafter, with "enlargement of the front legs due to gelatinous edema in the connective tissue."

ii. Mechanism of Action. The biochemical mechanisms responsible for the teratogenic effects of riboflavin deficiency have been the subject of investigation by several groups. Miller *et al.* (1962), noting the similarities in malformations induced by riboflavin and by folic acid deficiencies, investigated the effect of riboflavin deficiency on flavin and folate levels in whole embryos and in fetal and maternal livers. Their results suggested that galactoflavin teratogenicity may be due to a low level of flavin adenine dinucleotide (FAD) during differentiation. Maternal liver folate and citrovorum factor were reduced by 50% after 16 days of the riboflavin-deficient diet, but correlation of folate levels with the teratogenicity of riboflavin deficiency was not possible.

In embryos and fetuses from riboflavin-deficient, galactoflavin-treated rats, succinic and DPNH oxidase systems (the terminal electron transport systems) were markedly reduced from control values at 12 through 16 days of gestation, while placenta and maternal heart were unaffected (Aksu *et al.*, 1968). On the basis of these findings it was concluded that the teratogenic effects of riboflavin deficiency resulted from reduced activity of electron transport systems. In continuing work, Shepard and Bass (1970) concluded that basic mechanisms for limb defects occurred at the local cell level and not from generalized physiologic impairment of the embryo.

Potier de Courcy and colleagues (Potier de Courcy and Terroine, 1968a,b; Potier de Courcy and Desmettre-Miguet, 1970) have made an extensive study of chemical and biochemical changes in the maternal fetal organism resulting from riboflavin deficiency. Glycogen, protein, DNA, and RNA levels were reduced in placenta and fetus. Cell number and alkaline phosphatase were also reduced in placenta and fetus. The concentration of calcium was normal in 13-day embryos, but was reduced at 21 days, in contrast to that of sodium and potassium, which remained at normal levels of concentration. Zinc levels were also normal (Potier de Courcy *et al.*, 1970).

iii. Riboflavin in Human Pregnancy. A group of pregnant women found to have clinical riboflavin deficiency were studied with respect to the course of pregnancy, labor, condition of the newborn, and neonatal development of their infants. No effects were reported on incidence of toxemia, infections, or hemorrhagic complications, but vomiting and incidence of prematurity and stillbirths were increased. Riboflavin deficiency also depressed lactation. Birth weight and incidence of malformation appeared to be unaffected (Braun *et al.*, 1945; Brzezinski *et al.*, 1947).

b. Thiamine. Dietary deficiency of thiamine has been shown to have deleterious effects on reproduction in experimental animals. Nelson and Evans (1955a) gave pregnant rats thiamine-deficient diets beginning on the day of mating or for 1–3 weeks prior to pregnancy. This produced a very high rate of resorptions and stillbirths, and living young were of low birth weight. There were also marked losses in maternal body weight and increased maternal mortality. These effects seemed to be due largely to restriction of food intake induced by the vitamin deficiency. The effects on reproduction could be overcome by injection of estrone and progesterone, and under such conditions pregnancy was maintained even though the usual marked reduction in food intake occurred with the thiamine deficiency.

When female rats were fed from weaning to mating a diet low in thiamine, but were given adequate thiamine during pregnancy, they produced smaller litters than did controls (Brown and Snodgrass, 1965). In the pregnant sow, Ensminger and coworkers (1947) also found high rates of stillbirths, weak newborn young, and poor maintenance of pregnancy with a thiamine-deficient diet. In humans, congenital beriberi is seen in the newborn infants of pregnant women with thiamine deficiency (van Gelder and Darby, 1944).

c. **Niacin.** Most of the work on the importance of niacin (nicotinic acid) during prenatal development has been done with the use of an antimetabolite, 6-aminonicotinamide (6-AN). Chamberlain and Nelson (1963a) showed that multiple congenital abnormalities resulted when this antagonist was fed to pregnant rats. When the antimetabolite was given for 2- or 3-day periods during pregnancy, a high incidence of embryonic mortality and resorption and a low incidence of abnormal young was observed. Malformations affected the skeleton, central nervous system, eye, urinary system, trunk, thyroid, and thymus gland. Congenital defects were also produced by single injections of 6-AN during pregnancy in rats. These included a high incidence of hydrocephalus as well as ocular, urogenital, and vascular anomalies. Skeletal defects, cleft palate, cleft lip, and exomphalos were also observed (Chamberlain and Nelson, 1963b). 6-AN has also been used to study the development of cleft palate late in gestation (Chamberlain, 1966) and the effect of niacin directly on the rat embryo *in vitro* (Turbow and Chamberlain, 1968).

Studies of the effects of a nicotinic acid-deficient diet have also been made without use of an antagonist. Ruffo and Vescia (1941) found a marked reduction in the number of young born to rats given a nicotinic acid-deficient diet. Using a diet deficient in tryptophan as well as in niacin (since the rat can convert tryptophan to niacin), Fratta (1969; Fratta *et al.,* 1964) observed a high rate of fetal resorptions but did not see congenital malformations. Greengard *et al.* (1966), however, reported malformations in rats from such a diet although they did not describe the defects. They also showed that such effects could be prevented by adrenocortical steroids.

d. **Vitamin B$_6$**

i. *Deficiency.* The first report on the essential nature of pyridoxine in the pregnant rat came from the work of Nelson and Evans (1948) as part of their series of studies on the role of nutritional factors in reproduction. They produced pyridoxine deficiency by use of an antagonist, desoxypyridoxine, and found that the incidence of resorptions was very high. The percentage of young born dead was also high, and the number of young per litter and the birth weight of the young were low. Nelson and her group thought that the influence of pyridoxine deficiency on resorption was brought about through its effect on the gonadal function of the pregnant animal, since injection with estrone and progesterone (Nelson *et al.*, 1951) or of gonadotropic hormones could maintain pregnancy (Nelson *et al.*, 1953). Later, Davis *et al.* (1970) reported similar studies in which the amount of desoxypyridoxine was considerably greater than that used by Nelson and Evans. In this report congenital malformations resulted from vitamin B_6 deficiency in the pregnant rat. The malformations observed were digital defects, cleft palate, omphalocele, micrognathia, and exencephaly.

Pyridoxine deficiency in pregnant rats has also been studied without the use of a pyridoxine antagonist. Eberle and Eiduson (1968) concluded that the fetus was protected from B_6 deficiency during pregnancy. However, other workers found that feeding a pyridoxine-deficient diet to pregnant rats during the last two weeks of gestation and during lactation resulted in the birth of some young that had convulsions as early as 3 days of age. Plasma transaminases and pyridoxal phosphate of brain were low in B_6-deficient pups (Dakshinamurti and Stephens, 1969).

The pyridoxine level of the diet prior to pregnancy also appears to be important. Ross and Pike (1956) found that rats given a pyridoxine-deficient diet prior to the beginning of pregnancy but a supplemented diet thereafter gave birth to young with birth weight as low as or lower than those from mothers which received desoxypyridoxine during pregnancy. Körner and Nowak (1971) produced pyridoxine deficiency of rats in which they monitored the degree of deficiency by means of the tryptophan loading test and the erythrocytic glutamate oxaloacetate transaminase (EGOT) activation test. They measured the influence of the B_6 deficiency during gestation or gestation and lactation on the biochemical development of the brain of the fetuses and neonates. Animals which were B_6 deficient during gestation and lactation showed poor survival, retarded weight gain, reduced activity, and a higher incidence of errors in a maze test, or died with convulsive seizures. Brain tissues showed reduced concentration of GABA and reduced activity of glutamate decarboxylase and of dopa decarboxylase. Additional differences in chemical composition of the brains of pups from B_6-deficient dams included decreased concentrations of pyridoxine, cerebrosides, and protein (Pang and Kirksey, 1974). Moon and Kirksey (1973) studied cellular growth during prenatal and early postnatal periods in progeny of pyridoxine deficient rats, but they made no mention of examination of the newborn young for the possibility of congenital anomalies.

It should be noted that most of these studies have not differentiated between effects of pregnancy and lactation, that is, between nutrition of the fetus and the suckling.

ii. Excess. The effect of a high intake of pyridoxine during pregnancy was studied in rats. No effect was seen on the young, in litter size, growth to weaning, or requirement for pyridoxine after weaning (Schumacher *et al.*, 1965).

e. Folic Acid

i. Experimental Animals. The importance of folic acid in prenatal development has been studied primarily in experimental animals with the use of folate antagonists. The first experiments in which congenital malformations were caused by folate deficiency, however, used an intestinal antibiotic to suppress vitamin synthesis by gut bacteria in combination with a folate-deficient diet in pregnant rats (Nelson and Evans, 1949). Giroud and Lefèbvres-Boisselot (1951), with a relatively mild deficiency, produced a variety of malformations, including cleft lip, coloboma, hydrocephalus, ectopia of thoracic and abdominal organs, and eye abnormalities.

In an extensive series of publications, Nelson and her colleagues reported in detail a wide variety of congenital malformations caused by an inadequate supply of this vitamin in the pregnant rat. Using x-methyl pteroylglutamic acid (x-methyl PGA) as antagonist in combination with a folate-deficient diet, they showed that depending on timing, high rates of congenital abnormalities and resorptions resulted (Nelson *et al.*, 1952). The young showed marked edema and anemia, cleft palate, facial defects, syndactyly, skeletal abnormalities, and anomalies of urinary system, lungs, and eyes (Monie *et al.*, 1954; Asling *et al.*, 1955, 1960b; Asling, 1961). Defects of the cardiovascular system were also seen (Baird *et al.*, 1954; Monie *et al.*, 1957a). Even transitory use of this dietary regimen resulted in high incidences of multiple abnormalities (Nelson *et al.*, 1955, 1956a; Monie *et al.*, 1957b). For example, a deficiency of only 48 hr between day 7 and 12 resulted in malformations in 70–100% of the young. Malformations have also been produced in the mouse and the cat with x-methyl PGA (Tuchmann-Duplessis *et al.*, 1959).

The possibility that malformations in folate-deficient animals resulted from depression of nucleic acid synthesis was studied by several groups. Kinney and Morse (1964), using aminopterin as a folate antagonist, found that DNA levels in fetal liver were depressed, but RNA was unaffected. Potier de Courcy (1966b), however, found significant decreases in protein, DNA and RNA of whole body, liver, and brain. Placenta showed similar depressions. Administration of DNA, orotic acid, or ascorbic acid to pregnant rats subjected to the folate-deficiency regime had no influence on the deleterious effects to the fetuses. The diminution of DNA content did not correlate with DNase activity and therefore seemed to be due to inhibition of DNA synthesis rather than to increased catabolism (Potier de Courcy and Terroine, 1966).

Another approach to the study of biochemical alterations in embryos from folate-deficient rats has been used by Johnson and his colleagues. In one study, they found that fetuses from folate-deficient, 9-methyl-PGA-treated rats contained reduced levels of ATP, ADP, and AMP as compared with controls (Chepenik *et al.*, 1970). These observations might be related to the increased oxygen consumption observed in such embryos (Netzloff *et al.*,

1966, 1968). Alterations in nonspecific isoenzyme patterns for a number of enzymes were also seen (Johnson and Chepenik, 1966). Studies of phosphomonoesterases in limbs showed marked depression of activity in fetuses from 9-methyl-PGA-treated females. The changes in enzyme patterns could be correlated with anomalous chondrogenesis and osteogenesis (Jaffe and Johnson, 1973).

ii. Humans. Folic acid, an essential nutrient for humans, is also necessary for the developing human embryo. This is evidenced by clinical observations in which the folate antagonist aminopterin was used as an abortifacient agent. In certain cases, when given too late or in insufficient quantity, abortion was not produced, but instead the birth of a malformed child resulted (see Chapter 8). However, the possible relationship of folate deficiency in pregnant women to malformations or other abnormalities of pregnancy is still a subject of controversy (for reviews see Kitay, 1969; Rothman, 1970). Although there is general agreement that folic acid deficiency is not rare in pregnant women, with an incidence of 3–22% (Stone, 1968), clear evidence that malformations are associated with it are lacking.

Fraser and Watt (1964) reported that 5 of 17 women with megaloblastic anemia gave birth to infants with serious birth defects. Using elevated urinary excretion of formimino glutamic acid (FIGLU) following a histidine load as a sign of folate deficiency, the Hibbards (Hibbard, 1964; Hibbard and Smithells, 1965; Hibbard and Hibbard, 1966) found folate deficiency in 62% of mothers of malformed infants as compared with 17% of mothers of normal infants and concluded that there was an association between folate deficiency and malformations of the nervous system. They suggested that a genetic factor influencing folate metabolism might be involved. On the other hand, other investigators did not find a correlation between malformation and folate status as measured by plasma levels and erythrocyte morphology (Scott *et al.*, 1970; Hall, 1972).

f. Pantothenic Acid. The active metabolic form of pantothenic acid is coenzyme A and is important for embryonic development. Nelson and Evans (1946) first showed that dietary deficiency of the vitamin in pregnant rats resulted in resorptions and small litters with underdeveloped young. Teratogenic effects of pantothenate deficiency were demonstrated by Lefébvres-Boisselot (1951), who observed exencephaly, anophthalmia or microphthalmia, edema, and hemorrhage. Limb abnormalities also occurred (Giroud *et al.,* 1955). Malformations of the brain and of the renal system have also resulted from pantothenic acid deficiency in rats (Giroud *et al.,* 1957; Roux and Dupuis, 1961), and even fetuses that were not malformed at birth showed abnormal intrauterine growth (Lefebvres-Boisselot and Dupuis, 1956; Everson *et al.,* 1954).

These effects occurred in the absence of signs of deficiency in the mothers. Measurements of pantothenic acid levels in maternal and fetal livers under these conditions showed that when maternal liver pantothenate fell to

40% fetal death occurred, but with less severe depression of pantothenate levels malformations resulted (Giroud *et al.*, 1954). Indeed, on the 10th day of gestation, a time critical for development of brain and eye malformations, the level of liver pantothenate in panthothenate-deficient rats was only 16% lower than normal (Giroud *et al.*, 1961).

Biochemical studies of fetuses under these conditions showed that the content of protein, DNA, and RNA was low in liver, brain, and total body of deficient fetuses at term. The brain, where malformations were of high incidence, showed the least depression of these values. Placenta showed similar changes, but the maternal liver was normal with respect to these biochemical parameters (Potier de Courcy, 1966a).

Using a pantothenate antagonist, omega-methyl-pantothenic acid, in combination with a deficient diet, Nelson *et al.* (1957) observed cerebral and eye defects, digital hemorrhages and edema, and in addition, interventricular septal defects, anomalies of the aortic arch pattern, hydronephrosis and hydroureter, clubfoot, cleft palate, and tail and dermal defects. High doses of ascorbic acid were found to compensate partially for the effects of pantothenic acid deficiency (Everson *et al.*, 1954; Giroud *et al.*, 1956).

In rats the postnatal effects of maternal dietary pantothenate deficiency during the first 14 days of gestation were studied by Ershoff and Kruger (1962). Offspring of rats subjected to this regime showed motor incoordination, deficient righting reflexes, motor spasms, and poor head orientation. These abnormalities occurred in the absence of other teratogenic effects and persisted into adulthood.

In studies on the effect of pantothenic acid deficiency during pregnancy in the guinea pig, Hurley and her colleagues (1965) found that the deficiency during the ninth and tenth (last) weeks of gestation resulted in abortion or death of the mother. Liver fat and pantothenic acid levels of the newborn were also affected. The experiments suggested that the greatest need for pantothenic acid during fetal development of the guinea pig is in the period shortly before birth, a time at which concentration of pantothenic acid and coenzyme A in fetal liver rises sharply in this species (Hurley and Volkert, 1965).

Abnormal prenatal development, as well as infertility, resorptions, and abortions were seen in female swine given a pantothenic acid-deficient diet during pregnancy (Ullrey *et al.*, 1955).

g. Vitamin B$_{12}$. The influence of vitamin B$_{12}$ deficiency on the prenatal development of mammals has been studied in rats. The major effect of this deficiency is to produce hydrocephalus (for review, see Newberne and O'Dell, 1961). Because of the association of folic acid and vitamin B$_{12}$, much of the early work on this subject was confused by the inability to separate the two deficiencies (Richardson and Hogan, 1946; Hogan *et al.*, 1950). However, the hydrocephalus which occurs under these conditions appears to be caused by deficiency of vitamin B$_{12}$. Woodard and Newberne (1967) showed that the etiology of hydrocephalus produced by a vitamin B_{12}-deficient diet is the same with or without added x-methyl folic acid.

Newberne and Young (1973) have also shown that a marginal vitamin B_{12} intake during gestation in the rat has long-term effects on the offspring. In progeny of rats given a marginally deficient vitamin B_{12} diet, birth weight was low, and body weight after 21 days, three months, and one year remained significantly lower than normal as well. In addition, various liver enzyme activities were reduced, and B_{12} concentration in the liver was low in the newborns.

h. Choline. The importance of choline during pregnancy for development of the offspring was studied in sows by Ensminger and his colleagues (1947). They found that in sows given a choline-deficient diet most of the offspring failed to survive. Some abnormalities such as kinked tails were observed. Fatty livers, hemorrhages of the kidney, and other kidney abnormalities were seen on postmortum and microscopic examination of the newborns. In mice, choline deficiency during pregnancy resulted in the birth of offspring with high postnatal mortality (Mirone, 1954). The full-term fetuses of rats given a choline-deficient diet had fatty livers, as did their mothers (Meader, 1965).

The postnatal effects of a prenatal maternal dietary deficiency of choline were studied in rats by Kratzing and Perry (1971). Offspring of females given a choline deficient diet only during pregnancy were hypertensive during the period of 35–58 days of age. However, by 86–108 days of age, blood pressure in these animals had returned to normal. These findings may be related to alterations in kidney function or structure.

The influence of diets varying in methionine and choline content with and without vitamin B_{12} on development of the rat embryo was studied by Newberne *et al.* (1970). Marginally deficient newborn offspring appeared to be clinically normal, but biochemical measurements indicated smaller brain cells and irregularities of protein synthesis.

IV. MINERAL ELEMENTS

A. Major Elements

1. Calcium

The effects of calcium deficiency during pregnancy are similar to those of vitamin D deficiency. One of the earliest experiments demonstrating the importance of maternal nutrition for fetal development concerns calcium. Cows fed wheat alone produced stillborn or immature, weak calves that did not live

(Hart *et al.*, 1911). When the ration was supplemented with bone meal, the calves were normal (Hart *et al.*, 1924). If calcium deficiency is not prolonged, no abnormal effects are seen in the fetus. Fetal development and even calcification may be normal under conditions of calcium deficiency in the mother (Macomber, 1927). Removal of calcium from the maternal skeleton occurs under such conditions by stimulation from the parathyroid glands (Bodansky and Duff, 1941; Comar, 1956).

If calcium deficiency is severe and prolonged, normal development of the embryo is hampered (Boelter and Greenberg, 1943). Using tracer studies, Bawden and Osborne (1962) have shown that there was no depression of calcium concentration of the litter when pregnant rats were given a calcium-deficient diet, but there was a highly significant difference between the ash content of the femurs of calcium-deficient females and those of controls. Similar findings were reported in pregnant sows on normal and calcium-deficient rations (Itoh *et al.*, 1967). The transfer of calcium from maternal to fetal tissues has been studied by Comar (1956).

Not only is the amount of calcium in the diet of importance to the ossification and calcification of the fetal skeleton, but the ratio of calcium to phosphorus is also important for the development of normal skeletons in the offspring of pregnant animals (Cox and Imboden, 1936). In humans, good ossification of the fetus is dependent upon an adequate supply of dietary calcium and phosphorus (Tovarud and Tovarud, 1931; Mellanby and Coumoulus, 1944; Burke and Stuart, 1948).

Diets containing excessive amounts of calcium have also been deleterious to the young of pregnant animals (Cox and Imboden, 1936). Fairney and Weir (1970) found that the offspring of hypercalcemic rats (produced by feeding a high-calcium diet) produced small litters with offspring of low birth weight which grew poorly and had short sparse fur with patchy alopecia.

2. Magnesium

Magnesium deficiency is similar to zinc deficiency (see below) in the rapidity with which effects appear. With severe deficiency of magnesium in pregnant rats, embryonic death and malformation were observed (Hurley and Cosens, 1970). With milder deficiencies, some embryos survived but they were small and malformed, edematous and pale, with low plasma protein and severe anemia (Hurley and Cosens, 1971). In maternal animals under such conditions, packed cell volume (PVC) and hemoglobin were normal, but in the fetuses at term, hemoglobin, PCV, and red blood cell count were markedly reduced (Hurley, 1971). Cohlan *et al.* (1970) and Dancis *et al.* (1971) have also observed a fall in the maternal plasma magnesium, decreased magnesium concentration in the fetuses, and fetal anemia in magnesium-deficient rats during pregnancy. Possible relationships to human pregnancy have been reviewed (Hurley, 1971).

3. Sodium and Potassium

There have been surprisingly few studies on the effects of sodium deficiency during pregnancy on the development of the fetus. In early work, Orent-Keiles and coworkers (1937) studied the effect of a very low intake of sodium in rats. When sodium-depleted rats were mated, very few young were obtained and those that survived to term were extremely small in size. However, when pregnant rats were given a diet low in sodium for one week prior to mating and through pregnancy, litter size, resorptions, and fetal weight at term were not significantly different from controls (Kirksey and Pike, 1962). Neither were total sodium nor potassium levels of the fetus affected (Kirksey et al., 1962). High sodium intake by the mother also had no effect on these parameters.

Dancis and Springer (1970) have studied fetal homeostasis in rats in relation to potassium and sodium when these elements were deficient in the maternal diet. In both cases, the young at term were low in body weight, but potassium concentration in fetal plasma did not change, although it was decreased in fetal tissues. Sodium deficiency, on the other hand, produced proportional depression of plasma sodium in the fetus. Maternal hyperkalemia induced hyperkalemia in the fetus as well.

Sodium depletion of ewes brought about lower than normal levels of sodium in fetal plasma and amniotic fluid, and a higher than normal volume of allantoic fluid (Phillips and Sundaram, 1966).

Injection (subcutaneous) of sodium chloride at levels of 1900 and 2500 mg/kg at 10 or 11 days of gestation produced fetal death and malformation in Japanese dd strain mice (Nishimura and Miyamoto, 1969).

B. Trace Elements

1. Iron

The most noteworthy effect of iron deficiency in both animals and man is anemia characterized by low levels of hemoglobin. Iron deficiency during pregnancy produces a hypochromic microcytic anemia in the mother and results in the birth of young which also show low levels of hemoglobin. The degree of anemia in the offspring, as well as other effects, depends upon the degree of iron deficiency in the maternal diet.

Alt (1938), in studies of pregnant rats, showed that the degree of iron deficiency in offspring of female rats given a diet lacking in iron increased with the number of litters. In later work, using a diet more nearly complete than that of Alt but equally deficient in iron, O'Dell and coworkers (1961) found that the pregnant rats showed a mild anemia. Their offspring, however, were severely anemic, weak, and almost entirely nonviable although they were not grossly malformed. Similar observations were made in humans. The infants of anemic mothers frequently shared the iron deficiency and were also anemic. The more severe the anemia of the mother, the more severe was that of the infant at birth. Anemia in the newborn is important with regard to the development of anemia in later infancy (Sisson and Lund, 1958).

2. Copper

a. Deficiency

i. Lambs. Knowledge of the importance of copper for prenatal development arose from investigations of a disease in lambs in western Australia, called enzootic ataxia or "swayback." The disease is characterized by spastic paralysis especially of the hind limbs, severe incoordination, and in some cases blindness in newborn lambs. The brain of the affected animal is smaller than normal, with collapsed cerebral hemispheres and shallow convolutions. The cerebella are particularly small and throughout the brain there is marked demyelination (Bennetts, 1932; Innes, 1936; Innes and Shearer, 1940). Enzootic ataxia is primarily due to copper deficiency in the pregnant sheep, as was demonstrated by the low copper content of liver, blood, and milk of these ewes. The copper content of the livers of affected lambs was also below normal. This disease can be prevented by copper administration to ewes during gestation (Bennetts and Chapman, 1937; Bennetts and Beck, 1942).

A condition identical to enzootic ataxia was reported in lambs from ewes fed on seaweed (Palsson and Grimsson, 1953). More recently, another hypocuprotic disease similar but not identical to swayback has been reported in lambs (Roberts *et al.*, 1966a,b).

Ultrastructural studies of the central nervous system in lambs with swayback have shown an increased size of nerve cells, progressive lack of Nissl substance, abnormalities of the Golgi apparatus, and increase in neurofibrils. Alteration of neuronal and astroglial processes was also seen (Cancilla and Barlow, 1966a,b).

High intake of sulfate and molybdate also produce this disease by inducing a conditioned copper deficiency in the pregnant ewe (Mills and Fell, 1960). In such newborn lambs, there was a low level of copper in the brain and deficiency of cytochrome oxidase in the motor neurons (Fell *et al.*, 1961, 1965). It is not clear, however, that the activity of cytochrome oxidase can be correlated with incidence of swayback (Suttle and Field, 1968).

ii. Experimental Animals. Copper deficiency in rats resulted in a very high incidence of fetal death and resorption. In experiments in which rats were given a copper deficient diet for one month before mating, very few young survived to term (Dutt and Mills, 1960; Hall and Howell, 1969). In other work in which presumably the copper deficiency was not so severe, fetuses were able to survive to term (O'Dell *et al.*, 1961, 1968). However, the newborn young were anemic, nonviable, and about one fourth were edematous and had characteristic subcutaneous hemorrhages. In many cases there was localized ischemia of extremities such as feet and tip of the tail. The copper-deficient offspring showed a high incidence of skeletal anomalies and abdominal hernia. Marked changes were also apparent in histological sections of the skin, including a decreased number of hair follicles in the deficient animals.

Changes in the embryonic development of the nervous system similar to those in lambs with enzootic ataxia were produced by Everson and her colleagues in guinea pigs (1967). Newborn guinea pigs of females receiving a copper-deficient diet during growth and pregnancy showed a high incidence of ataxia, growth abnormalities, and gross abnormalities of the brain. Agenesis of the cerebellum was especially striking. Cerebral cortical tissue was often soft and translucent. Aneurisms of the aortic arch and the abdominal aorta were also seen. Liver copper values were abnormally low. Histological studies showed that cerebellar folia were missing or abnormal, and throughout the brain there was underdevelopment of myelin. Phospholipid determinations of whole brains were consistent with the histological evidence of depletion (Everson *et al.*, 1968). Skeletal anomalies were also seen (Asling and Hurley, 1963).

Recently an interaction between a mutant gene in mice (*crinkled*) and copper metabolism has been reported. Percentage of survival of mutant mice to 30 days of age was doubled by feeding their mothers a high-copper diet during pregnancy and lactation. High dietary copper also prevented the lag in pigment development, thinness of epidermis, and paucity of hair bulbs characteristic of the mutants (Hurley and Bell, 1974).

iii. Human. Mjølnerød and his colleagues (1971) reported a case in which a young woman was treated with penicillamine during pregnancy. She gave birth to a child with a connective tissue defect including lax skin, hyperflexibility of the joints, vein fragility, varicosities, and impaired wound healing. The authors concluded that the penicillamine was the cause of the defect. Since penicillamine is used to remove copper from patients with Wilson's disease, it is possible that the treatment produced copper deficiency in the fetus. However, since penicillamine also removes other trace elements such as zinc, the effect may involve other metals.

3. Iodine

a. Deficiency. The only known function of iodine is in relation to its role in the thyroid hormone. Thus, effects of iodine deficiency can be related to those of hypothyroidism. In the adult, this condition is associated with goiter and in the fetus with cretinism. Cretinism was recognized as early as the sixteenth century, and by the beginning of the nineteenth century its connection with goiter was realized. A connection between iodine and goiter was postulated soon thereafter, and the idea that endemic goiter and, by implication, endemic cretinism was due to iodine deficiency was proposed in the middle of the nineteenth century (Langer, 1960). Since then the causative connection between iodine deficiency and endemic goiter has become clear, but that between iodine deficiency and endemic cretinism is less so.

The cretinous child presents a clinical picture of mental and physical retardation with a potbelly, large tongue, and facies resembling those of Down's syndrome. The skin of the cretin is coarse and myxedematous. Shortness of stature, retarded dentition, and delayed epiphyseal development are characteristic. Deaf-mutism or impaired speech may also be present. Fierro-Benitez and his associates (1974) reported on the pattern of cretinism in an area of highland Ecuador where iodine deficiency is severe. Considerable variation was found in the various parameters of the condition that were measured.

The most convincing evidence that iodine is a causative factor in endemic cretinism is the fall in its incidence after the introduction of iodine prophylaxis. In certain regions where there was a high rate of endemic goiter, a high incidence of cretinism, and low iodine content of the soil, the incidence of endemic cretinism as well as endemic goiter fell markedly after the institution of iodized salt (Clements, 1960; Kelly and Snedden, 1960; Stanbury and Querido, 1956). Additional confirmatory evidence comes from a study in the New Guinea highlands where severe iodine deficiency was prevented by intramuscular iodized oil (Pharoah et al., 1971). To determine whether endemic cretinism was the result of iodine deficiency, alternate families were treated with either iodized oil or saline solution. Follow-up over 4 years revealed 26 endemic cretins out of a total of 534 children born to mothers who had not received iodized oil. In comparison, only 7 cases of endemic cretinism occurred among 498 children born to mothers who had been treated with iodized oil. In 6 of these 7 cases, the mother was pregnant when the trial began. In the mothers who had not used iodized oil, mothers of 5 of the cretins were pregnant at the beginning of the trial. These data suggest that severe iodine deficiency in the mothers produces neurological damage during fetal development.

Although it is clear that iodine deficiency is an important factor in the development of endemic cretinism, it is possible that other factors, genetic or environmental, may interact with iodine deficiency to produce this condition (Stanbury and Querido, 1956; 1957; Fierro-Benitez et al., 1974; Pharoah et al., 1971). Differential responses to low levels of intake may play a role (see also "Manganese" below).

In field animals, Hart and Steenbock reported in 1918 the birth of hairless pigs with iodine deficiency. The hairlessness was accompanied by hypertrophied thyroids. Stillborn or weak young also occurred.

b. Excess. An excessive amount of iodine taken by the pregnant woman as well as a deficiency of the element is deleterious to the developing fetus. Several cases of congenital goiter and hypothyroidism due to maternal ingestion of excess iodine have been reported (Parmelee et al., 1940). Neonatal deaths and mental retardation in the cretinous survivors were reported

(Carswell *et al.*, 1970). Similarly, cretinism resulting from maternal [131I]sodium iodide therapy during pregnancy has been well documented. In such cases, presumably, the condition developed through fetal thyroid damage by the radioactive iodide (Russell *et al.*, 1957; Green *et al.*, 1971). Similar observations have been made experimentally in dogs (Smith *et al.*, 1951).

4. Manganese

The first report that manganese is an essential element for animals also showed that it was necessary for the prenatal development of mammals. Orent and McCollum (1931) found that offspring of manganese-deficient rats were unable to survive the postnatal period, but cross-fostering experiments showed that the postnatal mortality was caused by debility of the offspring rather than by failure of lactation in the mother (Daniels and Everson, 1935). Offspring of manganese-deficient rats displayed a characteristic ataxia with incoordination, lack of equilibrium, and retraction of the head (Shils and McCollum, 1943). The ataxic condition has since been reported in a number of other mammalian species including the pig (Plumlee *et al.*, 1956), guinea pig (Everson *et al.*, 1959), and mouse (Erway *et al.*, 1970). The ataxia produced by prenatal manganese deficiency is irreversible, and the critical period for its production is between days 14 and 18 of gestation in the rat (Hurley *et al.*, 1958; Hurley and Everson, 1963). Development of body-righting reflexes in these animals was delayed (Hurley and Everson, 1969), presumably through abnormal development of the inner ear (Hurley *et al.*, 1960; Asling *et al.*, 1960a). The specific anatomical lesion apparently responsible for the abnormal equilibrium and body-righting reflexes of manganese-deficient offspring is defective development of the otoliths of the vestibular portion of the inner ear (Erway *et al.*, 1966, 1970), with missing or completely absent otoliths. Similar observations have been made in guinea pigs (Shrader and Everson, 1967) and rats (Hurley, 1968a).

The failure of otolith development resulting from manganese deficiency appears to be due to abnormal synthesis of mucopolysaccharides. *In vitro* studies of fetal cartilage showed reduced and delayed uptake of ^{35}S in tibias from manganese-deficient fetuses (Hurley *et al.*, 1968a). Incorporation of ^{35}S in the macular cells of the inner ear of manganese deficient fetuses was also reduced. Furthermore, a nonmetachromatic otolithic matrix which did not contain ^{35}S was observed (Shrader *et al.*, 1973).

Manganese deficiency during the prenatal period also results in abnormalities of the skeleton (Asling and Hurley, 1963) including disproportionate growth (Hurley *et al.*, 1961a,c) and tibial epiphyseal dysplasia (Hurley and Asling, 1963).

The ataxic condition of manganese-deficient offspring is also seen in mice homozygous for a mutant gene called *pallid* (Lyon, 1963), which is also caused by abnormal development of the otoliths. This congenital genetic ataxia can be prevented by giving a diet containing high levels of manganese to the pregnant female (Erway *et al.*, 1966, 1971). The failure of otolith development in *pallid* mice also seems to be due to abnormal synthesis of mucopolysaccharides as evidenced by uptake of ^{35}S (Shrader *et al.*, 1973) and may be related to ultrastructural changes (Bell and Hurley, 1973).

A mutant gene analogous to *pallid* also occurs in mink. In a recessive mutant, *pastel*, approximately 25% of the animals exhibit an ataxic condition as a pleiotropic effect of the gene. Erway and Mitchell (1973) have now demonstrated that this condition is caused by reduction or absence of otoliths and that it can be prevented by manganese supplementation of the pregnant female. The discovery of analogous genes interacting with manganese in at least two species, as well as the copper-*crinkled* example discussed above (Section IV. B. 2), suggests that similar interactions may occur in other species including man (Hurley, 1969).

The influence of the genetic background of mice on their response to manganese deficiency during prenatal development was studied with inbred strains (Hurley and Bell, 1974b). The results showed that although the usual normal level of dietary manganese was adequate for normal development in all strains studied (except of the mutant, *pallid*) the response to a low dietary intake of manganese varied considerably according to the genetic background. This suggests that what is considered to be the recommended dietary level of a nutrient will probably be sufficient to prevent signs of deficiency for most individuals. However, at low or borderline levels, responses of individuals will vary greatly depending in part on their genetic background. Similar relationships between genetic and nutritional factors may account for differences in incidence of cretinism in areas of low iodine intake (Section IV. B.3).

5. Zinc

Congenital malformations produced by zinc deficiency in mammals were first reported by Hurley and Swenerton (1966) in rats. When normal well-fed female rats were mated and given a zinc-deficient diet at the beginning of pregnancy, full-term young weighed about half as much as controls, half of the implantation sites were resorbed, and almost all fetuses showed gross congenital malformations. Food-intake controls had normal young (Hurley *et al.*, 1971). Teratogenic effects of dietary zinc deficiency have since been confirmed by several laboratories (Mills *et al.*, 1969; Warkany and Petering, 1972; Dreosti *et al.*, 1972). Shorter periods of deficiency were also teratogenic. When the zinc deficiency occurred from day 6 to 14 of pregnancy, about half the young were abnormal. Even when the deficiency was given for only the first 10 days of pregnancy, 22% of the full-term fetuses were malformed (Hurley *et al.*, 1971). Warkany and Petering (1973) have shown that with a deficiency period limited to 3 days (from the 10th through the 12th) a small but significant percentage of young had malformations including brain abnormalities.

Malformations developing from zinc deficiency affected every organ system (Hurley *et al.,* 1971), and abnormalities of the nervous system were of high incidence (Hurley and Shrader, 1972). A lesion characteristic of zinc deficiency in juvenile animals is hyperplasia of the esophageal mucosa, and zinc-deficient fetuses also showed this typical histological abnormality (Diamond and Hurley, 1970).

In order to investigate the mechanisms bringing about these disturbances of morphogenesis, a number of enzymes were assayed both histochemically and chemically in full-term fetuses and in severely deficient growing animals. In most cases no differences were seen between control and zinc-deficient animals. There were some differences in cellular localization of enzyme activity, but in general it appeared that intensity of enzyme activity was a function of the cell type present and that enzyme changes were probably not causative factors in producing the congenital malformations of zinc deficiency (Hurley *et al.,* 1968b; Swenerton and Hurley, 1968; Swenerton *et al.,* 1972).

The congenital malformations of zinc-deficient embryos appeared to be related to impaired synthesis of nucleic acid. In zinc deficient embryos at 12 days of gestation the uptake of tritiated thymidine was much lower than in normal embryos, suggesting that DNA synthesis was depressed (Swenerton *et al.,* 1969). Other evidence from different systems also indicates a requirement for zinc in DNA synthesis (Falchuk *et al.,* 1975).

The effects of zinc deficiency in rats appear very rapidly, because of the need for a constant intake of zinc to maintain plasma levels. Plasma zinc concentration in pregnant rats was observed to drop sharply. After only 24 hr of the deficiency, plasma zinc fell by approximately 40% (Dreosti *et al.,*1968). The pregnant rat apparently cannot mobilize zinc from maternal tissues in amounts sufficient to supply the need for normal fetal development (Hurley and Swenerton, 1971). Therefore, the hypothesis was developed that zinc could be released from the skeleton only under conditions in which there is a breakdown of bone, and it was tested by comparing the effects of a diet lacking calcium as well as zinc with the effects of a diet deficient in zinc alone. The teratogenic effects of zinc deficiency were alleviated by simultaneous lack of dietary calcium (Hurley and Tao, 1972) in intact but not in parathyroidectomized rats (Tao and Hurley, 1975), indicating that conditions bringing about resorption of bone increase the availability of skeletal zinc.

A similar effect occurred when increased breakdown of maternal tissues was brought about by giving pregnant females a diet deficient in protein as well as zinc. The percentage of implantation sites affected was lower in rats receiving a diet deficient in both protein and zinc than in those receiving a diet deficient in zinc alone (Hurley *et al.,* 1973).

The effect of milder states of zinc deficiency was also investigated by using various levels of dietary zinc. With diets containing less than 9 ppm zinc during pregnancy, the level of zinc correlated inversely with the incidence of fetal death and malformation. Both total litter weight and fetal weight at term varied with the level of dietary zinc up to 14 ppm, but no correlation was seen between the fetal zinc content or maternal plasma zinc at term and the incidence of malformation. Plasma zinc measured during the second trimester, however, did correlate with the incidence of malformation (Hurley and Cosens, 1974).

The results of mild, rather than severe, zinc deficiency were also studied by investigating the effects of transitory deficiency during prenatal life. When normal pregnant rats were given a zinc-deficient diet from days 6 to 14 of gestation, the maternal plasma zinc level fell rapidly but also returned quickly to original values. The offspring showed low birth weight, high incidence of congenital malformation, high rate of stillbirth, and very poor survival to weaning, with most of the postnatal mortality in the first week. The concentration of zinc in the plasma of the young was normal as was that of maternal plasma and milk, suggesting that postnatal zinc nutriture was adequate. These results showed that a short period of zinc deficiency during prenatal life caused an irreversible change which subsequently affected postnatal development (Hurley and Mutch, 1973).

The rapid effect of zinc deficiency on the developing embryo is apparent after only 3 days of zinc deficiency in the pregnant rat. When pregnant females were subjected to the zinc deficiency regime at the beginning of pregnancy (day 0), abnormal changes were seen in the preimplantation blastocysts. On day 4 of gestation, 97% of eggs from controls were characterized as normal blastocysts, while only 18% of eggs from zinc-deficient females could be so classified. Even on day 3 of gestation, abnormal cleavage was apparent in the preimplantation egg (Hurley and Shrader, 1975).

The observation of chromosomal aberrations in zinc-deficient animals is also consistent with the idea that zinc deficiency results in abnormal nucleic acid synthesis. In cytogenetic studies, chromosome spreads from both fetal liver and maternal bone marrow of zinc-deficient animals showed chromosomal aberrations, especially gaps and terminal deletions in significant incidence (Bell *et al.*, 1976).

The possibility that zinc deficiency may play a role in human congenital malformations has been proposed (Hurley, 1968b, 1974a,b; Warkany and Petering, 1972; Sever and Emanuel, 1973; Emanuel and Sever, 1973; Burch, 1976). The teratogenic effects of other metals have been reviewed by Ferm (1972) and in Chapter 9, this volume.

REFERENCES

Aksu, O., Mackler, B., Shepard, T. H., and Lemire, R. J., 1968, Studies of the development of congenital anomalies in embryos of riboflavin-deficient, galactoflavin-fed rats. II. Role of the terminal electron transport system, *Teratology* **1**:93.

Allan, J. D., and Brown, J. K., 1968, Maternal phenylketonuria and foetal brain damage, an attempt at prevention by dietary control, *in: Some Recent Advances in Inborn Errors of Metabolism* (K. Holt and V. Coffey, eds.), pp. 14–39, E. & S. Livingstone Ltd., Edinburgh.

Allen, L. H., and Zeman, F. J., 1971, Influence of increased postnatal food intake on body composition of progeny of protein-deficient rats, *J. Nutr.* **101:**1311.

Allen, L. H., and Zeman, F. J., 1973a, Influence of increased postnatal nutrient intake on kidney cellular development in progeny of protein-deficient rats, *J. Nutr.* **103:**929.

Allen, L. H., and Zeman, F. J., 1973b, Kidney function in the progeny of protein-deficient rats, *J. Nutr.* **103:**1467.

Alling, C., Bruce, A., Karlsson, I., Sapia, O., and Svennerholm, L., 1972, Effect of maternal essential fatty acid supply on fatty acid composition of brain, liver, muscle and serum in 21-day-old rats, *J. Nutr.* **102:**773.

Alt, H. L., 1938, Iron deficiency in pregnant rats, its effect on the young, *Am. J. Dis. Child.* **56:**975.

Andersen, D. H., 1949, Effect of diet during pregnancy upon the incidence of congenital hereditary diaphragmatic hernia in the rat. Failure to produce cystic fibrosis of the pancreas by maternal vitamin A deficiency, *Am. J. Pathol.* **25:**163.

Antonov, A. N., 1947, Children born during the seige of Leningrad in 1942, *J. Pediatr.* **30:**250.

Asling, C. W., 1961, Congenital defects of face and palate in rats following maternal deficiency of pteroylglutamic acid, *in: Congenital Anomalies of the Face and Associated Structures* (S. Pruzansky, ed.), pp. 173–187, Charles C Thomas, Springfield, Illinois.

Asling, C. W., and Hurley, L. S., 1963, The influence of trace elements on the skeleton. *Clin. Orthopaed.* **27:**213.

Asling, C. W., Nelson, M. M., Wright, H. V., and Evans, H. M., 1955, Congenital skeletal abnormalities in fetal rats resulting from maternal pteroylglumatic acid deficiency during gestation, *Anat. Rec.* **121:**775.

Asling, C. W., Hurley, L. S., and Wooten, E., 1960a, Abnormal development of the otic labyrinth in young rats following maternal dietary manganese deficiency, *Anat. Rec.* **136:**157.

Asling, C. W., Nelson, M. M., Dougherty, H. D., Wright, H. V., and Evans, H. M., 1960b, The development of cleft palate resulting from maternal pteroylglutamic (folic) acid deficiency during the latter half of gestation in rats, *Surg. Gynecol. Obstet.* **111:**19.

Baird, C. D. C., Nelson, M. M., Monie, I. W., and Evans, H. M., 1954, Congenital cardiovascular anomalies induced by pteroylglumatic acid deficiency during gestation in the rat, *Circ. Res.* **2:**544.

Bannon, S. L., Higginbottom, R. M., McConnell, J. M., and Kaan, H. W., 1945, Development of galactose cataract in the albino rat embryo, *Arch. Ophthalmol.* **33:**224.

Barnes, R. H., 1971, Nutrition and man's intellect and behavior, *Fed. Proc.* **30:**1429.

Bawden, J. W., and Osborne, J. W., 1962, Tracer study on the effect of dietary calcium deficiency during pregnancy in rats, *J. Dent. Res.* **41:**1349.

Bell, L. T., and Hurley, L. S., 1973, Ultrastructural effects of manganese deficiency in liver, heart, kidney and pancreas of mice. *Lab. Invest.* **29:**723.

Bell, L. T., Branstator, M., Roux, C., and Hurley, L. S., 1976, Chromosomal abnormalities in maternal and fetal tissues of magnesium or zinc deficient rats, *Teratology* **12:**221.

Bennetts, H. W., 1932, Enzootic ataxia of lambs in Western Australia, *Aust. Vet. J.* **8:**137.

Bennetts, H. W., and Chapman, F. E., 1937, Copper deficiency in sheep in Western Australia: A preliminary account of the aetiology of enzootic ataxia of lambs and an anaemia of ewes, *Aust. Vet. J.* **13:**138.

Bennetts, H. W., and Beck, A. B., 1942, Enzootic ataxia and copper deficiency of sheep in Western Australia, *Australian Council Sci. Ind. Res., Bull. No. 147.*

Berg, B. N., 1965, Dietary restriction and reproduction in the rat, *J. Nutr.* **87:**344.

Berg, B. N., Sigg, E. B., and Greengard, P., 1967, Maintenance of pregnancy in protein-deficient rats by adrenocortical steroid or ACTH administration, *Endocrinology* **80:**829.

Bergner, L., and Susser, M. W., 1970, Low birth weight and prenatal nutrition: An interpretative review, *Pediatrics* **46:**946.

Bernhardt, I. B., and Dorsey, D. J., 1974, Hypervitaminosis A and congenital renal anomalies in a human infant, *Obstet. Gynecol.* **43:**750.

Bessman, S. P., 1972, Genetic failure of fetal amino acid "justification": A common basis for many forms of metabolic, nutritional, and "nonspecific" mental retardation, *J. Pediatr.* **81:**834.

Blackwell, B. N., Blackwell, R. Q., Yu, T. T. S., Weng, Y. S., and Chow, B. F., 1969, Further studies on growth and feed utilization in progeny of underfed mother rats, *J. Nutr.* **97**:79.

Bodansky, M., and Duff, V. B., 1941, Relation of parathyroid function and diet to the mineral composition of the bones in rats at the conclusion of pregnancy, *J. Nutr.* **21**:235.

Boelter, M. D. D., and Greenberg, D. M., 1943, Effect of severe calcium deficiency on pregnancy and lactation in the rat, *J. Nutr.* **26**:105.

Bologna, U., and Piccioni, V., 1956, Effetti dell'ipovitaminosi B₂ sulla gravidanza e sui nati del ratto albino, *Soc. Ital. Biol. Speriment.* **26**:219.

Brasel, J. A., and Winick, M., 1972, Maternal nutrition and prenatal growth. Experimental studies of effects of maternal undernutrition on fetal and placental growth, *Arch. Dis. Child.* **47**:479.

Braun, K., Bromberg, Y. M., and Brzezinski, A., 1945, Riboflavin deficiency in pregnancy, *J. Obstet. Gynecol.* **52**:43.

Brimblecombe, F. S. W., Ashford, J. R., and Fryer, J. G., 1968, Significance of low birth weight in perinatal mortality, *Br. J. Prev. Soc. Med.* **22**:27.

Brown, M. L., and Snodgrass, C. H., 1965, Effect of dietary level of thiamine on reproduction in the rat, *J. Nutr.* **85**:102.

Brzezinski, A., Bromberg, Y. M., and Braun, K., 1947, Riboflavin deficiency in pregnancy, its relationship to the course of pregnancy and to the condition of the foetus, *J. Obstet. Gynecol.* **54**:182.

Burch, R. E., and Sullivan, J. F., 1976, Clinical and nutritional aspects of zinc deficiency and excess, *in: Symposium on Trace Elements,* (R. E. Burch and J. F. Sullivan, eds.) *Med. Clin. N. Amer.* **60**: 675–685.

Burke, B. S., and Stuart, H. C., 1948, Nutritional requirements during pregnancy and lactation, *J. Am. Med. Assoc.* **137**:119.

Burke, B. S., Beal, V. A., Kirkwood, S. B., and Stuart, H. C., 1943, The influence of nutrition during pregnancy upon the condition of the infant at birth, *J. Nutr.* **26**:569.

Burr, G. O., and Burr, M. M., 1930, On the nature and role of the fatty acids essential in nutrition, *J. Biol. Chem.* **86**:587.

Butcher, R. E., Brunner, R. L., Roth, T., and Kimmel, C. A., 1972, A learning impairment associated with maternal hypervitaminosis-A in rats, *Life Sci. I* **11**(3):141.

Cancilla, P. A., and Barlow, R. M., 1966a, Structural changes of the central nervous system in swayback (enzootic ataxia) of lambs. II. Electron microscopy of the lower motor neuron, *Acta Neuropathol.* **6**:251.

Cancilla, P. A., and Barlow, R. M., 1966b, Structural changes of central nervous system in swayback (enzootic ataxia) of lambs. III. Electron microscopy of the cerebral lesion, *Acta Neuropathol.* **6**:260.

Carswell, F., Kerr, M. M., and Hutchison, J. H., 1970, Congenital goitre and hypothyroidism produced by maternal ingestion of iodides, *Lancet* **1**:1241.

Chamberlain, J. G., 1966, Development of cleft palate induced by 6-aminonicotinamide late in rat gestation, *Anat. Rec.* **156**:31.

Chamberlain, J. G., and Nelson, M. M., 1963a, Congenital abnormalities in the rat resulting from single injections of 6-aminonicotinamide during pregnancy, *J. Exp. Zool.* **153**:285.

Chamberlain, J. G., and Nelson, M. M., 1963b, Multiple congenital abnormalities in the rat resulting from acute maternal niacin deficiency during pregnancy, *Proc. Soc. Exp. Biol. Med.* **112**:836.

Chase, H. P., 1973, The effects of intrauterine and postnatal undernutrition on normal brain development, *Ann. N.Y. Acad. Sci.* **205**:231.

Cheng, D. W., Chang, L. F., and Bairnson, T. A., 1957, Gross observations on developing abnormal embryos induced by maternal vitamin E deficiency, *Anat. Rec.* **129**:167.

Cheng, D. W., Bairnson, T. A., Rao, A. N., and Subbammal, S., 1960, Effect of variations of rations on the incidence of teratogeny in vitamin E deficient rats, *J. Nutr.* **71**:54.

Cheng, D. W:, Braun, K. G., Braun, B. J., and Udani, K. H., 1961, Tocopherol content of maternal and fetal rat tissues as related to vitamin E intake during gestation, *J. Nutr.* **74**:111.

Chepenik, K. P., Johnson, E. M., and Kaplan, S., 1970, Effects of transitory maternal pteroylglutamic acid (PGA) deficiency on levels of adenosine phosphates in developing rat embryos. *Teratology* **3**:229.

Chow, B. F., 1964, Growth of rats from normal dams restricted in diet in previous pregnancies, *J. Nutr.* **83**:289.

Chow, B. F., and Lee, C. J., 1964, Effect of dietary restriction of pregnant rats on body weight gain of the offspring, *J. Nutr.* **82**:10.

Christian, J. R., and Diamond, E. F., 1961, Galactosemia, a family study, *Am. J. Dis. Child.* **101**:75.

Clark, G. M., Zamenhof, S., van Marthens, E., Grauel, L., and Kruger, L., 1973, The effect of prenatal malnutrition on dimensions of cerebral cortex, *Brain Res.* **54**:397.

Clausen, J., 1969, The effect of vitamin A deficiency on myelination in the central nervous system of the rat, *Eur. J. Biochem.* **7**:575.

Clements, F. W., 1960, Health significance of endemic goitre and related conditions. *in: Endemic Goitre*, pp. 245–260, World Health Organization, Geneva.

Cohlan, S. Q., 1953, Excessive intake of vitamin A as a cause of congenital anomalies in the rat, *Science* **117**:535.

Cohlan, S. Q., and Stone, S. M., 1961, Effects of dietary and intraperitoneal excess of L-lysine and L-leucine on rat pregnancy and offspring, *J. Nutr.* **74**:93.

Cohlan, S. Q., Jansen, V., Dancis, J., and Piomelli, S., 1970, Microcytic anemia with erythroblastosis in the offspring of magnesium deprived rats. *Blood* **36**:500.

Comar, C. L., 1956, Radiocalcium studies in pregnancy, *Ann. N.Y. Acad. Sci.* **64**:281.

Committee on Maternal Nutrition/Food and Nutrition Board, National Research Council, 1970, *Maternal Nutrition and the Course of Pregnancy*, National Academy of Sciences, Washington, D.C.

Cox, W. M., Jr., and Imboden, M., 1936, The role of calcium and phosphorus in determining reproductive success, *J. Nutr.* **11**:147.

Dakshinamurti, K., and Stephens, M. C., 1969, Pyridoxine deficiency in the neonatal rat, *J. Neurochem.* **16**:1515.

Dancis, J., and Springer, D., 1970, Fetal homeostatsis in maternal malnutrition: Potassium and sodium deficiency in rats, *Pediatr. Res.* **4**:345.

Dancis, J., Springer, D., and Cohlan, S. Q., 1971, Fetal homeostasis in maternal malnutrition. II. Magnesium deprivation. *Pediatr. Res.* **5**:131.

Daniels, A. L., and Everson, G. J., 1935, The relation of manganese to congenital debility. *J. Nutr.* **9**:191.

Danks, D. M., Campbell, P. E., Stevens, B. F., Mayne, V., and cartwright, E., 1972, Menkes's kinky hair syndrome. An inherited defect in copper absorption with widespread effects. *Pediatrics* **50**:188.

Davis, S. D., Nelson, T., and Shepard, T. H., 1970, Teratogenicity of vitamin B_6 deficiency: Omphalocele, skeletal and neural defects, and splenic hypoplasia, *Science* **169**:1329.

DeLuca, H. F., 1974, Vitamin D: The vitamin and the hormone, *Fed. Proc.* **33**:2211.

Deuel, H. J., Jr., Martin, C. R., and Alfin-Slater, R. B., 1954, The effect of fat level of the diet on general nutrition. XII. The requirement of essential fatty acids for pregnancy and lactation, *J. Nutr.* **54**:193.

Diamond, I., and Hurley, L. S., 1970, Histopathology of zinc-deficient fetal rats. *J. Nutr.* **100**:325.

Dickerson, J. W. T., and Hughes, P. C. R. 1972, Growth of the rat skeleton after severe nutritional intrauterine and post-natal retardation. *Resuscitation* **1**:163.

Dingle, J. T., and Lucy, J. A., 1965, Vitamin A. carotenoids and cell formation, *Biol. Rev.* **40**:(3):422.

DiSaia, P. J., 1966, Pregnancy and delivery of a patient with a Starr–Edwards mitral valve prosthesis, *Obstet. Gynecol.* **28**:469.

Dobbing, J., 1972, Vulnerable periods of brain development. *In: Lipids, Malnutrition and the Developing Brain* (A. Von Muralt, ed.), p. 9, Assoc. Sci. Pub., Amsterdam.

Dreosti, I. E., Tao, S., and Hurley, L. S., 1968, Plasma zinc and leukocyte changes in weanling and pregnant rats during zinc deficiency. *Proc. Soc. Exp. Biol. Med.* **127**:169.

Dreosti, I. E., Grey, P. C., and Wilkens, D. J., 1972, Deoxyribonucleic acid synthesis, protein synthesis and teratogenesis in zinc-deficient rats. *S. Afr. Med. J.* **46:**1585.

Dutt, B., and Mills, C. F., 1960, Reproductive failure in rats due to copper deficiency, *J. Comp. Pathol.* **70:**120.

Eberle, E. D., and Eiduson, S., 1968, Effect of pyridoxine deficiency on aromatic L-amino acid decarboxylase in the developing rat liver and brain, *J. Neurochem.* **15:**1071.

Emanuel, I., and Sever, L. E., 1973, Questions concerning the possible association of potatoes and neural-tube defects, and an alternative hypothesis relating to maternal growth and development. *Teratology* **8:**325.

Ensminger, M. E., Bowland, J. P., and Cunha, T. J., 1947, Observations on the thiamine, riboflavin, and choline needs of sows for reproduction *J. Anim. Sci.* **6:**409.

Ershoff, B. H., and Kruger, L., 1962, A neurological defect in the offspring of rats fed a pantothenic acid-deficient diet during pregnancy. *Exp. Med. Surg.* **20:**180.

Erway, L. C., and Mitchell, S. E., 1973, Prevention of otolith defect in pastel mink by manganese supplementation. *J. Hered.* **64:**110.

Erway, L., Hurley, L. S., and Fraser, A., 1966, Neurological defect: Manganese in phenocopy and prevention of a genetic abnormality of inner ear. *Science* **152:**1766.

Erway, L., Hurley, L. S., and Fraser, A., 1970, Congenital ataxia and otolith defects due to manganese deficiency in mice. *J. Nutr.* **100:**643.

Erway, L., Fraser, A., and Hurley, L. S., 1971, Prevention of congenital otolith defect in *Pallid* mutant mice by manganese supplementation. *Genetics* **67:**97.

Evans, H. M., and Bishop, K. S., 1922, On the existence of a hitherto unrecognized dietary factor essential for reproduction, *Science* **56:**650.

Evans, H. M., Lepkovsky, S., and Murphy, E. A., 1934a, Vital need of the body for certain unsaturated fatty acids. IV. Reproduction and lactation upon fat-free diet, *J. Biol. Chem.* **106:**431.

Evans, H. M., Lepkovsky, S., and Murphy, E. A., 1934b, Vital need of the body for certain unsaturated fatty acids. V. Reproduction and lactation upon diets containing saturated fatty acids as their sole source of energy, *J. Biol. Chem.* **106:**441.

Everson, G., Northrop, L., Chung, N. Y., Getty, R., and Pudelkewicz, C., 1954, Effect of varying the intake of calcium pantothenate of rats during pregnancy. I. Chemical findings in the young at birth, *J. Nutr.* **53:**341.

Everson, G. J., Hurley, L. S., and Geiger, J. F., 1959, Manganese deficiency in the guinea pig. *J. Nutr.* **68:**49.

Everson, G. J., Tsai, H.-C. C., and Wang, T., 1967, Copper deficiency in the guinea pig, *J. Nutr.* **93:**533.

Everson, G. J., Shrader, R. E., and Wang, T., 1968, Chemical and morphological changes in the brains of copper-deficient guinea pigs, *J. Nutr.* **96:**115.

Fairney, A., and Weir, A. A., 1970, The effect of abnormal maternal plasma calcium levels on the offspring of rats, *J. Endocrinol.* **48:**337.

Falchuk, K. H., Fawcett, D. W., and Vallee, B. L., 1975, Role of zinc in cell division of *Euglena Gracilis, J. Cell Sci.* **17:**57.

Felig, P., and Lynch, V., 1970, Starvation in human pregnancy: Hypoglycemia, hypoinsulinemia, and hyperketonemia, *Science* **170:**990.

Fell, B. F., Williams, R. B., and Mills, C. F., 1961, Further studies of nervous-tissue degeneration resulting from "conditioned" copper deficiency in lambs. *Proc. Nutr. Soc.* **20:**xxvii.

Fell, B. F., Mills, C. F., and Boyne, R., 1965, Cytochrome oxidase deficiency in the motor neurones of copper-deficient lambs: A histochemical study, *Res. Vet. Sci.* **6:**170.

Fellers, F. X., and Schwartz, R., 1958, Etiology of the severe form of idiopathic hypercalcemia of infancy, *N. Engl. J. Med.* **259:**1050.

Ferm, V. H., 1972, The teratogenic effects of metals on mammalian embryos, *Adv. Teratol.* **5:**51.

Ferm, V. H., 1967, Potentiation of the teratogenic effect of vitamin A with exposure to low environmental temperature, *Life Sci.* **6:**493.

Fierro-Benitez, R., Ramirez, I., Garces, J., Jaramillo, C., Moncayo, F., and Stanbury, J. B., 1974, The clinical pattern of cretinism as seen in highland Ecuador, *Am. J. Clin. Nutr.* **27**:531.

Fillmore, S. J., and McDevitt, E., 1970, Effects of coumarin compounds on the fetus, *Ann. Intern. Med.* **73**:731.

Finnerty, J. J., and Mackay, B. R., 1962, Antepartum thrombophlebitis and pulmonary embolism, *Obstet. Gynecol.* **19**:405.

Fisch, R. O., Doeden, D., Lansky, L. L., and Anderson, J. A., 1969, Maternal phenylketonuria, *Am. J. Dis. Child.* **118**:847.

Forfar, J. O., Tompsett, S. L., and Forshall, W., 1959, Biochemical studies in idiopathic hypercalcaemia of infancy, *Arch. Dis. Child.* **34**:525.

Fraser, J. L., and Watt, H. J., 1964, Megaloblastic anemia in pregnancy and the puerperium, *Am. J. Obstet. Gynecol.* **89**:532.

Fratta, I. D., 1969, Nicotinamide deficiency and thalidomide: Potential teratogenic disturbances in Long–Evans rats, *Lab. Anim. Care* **19**:727.

Fratta, I., Zak, S. B., Greengard, P., and Sigg, E. B., 1964, Fetal death from nicotinamide-deficient diet and its prevention by chlorpromazine and imipramine, *Science* **145**:1429.

Friedman, W. F., and Mills, L. F., 1969, The relationship between vitamin D and the craniofacial and dental anomalies of the supravalvular aortic stenosis syndrome, *Pediatrics* **43**:12.

Friedman, W. F., and Roberts, W. C., 1966, Vitamin D and the supravalvular aortic stenosis syndrome, the transplacental effects of vitamin D on the aorta of the rabbit, *Circulation* **34**:77.

Gal, I., Sharman, I. M., and Pryse-Davies, J., 1972, Vitamin A in relation to human congenital malformations. *Adv. Teratol.* **5**:143.

Giroud, A., and Boisselot, J., 1947, Répercussions de l'avitaminose B$_2$ sur l'embryon du rat, *Arch. Fr. Pediatr.* **4**:319.

Giroud, A. and Lefèbvres-Boisselot, J., 1951, Anomalies provoquées chez le foetus en l'absence d'acide folique, *Arch. Fr. Pediatr.* **8**:648.

Giroud, A., and Martinet, M., 1954, Fentes du palais chez l'embryon de rat par hypervitaminose A. *C.R. Soc. Biol.* **148**:1742.

Giroud, A., and Martinet, M., 1955a, Hypervitaminose A et anomalies chez le foetus de rat, *Rev. Intern. Vitaminol.* **26**:10.

Giroud, A., and Martinet, M., 1955b, Production d'un excès de liquide amniotique dan l'anencéphalie, *C.R. Soc. Biol.* **149**:452.

Giroud, A., and Martinet, M., 1955c, Malformations diverses du foetus de rat suivant les stades d'administration de vitamine A en excès, *C.R. Soc. Biol.* **149**:1088.

Giroud, A., and Martinet, M., 1956a, Tératogenèse par hautes doses de vitamine A en fonction des stades du développement, *Arch. Anat. Microsc. Morphol. Exp.* **45**:7.

Giroud, A., and Martinet, M., 1956b, Malformations de la face et hypervitaminose A, *Rev. Stomatol.* **57**:454.

Giroud, A., and Martinet, M., 1956c, Hypervitaminose A et malformations congénitales, *Ann. Pediatr. (Semaine des Hopitaux)* **32**:1.

Giroud, A., and Martinet, M., 1956d, Hypovitaminose et hypervitaminose A chez le jeune et chez l'embryon, *Etud. Néo-Natales* **5**:55.

Giroud, A., and Martinet, M., 1959a, Tératogénèse par hypervitaminose A chez lè rat, la souris, le cobaye et le lapin, *Arch. Fr. Pediatr.* **16**:1.

Giroud, A., and Martinet, M., 1959b, Anencéphalie chez la souris par hypervitaminose A, *C.R. Assoc. Anat.*, XLVI Réunion, Montpellier, March 22–26, p. 288.

Giroud, A., and Martinet, M., 1959c, Malformations oculaires avec fibrose du vitré chez des embryons de lapin soumis à l'hypervitaminose A, *Bull. Soc. Ophtalmol.* **1959**(3):1.

Giroud, A., and Martinet, M., 1962, Légèreté de la dose tératogène de la vitamine A, *C.R. Soc. Biol.* **156**:449.

Giroud, A., Lévy, G., Lefèbvres, J., and Dupuis, R., 1952, Chute du taux de la riboflavine au stade ou se déterminent les malformations embryonnaires, *Intern. Z. Vitaminf.* **23**:490.

Giroud, A., Lévy, G., and Lefèbvres, J., 1954, Recherches sur le taux de l'acide pantothénique chez les mères et les foetus normaux et chez les mères carencées, *Rev. Intern. Vitaminol.* **25**:148.

Giroud, A., Lefèbvres, J., Prost, H., and Dupuis, R., 1955, Malformations des membres dues à des lésions vasculaires chez le foetus de rat déficient en acide pantothénique, *J. Embryol. Exp. Morphol.* **3**:1.

Giroud, A., Lefèbvres, J., and Dupuis, R., 1956, Action partiellement compensatrice de doses élevées d'acide ascorbique au cours de la gestation, chez des rattes déficientes en acide pantothénique, *C.R. Soc. Biol.* **150**:1735.

Giroud, A., Delmas, A., Prost, H., and Lefèbvres, J., 1957, Malformations encéphaliques par carence en acide pantothénique et leur interprétation, *Acta Anat.* **29**:17.

Giroud, A., Martinet, M., and Roux, C., 1958a, Anencéphalie, encéphalocèles, ménigocèles par hypervitaminose A, *Arch. Fr. Pediatr.* **15**:(6):1.

Giroud, A., Martinet, M., and Roux, C., 1958b, Urétéro-hydronéphrose expérimentale chez l'embryon par hypervitaminose A, *Arch. Fr. Pediatr.* **15**:(4):1.

Giroud, A., Martinet, M., and Solère, M., 1958c, Encéphalocèles, méningocèles par hypervitaminose A et considérations cliniques, *Rev. Neurol.* **98**:13.

Giroud, A., Delmas, A., and Martinet, M., 1959, Etude morphogénétique sur des embryons anencéphales, *Arch. Anat. Histol. Embryol. Norm. Exper.* **42**:203.

Giroud, A., Boisselot-Lefèbvres, J., and Dupuis, R., 1961, Carence tératogène en acide pantothénique. Légèreté de la carence, *Bull. Soc. Chim. Biol.* **43**:859.

Goodenday, L. S., and Gordan, G. S., 1971, No risk from vitamin D in pregnancy, *Ann. Intern. Med.* **75**:807.

Gortner, R. A., Jr., and Ekwurtzel, J. B., 1965, Incidence of teratogeny induced by vitamin E deficiency in the rat, *Proc. Soc. Exp. Biol. Med.* **119**:1069.

Grainger, R. B., O'Dell, B. L., and Hogan, A. G., 1954, Congenital malformations as related to deficiencies of riboflavin and vitamin B_{12}, source of protein, calcium to phosphorus ratio and skeletal phosphorus metabolism, *J. Nutr.* **54**:33.

Green, H. G., Gareis, F. J., Shepard, T. H., and Kelley, V. C., 1971, Cretinism associated with maternal sodium iodide I[131] therapy during pregnancy, *Am. J. Dis. Child.* **122**:247.

Greengard, P., Sigg, E. B., Fratta, I., and Zak, S. B., 1966, Prevention and remission by adrenocortical steroids of nicotinamide deficiency disorders and of 6-aminonicotinamide toxicity in rats and dogs, *J. Pharmacol. Exp. Ther.* **154**:624.

Hale, F., 1933, Pigs born without eye balls, *J. Hered.* **24**:105.

Hale, F., 1934, The relation of vitamin A to the eye development in the pig, Record of Proceedings, Annual Meeting, 1934, American Society of Animal Production, p. 126.

Hale, F., 1935, The relation of vitamin A to anophthalmos in pigs, *Am. J. Ophthalmol.* **18**:1087.

Hale, F., 1937, The relation of maternal vitamin A deficiency to microphthalmia in pigs, *Tex. State J. Med.* **33**:37.

Hall, G. A., and Howell, J. McC., 1969, The effect of copper deficiency on reproduction in the female rat, *Br. J. Nutr.* **23**:41.

Hall, M. H., 1972, Folic acid deficiency and congenital malformation, *J. Obstet. Gynaecol. Br.* **79**:159.

Hall, S. M., and Zeman, F. J., 1968, Kidney function of the progeny of rats fed a low protein diet. *J. Nutr.* **95**:49.

Hart, E. B., and Steenbock, H., 1918, Thyroid hyperplasia and the relation of iodine to the hairless pig malady, *J. Biol. Chem.* **33**:313.

Hart, E. B., McCollum, E. V., Steenbock, H., and Humphrey, G. C., 1911, Physiological effect on growth and reproduction of rations balanced from restricted sources, *Wis. Exp. Stn. Res. Bull.* **17**:160.

Hart, E. B., Steenbock, H., Humphrey, G. C., and Hulce, R. S., 1924, New observations and a reinterpretation of old observations on the nutritive value of the wheat plant, *J. Biol. Chem.* **62**:315.

Haworth, J. C., Ford, J. D., and Younoszai, M. K., 1969, Effect of galactose toxicity on growth of the rat fetus and brain, *Pediatr. Res.* **3**:441.

Hibbard, B. M., 1964, The role of folic acid in pregnancy, *J. Obstet. Gynecol.* **71**:529.

Hibbard, B. M., and Hibbard, E. D., 1966, Recurrence of defective folate metabolism in successive pregnancies, *J. Obstet. Gynaecol.* **73**:428.

Hibbard, E. D., and Smithells, R. W., 1965, Folic acid metabolism and human embryopathy, *Lancet* **1**:1254.

Hirsh, J., Cade, J. F., and O'Sullivan, E. F., 1970, Clinical experience with anticoagulant therapy during pregnancy, *Br. Med. J.* **1**:270.

Hogan, A. G., O'Dell, B. L., and Whitley, J. R., 1950, Maternal nutrition and hydrocephalus in newborn rats, *Proc. Soc. Exp. Biol. Med.* **74**:272.

Hsueh, A. M., Blackwell, R. Q., and Chow, B. F., 1970, Effect of maternal diet in rats on feed consumption of the offspring, *J. Nutr.* **100**:1157.

Hurley, L. S., 1968a, Approaches to the study of nutrition in mammalian development, *Fed. Proc.* **27**:193.

Hurley, L. S., 1968b, The consequences of fetal impoverishment, *Nutr. Today* **3**:3.

Hurley, L. S., 1969, Nutrients and genes: Interactions in development. *Nutr. Rev.* **27**:3.

Hurley, L. S., 1971, Magnesium deficiency during pregnancy and its effect on the offspring. A comprehensive review. *Proceedings of First International Symposium on Magnesium Deficiency in Human Pathology*, pp. 481–492, May, Vittel, France, Imprimerie Amelot, 27 Brionne, France.

Hurley, L. S., 1974a, Aspects of mineral metabolism in the perinatal period. *in: Perinatal Pharmacology: Problems and Priorities*, (J. Dancis and J. C. Hwang, eds.), pp. 149–158, Raven Press, New York.

Hurley, L. S., 1974b, Zinc deficiency, potatoes, and congenital malformations in man, *Teratology* **10**:205.

Hurley, L. S., and Asling, C. W., 1963, Localized epiphyseal dysplasia in offspring of manganese-deficient rats, *Anat. Rec.* **145**:25.

Hurley, L. S., and Bell, L. T., 1974a, Amelioration by copper supplementation of mutant gene crinkled in mice, *Teratology* **9**:A22.

Hurley, L. S., and Bell, L. T., 1974b, Genetic influence on response to dietary manganese deficiency, *J. Nutr.* **104**:133.

Hurley, L. S., and Cosens, G., 1970, Teratogenic magnesium deficiency in pregnant rats, *Teratology* **3**:202.

Hurley, L. S., and Cosens, G., 1971, Congenital malformations and fetal anemia resulting from magnesium deficiency in rats, *Fed. Proc.* **30**:516.

Hurley, L. S., and Cosens, G., 1974, Reproduction and prenatal development in relation to dietary zinc level, *in: Trace Element Metabolism in Animals*, Vol. 2 (W. G. Hoekstra, J. W. Suttie, H. E. Ganther, and W. Mertz, eds.), pp. 516–518, University Park Press, Baltimore.

Hurley, L. S., and Everson, G. J., 1959, Delayed development of righting reflexes in offspring of manganese-deficient rats, *Proc. Soc. Exp. Biol. Med.* **102**:360.

Hurley, L. S., and Everson, G. J., 1963, Influence of timing of short-term supplementation during gestation on congenital abnormalities of manganese deficient rats, *J. Nutr.* **79**:23.

Hurley, L. S., and Mutch, P. B., 1973, Prenatal and postnatal development after transitory gestational zinc deficiency in rats. *J. Nutr.* **103**:649.

Hurley, L. S., and Shrader, R. E., 1972, Congenital malformations of the nervous system in zinc-deficient rats, *in: Neurobiology of the Trace Metals Zinc and Copper*, (C. C. Pfeiffer, ed.), pp. 7–51, International Review of Neurobiology, Suppl. 1, Academic Press, New York.

Hurley, L. S. and Shrader, R. E., 1975, Abnormal development in preimplantation eggs after four days of maternal dietary zinc deficiency, *Nature* **254**:427.

Hurley, L. S., and Swenerton, H., 1966, Congenital malformation resulting from zinc deficiency in rats. *Proc. Soc. Exp. Biol. Med.* **123**:692.

Hurley, L. S., and Swenerton, H., 1971, Lack of mobilization of bone and liver zinc under teratogenic conditions of zinc deficiency in rats. *J. Nutr.* **101**:597.

Hurley, L. S., and Tao, S., 1972, Alleviation of teratogenic effects of zinc deficiency by simultaneous lack of calcium. *Am. J. Physiol.* **222**:322.

Hurley, L. S., and Volkert, N. E., 1965, Pantothenic acid and coenzyme A in the developing guinea pig liver, *Biochim. Biophys. Acta.* **104:**372.

Hurley, L. S., Everson, G. J., and Geiger, J. F., 1958, Manganese deficiency in rats: Congenital nature of ataxia, *J. Nutr.* **66:**309.

Hurley, L. S., Wooten, E., Everson, G. J., and Asling, C. W., 1960, Anomalous development of ossification in the inner ear of offspring of manganese deficient rats, *J. Nutr.* **71:**15.

Hurley, L. S., Wooten, E., and Everson, G. J., 1961a, Disproportionate growth in offspring of manganese-deficient rats. II. Skull, brain, and cerebrospinal fluid pressure, *J. Nutr.* **74:**282.

Hurley, L. S., Everson, G. J., Wooten, E., and Asling, C. W., 1961b, Disproportionate growth in offspring of manganese-deficient rats. I. The long bones, *J. Nutr.* **74:**274.

Hurley, L. S., Volkert, N. E., and Eichner, J. T., 1965, Pantothenic acid deficiency in pregnant and non-pregnant guinea pigs, with special reference to effects on the fetus, *J. Nutr.* **86:**201.

Hurley, L. S., Gowan, J., and Shrader, R., 1968a, Genetic–nutritional interaction in relation to manganese and calcification. *in: Les tissues calcifiés,* V°Symposium européen, Société d'enseignement supérieur, Paris, pp. 101–104.

Hurley, L. S., Dreosti, I. E., and Swenerton, H., 1968b, Studies on zinc enzymes and nucleic acid synthesis in relation to congenital malformations in zinc deficient rats, *Proc. West. Hemisphere Nutr. Congr. II,* San Juan, Puerto Rico, p. 39.

Hurley, L. S., Gowan, J., and Swenerton, H., 1971, Teratogenic effects of short-term and transitory zinc deficiency in rats, *Teratology* **4:**199.

Hurley, L. S., Sucher, K., Story, D., and Cosens, G., 1973, Interaction of dietary protein and zinc in the pregnant rat, *J. Nutr.* **103:**xxv.

Hutchings, D. E., and Gaston, J., 1974, The effects of vitamin A excess administered during the mid-fetal period on learning and development in rat offspring, *Dev. Psychobiol.* **7:**225.

Hutchings, D. E., Gibbon, J., and Kaufman, M. A., 1973, Maternal vitamin A excess during the early fetal period: Effects on learning and development in the offspring, *Dev. Psychobiol.* **6:**445.

Hytten, F. E., and Leitch, I., 1971, *The Physiology of Human Pregnancy,* Blackwell Scientific Publications, Oxford, U.K.

Innes, J. R. M., 1936, "Swayback"—a demyelinating disease of lambs with affinities to Schilder's encephalitis in man, *Vet. Rec.* **48:**1539.

Innes, J. R. M., and Shearer, G. D., 1940, "Swayback"; a demyelinating disease of lambs with affinities to Schilder's encephalitis in man *J. Comp. Pathol. Ther.* **53:**1.

Isselbacher, K. J., 1959, Galactose metabolism and galatosemia, *Am. J. Med.* **26:**715.

Itoh, H., Hansard, S. L., Glenn, J. C., Hoskins, F. H., and Thrasher, D. M., 1967, Placental transfer of calcium in pregnant sows on normal and limited-calcium rations, *J. Anim. Sci.* **26:**335.

Jaffe, N. R., and Johnson, E. M., 1973, Alterations in the ontogeny and specific activity of phosphomonoesterases associated with abnormal chondrogenesis and osteogenesis in the limbs of fetuses from folic acid-deficient pregnant rats, *Teratology* **8:**33.

Jeans, P. C., Smith, M. B., and Stearns, G., 1944, Incidence of prematurity in relation to maternal nutrition, *J. Am. Dent. Assoc.* **31:**576.

Johnson, E. M., and Chepenik, K. P., 1966, Effects of folic acid deficiency on enzymatic differentiation in embryonic tissues and organs. *Anat. Rec.* **154:**362.

Kalter, H., and Warkany, J., 1957, Congenital malformations in inbred strains of mice induced by riboflavin-deficient, galactoflavin-containing diets, *J. Exp. Zool.* **136:**531.

Kelly, F. C., and Snedden, W. W., 1960, Prevalence and geographical distribution of endemic goitre, *in: Endemic Goitre,* pp. 27–233, World Health Organization, Geneva.

Kenney, M. A., and Roderuck, C. E., 1963, Fatal syndrome associated with vitamin E status of pregnant rats, *Proc. Soc. Exp. Biol. Med.* **114:**257.

Kerber, I. J., Warr, O. S., III, and Richardson, C., 1968, Pregnancy in a patient with a prosthetic mitral valve, *J. Am. Med. Assoc.* **203:**223.

Kerr, G. R., Chamove, A. S., Harlow, H. F., and Waisman, H. A., 1968, "Fetal PKU": The effect of maternal hyperphenylalaninemia during pregnancy in the Rhesus monkey (*Macaca mulatta*), *Pediatrics* **42:**27.

Kinney, C. S., and Morse, L. M., 1964, Effect of a folic acid antagonist, aminopterin, on fetal development and nucleic acid metabolism in the rat, *J. Nutr.* **84:**288.

Kirksey, A., and Pike, R. L., 1962, Some effects of high and low sodium intakes during pregnancy in the rat. I. Food consumption, weight gain, reproductive performance, electrolyte balances, plasma total protein and protein fractions in normal pregnancy, *J. Nutr.* **77:**33.

Kirksey, A., Pike, R. L., and Callahan, J. A., 1962, Some effects of high and low sodium intakes during pregnancy in the rat. II. Electrolyte concentrations of maternal plasma, muscle, bone and brain and of placenta, amniotic fluid, fetal plasma and total fetus in normal pregnancy, *J. Nutr.* **77:**43.

Kitay, D. Z., 1969, Folic acid deficiency in pregnancy, *Am. J. Obstet. Gynecol.* **104:**1067.

Kochhar, D. M., 1967, Teratogenic activity of retinoic acid, *Acta Pathol. Microbiol. Scand.* **70:**398.

Kochhar, D. M., 1968, Studies of vitamin A-induced teratogenesis: Effects on embryonic mesenchyme and epithelium, and on incorporation of H^3-thymidine, *Teratology* **1:**299.

Kochhar, D. M., 1973, Limb development in mouse embryos, I. Analysis of teratogenic effects of retinoic acid, *Teratolgy* **7:**289.

Kochhar, D. M., and Johnson, E. M., 1965, Morphological and autoradiographic studies of cleft palate induced in rat embryos by maternal hypervitaminosis A, *J. Embryol. Exp. Morphol.* **14:**223.

Körner, W. F., and Nowak, H., 1971, The influence of vitamin B_6 deficiency during pregnancy on the fetal postnatal development of the rat, *in: Malformations congénitales des Mammifères* (H. Tuchmann-Duplessis, ed.), pp. 243–253, Masson et Cie, Paris.

Kratzing, C. C., and Perry, J. J., 1971, Hypertension in young rats following choline deficiency in maternal diets, *J. Nutr.* **101:**1657.

Kraus, A. P., Perlow, S., and Singer, K., 1949, Danger of Dicurmarol[R] treatment in pregnancy, *J. Am. Med. Assoc.* **139:**758.

Lamming, G. E., Salisbury, G. W., Hays, R. L., and Kendall, K. A., 1954a, The effect of incipient vitamin A deficiency on reproduction in the rabbit. I. Decidua, ova and fertilization, *J. Nutr.* **52:**217.

Lamming, G. E., Woollam, D. H. M., and Millen, J. W., 1954b, Hydrocephalus in young rabbits associated with maternal vitamin A deficiency, *Br. J. Nutr.* **8:**363.

Langer, P., 1960, History of goitre, *in: Endemic Goitre*, pp. 9–25, World Health Organization, Geneva.

Langman, J., and Welch, G. W., 1966, Effect of vitamin A on development of the central nervous system, *J. Comp. Neurol.* **128:**1.

Langman, J., and Welch, G. W., 1967, Excess vitamin A and development of the cerebral cortex, *J. Comp. Neurol.* **131:**15.

Latham, M. C., and Cobos, F., 1971, The effects of malnutrition on intellectual development and learning, *Am. J. Public Health* **61:**1307.

Lee, Y. C. P., King, J. T., and Visscher, M. B., 1953, Strain difference in vitamin E and B_{12} and certain mineral trace-element requirements for reproduction in A and Z mice, *Am. J. Physiol.* **173:**456.

Lefèbvres-Boisselot, J., 1951, Role tératogène de la déficience en acide pantothénique chez le rat, *Ann. Med.* **52:**225.

Lefèbvres-Boisselot, J., and Dupuis, R., 1956, Réduction de la croissance pondérale et trouble du métabolisme de l'eau, chez les foetus de rates déficientes en acide pantothénique, *J. Physiol.* **48:**598.

Loh, K. R. W., Shrader, R. E., and Zeman, F. J., 1971, Effect of maternal protein deprivation on neonatal intestinal absorption in rats, *J. Nutr.* **101:**1663.

Luse, S. A., Rhys, A., and Lessey, R., 1970, Effects of maternal phenylketonuria on the rat fetus, *Am. J. Obstet. Gynecol.* **108:**387.

Lyon, M. F., 1953, Absence of otoliths in the mouse: An effect of the pallid mutant, *J. Genet.* **51:**638.

Macomber, D., 1927, Effect of a diet low in calcium on fertility, pregnancy and lactation in the rat, *J. Am. Med. Assoc.* **88**:6.

Marin-Padulla, M., and Ferm, V. H., 1965, Somite necrosis and developmental malformations induced by vitamin A in the golden hamster, *J. Embryol. Exp. Morphol.* **13**(1):1.

Martin, M. P., Bridgforth, E., McGanity, W. J., and Darby, W. J., 1957, The Vanderbilt cooperative study of maternal and infant nutrition. X. Ascorbic acid, *J. Nutr.* **62**:201.

Masi, P. L., Frumento, F., and Parrini, C., 1966, Malformazioni congenite maxillo-dentali nella prole di topi trattati con ipervitaminosi A, *Riv. Ist. Sieroter. Ital.* **21**:1129.

Maxwell, J. P., 1934, Osteomalacia and diet, *Nutr. Abstr. Rev.* **4**:1.

McClure, T. J., 1961, Pathogenesis of early embryonic mortality caused by fasting pregnant rats and mice for short periods, *J. Reprod. Fertil.* **2**:381.

McLaren, D. S., 1968, To eat to see, *Nutr. Today* **3**:2.

Meader, R. D., 1965, Livers of choline-deficient pregnant and fetal rats. *Anat. Rec.* **153**:407.

Mellanby, M., and Coumoulos, H., 1944, The improved dentition of 5-year-old London schoolchildren, a comparison between 1943 and 1929, *Br. Med. J.* **1**:837.

Millen, J. W., and Woollam, D. H. M., 1956, The effect of the duration of vitamin A deficiency in female rabbits upon the incidence of hydrocephalus in their young, *J. Neurol. Psychiatr.* **19**:17.

Millen, J. W., Woollam, D. H. M., and Lamming, G. E., 1953, Hydrocephalus associated with deficiency of vitamin A, *Lancet* **2**:1234.

Millen, J. W., Woollam, D. H. M., and Lamming, G. E., 1954, Congenital hydrocephalus due to experimental hypovitaminosis A, *Lancet* **2**:679.

Miller, J. R., 1962, Strain difference in response to the teratogenic effect of maternal fasting in the house mouse, *Can. J. Genet. Cytol.* **4**:69.

Miller, Z., Poncet, I., and Takacs, E., 1962, Biochemical studies on experimental congenital malformations: flavin nucleotides and folic acid in fetuses and livers from normal and riboflavin-deficient rats, *J. Biol. Chem.* **237**:968.

Mills, C. F., and Fell, B. F., 1960, Demyelination in lambs born of ewes maintained on high intakes of sulphate and molybdate, *Nature* **185**:20.

Mills, C. F., Quarterman, J., Chesters, J. K., Williams, R. B., and Dalgarno, A. C., 1969, Metabolic role of zinc, *Am. J. Clin. Nutr.* **22**:1240.

Mirone, L., 1954, Effect of choline-deficient diets on growth, reproduction and mortality of mice. *Am. J. Physiol.* **179**:49.

Mjølnerød, O. K., Dommerud, S. A., Rasmussen, K., and Gjeruldsen, S. T., 1971, Congenital connective-tissue defect probably due to D-penicillamine treatment in pregnancy, *Lancet* **1**:673.

Monie, I. W., Nelson, M. M., and Evans, H. M., 1954, Abnormalities of the urinary system of rat embryos resulting from maternal pteroylglutamic acid deficiency, *Anat. Rec.* **120**:119.

Monie, I. W., Nelson, M. M., and Evans, H. M., 1957a, Persistent right umbilical vein as a result of vitamin deficiency during gestation, *Circ. Res.* **5**:187.

Monie, I. W., Nelson, M. M., and Evans, H. M., 1957b, Abnormalities of the urinary system of rat embryos resulting from transitory deficiency of pteroylglutamic acid during gestation, *Anat. Rec.* **127**:711.

Moon, W.-H. Y., and Kirksey, A., 1973, Cellular growth during prenatal and early postnatal periods in progeny of pyridoxine-deficient rats, *J. Nutr.* **103**:123.

Moore, R. A., Bittenger, I., Miller, M. L., and Hellman, L. M., 1942, Abortion in rabbits fed a vitamin K deficient diet, *Am. J. Obstet. Gynecol.* **43**:1007.

Morriss, G. M., 1972, Morphogenesis of the malformations induced in rat embryos by maternal hypervitaminosis A, *J. Anat.* **113**:241.

Mouriquand, G., and Edel, V., 1953, Sur l'hypervitaminose C, *C.R. Soc. Biol.* **147**:1432.

Mouriquand, G., Gillet, R., and Coeur, A., 1935, La phase "d'avortement" et la phase de protection maternelle et foetale dans le scorbut expérimental, *C.R. Soc. Biol.* **119**:1230.

Naeye, R. L., 1965, Cardiovascular abnormalities in infants malnourished before birth, *Biol. Neonate* **8**:104.

Naeye, R. L., Diener, M. M., and Dellinger, W. S., 1969, Urban poverty: Effects on prenatal nutrition, *Science* **166:**1026.

Naeye, R. L., Diener, M. M., Harcke, H. T., Jr., and Blanc, W. A., 1971, Relation of poverty and race to birth weight and organ and cell structure in the newborn, *Pediatr. Res.* **5:**17.

Nanda, R., 1971, Tritiated thymidine labelling of the palatal processes of rat embryos with cleft palate induced by hypervitaminosis A. *Arch. Oral Biol.* **16:**435.

Nanda, R., 1974, Effect of vitamin A on the potentiality of rat palatal processes to fuse *in vivo* and *in vitro, Cleft Palate J.* **11:**123.

Nelson, M. M., and Evans, H. M., 1946, Pantothenic acid deficiency and reproduction in the rat, *J. Nutr.* **31:**497.

Nelson, M. M., and Evans, H. M., 1948, Effect of desoxypyridoxine on reproduction in the rat, *Proc. Soc. Exp. Biol. Med.* **68:**274.

Nelson, M. M., and Evans, H. M., 1949, Pteroylglutamic acid and reproduction in the rat, *J. Nutr.* **38:**11.

Nelson, M. M., and Evans, H. M., 1953, Relation of dietary protein levels to reproduction in the rat, *J. Nutr.* **51:**71.

Nelson, M. M., and Evans, H. M., 1954, Maintenance of pregnancy in the absence of dietary protein with estrone and progesterone, *Endocrinology* **55:**543.

Nelson, M. M., and Evans, H. M., 1955a, Relation of thiamine to reproduction in the rat, *J. Nutr.* **55:**151.

Nelson, M. M., and Evans, H. M., 1955b, Maintenance of pregnancy in absence of dietary protein with progesterone, *Proc. Soc. Exp. Biol. Med.* **88:**444.

Nelson, M. M., Lyons, W. R., and Evans, H. M., 1951, Maintenance of pregnancy in pyridoxine-deficient rats when injected with estrone and progesterone, *Endocrinology* **48:**726.

Nelson, M. M., Asling, C. W., and Evans, H. M., 1952, Production of multiple congenital abnormalities in young by maternal pteroylglutamic acid deficiency during gestation, *J. Nutr.* **48:**61.

Nelson, M. M., Lyons, W. R., and Evans, H. M., 1953, Comparison of ovarian and pituitary hormones for maintenance of pregnancy in pyridoxine-deficient rats, *Endocrinology* **52:**585.

Nelson, M. M., Wright, H. V., Asling, C. W., and Evans, H. M., 1955, Multiple congenital abnormalities resulting from transitory deficiency of pteroylglutamic acid during gestation in the rat, *J. Nutr.* **56:**349.

Nelson, M. M., Wright, H. V., Baird, C. D. C., and Evans, H. M., 1956a, Effect of 36-hour period of pteroylglutamic acid deficiency on fetal development in the rat, *Proc. Soc. Exp. Biol. Med.* **92:**554.

Nelson, M. M., Baird, C. D. C., Wright, H. V., and Evans, H. M., 1956b, Multiple congenital abnormalities in the rat resulting from riboflavin deficiency induced by the antimetabolite galactoflavin, *J. Nutr.* **58:**125.

Nelson, M. M., Wright, H. V., Baird, C. D. C., and Evans, H. M., 1957, Teratogenic effects of pantothenic acid deficiency in the rat, *J. Nutr.* **62:**395.

Netzloff, M., Johnson, E., and Kaplan, S., 1968, Respiratory changes observed in abnormally developing rat embryos, *Teratology* **1:**375.

Neuweiler, W., 1951, Die Hypervitaminose und ihre Beziehung zur Schwangerschaft, *Intern Z. Vitaminforsch* **22:**392.

Newberne, P. M., and O'Dell, B. L., 1961, Vitamin B_{12} deficiency and hydrocephalus, *in: Disorders of the Developing Nervous System* (W. S. Fields and M. M. Desmond, eds.), pp. 123–143, Charles C Thomas, Springfield, Illinois.

Newberne, P. M., and Young, V. R., 1973, Marginal vitamin B_{12} intake during gestation in the rat has long term effects on the offspring, *Nature* **242:**263.

Newberne, P. M., Ahlstrom, A., and Rogers, A. E., 1970, Effects of maternal dietary lipotropes on prenatal and neonatal rats, *J. Nutr.* **100:**1089.

Niiyama, Y., Kishi, K., and Inoue, G., 1970, Effect of ovarian steroids on maintenance of pregnancy in rats fed diets devoid of one essential amino acid, *J. Nutr.* **100:**1461.

Niiyama, Y., Kishi, K., Endo, S., and Inoue, G., 1973, Effect of diets devoid of one essential amino acid on pregnancy in rats maintained by ovarian steroids, *J. Nutr.* **103:**207.

Nishikawa, M., 1963, Influence of maternal fasting for a short period upon the development of mouse embryos, *Kaibogaku Zasshi* **38:**181.

Nishimura, H., and Miyamoto, S., 1969, Teratogenic effects of sodium chloride in mice, *Acta Anat.* **74:**121.

O'Dell, B. L., 1968, Trace elements in embryonic development, in: Symposium on Nutrition and Prenatal Development, *Fed. Proc.* **27:**199.

O'Dell, B. L., Hardwick, B. C., and Reynolds, G., 1961, Mineral deficiencies of milk and congenital malformation in the rat, *J. Nutr.* **73:**151.

Olwin, J. H., and Koppel, J. L., 1969. Anticoagulant therapy during pregnancy, *Obstet. Gynecol.* **34:**847.

Orent, E. R., and McCollum, E. V., 1931, Effects of deprivation of manganese in the rat, *J. Biol. Chem.* **92:**651.

Orent-Keiles, E., Robinson, A., and McCollum, E. V., 1937, The effects of sodium deprivation on the animal organism, *Am. J. Physiol.* **119:**651.

Ornoy, A., Menczel, J., and Nebel, L., 1968, Alterations in the mineral composition and metabolism of rat fetuses and their placentas induced by maternal hypervitaminosis D_2, *Isr. J. Med. Sci.* **4:**827.

Ornoy, A., Nebel, L., and Menczel, Y., 1969, Impaired osteogenesis of fetal long bones induced by maternal hypervitaminosis D_2, *Arch. Pathol.* **87:**563.

Orony, A., Kaspi, T., and Nebel, L., 1972, Persistent defects of bone formation in young rats following maternal hypervitaminosis D_2, *Isr. J. Med. Sci.* **7:**943.

Palacios-Macedo, X., Díaz-Devis, C., and Escudero, J., 1969, Fetal risk with the use of coumarin anticoagulant agents in pregnant patients with intracardiac ball valve prosthesis, *Am. J. Cardiol.* **24:**853.

Palludan, B., 1966a, *A-Avitaminosis in Swine,* Munksgaard, Copenhagen.

Palludan, B., 1966b, Swine in teratological studies, *in: Swine in Biomedical Research* (L. K. Bustad and R. O. McClella, eds.), pp. 51–78, Frayn Printing, Seattle, Washington.

Palsson, P. A., and Grimsson, H., 1953, Demyelination in lambs from ewes which feed on seaweeds, *Proc. Soc. Exp. Biol. Med.* **83:**518.

Pang, R. L., and Kirksey, A., 1974, Early postnatal changes in brain composition in progeny of rats fed different levels of dietary pyridoxine, *J. Nutr.* **104:**111.

Parmelee, A. H., Allen, E., Stein, I. F., and Buxbaum, H., 1940, Three cases of congenital goiter, *Am. J. Obstet. Gynecol.* **40:**145.

Persaud, T. V. N., and Tiesj, D., 1969, Developmental abnormalities in rat induced by amino acid leucine, *Naturwissenschaften* **56:**37.

Persaud, T. V. N., and Kaplan, S., 1970, The effects of hypoglycin-A, a leucine analogue, on the development of rat and chick embryos, *Life Sci. (II)* **9:**1305.

Pharoah, P.O.D., Buttfield, I. H., and Hetzel, B. S., 1971, Neurological damage to the fetus resulting from severe iodine deficiency during pregnancy, *Lancet* **1:**308.

Phillips, G. D., and Sundaram, S. K., 1966, Sodium depletion of pregnant ewes and its effects on foetuses and foetal fluids, *J. Physiol.* **184:**889.

Pike, R. L., 1951, Congenital cataract in albino rats fed different amounts of tryptophan and niacin, *J. Nutr.* **44:**191.

Plumlee, M. P., Thrasher, D. M., Beeson, W. M., Andrews, F. N., and Parker, H. E., 1956, The effects of a manganese deficiency upon the growth, development, and reproduction of swine, *J. Anim. Sci.* **15:**352.

Pond, W. G., Wagner, W. C., Dunn, J. A., and Walker, E. F., Jr., 1968, Reproduction and early postnatal growth of progeny in swine fed a protein-free diet during gestation, *J. Nutr.* **94:**309.

Popjak, G., 1946, Maternal and foetal tissue- and plasma-lipids in normal and cholesterol-fed rabbits, *J. Physiol.* **105:**236.

Potier de Courcy, G., 1966a, Caractères globaux des métabolismes nucléique et protéolique chez le foetus de rat carencé en acide pantothénique, *Arch. Sci. Physiol.* **20:**43.

Potier de Courcy, G., 1966b, Conséquences de la carènce folique sur les caractères généraux des métabolismes nucléique et protéique chez le foetus de rat, *Arch. Sci. Physiol.* **20**:189.

Potier de Courcy, G., and Desmettre-Miguet, S., 1970, Effects de la carence en riboflavine sur quelques éléments minéraux des tissus foeto-maternels du rat, *Arch. Sci. Physiol.* **24**:181.

Potier de Courcy, G., and Terroine, T., 1966, Les désoxyribonucléases I et II dans la vie pré- et post-natale du rat normal et carencé en acide folique, *Arch. Sci. Physiol.* **20**:1.

Potier de Courcy, G., and Terroine, T., 1968a, Influence de la carence en riboflavine sur la fonction phosphatasique alcaline des organes maternels et foetaux à différents stades de la gestation, *Ann. Nutr. Aliment.* **22**:95.

Potier de Courcy, G., and Terroine, T., 1968b, Conséquences chez le rat de la carence en riboflavine sur la composition globale de certains tissus maternels et foetaux, *Arch. Sci. Physiol.* **22**:1.

Potier de Courcy, G., Susbielle, H., and Terroine, T., 1970, Etude du zinc dans l'ariboflavinose tératogéne chez le rat, *Arch. Sci. Physiol.* **24**:409.

Potvliege, P. R., 1972, Hypervitaminosis D_2 in gravid rats, *Arch. Pathol.* **73**:29.

Primrose, T., and Higgins, A., 1971, A study in human antepartum nutrition, *J. Reprod. Med.* **7**:257.

Purvis, R. J., MacKay, G. S., Cockburn, F., Barrie, W. J. McK., Wilkinson, E. M., Belton, N. R., and Forfar, J. O., 1973, Enamel hypoplasia of the teeth associated with neonatal tetany: A manifestation of maternal vitamin-D deficiency, *Lancet* **2**:811.

Quick, A. J., 1946, Experimentally induced changes in the prothrombin level of the blood. III. Prothrombin concentration of new-born pups of a mother given dicumarol before parturition, *J. Biol. Chem.* **164**:371.

Richardson, L. R., and Hogan, A. G., 1946, Diet of mother and hydrocephalus in infant rats, *J. Nutr.* **32**:459.

Rivers, J. M., Krook, L., and Cormier, A., 1970, Biochemical and histological study of guinea pig fetal and uterine tissue in ascorbic acid deficiency, *J. Nutr.* **100**:217.

Roberts, H. E., Williams, B. M., and Harvard, A., 1966a, Cerebral oedema in lambs associated with hypocuprosis, and its relationship to swayback. I. Field, clinical gross anatomical and biochemical observations, *J. Comp. Pathol.* **76**:279.

Roberts, H. E., Williams, B. M., and Harvard, A., 1966b, Cerebral oedema in lambs associated with hypocuprosis, and its relationship to swayback. II. Histopathological findings, *J. Comp. Pathol.* **76**:285.

Roderick, L. M., 1929, The pathology of sweet clover disease in cattle, *J. Am. Vet. Med. Assoc.* **75**:314.

Roeder, L. M., and Chow, B. F., 1972, Maternal undernutrition and its long-term effects on the offspring, *Am. J. Clin. Nutr.* **25**:812.

Rosa, F. W., and Turshen, M., 1970, Fetal Nutrition, *Bull. W. H. O.* **43**:785.

Rosen, J. F., Roginsky, M., Nathenson, G., and Finberg, L., 1974, 25-hydroxyvitamin D, plasma levels in mothers and their premature infants with neonatal hypocalcemia, *Am. J. Dis. Child.* **127**:220.

Ross, M. L., and Pike, R. L., 1956, The relationship of vitamin B_6 to protein metabolism during pregnancy in the rat, *J. Nutr.* **58**:251.

Rothman, D., 1970, Folic acid in pregnancy, *Am. J. Obstet. Gynecol.* **108**:149.

Roux, C., and Aubry, M., 1966, Teratogenic effect of an inhibition of cholesterol synthesis (AY9944) in the rat, *C.R. Soc. Biol.* **160**:1353.

Roux, C., and Dupuis, R., 1961, Urohydronéphroses par carence pantothénique, *Arch. Fr. Pediatr.* **18**:1337.

Roux, J. F., and Yoshioka, T., 1970, Lipid metabolism in the fetus during development, *Clin. Obstet. Gynecol.* **13**:595.

Roux, C., Fournier, P., Dupuis, Y., and Dupuis, R., 1962a, Anomalies oculaires congénitales dans la carence en vitamine A, *Bull. Soc. Ophthalmol. Fr.* **1962**:1.

Roux, C., Fournier, P., Dupuis, Y., and Dupuis, R., 1962b, About teratogenic vitamin A deficiency, *Biol. Neonate* **4**:371.

Ruffo, A., and Vescia, A., 1941, Importanza dell'acido nicotinico per il ratto, *Boll. Soc. Ital. Biol. Sper.* **16**:185.

Runner, M. N., and Miller, J. R., 1956, Congenital deformity in the mouse as a consequence of fasting, *Anat. Rec.* **124**:437.

Russell, K. P., Rose, H., and Starr, P., 1957, The effects of radioactive iodine on maternal and fetal thyroid function during pregnancy, *Surg. Gynecol. Obstet.* **104**:560.

Samborskaya, E. P., and Ferdman, T. D., 1966, The mechanism of termination of pregnancy by ascorbic acid, *Bull. Exp. Biol. Med.* **62**:934(in Russian).

Sarma, V., 1959, Maternal vitamin A deficiency and fetal microcephaly and anophthalmia, *Obstet. Gynecol.* **13**:299.

Schofield, F. W., 1923–1924, Damaged sweet clover: The cause of a new disease in cattle simulating hemorrhagic septicemia and blackleg, *J. Am. Vet. Med. Assoc.* **64**:553.

Schumacher, M. F., Williams, M. A., and Lyman, R. L., 1965, Effect of high intakes of thiamine, riboflavin and pyridoxine on reproduction in rats and vitamin requirements of the offspring, *J. Nutr.* **86**:343.

Scott, K. E., and Usher, R., 1966, Fetal malnutrition: Its incidence, causes, and effects, *Am. J. Obstet. Gynecol.* **94**:951.

Scott, D. E., Whalley, P. J., and Pritchard, J. A., 1970, Maternal folate deficiency and pregnancy wastage. II. Fetal malformation, *Obstet. Gynecol.* **36**:26.

Scrimshaw, N. S. and Gordon, E., 1968, *Malnutrition, Learning and Behavior,* M.I.T. Press, Cambridge, Mass.

Scriver, C. R., 1970, Vitamin D dependency, *Pediatrics* **45**:361.

Scriver, C. R., 1974, Inborn errors of metabolism: A new frontier of nutrition, *Nutr. Today* **9**:4.

Segal, S., and Bernstein, H., 1963, Observations on cataract formation in the newborn offspring of rats fed high-galactose diet, *J. Pediatr.* **62**:363.

Sever, L. E., and Emanuel, I., 1973, Is there a connection between maternal zinc deficiency and congenital malformations of the central nervous system in man? *Teratology* **7**:117.

Shepard, T. H., and Bass, G. L., 1970, Organ culture of limb buds from riboflavin-deficient and normal rat embryos in normal and riboflavin deficient media, *Teratology* **3**:163.

Shepard, T. H., Lemire, R. J., Aksu, O., and Mackler, B., 1968, Studies of the development of congenital anomalies in embryos of riboflavin deficient, galactoflavin fed rats. I. Growth and embryologic pathology, *Teratology* **1**:75.

Shils, M. E., and McCollum, E. V., 1943, Further studies on the symptoms of manganese deficiency in the rat and mouse, *J. Nutr.* **26**:1.

Shrader, R. E., and Everson, G. J., 1967, Anomalous development of otoliths associated with postural defects in manganese-deficient guinea pigs, *J. Nutr.* **91**:453.

Shrader, R. E., and Zeman, F. J., 1969, Effect of maternal protein deprivation on morphological and enzymatic development of neonatal rat tissue, *J. Nutr.* **99**:401.

Shrader, R. E., and Zeman, F. J., 1973a, Skeletal development in rats as affected by maternal protein deprivation and postnatal food supply, *J. Nutr.* **103**:792.

Shrader, R. E., and Zeman, F. J., 1973b, *In vitro* synthesis of anterior pituitary growth hormone as affected by maternal protein deprivation and postnatal food supply, *J. Nutr.* **103**:1012.

Shrader, R. E., Erway, L., and Hurley, L. S., 1973, Mucopolysaccharide synthesis in the developing inner ear of manganese-deficient and pallid mutant mice, *Teratology* **8**:257.

Sima, A., and Sourander, P., 1973, The effect of perinatal undernutrition on perineurial diffusion barrier to exogenous protein, *Acta Neuropathol. (Berlin)* **24**:263.

Sisson, T. R. C., and Lund, C. J., 1958, The influence of maternal iron deficiency on the newborn, *Am. J. Clin. Nutr.* **6**:376.

Smith, C. A., 1947a, Effects of maternal undernutrition upon the newborn infant in Holland (1944–1945), *J. Pediatr.* **30**:229.

Smith, C. A., 1947b, The effect of wartime starvation in Holland upon pregnancy and its product, *Am. J. Obstet. Gynecol.* **53**:599.

Smith, C. A., Oberhelman, H. A., Jr., Storer, E. H., Woodward, E. R., and Dragstedt, L. R., 1951,

Production of experimental cretinism in dogs by the administration of radioactive iodine, *Arch. Surg.* **63**:807.

Solursh, M., and Meier, S., 1973, The selective inhibition of mucopolysaccharide synthesis by vitamin A treatment of cultured chick embryo chondrocytes, *Calcif. Tissue Res.* **13**:131.

Sontag, L. W., Munson, P., and Huff, E., 1936, Effects on the fetus of hypervitaminosis D and calcium and phosphorus deficiency during pregnancy, *Am. J. Dis. Child.* **51**:304.

Stanbury, J. B., and Querido, A., 1956, Genetic and environmental factors in cretinism: A classification, *J. Clin. Endocrinol. Metab.* **16**:1522.

Stanbury, J. B., and Querido, A., 1957, On the nature of endemic cretinism, *J. Clin. Endocrinol. Metab.* **17**:801.

Stein, Z., Susser, M., Saenger, G., and Marolla, F., 1972, Nutrition and mental performance, *Science* **178**:708.

Stone, M. L., 1968, Effects on the fetus of folic acid deficiency in pregnancy, *Clin. Obstet. Gynecol.* **11**:1143.

Suomalainen, P., 1950, Effect of E-avitaminosis on the histiotrophic nutrition of the mouse embryo, *Nature* **165**:364.

Suttle, N. F., and Field, A. C., 1968, Effect of intake of copper, molybdenum and sulphate on copper metabolism in sheep. II. Copper status of the newborn lamb, *J. Comp. Pathol.* **78**:363.

Swanson, W. W., and Iob, V., 1939, The growth of fetus and infant as related to mineral intake during pregnancy, *Am. J. Obstet. Gynecol.* **38**:382.

Swenerton, H., and Hurley, L. S., 1968, Severe zinc deficiency in male and female rats, *J. Nutr.* **95**:8.

Swenerton, H., Shrader, R., and Hurley, L. S., 1969, Zinc-deficient embryos: Reduced thymidine incorporation, *Science* **166**:1014.

Swenerton, H., Shrader, R., and Hurley, L. S., 1972. Lactic and malic dehydrogenases in testes of zinc-deficient rats, *Proc. Soc. Exp. Biol. Med.* **141**:283.

Takekoshi, S., 1964, The mechanism of vitamin A induced teratogenesis, *J. Embryol. Exp. Morphol.* **12**:263.

Tao, S. H., and Hurley, L. S., 1975, Effect of dietary calcium deficiency during pregnancy on zinc mobilization in intact and parathyroidectomized rats, *J. Nutr.* **105**:220.

Terada, M., 1970, Effect of intermittent fasting before pregnancy upon maternal fasting as a teratogen in mice, *J. Nutr.* **100**:767.

Toverud, K. U., and Toverud, G., 1931, Studies on the mineral metabolism during pregnancy and lactation and its bearing on the disposition to rickets and dental caries, *Acta Paediatr. Suppl. 2* **12**:1.

Tuchmann-Duplessis, H., and Mercier-Parot, L., 1964, Avortements et malformations sous l'effet d'un agent provoquant une hyperlipémie et une hypercholestérolémie, *Bull. Acad. Natl. Med. (Paris)* **148**:392.

Tuchmann-Duplessis, H., Lefébvres-Boisselot, J., and Mercier-Parot, L., 1959, L'action tératogène de l'acide x-methyl-folique sur diverses espéces animales, *Arch. Fr. Pediatr.* **15**(4):1. **15**(1):1.

Tuncer, M., 1970, Bone development, incidence of hypoglycemia and effect of maternal and fetal factors in low birth weight infants, *Turk. J. Pediatr.* **12**:59.

Turbow, M. M., and Chamberlain, J. G., 1968, Direct effects of 6-aminonicotinamide on the developing rat embryo *in vitro* and in *in vivo*, *Teratology* **1**:103.

Ullrey, D. E., Becker, D. E., Terrill, S. W., and Notzold, R. A., 1955, Dietary levels of pantothenic acid and reproductive performance of female swine, *J. Nutr.* **57**:401.

van den Berg, B. J., and Yerushalmy, J., 1966, The relationship of the rate of intrauterine growth of infants of low birth weight to mortality, morbidity, and congenital anomalies, *J. Pediatr.* **69**:531.

van Gelder, D. W., and Darby, F. U., 1944, Congenital and infantile beriberi, *J. Pediatr.* **25**:226.

Wallis, G. C., 1938, Some effects of vitamin D deficiency on mature dairy cows, *J. Dairy Sci.* **21**:315.

Warkany, J., 1943, Effect of maternal rachitogenic diet on skeletal development of young rat, *Am. J. Dis. Child.* **66:**511.

Warkany, J., and Deuschle, F. M., 1955, Congenital malformations induced in rats by maternal riboflavin deficiency: Dentofacial changes, *J. Am. Dent. Assoc.* **51:**139.

Warkany, J., and Petering, H. G., 1972, Congenital malformations of the central nervous system in rats produced by maternal zinc deficiency, *Teratology* **5:**319.

Warkany, J., and Petering, H. G., 1973, Congenital malformations of the brain produced by short zinc deficiencies in rats, *Am. J. Ment. Defic.* **77:**645.

Warkany, J., and Roth, C. B., 1948, Congenital malformations induced in rats by maternal vitamin A deficiency. II. Effect of varying the preparatory diet upon the yield of abnormal young, *J. Nutr.* **35:**1.

Warkany, J., and Schraffenberger, E., 1943, Congenital malformations induced in rats by maternal nutritional deficiency. V. Effects of purified diet lacking riboflavin, *Proc. Soc. Exp. Biol. Med.* **54:**92.

Warkany, J., and Schraffenberger, E., 1944, Congenital malformations of the eyes induced in rats by maternal vitamin A deficiency, *Proc. Soc. Exp. Biol. Med.* **57:**49.

Warkany, J., and Schraffenberger, E., 1946, Congenital malformation induced in rats by maternal vitamin A deficiency, *Arch. Ophthalmol.* **35:**150.

Wilson, J. G., and Warkany, J., 1947, Epithelial keratinization as evidence of fetal vitamin A deficiency, *Proc. Soc. Exp. Biol. Med.* **64:**419.

Wilson, J. G., and Warkany, J., 1948, Malformations in the genito-urinary tract induced by maternal vitamin A deficiency in the rat, *Am. J. Anat.* **83:**357.

Wilson, J. G., and Warkany, J., 1950, Cardiac and aortic arch anomalies in the offspring of vitamin A deficient rats correlated with similar human anomalies, *Pediatrics* **5:**708.

Wilson, J. G., Roth, C. B., and Warkany, J., 1953, An analysis of the syndrome of malformations induced by maternal vitamin A deficiency. Effects of restoration of vitamin A at various times during gestation, *Am. J. Anat.* **92:**189.

Wilson, M. G., Meyers, H. I., and Peters, A. H., 1967, Postnatal bone growth of infants with fetal growth retardation, *Pediatrics* **40:**213.

Winick, M., 1969, Malnutrition and brain development, *J. Pediatr.* **74:**667.

Winick, M., and Noble, A., 1966, Cellular response in rats during malnutrition at various ages, *J. Nutr.* **89:**300.

Winick, M., Brasel, J. A., and Velasco, E. G., 1973, Effects of prenatal nutrition upon pregnancy risk, *Clin. Obstet. Gynecol.* **16:**184.

Woodard, J. C., and Newberne, P. M., 1967, The pathogenesis of hydrocephalus in newborn rats deficient in vitamin B_{12}, *J. Embryol. Exp. Morphol.* **17:**177.

Wooley, D. W., and van der Hoven, T., 1964, Serotonin deficiency in infancy as one cause of a mental defect in phenylketonuria, *Science* **144:**883.

Yasuda, M., Nanjo, H., and Suzuki, M., 1966, The effect of fasting upon the development of embryos in elderly pregnant mice, *Kaibogaku Zasshi* **41:**43.

Zamenhof, S., and van Marthens, E., 1974, Study of factors influencing prenatal brain development, *Mol. Cell. Biochem.* **4:**157.

Zamenhof, S., Hall, S. M., Grauel, L., van Marthens, E., and Donahue, M. J., 1974, Deprivation of amino acids and prenatal brain development in rats, *J. Nutr.* **104:**1002.

Zeman, F. J., 1967, Effect on the young rat of maternal protein restriction, *J. Nutr.* **93:**167.

Zeman, F. J., 1968, Effects of maternal protein restriction on the kidney of the newborn young of rats, *J. Nutr.* **94:**111.

Zeman, F. J., 1970, Effect of protein deficiency during gestation on postnatal cellular development in the young rat, *J. Nutr.* **100:**530.

Zeman, F. J., and Stanbrough, E. C., 1969, Effect of maternal protein deficiency on cellular development in the fetal rat, *J. Nutr.* **99:**274.

Zeman, F. J., Shrader, R. E., and Allen, L. H., 1973, Persistent effects on maternal protein deficiency in postnatal rats, *Nutr. Rep. Int.* **7:**421.

Zilversmit, D. B., Remington, M., and Hughes, L. B., 1972, Fetal growth and placental permeability in rabbits fed cholesterol, *J. Nutr.* **102:**1681.

Embryotoxicity of Drugs in Man 8

JAMES G. WILSON

I. INTRODUCTION

The subject of possible adverse effects on development of drugs and environmental chemicals has been reviewed with some regularity (Cahen, 1966; Cohlan, 1964; Lenz, 1964; Nishimura, 1964; Palmisano and Polhill, 1972; Slater, 1965; Tuchmann-Duplessis, 1965; and others). Until more effective systems for prior testing and later surveillance for possible teratogenic effects of chemicals are in operation, however, frequent scrutiny in the form of surveys and reviews may serve to maintain a degree of alertness, if not provide the needed safeguard. Table 1 summarizes available information on known causes in man. It is apparent that only a very small percentage of the total burden of human developmental defects can at present be attributed to drugs and chemicals. Such percentages, however, must not be taken literally because they are only crude estimates based on the relatively few individual entities positively identified as human teratogens to date. Nevertheless, it appears unlikely that a majority of infants born with or diagnosed as having developmental defects will ever be assigned to any single category of environmental causation.

Some investigators have suggested that much of the presently unknown category (Table 1) may eventually be explainable in terms of *combinations and interactions* of two or more agents, either within and among categories listed in the table, or involving factors not now suspected. Whether or not this is true, and the present author believes that it may be, there is reason to think that chemicals, including drugs, will rank high in the lists of interacting substances.

JAMES G. WILSON · Children's Hospital Research Foundation, Elland and Bethesda Avenues, Cincinnati, Ohio 45229.

**Table 1. Causes of Developmental Defects in Man—
Estimates Based on Surveys and
Case Reports in the Medical Literature[a]**

Known genetic transmission		20%
Chromosomal aberration		3–5%
Environmental causes		
Ionizing radiations		<1%
Therapeutic	Nuclear	
Infections		2–3%
Rubella virus	Varicella virus(?)	
Cytomegalovirus	Toxoplasma	
Herpes virus hominis	Syphilis	
Maternal metabolic imbalance		1–2%
Endemic cretinism	Phenylketonuria	
Diabetes	Virilizing tumors	
Drugs and environmental chemicals		4–5%
Androgenic hormone	Anticonvulsants	
Folic antagonists	Oral hypoglycemics(?)	
Thalidomide	Few neurotropic-anorectics(?)	
Oral anticoagulants	Organic mercury	
Maternal alcoholism		
Combinations and interactions		?
Unknown		65–70%

[a]Modified from Wilson (1973b).

Support for this opinion is found in the tremendous numbers of drugs and chemicals already known to be teratogenic in laboratory animals (Table 2). Striking degrees of potentiative interactions have been demonstrated in laboratory animals when two or more compounds, several included in the accompanying list, were used simultaneously at doses at or below the threshold level for teratogenicity of the same compounds when applied singly (Wilson, 1964). It should be emphasized that most of the substances listed in Table 2 were teratogenic only at doses well above human therapeutic or likely exposure levels. One can only guess at the extent to which pregnant women are exposed simultaneously to two or more of such agents.

The implications of the thalidomide catastrophe notwithstanding, drugs may still be used in appreciable numbers during human pregnancy. A prospective study a few years ago revealed that a mean of 3.7 drugs regarded by the authors (Nora *et al.*, 1967a) as potentially teratogenic were taken by 240 women during the first trimester of their pregnancies. In Scotland over 97% of 1369 mothers were found to have taken prescribed drugs and 65% self-administered drugs during pregnancy (Nelson and Forfar, 1971). Although not proving a cause–effect relationship, it was found that significantly more of the mothers in the latter study who gave birth to infants with developmental abnormality took aspirin, antacid, dextroamphetamine, phenobarbitone, sodium amytal, other barbiturates, cough medicines, iron, sulfonamides, and nicotinamide, than did mothers in the control group. The real meaning of such observations is uncertain because it is impossible, without suitable controls, to separate influences on development caused by a drug from those caused by the condition for which it was taken. An untreated control group with identical indications for taking the drug at comparable times in gestation would be needed to evaluate these situations, but such data are rarely, if ever, available. Furthermore, the time during gestation when a drug is taken is often an important determinant of whether and what type of embryotoxicity will result, but reliable information on level and duration of dosage are sometimes difficult to obtain.

If, after animal tests, a new drug were approved for use, how might an undetected potential to cause human teratogenicity be recognized? There are four criteria by which such recognition might be possible, assuming widespread use of or exposure to the new agent (Table 3). These, in essence, were used to identify the agent in two previous human teratologic catastrophes, namely, those caused by rubella and thalidomide. This method, however, depends on the presence of a distinctive defect or pattern of defects and of sufficient numbers of cases to ensure recognition of an association between cause and effect. Recognition also requires a sufficient number of exposures and a relatively high risk of defects per exposure. Lacking any of these conditions, the criteria listed would not be dependable in identifying new human teratogens. It is well known that in animal experiments some teratogenic agents such as folate antagonists do not produce discrete, easily recognized patterns or syndromes, but are capable of affecting almost any organ in the

Table 2. Some Types of Drugs and Environmental Chemicals That Have Been Shown to Be Teratogenic[a] in One or More Species of Mammals[b]

Salicylates (e.g., aspirin, oil of wintergreen)
Certain alkaloids (e.g., caffeine, nicotine, colchicine)
Tranquilizers (e.g., meprobamate, chlorpromazine, reserpine, diazepam)
Antihistamines (e.g., buclizine, meclizine, cyclizine)
Antibiotics (e.g., chloramphenacol, streptonigrin, penicillin)
Hypoglycemics (e.g., carbutamide, tolbutamide, hypoglycins)
Corticoids (e.g., triamcinolone, cortisone)
Alkylating agents (e.g., busulfan, chlorambucil, cyclophosphamide, TEM)
Antimalarials (e.g., chloroquine, quinacrine, pyrimethamine)
Anesthetics (e.g., halothane, urethan, nitrous oxide, pentobarbital)
Antimetabolites (e.g., folic acid, purine and pyrimidine analogs)
Solvents (e.g., benzene, dimethylsulfoxide, propylene glycol)
Pesticides (e.g., aldrin, malathion, carbaryl, 2,4,5-T, captan, folpet)
Industrial effluents (e.g., some compounds of Hg, Pb, As, Li, Cd)
Plants (e.g., locoweed, lupins, jimsonweed, sweet peas, tobacco stalks)
Miscellaneous (e.g., trypan blue, triparanol, diamox, etc.)

[a]Teratogenic effects were usually seen only at doses well above therapeutic levels for the drugs, or above likely exposure levels for the environmental chemicals.
[b]From Wilson (1973b).

body, depending on when in gestation they are given. The same also holds for man, as will be evident later in this chapter. The possibility that some portion of human developmental defects very likely result from the interaction of two or more causative factors compounds the difficulty of relating effect to cause(s). These complications, together with incomplete surveillance and reporting, and lack of appropriate control groups for comparison, doubtless will continue to hamper efforts to identify and better define conditions of exposure to drugs and chemicals that may involve risk to the human embryo, fetus, or postnatal individual.

Published information on the effects on the offspring of drugs known to have been taken during the first two or three months of pregnancy has been carefully reviewed and the drugs classified into four categories. The categories and the criteria used are presented in Table 4. It should be emphasized that placing a drug in one or another of these classifications in the following discussions often includes elements of judgment and bias on the part of the present as well as the cited authors. In other words, the categorization of a drug here is not always based on uncontrovertible fact, but it does reflect a considerable effort to accumulate and evaluate all information available at the time of writing.

It should also be noted that the broader concept of *developmental toxicity* is used here instead of the traditional, more restrictive, and sometimes misleading one of *teratogenicity*. The reason for this nonconformity with custom is logical, namely, that drugs and other potentially deleterious agents applied during the developmental span are now known to be capable of producing several different kinds of developmental defects, namely: (1) prenatal or perinatal death, (2) malformation, (3) prenatal or postnatal growth retardation, and (4) postnatal functional aberration. (A fifth kind of developmental problem, congenital neoplasms, is dealt with in the chapter by Bolande, Vol. 2.) Thus the following classifications were made in consideration of all available information on intrauterine and perinatal death, on growth retardation and functional abnormality, as well as on the more commonly reported structural abnormalities. Doubtless, the human conceptus is most vulnerable to the adverse effects of drugs during the first two or three months after fertilization. Animal studies at comparable stages of development have amply demon-

---

Table 3. Criteria for Recognizing a New Teratogenic Agent in Man[a]

1. An abrupt increase in the incidence of a particular defect or association of defects (syndrome).
2. Coincidence of this increase with a known environmental change, e.g., widespread use of a new drug.
3. Known exposure to the environmental change early in pregnancies yielding characteristically defective infants.
4. Absence of other factors common to all pregnancies yielding infants with the characteristic defect(s).

---

[a]From Wilson (1973b).

**Table 4. Categories and Criteria Used in Classifying Drugs
According to Their Embryotoxic Potential in Man**[a]

1. Established as being embryotoxic.
   a. Unquestionably produces a higher percentage of developmental defects or intrauterine death in infants of exposed than otherwise similar, nonexposed women.
2. Suspected of being embryotoxic.
   a. Apparent increase above background level of developmental defects or intrauterine death among infants of exposed women, based on large numbers of accumulated case reports.
   b. Retrospective surveys indicate significant increase of defects or death among infants of exposed women, but "memory bias" of mothers introduces uncertainty.
   c. Prospective surveys indicate significant increase among infants of exposed women, but total number of exposed cases relatively small in consecutive series.
3. Possibly embryotoxic, under unusual conditions of dosage or in combination with other unidentified etiologic factors.
   a. Questionable increase of defects or death among infants of exposed woman based on large number of accumulated cases or on extensive epidemiologic surveys.
   b. Strong indication from studies in more than one animal species that drug has definite embryotoxic potential, especially at high dosage.
4. Not embryotoxic under any known conditions of human usage.
   a. No increase in defects or death among infants born to exposed women in well-controlled epidemiologic surveys.
   b. Extensively used by pregnant women over many years without substantiation of cause–effect relationship between use of the drug and adverse effects on the offspring.

[a]Assuming exposure to therapeutic or higher dose during 1st trimester.

strated that death and malformation are most likely to be induced during this period of organogenesis. When no mention is made below of growth retardation or of specific functional aspects, it may mean that none occurred, but more likely it means that the original authors did not evaluate or report on them.

II. DRUGS ESTABLISHED AS EMBRYOTOXIC IN MAN

A. Thalidomide

In the years since the first recognition of its devastating effect on the human embryo, thalidomide has assumed the status of a classic teratogenic agent. It was shown to be almost invariably effective when taken in adequate dosage on a very few or a single day(s) during the susceptible period of the human embryo, from the 20th to the 35th day postconception (Lenz, 1964). It produced a well-defined pattern of musculoskeletal deformities, affecting mainly the extremities and face, not precisely duplicated by any known syndrome, although it bore considerable resemblance to certain infrequently seen hereditary or familial conditions (Appelt *et al.*, 1966; Herrmann *et al.*,

1969; Holmes and Borden, 1974), some of which are mimicked in children exposed to thalidomide at specific times in gestation (Lenz, 1972). It was virtually nontoxic to young adults and consequently entailed little risk of overdosage.

At the expense of several thousand crippled children and the loss of an otherwise safe and effective sedative, thalidomide has taught teratologists some valuable lessons. The most important of these is that the human embryo may be inordinately sensitive to a substance that causes little or no toxicity in human adults or in certain test animals. It is a disquieting fact that the animals most widely used in teratogenicity testing, i.e., mouse, rat, rabbit, are relatively insensitive to this most potent human teratogen. The only animals showing substantially the same teratogenic sensitivity as does man, in terms of types of defects as well as dosage levels, are seven species of macaque monkeys, baboons, and marmosets (Wilson, 1973a). This has resulted in a tendency to consider the simian primate as more valid test animals for the evaluation of human teratological risk than are the widely used rodents and rabbits, but it must be emphasized that the close parallel between man and monkey as regards to teratogenicity of thalidomide has not been demonstrated for other drugs except those with androgenic properties.

B. Androgenic Hormones

It has been known for more than 30 years that treatment of pregnant mammals, including monkeys (van Wagenen and Hamilton, 1943), with male sex hormones produced pseudohermaphroditism in the female offspring similar to that occasionally seen in human infants. The earlier practice of injecting androgenic hormones for treatment of breast cancer resulted in the masculinization of a number of female fetuses when injections were begun prior to the 12th week of gestation (Grumbach and Ducharme, 1960). The use of progesterone from natural sources for treatment of threatened abortion led to the widespread use between 1950 and 1960 of synthetic progestins for the same purpose. More than 600 female infants with equivocal or frankly masculinized external genitalia were born before it was realized that some of these synthetic compounds had appreciable androgenic activity (Cahen, 1966; Voorhess, 1967; Wilkins, 1960). Differentiation of external genitalia is largely completed by the 12th week of gestation; therefore, hormone treatment after this time is no longer able to cause irreversible modification of these organs.

C. Folic Acid Antagonists

Early reports of the use of folic acid antagonists in laboratory animals indicated that these compounds were almost invariably embryolethal. This knowledge was the justification for their trial as abortifacients in human pregnancies scheduled for therapeutic abortion (Thiersch, 1952, 1956). Of 24 pregnancies so treated, 8 failed to abort and 2 of these yielded malformed infants. In addition, 2 of those that died *in utero* also appear to have been malformed. In another series of 20 cases treated with aminopterin during pregnancy (Goetsch, 1962), there was subsequent abortion in 15 and malformation in 1 of the 5 survivors. Combining these reports, 70% of treated pregnancies aborted and 23% of the surviving infants were malformed. At least 8 additional cases of malformed liveborn infants, making a total of 11, have now been attributed to antifolic compounds, mainly aminopterin (de Alvarez, 1962; Brandner and Nusslé, 1969; Emerson, 1962; Meltzer, 1956; Milunsky *et al.,* 1968; Shaw and Steinbach, 1968; Warkany *et al.,* 1959).

It is thus apparent that treatment with folic acid antagonists during the early months of human pregnancy usually results in intrauterine death, but 20–30% of the fetuses that are not aborted and survive to term display a variety of structural or functional defects. The defects observed in these infants have been so varied as not to comprise a recognizable syndrome, a fact in keeping with studies in rats which have shown that almost any organ can be the site of abnormality depending on the time in gestation at which the antagonist is applied (Nelson, 1963). Attempts to duplicate the high level of embryotoxicity of folate antagonists in rhesus monkeys have shown this species to be more resistant than man to comparable doses (Wilson, 1973b).

III. DRUGS SUSPECTED TO BE EMBRYOTOXIC IN MAN

Several drugs are suspected of having a degree of teratogenic potential because they have with some frequency been associated with malformations in the offspring of women treated during early pregnancy. Nevertheless, these reports constitute only a small percentage of the total number of women known or thought to have been treated, and consequently are subject to alternative interpretations, for example: (1) They could be spontaneous occurrences only coincidentally related to drug treatment. (2) They may in part be attributable to the maternal disease for which the drug was given. (3) They may in fact reflect a low order of teratogenic potential of the drug. Voluntary reporting of individual cases of anomalies thought to be related to the taking of drugs generally does not provide the type or quantity of data needed to choose among these alternatives. Prospective epidemiologic surveys have not identified any new teratogenic drugs, although they have helped to rule out the possibility of drug-related causation, as was the case with meclizine (Mellin and Katzenstein, 1964).

A. Maternal Alcoholism

Although previous authors had suggested that such defects as mental retardation and certain encephalopathies were a consequence of maternal alcoholism, after reviewing the literature on the subject Giroud and

Tuchmann-Duplessis (1962) concluded that alcoholism had not caused true embryopathies, but noted that there was an increased frequency of prematurity, infant mortality, mental retardation, and neurological disorders among children of alcoholics. The question was again raised by Lemoine *et al.* (1968), but only recently has the matter come under intensive scrutiny. Several authors have now accumulated data indicating that a history of severe chronic alcoholism in the mother is often associated with the birth of children which exhibit a recurring syndrome of defects (Christoffel and Salafsky, 1975; Green, 1974; Jones, 1974; Jones and Smith, 1973, 1975; Jones *et al.*, 1973, 1974; Palmer *et al.*, 1974, Tenbruick and Buchin, 1975).

Although some variations would be expected, these authors generally agree on the composition of a syndrome as follows: prenatal growth retardation, microcephaly, reduced palpebral fissures, borderline mental deficiency, and less frequently joint anomalies, cardiovascular defects, perinatal mortality, and postnatal failure to thrive. The authors also tend to agree that this syndrome of abnormalities is not attributable solely to maternal nutritional deficiency during pregnancy, because several of the mothers were judged to have had adequate or near-adequate nutrition during pregnancy. The full magnitude of the problem is not yet defined, but some idea of its possible extent is suggested by the observation of Forfar and Nelson (1973) that 14% of a sample of 911 pregnant women in Edinburgh were considered to be heavy drinkers of alcohol. Final recognition of the association of alcohol with what is admittedly a loosely defined and variable set of defects may be attributable to the recently intensified interest in all syndromes and repetitive patterns of abnormalities.

Unlike many established or suspected associations between drugs and developmental defects, in this instance the possibility has not been highlighted by clearcut animal studies. Alcohol given to pregnant mammals, even in large doses, has produced variable effects on intrauterine growth and litter size, and usually few or no malformations (Chernoff, 1975; Papara-Nicholson and Telford, 1957; Sandor and Amels, 1971; Tze and Lee, 1975). In summary, there no longer seems reason to doubt that defective children are born to chronically alcoholic women with some frequency, but uncertainty remains about such things as the duration and level of alcohol consumption and the possible involvement of other factors (nutrition?) which may account for the observed variability in the outcome of alcoholic pregnancies.

B. Anticonvulsants

These drugs have been suspected of having a low level of teratogenic potential in man for a number of years, beginning with the reports of Massey (1966), Melchoir *et al.* (1967), and Meadow (1968) of association between the birth of defective infants and the taking of the drugs ordinarily used to control epilepsy. Meadow (1970) later reported a total of 38 cases of cleft lip and/or cleft palate in infants born to epileptic women on anticonvulsant therapy throughout pregnancy. A syndrome consisting mainly of facial clefts and congenital heart defects was postulated. Despite the accumulating evidence of an association between the epileptic state and an increased rate of malformations, however, many investigators hesitated to assign an etiologic role to drugs for three reasons. (1) The mothers of defective infants usually had taken one or more of several different anticonvulsants for variable times and doses during their pregnancies. (2) The reports were of such nature as to make it virtually impossible to separate the effects of the taking of drugs from the physiologic imbalances that accompanied or followed convulsive seizures. (3) There was a possibility that whatever genetic factors predispose to epilepsy also introduce elements of instability into embryonic development.

Reports of an association between the taking of anticonvulsive drugs and the birth of defective babies have continued at an increasing rate in the last few years. Approximately 35 such reports now in the medical literature cover more than 1700 livebirths to women on therapy to control seizures during all or a substantial part of their pregnancies. The authors of several surveys have concluded that the malformation rate among births to epileptic women is at least twice that expected in comparable nonepileptic women (Barry and Danks, 1974; Elshove and van Eck, 1971; Koppe *et al.*, 1973; Lowe, 1973; Monson *et al.*, 1973; Niswander and Wertelecki, 1973; Speidel and Meadow, 1972). It also has been noted that there are many women on anticonvulsant therapy throughout pregnancy without untoward effects, or in whom the malformation rate was only questionably elevated (Bird, 1969; Janz and Fuchs, 1964). Nevertheless, the weight of evidence favoring some cause-and-effect relation between some aspect of the epileptic state in pregnant women and an increased risk of defective offspring now seems sufficient to warrant accepting this relation as a causative factor in human teratology.

It is still not possible, however, to identify a single primary agent, i.e., whether the disease or one or more of the drugs used in its therapy is at fault. Koppe *et al.* (1973) in a moderately large survey in Holland found no statistical basis for ascribing the teratogenic effects seen in 197 infants born to epileptic women to the taking of anticonvulsants. On the other hand, Lowe (1973) in a similar survey in Wales found that epileptic mothers taking anticonvulsants during the first trimester produced 6.7% malformed infants, whereas 111 women with a history of epilepsy but not taking anticonvulsant drugs during pregnancy produced no more defective babies (2.7%) than did the controls. Monson *et al.* (1973) reported that the malformation rate among children of epileptic women not taking drugs or exposed only occasionally or late in pregnancy was intermediate between that of women taking diphenylhydantoin daily (61/1000) and that of nonepileptic women (25/1000). Most authors have failed to segregate epileptic women who are on anticonvulsant drugs from those not on any therapy, as regards malformations among their offspring. Many have undoubtedly assumed, as did Niswander and Wertelecki (1973), that the drug-treated cases were the more severe ones, that is, more prone to seizures. Certainly to define the role of drugs precisely, more attention must be given to the rate of malformation among infants born to epileptic women not on anticonvulsant therapy during pregnancy and whether or not seizures occurred. Nevertheless, there now seems to be little reason to doubt that women whose convulsive disease is sufficiently severe to require treatment with anticonvulsant drugs are about twice as likely to produce defective offspring as are women in the general population.

It is even more difficult to indict any one drug of the several in use. The most widely used ones are of the hydantoin (e.g., diphenylhydantoin) or the barbiturate (e g., phenobarbital, primidone) classes, but others are in current use and it is by no means unusual for epileptics to take two or more drugs concurrently. Speidel and Meadow (1972) in a retrospective study of 427 pregnancies in epileptic women found that 10 defective babies were born to the 240 mothers who had taken phenobarbitone, 9 to the 192 who had taken phenatoin, 3 to the 65 who had taken primidone, and 2 to the 10 who had taken diazepam. Such data are consistent with the view that the severity of the disease may be the primary factor. In other words disease of a degree such as to require the taking of large doses of anticonvulsant drugs, or perhaps this and unidentified physiologic imbalances associated with this level of disease, may be more important than which drug is taken. This possibility, together with measurements of plasma levels of unbound drug and greater diligence in identifying maternal toxic levels of dosage, are much in need of attention in future studies on this subject.

It was stressed in early reports that facial clefts were the predominant type of defect in affected children (Elshove and van Eck, 1971; Meadow, 1970; Melchoir et al., 1967; Mirkin, 1971; Pashayan et al., 1971). Subsequent studies, however, have shown that many other types of defects may be involved, ranging through abnormal digits (Loughnan et al., 1973; Barr et al., 1974; Barry and Danks, 1974), chromosomal aberrations (Grosse et al., 1972), a variety of visceral defects (Starreveld-Zimmerman et al., 1973), and mental subnormality (Speidel and Meadow, 1972). Several recent studies have made it clear that facial clefts were not necessarily the predominant defects (Koppe et al., 1973; Kuenssberg and Knox, 1973; Lowe, 1973), and most consisting of more than a very few cases have usually included some with congenital heart defects. Although facial clefts and heart defects have frequently been reported, it can be questioned whether they occur with sufficient regularity to form the basis of a syndrome. Hanson and Smith (1975) found that five unrelated children born to epileptic women treated with hydantoin anticonvulsants displayed a broad array of defects including craniofacial anomalies, nail and digit hypoplasia, growth retardation, and mental deficiency; but none had heart defects and only one had cleft palate. Thus, the presence of a discrete syndrome may still be questioned, and the lack of such does not preclude assigning a low level of teratogenicity to anticonvulsant drugs. It should be recalled that the folic acid antagonists produce a broad spectrum of developmental defects in both animal and man (Nelson et al., 1955; Wilson, 1973b).

Unlike other anticonvulsant drugs, diphenylhydantoin has been repeatedly shown to be teratogenic in mice and rats (Elshove, 1969; Gabler, 1968; Gibson and Becker, 1968; Kernis *et al.*, 1973; Marsh and Fraser, 1973; Mercier-Parot and Tuchmann-Duplessis, 1974; Murray, 1974). Furthermore, Murray (1974) has shown that the level of unbound drug in maternal plasma that is associated with teratogenicity in animals is only two or three times higher than the plasma levels ordinarily associated with the therapeutic dose in man. This author also noted that teratogenic effects in animals were observed at or near maternal toxic doses and suggests that the occasionally observed teratogenic effects in man might also be at doses approaching maternal toxic levels. The present author (Wilson, 1974) reported that of 17 pregnant rhesus monkeys treated with large doses of diphenylhydantoin during organogenesis, 3 produced fetuses with the same unusual urinary tract anomaly, several had fetuses with minor skeletal defects, and 2 aborted (Table 5). The later study, despite the limited number of animals, suggests that this anticonvulsant has a low teratogenic potential in monkeys. Thus animal studies lend some support to the possibility that diphenylhydantoin alone has a teratogenic potential under certain conditions of dosage, regardless of maternal disease.

Other anticonvulsants, *trimethadione* and *paramethadione*, ordinarily used in the treatment of petit mal epilepsy, have also been associated with possible teratogenic effects in man. German *et al.* (1970) reported that 8 of 14 infants from 4 women given trimethadione during the early months of pregnancy had facial abnormality and 4 had cardiac defects. Rutman (1973) described 2 defective infants whose mothers had taken paramethadione in combination with diphenylhydantoin, and one of these women also had 2 miscarriages and a premature infant who did not thrive postnatally. Nichols (1973) reported the birth of one infant with severe facial, abdominal, cardiac and spinal defects to a mother who had taken trimethadione for several years; and Tsun-Yu and Stiles (1975) observed a rare persistent fifth aortic arch in the child of a woman who had a seizure during the fourth month of pregnancy while on thrimethadione medication. On the basis of 7 malformed infants born to 3 mothers on trimethadione during their pregnancies, together with information in the medical literature, Zachai *et al.* (1975) proposed a "fetal trimethadione syndrome" consisting of mild mental retardation, speech difficulties, V-shaped eyebrows, epicanthus, low set ears, palatal anomalies and abnormal teeth, and less commonly, growth retardation, microcephaly, ocular anomalies, congenital heart disease, etc. The comment seems warranted that, if syndromes exist for either trimethadione or other anticonvulsants, they are remarkably broad in scope and they overlap to the point of being indistinguishable.

In subhuman primates Poswillo (1972) was unable to produce malformations in 6 cynomolgus monkeys treated during early pregnancy with large doses of trimethadione or paramethadione, and Wilson (1974) observed only one frankly malformed fetus from 16 pregnant rhesus monkeys treated with

Table 5. Developmental Defects Observed in 17 Fetuses from Rhesus Monkeys Treated with Diphenylhydantoin During Organogenesis, and of 4 Fetuses from Females Treated with Both Diphenylhydantoin and Phenobarbital[a]

| mg/kg (b.i.d.) | Dosage[b] | | Number Rx | Condition of 100-day fetus |
	Number of days dosed	Gestation period (days)		
5	21	25–45	4	2 normal
				1 renal ectopia, retrocaval ureter[c]
				1 retrocaval rt. ureter
10	21	20–45	4	2 normal
				1 webbing of neck (?)
				1 slight growth retardation
20	21	20–44	4	1 bilateral cervical ribs
				1 retrocaval ureter, retarded
				1 3 cusps on mitral valve
				1 aborted d. 44
30	21	22–42	1	1 rudimentary 13th ribs
	10	35–44	4	2 normal
				1 bowed femora
				1 aborted d. 55
Diphenylhydantoin (30 mg/kg) plus phenobarbital (2 mg/kg b.i.d.)				
	10	35–44	4	1 normal
				1 dextrocardia
				1 malrotation of gut
				1 6 cervical 8 lumbar vert.

[a]From Wilson (1974).
[b]Human dosage ranges upwards from 1.5 mg/kg/d. A dose of 50 mg/kg/d caused severe maternal toxicity in monkeys.
[c]This defect, which was observed in three instances here, is not known to have occurred spontaneously in rhesus monkeys.

large doses of trimethadione (Table 6). These human and simian data are too limited to warrant a final conclusion, but some teratogenic potential of these oxyzolidinedione anticonvulsants must be regarded as a possibility.

C. Neurotropic–Anorectic Drugs

Although such drugs as *amphetamines* and *phenmetrazine* are usually thought of in connection with their effects on the autonomic and central nervous systems, they are anorectics and as such are sometimes used in weight control during pregnancy. Concern about their possible teratogenic potential grew originally from repeated demonstrations that such drugs were teratogenic in mice (Tuchmann-Duplessis and Mercier-Parot, 1964; Nora *et al.,* 1965, 1968) and rabbits (Kasirsky and Tansy, 1971). In an early retrospective analysis of 219 cases and prospective analysis of 52 cases of women using dextroamphetamine during pregnancy, Nora *et al.* (1967b) were unable to demonstrate a teratogenic effect of this drug in man, but found in a later study that 184 mothers of children with congenital heart defects had a higher incidence of amphetamine use than did mothers of the control group (Nora *et al.,* 1970). Other authors have reported that amphetamine use during early pregnancy has been associated with the birth of infants with such diverse malformations as exencephaly and biliary atresia (Levin, 1971; Matera *et al.,* 1967). In a retrospective survey of 458 mothers of children with various malformations Nelson and Forfar (1971) found that a higher proportion of these women had taken dextroamphetamine to suppress appetite during early pregnancy than had the control mothers of normal children. Women known to have taken large doses of amphetamine throughout pregnancy, however, do not always have abnormal babies (Briggs *et al.,* 1975). Nevertheless, there is reason to entertain suspicion that amphetamines may occasionally have a low level of teratogenic activity in man, but the evidence to date is by no means conclusive. Adding to the uncertainty is the fact that rats and some strains of rabbits appear to be resistant to the teratogenic action of these drugs (Jordan and Shiel, 1972).

Phenmetrazine has been reported occasionally to be associated with the birth of defective infants (Lenz, 1962; Moss, 1962; Powell and Johnstone, 1962), although no anomalies were found among infants born to mothers taking this drug during pregnancy in another study (Notter and Delande, 1962). Animal studies on the teratogenic potential of this drug have not been

Table 6. Effects of Trimethadione on the Fetuses of Rhesus Monkeys Treated at Various Times During Organogenesis[a]

| mg/kg (i.d.) | Dosage[b] | | | Condition of 100-day fetus |
	Number of days dosed	Gestation period (days)	Number R_x	
30	4	20–45	5	4 normal 1 extra T vert. & ribs
60	4	20–45	5	4 normal 1 growth retarded
60	20	21–47	6	1 normal 1 11 T vert. & ribs, 8 C vert. 1 growth retarded 1 colonic atresia, short tail, 6 L vert., no caudal vert. 2 aborted ds. 34 and 36

[a]From Wilson (1974).
[b]Human adult dosage 1–2 g/day = 16–32 mg/kg/day.

reported. A drug of similar action, *diethylaminopropiophenone*, was found not to be embryotoxic in either rats or mice (Cahen *et al.*, 1964).

Other neurotropic substances infrequently mentioned as possibly teratogenic are *dominal* (Tuchmann-Duplessis and Mercier-Parot, 1964a) and *bromine-containing compounds* (Opitz *et al.*, 1972; Rossiter and Rendle-Short, 1972), but supporting evidence is meager.

D. Oral Anticoagulants—Warfarin

There is an understandable scarcity of animal studies (Kronick *et al.*, 1974) on the embryotoxicity of warfarin (Coumadin) in view of the fact that it is an effective rodenticide. Reported associations between the use of this compound during human pregnancy and abnormality in the newborn child apparently began when Kerber *et al.* (1968) called attention to the fact that a mother on warfarin therapy because of a prosthetic heart valve gave birth to a child with nasal hypoplasia. Holmes *et al.* (1972) also ascribed an earlier reported case of an infant with nasal hypoplasia, optic atrophy, stippled epiphyses, and mental retardation to the taking of warfarin during pregnancy by the mother who had a mitral valve prosthesis. To these two early cases of malformation have recently been added several additional published ones and reference made to some unpublished ones (Anon., 1975; Becker *et al.*, 1975; Pettifor and Benson, 1975; Shaul *et al.*, 1975; Warkany, 1975; Warkany and Bofinger, 1975). A syndrome consisting of nasal hypoplasia and stippled epiphyses and sometimes including optic atrophy, mental retardation, brachydactyly, and a scattering of other defects has recurred with some regularity. This bears a resemblance to the genetically determined Conradi–Hünermann type of chondrodystrophia calcificans congenita. In some of the above cited cases the mothers had also been exposed during the first trimester to other suspected embryotoxic factors, but warfarin was the only factor common to all.

Oral anticoagulants have been used during pregnancy for a number of years for the treatment of thrombophlebitis or other potentially embolic conditions, and although increased fetal and perinatal death have been reported, malformations have not generally been observed to increase (Fillmore and McDevitt, 1970; Finnerty and Mackay, 1962; Hirsch *et al.*, 1970; Olwin and Koppel, 1969). Not all women who have more recently given birth to deformed children have had intracardiac prostheses (Palacios-Macedo *et al.*, 1969); but several have, or have had other indications for the use of large dosage of anticoagulants throughout pregnancy. This suggests that there may be a dosage factor in determining whether malformations will occur. This and other possible complicating factors, e.g., other potentially teratogenic drugs taken concurrently, must be further examined before the true teratogenic status of anticoagulants can be established.

E. Alkylating Agents

Owing to their polyfunctional action in suppressing growth, alkylating agents are immediately suspect as teratogens. In addition to being radiomimetic and cytotoxic, they are known to be highly teratogenic in laboratory animals (Chaube and Murphy, 1968; DiPaolo and Kotin, 1966; Tuchmann-Duplessis, 1969; Duke, 1975). There are, however, surprisingly few instances of human maldevelopment having been attributed to use of these compounds during pregnancy (Sokal and Lessman, 1960), although intrauterine death and abortion regularly occur with high dosage (Nicholson, 1968). One woman alternately given 6-mercaptopurine and busulfan delivered a baby with multiple malformations (Diamond *et al.,* 1960), another given chlorambucil delivered an infant with urinary tract anomalies (Shotten and Monie, 1963), and one given cyclophosphamide produced a child with multiple musculoskeletal defects (Greenberg and Tanaka, 1964). A number of cytotoxic, cancer chemotherapeutic drugs have been administered to pregnant monkeys and found to produce high rates of intrauterine death but relatively few outright malformations (Wilson, 1972). The paucity of human cases attributable to the therapeutic use of alkylating agents may relate to three factors: (1) limited use during pregnancy because of high teratogenicity known to have occurred in laboratory animals (Hoskins, 1948; Murphy *et al.,* 1957; Thiersch, 1957), (2) easy justification of elective abortion because of the life-saving role of these drugs for the mother, and (3) high rate of treatment-induced intrauterine death and abortion. In other words, the known biological effects of alkylating agents contraindicate their use during pregnancy except in treatment of a life-threatening neoplasm. Apart from their direct embryotoxic effects, many of these compounds are known to be mutagenic in microorganisms and in mammalian cell cultures (Fishbein *et al.,* 1970), and to a lesser extent, in intact mammals (Bateman and Epstein, 1971).

F. Oral Hypoglycemics

Certain of these drugs are in the suspect category on the basis of infrequent but recurring reports of association between their use during pregnancy and the birth of defective children. The first two cases, described by Campbell (1961) and Larsson and Sterky (1961), were malformed infants from mothers who had taken *tolbutamide* or other sulfonylurea compounds throughout pregnancy to control diabetes; but a contemporaneous report of 20 pregnant women on sulfonylurea drugs, after lowered glucose tolerance was observed, included none with defective children (Ghanem, 1961). Since that time, however, at least a dozen additional cases of association have been reported (Caldera, 1970; Campbell, 1963; Coopersmith and Kerbel, 1962; Dolger *et al.,* 1969; Malius *et al.,* 1964; Schiff *et al.,* 1970). Generally the defective children have occurred in a ratio of about 1 in 20 among all births to women taking tolbutamide during pregnancy, which is slightly higher than the 2–3% usually reported for the general population when no extraordinary diagnostic procedures are used. On the other hand there have been reports of fairly large series of cases in which no abnormality was observed (Sterne, 1963). The infrequent occurrence of a positive association between use during pregnancy and abnormal births, together with the absence of an identifiable syndrome of defects, leaves the teratogenic status of the sulfonylurea hypoglycemics in question at present. Further confusing the issue is the fact that diabetic women give birth to more defective infants, apparently without regard to the degree to which their blood glucose levels are controlled, than do nondiabetic women. There is no doubt that these compounds have direct teratogenicity in mice, rats, and rabbits (Tuchmann-Duplessis and Mercier-Parot, 1963).

IV. DRUGS POSSIBLY EMBRYOTOXIC IN MAN

In large part this category includes drugs which have been very widely used and only occasionally implicated in human developmental toxicity. In other instances acceptable data suggesting embryotoxicity in man are extremely meager, but the fact that several species of animals have shown the drug to have definite teratogenic potential has been the basis of their inclusion here.

A. Female Sex Hormones

As discussed earlier the synthetic progestogens used in large doses during the 1950s and early 1960s as a preventive measure in cases of threatened abortion caused pseudohermaphroditism in a large number of genetic females whose mothers were treated prior to the 12th week of gestation. This unequivocally teratogenic effect, however, was due to the androgenic properties of these substances and presumably not to their progestational action. Neither natural nor synthetic estrogenic or progestational hormones were thought to pose a significant embryotoxic hazard to the human conceptus until relatively recently, although several years ago Uhlig (1959) did describe three severely malformed infants born to women who took high doses of dienoestrol in attempting to induce abortion. Also in abortion attempts, large doses of norethindrone–mestranone contraceptive hormones have now been associated with the birth of two defective children, one with multiple skeletal malformations (Papp and Gardo, 1971) and one that was a pseudohermaphroditic female (46XY chromosome count) with cleft palate, retrognathia, kyphoscoliosis, and multiple appendicular defects (Gardner et al., 1970).

Concern about the possible adverse effects of hormonal pregnancy tests prompted a survey of 100 mothers of babies with meningomyelocele or hydrocephaly, and it was found that 19 mothers of defective babies had had such tests, as compared with only 4 of 100 mothers of normal babies (Gal *et al.,* 1967). In a larger survey of 149 children with varied but mainly nervous system malformations, Greenberg *et al.* (1975) reported that 23 of the mothers had had hormonal pregnancy tests, whereas only 8 of 149 mothers of normal children from matched pregnancies had been exposed to such tests. Combining the results of these somewhat similar studies, 16.9% of mothers of defective children had taken the tests but only 4.8% of mothers of normal children had done so, strongly suggesting that the tests increased the risk of abnormal development *in utero.* Laurence *et al.* (1971), however, were unable to find a significant difference between mothers of children with neural tube defects and those of normal children as regards hormonal pregnancy tests.

The use of oral contraceptives in large dosage for treatment of threatened abortion has also been associated with the birth of two masculinized female infants (Voorhess, 1967) and of one with multiple defects (Kaufman, 1973). Several types of malformations or syndromes of defects have in recent surveys been variously associated with the taking of contraceptive pills or other use of estrogen and/or progestin preparations during pregnancy. Levy *et al.* (1973) found that these hormones had been taken by 7 of 76 mothers of babies with transposition of the great vessels whereas none of 76 mothers of children with other types of defects had taken the hormones; this was interpreted to mean that exogenous sex hormones may be part of a multifactorial causation of congenital heart disease. Mulvihill *et al.* (1974), however, were unable to show that children with transposition were more likely to be born to mothers exposed prenatally to female sex hormones than were children with ventricular septal defects or functional murmurs. Nishimura *et al.* (1974) also failed to find more congenital heart defects or other malformations in fetuses obtained by induced abortion in women exposed to sex hormones, as opposed to those obtained from women not exposed. On the other hand, Robertson-Rintoul (1974) reported that four of five women known to have continued taking oral contraceptives during pregnancy produced infants with serious malformations, two with congenital heart defects, one with exomphalos, and one with skeletal malformations.

Exposure to estrogen–progestogen contraceptive preparations during the early weeks of gestation was associated by Nora and Nora (1973, 1974) with congenital heart disease in 20 cases in a retrospective study of 224 patients. Of these 10 had a syndrome-like clustering of vertebral, cardiac, tracheal, esophogeal, renal, and limb anomalies. In their latest report these authors had accumulated 19 patients with this syndrome, and 13 of these gave a history of having taken progestogen alone or progestogen–estrogen compounds during the period of organogenesis (Nora and Nora, 1975). It should be noted that a similar association of malformations is known to occur in the absence of female sex hormone dosage during pregnancy (Balci *et al.*, 1973; Yasuda and Poland, 1974). Janerich *et al.* (1974) found that among 108 mothers of children with congenital limb-reduction defects, 15 (14%) had a history of exposure to sex steroids whereas only 4 (4%) mothers of normal children had such exposure. Harlap *et al.* (1975) reported that of 11,468 pregnant women interviewed, appreciably more infants with major and minor malformations were born to women given estrogens and progesterones during pregnancy than to those receiving no hormones.

Other investigators have found less-pronounced or only marginal indications of a relationship between the use of oral contraceptives and developmental abnormalities in the offspring. Robinson (1971) failed to find a statistically significant difference in fetal outcome between 1250 women who had taken estrogen–progestin pills and an equal number of paired control women, although children of mothers on contraceptives had more life-threatening anomalies by a ratio of 12:5 and more congenital heart defects by a ratio of 7:2 ($P > 0.05$) than did control mothers. In a prospective survey of 50,282 pregnancies involving maternal exposure to oral contraceptives or other female sex hormones, Hook *et al.* (1974) noted a suggestive association between exposure of extraneous female sex hormones during pregnancy and infant malformation; but this association was present after both early and late exposure, which tends to weaken the likelihood of a direct cause–effect relationship with developmental defects. Also of uncertain significance, Cervantes *et al.* (1973) made clinical examinations of the offspring of 32 women who were on hormonal contraceptives for 1 to 5 cycles during their pregnancies and found that one had major defects, 3 had minor defects, 7 aborted, 4 were lost to the study, and the remaining 17 were normal.

Of possible relevance to the question of hormonal influences during pregnancy are other reports such as that of Yalom *et al.* (1973) who, after studying women with elevated estrogen levels during pregnancy, felt there was strong reason to believe that the postnatal psychological development of male children was modified in that they were less aggressive and athletic and generally had diminished "masculinity," and two were hypospadiac. Poland and Ash (1973) examined abortuses from women who had been on oral contraceptive pills before pregnancy and found that 54.7% of these women aborted specimens judged to be developmentally abnormal, whereas only 42.2% of abortuses from non-pill-users were abnormal, most of the difference being accounted for by the larger numbers of "disorganized" fetuses from pill-using mothers. A high frequency of polyploid chromosomes was noted among abortuses from women on steroid-contraceptive treatment by Carr (1970), but this was not confirmed by Boué *et al.* (1975).

Several other studies have failed to find any correlation between increased embryotoxicity and the taking of sex hormones during early pregnancy (Bačič *et al.*, 1970; Spira *et al.*, 1972; Tuchmann-Duplessis, 1973; Wild *et al.*, 1974) or prior to pregnancy (Girotti and Schaer, 1973; Peterson, 1969). Because of the conflicting nature of evidence on the effects of extraneous sex hormones on intrauterine development, and the lack of consistency as to types of defects reported, the question of possible adverse effects on development cannot at this time be resolved. Despite their inconsistency, however, the persistence of reports of such effects requires that the continued use of female sex hormones during pregnancy, for whatever purpose or in whatever dosage, be kept under close scrutiny.

The fact that laboratory animals have been reported to show increased rates of malformations not related to the genital system following treatment of pregnant dams with contraceptive hormones (Takano, 1964; Takano *et al.*, 1966; Christenson *et al.*, 1971; Andrew and associates, 1972, 1973, 1974), adds to the disquiet about the complete safety of these compounds during early human pregnancy.

The nonsteroidal estrogenic substance, *diethylstilbesterol* (DES), has attracted much attention since 1970 when it was first reported to be associated with adenocarcinoma of the vagina in girls and young women whose mothers had been given this drug during the first trimester of pregnancy for threatened abortion (Herbst and Scully, 1970; Herbst *et al.*, 1971). Subsequently the number of such cases reported, some involving the cervix as well as the vagina, has risen to more than 200 (Greenwald *et al.*, 1971; Herbst *et al.*, 1974). This is estimated to account for only a small percentage of the total number of pregnant women treated with this drug during the 1950s and early 1960s. It has recently been shown that treatment with DES causes a high rate (97%) of vaginal adenosis and other benign changes in epithelia of the vagina and cervix among the female offspring of women so treated during pregnancy, but that in only a very small fraction of the cases does the abnormal epithelium become malignant (Stafl and Mattingly, 1974; Herbst *et al.*, 1975). Thus, it appears that this drug with some regularity causes abnormal histogenesis in the lower end of the mullerian duct system from which the upper vagina and cervix are derived, and under as yet unidentified conditions a small proportion of these later become neoplastic. If this interpretation is correct, the abnormal extension of columnar epithelium into the vagina as a result of DES treatment during gestation results in a high incidence of a minor developmental abnormality. The subsequent metaplasia and occasional neoplasia are not teratogenic manifestations themselves but are secondary to transformation of an ectopic epithelial type in an unfavorable environment, i.e., the vagina.

Paradoxically, treatment with DES during pregnancy has been associated with masculinization of the female fetus in four instances (Bongiovanni *et al.*, 1959). One male fetus is reported to have been feminized by this synthetic compound as a result of treatment of the mother from the 6th week until term (Kaplan, 1959).

B. Tranquilizers

Until very recently drugs of this loose pharmacological and chemically diverse class have not been implicated in human developmental toxicology, despite the fact that some of these are among the most widely used prescription drugs sold today. Two so-called minor tranquilizers used for treatment of anxiety, *meprobamate* and *chlordiazepoxide,* have now been reported to be associated with the birth of increased numbers of defective children when taken during the first 42 days of gestation (Milkovich and van den Berg, 1974). This carefully controlled prospective study found that the rate of congenital anomalies after taking meprobamate during early weeks of pregnancy was 2½ times the rate after other drugs ($P = 0.045$) and more than 4½ times the rate after no drugs ($P = 0.027$) during the same period. The rate of anomalies after chlordiazepoxide during early pregnancy was lower than the above but was, nevertheless, 2½ times the rate after other drugs ($P = 0.125$) and more than 4 times the rate after no drugs ($P = 0.075$) during the same period. Although the significance levels of difference between rates associated with the taking of chlordiazepoxide and the rates for other drugs or for no drugs were $P > 0.05$, there was strong suggestion that this drug as well as meprobamate may adversely affect development during early gestation. This possibility was enhanced by the observation that the taking of either of these tranquilizers during early pregnancy seemed to have increased the likelihood of fetal death, recalling the fact that increased embryolethality and increased malformations tend to occur concurrently in experimental teratology. Further strengthening this study were the facts that it included patients from a broad range of socioeconomic backgrounds and that there was a 5-year follow-up during which additional diagnoses were made and previous ones confirmed.

Neither drug was associated with a clearly defined pattern of defects, although congenital heart defects predominated among children from mothers on meprobamate and nervous system defects were most numerous among those from mothers on chlordiazepoxide. The existence of a recognizable syndrome is not a necessary characteristic of human teratogens, however, as discussed earlier in this chapter.

Contrary evidence was reported in a prospective study (Hartz *et al.,* 1975) based on follow-up of children from 50,282 pregnancies in 12 hospitals over the period 1958–1966. The children exposed *in utero* to either meprobamate or chlordiazepoxide during any of the first four months of gestation showed no significant difference from children not exposed as regards malformations, mental development or childhood mortality between birth and the fourth birthday. It should be noted that much of the information on drug consumption during pregnancy in this study was based on interviews with mothers and not on physicians' records, although medication history was confirmed in most cases by the attending physician or from hospital records.

An association between prenatal exposure to *diazepam,* another benzodiazepin tranquilizer closely related to chlordiazepoxide, and the birth of defective children has very recently been indicated. Three retrospective studies (Safra and Oakley, 1975; Saxén and Saxén, 1975; Aarskog, 1975) have all reported a higher incidence of oral clefts from pregnancies in which diazepam was taken during the first trimester. Although there were possible complicating factors in all of these surveys, the consistency of the findings suggests a low level of causal relationship between this drug and these defects. Animal data do not help to resolve the issue. Of two studies in mice one indicates that diazepam is embryotoxic (Stenchever and Parks, 1975) and the other that it is not (Miller and Becker, 1975).

More prospective studies of the type reported by Milkovich and van den Berg and by Hartz and colleagues are needed to determine the true teratogenic status of meprobamate, chlordiazepoxide, and diazepam. The main limitation of such studies is that inherent in all prospective surveys, namely, a relatively small number of pertinent cases in which all conditions necessary to ascertain an association are met. Nevertheless, until more data are available it would seem prudent to regard these compounds as having a low level of teratogenic potential under as yet unidentified conditions associated with their use during the early weeks of pregnancy.

Haloperidol, a so-called major tranquilizer because of its use in treatment of psychoses, has also recently been implicated in human developmental toxicology, but in this instance supporting evidence is meager in the extreme. Only two cases of association between drug ingestion during pregnancy and the birth of defective children are known: one of multiple digital and long-bone malformations (Kopelman *et al.,* 1975), and one of ectrophocomelia (Dieulangard *et al.,* 1966). Both infants had reduction defects of the limbs but of somewhat different types and degrees. Of considerable importance in evaluating these case reports is the fact that both mothers also took during their pregnancies other drugs that have been under question as having some teratogenicity, i.e., diphenylhydantoin in one instance and hydroxyprogesterone in the other (see above). No data are available on the extent of use of haloperidol in the United States, but it is said to have been fairly widely used in Europe since 1957. In one retrospective study of 98 women given haloperidol for hyperemesis gravidarum, no malformations or other untoward effects were found in the newborn infants (Van Waes and Van de Velde, 1969). It has been shown to have variable teratogenicity in mice and virtually none in other animals (Bertelli *et al.,* 1968; Einer-Jensen and Secher, 1970; Seay and Field, 1963; Vichi, 1969; Vichi and Pierleoni, 1970; Tuchmann-Duplessis and Mercier-Parot, 1967). Although possibly embryotoxic in man, there is little evidence to support such a contention at this time.

C. Salicylates (Aspirin)

From time to time reports in the medical literature have suggested that there was increased pregnancy wastage, usually in the form of abortion, following repeated or excessive use of salicylates during pregnancy (Jackson, 1948; Carter and Wilson, 1963), but causal relationships were never clearly shown. In a retrospective survey of 833 pregnancies yielding malformed infants, however, Richards (1969) found that significantly more mothers ($P <$ 0.001) took salicylates during the first trimester than did matched control mothers of normal infants. In a similar survey of 458 mothers of defective children, Nelson and Forfar (1971) also found that aspirin was taken by a significantly higher proportion of these mothers than by control mothers during the first trimester. In a partly prospective study of 599 children with oral clefts, I. Saxén (1975) reported that 14.9% of the mothers had taken salicylates during the first trimester, whereas only 5.6% of the control mothers of normal children had used these compounds ($P < 0.001$).

A contrary indication, however, has come from another survey which inquired into the types of drugs prescribed during early pregnancy to mothers of malformed children. Significantly fewer aspirin-containing preparations were issued to such mothers than to women delivering normal babies (Crombie et al., 1970). In Australia 144 regular users of salicylate powders were found to deliver more infants with reduced birth weight and higher perinatal mortality rates than were delivered by nonsalicylate users, but the incidence of congenital anomalies was not significantly raised (Turner and Collins, 1975).

Eight cases of a suggestive association between the taking of salicylate-containing preparations during early pregnancy and the birth of variously malformed children have been reported (McNiel, 1973). This paucity of case reports is surprising, considering that these compounds probably comprise the most widely used class of therapeutic agents in existence. But in spite of the fact that salicylates have not been positively implicated as causing developmental toxicity in man, there is increasing epidemiologic evidence of a low level of teratogenicity.

Since Warkany and Takacs (1959) in rat and Larsson et al. (1963) in mouse demonstrated that large doses of salicylates were highly teratogenic in animals, these compounds have been widely used in experimental studies. Aspirin in doses five or six times larger than those effective in rats was also found to be embryotoxic in rhesus monkeys (Wilson, 1971). Of 8 pregnant monkeys treated, 2 aborted a few weeks after treatment, 2 produced growth retarded fetuses, 3 produced malformed fetuses (2 with heart defects) and only 1 yielded a normal fetus. It should be stressed that the daily dose (500 mg/kg) was well in excess of that likely to be used therapeutically in pregnant women. This differential of dosage, with its implied margin of safety, however, is diminished by the observation of Kimmel et al. (1971) that the teratogenic potential of a given dose of aspirin in rats can be appreciably increased by concurrent administration of benzoic acid, a widely used food preservative. Similar metabolic interaction of benzoic acid and salicylates has been shown to occur in man (Levy, 1965). Thus, even though the primary dosage of salicylates ingested by pregnant women in unlikely to reach the level shown to be embryotoxic in animals, the possibility exists that other extraneous chemicals may interfere with the excretion of salicylic acid, the presumed active metabolite, to effect a higher and more prolonged plasma concentration, which could reach embryotoxic levels.

D. Antibiotics

These bacteriostatic and bacteriocidal agents have been implicated in human teratology mainly by the report of Carter and Wilson (1965), who made a retrospective analysis of 85 case reports of women given such a drug during the first 12 weeks of pregnancy. There were 12 malformed children, as against 2 or 3 that might be expected in equal numbers of untreated pregnancies; and 13 aborted, whereas the expected number would be 8. An additional 67 cases involving antibiotic administration during the first 12 weeks of gestation showed approximately 25% of pregnancy wastage (the sum of abortion, malformation, stillbirth, and neonatal death) as compared with 14% in controls (Carter, 1965). Deafness has occasionally been reported in young children after streptomycin therapy during pregnancy (Boletti and Croatto, 1958; Robinson and Cambon, 1964). Recent retrospective studies (Nelson and Forfar, 1971; Richards, 1969) have failed to support the above claims of an increase in developmental defects after antibiotics given during pregnancy; but a partially prospective survey of 599 children with oral cleft showed an excess above controls when the penicillins ($P < 0.01$) and tetracyclines and/or chloramphenicol ($P < 0.05$) were taken during the first trimester of pregnancy (I. Saxén, 1975).

Several antibiotics are unquestionably teratogenic in rodents, e.g., tetracycline (Filippi, 1967), streptonigrin (Warkany and Takacs, 1965), hadicidin (Lejour-Jeanty, 1966), penicillin–streptomycin mixtures (Filippi, 1967), actinomycin D (Tuchmann-Duplessis and Mercier-Parot, 1960; Wilson, 1966), and mitomycin C (Nishimura, 1963). Not all antibiotics are teratogenic in animals, however. Those acting primarily by inhibiting protein synthesis, such as puromycin, lincomycin, and streptomycin, have been found at high doses to be only slightly if at all teratogenic but highly embryolethal (unpublished data, author's laboratory).

The widespread use of these drugs during pregnancy over the past 25 years, and the fact that well-substantiated embryotoxicity for man has not been reported, probably means that the teratogenic potential of these compounds is quite low under usual conditions of usage. Nevertheless, their known effects on rapidly proliferating cells and organisms in culture, together with demonstrated teratogenicity of some antibiotics on mammalian embryos *in vivo,* suggests that such potential under some conditions be regarded as a possibility in man.

E. Antituberculous Drugs

A survey of children born to 61 women with pulmonary tuberculosis revealed that four had major congenital defects (2 were anencephalic, 1 had agenesis of kidneys, and 1 had a heart defect), which the author (McDonald, 1961) felt might represent an excess over the expected rate in healthy women. To answer the question of whether one of the drugs used in treatment of tuberculosis was responsible, 253 pregnancies in women known to have been under supervision or treatment for the disease during pregnancy have been analyzed (Lowe, 1964). Various combinations of the drugs, *isoniazid, p-aminosalicylic acid,* and *streptomycin,* were given to 74 of the women during early gestation, and two (2.8%) delivered defective infants. A group of 173 tuberculous women not given these drugs during pregnancy delivered 4.1% of defective infants. An even larger group of tuberculous pregnant women (726) were given various antitubercular drugs during pregnancy and found not to have significantly more malformed babies than untreated women (Ganguin, 1971). In another study of 123 tuberculous women treated during pregnancy with the same drugs, however, it was reported that malformations among the offspring were increased by a factor of two or three compared with controls (Varpela, 1964). *Ethionamide* alone was reported to be teratogenic in a study based on 22 pregnant women who received the drug during pregnancy (Potworowska *et al.,* 1966). Of the five mothers known to have taken the drug during the early months of pregnancy four produced malformed children. There was no prevalent defect or syndrome but the central nervous system was involved in all. In a survey of 110 phthisiologists, 9 cases of ethionamide treatment during pregnancy were found but none involved any teratogenic effects in the offspring (Jentgens, 1968).

A study on the use of isoniazid alone during 125 human pregnancies reported no increase in malformations but noted 5 cases of functional disorders of the nervous system in the offspring (Monnet *et al.,* 1967); in mice doses of isoniazid that were highly toxic to the pregnant females failed to induce abnormal development in the offspring (Kalter, 1972a). Other workers (Dluzniewski and Gastol-Lewinska, 1971) failed to find significant teratogenicity in either rats or chinchilla rabbits after any of several tuberculostatic drugs, ethionamide, isoniazid, *p*-aminosalicylic acid or streptomycin, given to pregnant dams; although increased skeletal variations were observed in rats. In view of these conflicting reports, whether any of the drugs used in treatment of tuberculosis possess embryotoxic potential remains uncertain, but at present there is no reason to consider this as more than a possibility.

F. Quinine and Other Antimalarials

Several years ago quinine was said to have been used in large doses as an abortifacient and to have entailed a certain teratogenic risk when attempted abortion was unsuccessful (Grebe, 1952; Sylvester and Hughes, 1954; Uhlig, 1957; Windorfer, 1953; Zolcinski *et al.,* 1965). Reliable data on this illicit use of quinine are understandably scarce, but Tanimura (1972) was able to find 21 cases since 1947 of human malformations thought to be associated with large doses of this drug during pregnancy. Winckel (1948) collected 17 cases in which visual and auditory defects in children were associated with the taking of quinine during pregnancy, not always in abortion attempts, but the evidence of a causal relationship was tenuous in several of these, and no information was available on the total use during pregnancy. Tanimura (1972) reviewed the results of published animal studies and found that large doses given during pregnancy to several species caused intrauterine death and growth retardation but relatively few malformations. This author also treated six pregnant macaque monkeys and observed no embryotoxic effects other than retardation of growth in two. Savini *et al.* (1971) observed no increase in malformations when quinine was given to pregnant rats, rabbits, and dogs at or near maternal toxic levels, although there was increased embryolethality in rabbits and dogs at these doses.

Quinacrine has been associated with one case of human malformation (Vevera and Zatloukal, 1964), but in mice this antimalarial caused only an increase in resorptions (de Rothschild and Levy, 1950).

G. Anesthetics

The chemical agents used in surgical anesthesia have generally been thought to represent no particular hazard to mammalian embryos during the early stages of pregnancy (Smith, 1968). Urethane is unquestionably teratogenic in mice (Sinclair, 1950; Nishimura and Kuginuki, 1958), but it is no longer used as an anesthetic in man. Anesthetic agents such as nitrous oxide and halothane produce no or very low levels of developmental abnormality in rats, and these have consisted mainly of minor skeletal variations (Basford and Fink, 1968; Shepard and Fink, 1968). Smith *et al.* 1971), however, produced an increase in the incidence of cleft palate in C-57 black mice using halothane, although Chang *et al.* (1974) using the same agent in rats obtained no gross malformations but did find ultrastructural changes in the brain, kidney, and liver in neonatal animals exposed *in utero*.

Some concern has arisen following several reports that anesthetists and operating room nurses, who are subjected to repeated exposure to inhalation of anesthetics, may have higher-than-normal rates of miscarriages and possibly some increase in the birth of defective children (Askrog and Harvald, 1970; Corbett, 1972; Cohen *et al.*, 1971; Knill-Jones *et al.*, 1972). Corbett *et al.* (1974a) in a survey of 621 nurse–anesthetists found that of those who continued to work as anesthetists during pregnancy 16.4% had defective children, whereas of those who did not work during their pregnancies only 5.7% had defective children ($p < 0.005$); but the total number of pregnancies of both types was relatively small and other environmental factors were possibly contributory. One of these agents, methoxyflurane, has been found to have no embryotoxicity in rats exposed 8–16 hr/day to operating-room concentrations and higher (100 ppm) (Corbett *et al.*, 1974b). There has been an inclination to relate these problems to the widespread use of halothane; however, the matter is still under investigation, and no conclusion is warranted until additional data are available.

H. Insulin

This widely used drug is of less concern today than formerly, because it is no longer used in shock treatment of psychiatric disorders. Of 14 women subjected to insulin coma in psychiatric therapy during the first 14 weeks of gestation, 4 were reported to have delivered dead and 2 to have delivered malformed fetuses (Sobel, 1960). Two mentally defective children were associated with similar maternal therapy during the third month of pregnancy (Wickes, 1954). The occasional malformed infant born to a diabetic woman maintained on insulin is probably coincidental (Pettersson *et al.*, 1970), although the fact that diabetic women are known to produce about twice as many defective children as healthy women should be kept in mind.

I. Lithium Carbonate

This compound, which is increasingly used in the treatment of manic–depressive illness, is the subject of some disagreement as regards its embryotoxic potential. It has been known for many years to interfere with embryonic development in submammalian forms. In pregnant mammals, however, reports vary widely in the extent to which they implicate lithium salts in teratogenicity. At doses producing blood levels said to be in the range of those observed during human therapy, it caused low to moderate embryotoxicity in mice (Szabo, 1970; Matsumoto *et al.*, 1974) and in Sprague–Dawley rats (Wright *et al.*, 1971) but failed to do so in other strains of rats or in rabbits and monkeys (Johansen, 1971; Gralla and McIlhenny, 1972; Trautner *et al.*, 1958). Thus this compound can hardly be said to be highly teratogenic in laboratory mammals. Nevertheless, there is some reason for concern as to its use during the first third of human pregnancy.

On the basis of the earlier reports of embryotoxicity in animals, a Registry of Lithium Babies was set up to collect data on the outcome of human pregnancies in which lithium salts had been administered during the first trimester. The first report of the registry (Schou *et al.*, 1973) provided information on 118 children born to women so treated. Of these, 5 infants were stillborn of which 1 was malformed; 7 died within the first week, of which 5 were malformed. Including those just mentioned, a total of 9 babies (7.6%) were judged to have macroscopic abnormalities of structure attributable to faulty development. Other reports of embryotoxicity after lithium carbonate therapy during pregnancy (Goldfield and Weinstein, 1973) are based on cases presumably included in the above-mentioned registry. Nora *et al.* (1974) have reported two additional cases, both of which were demonstrated to have the relatively rare (1/20,000 births) Ebstein's anomaly of the heart. The latter authors emphasize that 8 of the 11 known malformed infants delivered by mothers on lithium therapy had cardiac defects and that 4 of these 8 had the Ebstein's anomaly, which was noted as representing an approximately 400-

fold increase over expected frequency. In any event, there appears to be ample reason for continuing to regard lithium salts as possibly having teratogenic potential under as yet undefined conditions during human pregnancy.

V. DRUGS BELIEVED NOT TO BE EMBRYOTOXIC IN MAN UNDER LIKELY CONDITIONS OF USAGE

To maintain that any drug or other chemical agent is devoid of embryotoxic potential in man is foolhardy. In the first place, current methods for evaluating embryotoxicity in man are far from exact. Preclinical tests in pregnant animals, even when the test animal is known to absorb, metabolize, and eliminate a drug as does man, provide only a crude measure of possible effects on the human conceptus and often involve insignificant numbers compared with the numbers of pregnant women that may use the drug. The difficulties often encountered in judging the embryotoxicity of a drug on the basis of accumulated case reports or of planned epidemiologic surveys are evident in the preceding sections. To further complicate the situation, there are the unpredictable possibilities of accidental overdosage, unusual individual susceptibility, and interactions with other chemical agents or pathological conditions.

Nevertheless, the following drugs have for the most part been extensively used during human pregnancy without verified increase in the numbers of dead or defective infants born to exposed women. In some instances, the drug may not have been widely used; but owing to a clustering of reported adverse effects on development, intensive efforts were made to examine large numbers of cases involving use during pregnancy. This was particularly true of meclizine and, to a lesser extent, of imipramine and LSD. Many other drugs not mentioned here have never been suggested as being causally associated with developmental toxicity. Coincidental associations are inevitable when it is considered that 3–5% of human births yield structurally or functionally defective children, and that this incidence may double during a 1-year follow-up.

Animal studies that report a high teratogenic susceptibility to a given drug in one species, especially if a particular defect is involved, sometimes lead to a flurry of reported associations between that drug and that defect in man. A good example is the glucocorticoid, cortisone, which is widely known to cause a high incidence of cleft palate in certain strains of mice. Most other animal species are much less sensitive to this treatment, however, and such lack of uniformity in response is a useful reminder that interspecies variations may be great. Several of the agents mentioned below have been found to cause developmental toxicity in one or a few species but not in others. Accordingly, relatively little significance has been placed on reports of teratogenicity in a single species as an indicator of potential risks in man.

A. Imipramine

Of the drugs used in the treatment of depression, two dibenzazepine derivatives, imipramine and amitriptyline, were briefly regarded as possibly teratogenic in man. Reports from Australia in 1972 indicated that positive correlation had been established between the taking of imipramine during the first third of pregnancy and the birth of infants with reduction deformities of the upper limb, resembling those seen after thalidomide ingestion (McBride, 1972). These claims later were largely withdrawn, but the fact that this and related tricyclic antidepressants had been found to show a variable level of teratogenicity in rabbits and rats (Aeppli, 1969; Harper *et al.*, 1965; Robson and Sullivan, 1963) required that the possibility of human teratogenicity be considered further. A search through recent literature yielded some 225 published cases in which one of these drugs was taken during the first trimester of pregnancy (Barson, 1972; Crombie *et al.*, 1972; Grabowsky, 1966; Kuenssberg and Knox, 1972; Scanlon, 1969; Sim, 1972; Report of the New Zealand Committee on Adverse Drug Reactions, 1971). This group included 7 children (3.1%) with various malformations, and interestingly, none had limb defects such as were characteristic of McBride's cases. No data on the total number of pregnant women taking such drugs are available. The Metropolitan Atlanta Congenital Malformations Report (Jan.–Feb., 1972) examined 120 cases involving reduction deformity of the limbs similar to those described by McBride and were unable to establish that any of the mothers had taken imipramine or related antidepressants. Thus, the tricyclic antidepressant drugs have not been substantiated as having teratogenic potential in man when used at recommended therapeutic doses. Attempts in the author's laboratory to induce malformations with large doses of imipramine given to pregnant monkeys were negative (Table 7). Hendrickx (1975) gave 21 bonnet and rhesus monkeys daily doses up to 10 times the recommended therapeutic level for varying periods during organogenesis and observed no teratogenic effects, although abortion rate was possibly elevated above that of controls.

B. LSD

There have been several claims of cytogenetic and teratogenic effects of use of lysergic acid diethylamide (LSD), and the extensive literature cannot be fully discussed here. A careful review of more than 100 publications on the subject by Dishotsky *et al.* (1971) has done much to place the matter in perspective. Briefly, these authors concluded that: (1) when chromosomal damage has been found, it was among illicit drug users and appeared to be related to drug abuse generally; (2) pure LSD ingested in moderate dosage does not produce chromosomal damage that can be detected by methods current at the time studies were made; (3) there was some evidence that LSD may be a weak mutagen, but unlikely to be mutagenic in concentrations used

Table 7. Effects of Imipramine on the Fetuses of Rhesus Monkeys Treated at Various Times During Organogenesis[a]

mg/kg (i.d.)	Number of days dosed	Gestation period (days)	Number of R_x	Condition of 100-day fetus
	Dosage[b]			
2–4	20	21–48	1	1 normal
5–7	20	22–49	1	1 normal
10	24	24–48	1	1 normal
15	24	21–49	4	1 normal
				3 growth retarded
20	24	20–45	2	1 normal
				1 extra thor. vert. & ribs

[a]From Wilson (1974).
[b]Range of human dosage = 1.5–3.5 mg/kg/day.

by human beings; (4) early reports of teratogenic effects in rats and hamsters were, for the most part, not confirmed in later studies and results in mice have been highly inconsistent; (5) the several instances of malformations among infants born to women having ingested illicit LSD during pregnancy could be attributed either to coincidence or to some other aspect of the drug-abuse problem; (6) there was no reported instance of a woman delivering a malformed child after taking pure LSD during pregnancy. Long (1972) reviewed 161 cases of children born to parents who took LSD before and/or during pregnancy and found five instances of limb malformation; but uncertainty about the number of other drugs taken, the time in pregnancy when taken, and other variables inherent in drug-abuse situations led this author to doubt that LSD was a causal factor. Dumars (1971) reported that parental use of illicit drugs does not cause differences in the incidence of chromosomal breaks or rearrangements from those seen in non-drug-users.

Kato *et al.* (1970) injected 4 pregnant monkeys with very large doses of LSD (0.125–1.0 mg/kg, approximately 90 times the human hallucinogenic dose) repeatedly at times presumed to be in the third or fourth month of gestation. The offspring included 1 normal infant, 2 stillborn infants which "on gross inspection showed facial deformity," and 1 which died one month postnatally. In the present author's laboratory 8 pregnant rhesus monkeys were treated orally three times per week for four weeks between days 20 and 45 of gestation with doses of 200 μg (20 times human "trip" dose on mg/kg basis). One female aborted, but 7 had normal 100-day fetuses at hysterotomy. Chromosomal abnormalities were not significantly increased in either the mothers or the fetuses. Thus there is little reason to believe that LSD has either teratogenic or cytogenetic effects at ordinary use levels, either in man or in laboratory animals.

C. Marihuana

Several studies on the teratogenic and cytogenetic effects of marihuana or its active principle, tetrahydrocannabinol, in laboratory animals have occasionally found teratogenic effects at high doses (Persaud and Ellington, 1968; Geber and Schramm, 1969a; Joneja, 1975), but more often have found only variable amounts of intrauterine death and some growth retardation (Borgen *et al.*, 1971; Banerjee *et al.*, 1975; Fleischman *et al.*, 1975; Harbison and Mantilla-Plata, 1972; Pace *et al.*, 1971; Persaud and Ellington, 1967). No substantiated cases of human malformations have been attributed to the use of marihuana during pregnancy. *In vivo* studies on chromosomal damage in users of this drug indicate that such damage is minimal and of undetermined significance (Martin *et al.*, 1974; Stenchever *et al.*, 1974).

D. Sulfonamides

Theoretically these antimicrobial agents would seem to be likely teratogens by the nature of their biological action, but such compounds have been found to have little effect on early human or other mammalian embryos. One retrospective survey found no increase in malformations among children of women taking these drugs during pregnancy, as compared with controls (Richards, 1969); another reported a marginally significant increase in major malformations ($p < 0.05$) in treated as compared with untreated pregnancies (Nelson and Forfar, 1971); and a partially prospective study found no significant increase of oral clefts in children of women taking sulfonamides during any trimester of their pregnancies (I. Saxén, 1975). Despite widespread use for many years, there are no verified case reports of association between the taking of sulfonamides during early pregnancy and subsequent delivery of a defective child.

Although useful in studying teratogenic mechanisms in chicks (Landauer and Wakasugi, 1968), sulfonamides have usually been found to have few embryotoxic effects in laboratory mammals (Giroud and Martinet, 1950; Paget and Thorpe, 1964; Kato and Kitagawa, 1973, 1974). An exception was the observation that the sulfonamides which are potent carbonic anhydrase inhibitors, acetazolamide in particular, are teratogenic in rodents (Wilson *et al.*, 1968), causing a uniquely localized malformation of the forelimbs.

E. Antihistamines

Of this diverse group of drugs only one class, the piperazines (meclizine, cyclizine, etc.) has been seriously questioned as regards embryotoxicity. *Meclizine* and related antihistamines were purported to have adverse effects on human development when taken as antiemetics during pregnancy in a number of publications soon after thalidomide was revealed to be teratogenic. However, several subsequent prospective studies involving large numbers of cases (Mellin, 1963; Smithells and Chinn, 1964; Yerushalmy and Milkovich, 1965) indicated that pregnant women taking these drugs produced no more defective babies than other women. Fifteen of the early reports of positive association were pooled by Lenz (1966), who noted that of 3333 infants from mothers given meclizine in the first trimester, 12 had cleft palate or cleft lip or both, an incidence only slightly higher than the expected number in a nonexposed group this size. No association between increased malformation rate and the taking of meclizine was noted in the retrospective survey of Nelson and Forfar (1971). Meclizine and several related antihistamines, however, are known to cause a high incidence of cleft palate in rats (King *et al.*, 1965; Stalsberg, 1965), but the teratogenic dose in rats is much higher than the antiemetic dose in pregnant women. Studies with large doses of meclizine in pregnant monkeys have not shown embryotoxic effects (Courtney and Valerio, 1968; Little, 1966).

Diphenhydramine has been studied in several animal experiments and found to have little effect on reproduction except for some reduced litter size and birth weight with large doses (Naranjo and de Naranjo, 1968; Schardein *et al.*, 1971; Shelesnyak and Davies, 1955). In a recent survey of Finnish children with cleft palate, however, it was reported that the facial clefts were significantly higher in the offspring of mothers having taken diphenhydramine during pregnancy than of control mothers (L. Saxén, 1974). In the same study clefts were no higher in children from women who took a cyclizine derivative than from control women.

F. Adrenocortical Steroids

These drugs, *cortisone* in particular, have produced cleft palate in the offspring of pregnant mice (Fraser and Fainstat, 1951; and many others) with such regularity as to cause concern about their use during human pregnancy. Published reports on the administration of these substances to pregnant women indicate some increase in abortion, stillbirth, and prematurity above background levels (Reilly, 1958). Some degree of oral cleft was observed in 5 infants from more than 300 women treated for a variety of reasons (Bongiovanni and McFadden, 1960; Doig and Coltman, 1956; Harris and Ross, 1956; Popert, 1962), but the known hereditary factor in the etiology of some types of clefts makes evaluation of such data difficult. A few malformations other than cleft palate also have been associated with cortisone treatment during pregnancy (Guilbeau, 1953; Warrell and Taylor, 1968; Wells, 1953). Although most such treatment has resulted in no frank embryotoxicity, transitory lymphopenia has sometimes been observed (Coté *et al.*, 1974). Thus, considering the extensive use of corticoids to treat a variety of diseases during human pregnancy and the failure of epidemiologic surveys to demonstrate statistical correlation, it is probable that these compounds under ordinary conditions of use do not appreciably increase the rate of spontaneously occurring developmental defects (De Costa and Abelman, 1952).

G. Narcotics

These drugs have generally been assumed to be nonteratogenic in animals (Cahen, 1964) as well as in man. Studies with large doses of *morphine* and *heroin* have caused some prenatal death and malformations in rodents (Geber and Schramm, 1969b,c; Harpel and Gautieri, 1968; Iuliucci and Gautieri, 1971), but in most such experiments interpretation of results was complicated by the alterations in maternal behavior and food intake caused by the drugs. *Methadone* has shown little or no embryotoxicity until doses at or near maternal toxic levels were used in mice, rats, and rabbits (Jurand, 1973; Markham *et al.*, 1971). No well-documented case reports are known of embryotoxic association between the taking of narcotic drugs during human pregnancy and the birth of malformed infants, although I. Saxén (1975) noted a significant increase above controls of infants with oral cleft when the mothers had taken opiates (mainly codeine) during the first trimester.

H. Barbiturates

The extensive experimental literature on effects of these drugs in laboratory animals (reviewed by Kalter, 1972b) contains little indication of teratogenicity, with very few exceptions (Persaud and Henderson, 1969; Setälä and Nyyssönen, 1964). Tuchmann-Duplessis and Mercier-Parot (1964) felt that the vast medical experience with these drugs in the absence of reports of adverse effects on the human fetus was indicative of their safety in this regard. This view was to a degree confirmed by Richards (1969) who, in a retrospective survey of the environmental factors thought likely to have been etiologic in 833 cases of malformation, found no association with these drugs. On the other hand, a similar study of 458 of defective children by Nelson and Forfar (1971) found that significantly more of the mothers took barbiturate-containing drugs than did control mothers of normal infants. The same was also true for vitamins, iron, antacids, and a number of other drugs, therefore, barbiturates can hardly be implicated on this basis alone.

I. Miscellaneous Sedatives, Antiemetics, and Tranquilizers

Glutethemide is chemically similar to thalidomide and for this reason was thoroughly investigated for teratogenic potential by Tuchmann-Duplessis and Mercier-Parot (1964) in mice, rats, and rabbits. Some dose-related resorptions were noted but none of the three species showed increased malformations. Similar results have been obtained by several other investigators (Takano *et al.*, 1963; see Kalter, 1972b, for review).

Chlorpromazine has been studied extensively for effects in pregnant animals and on reproduction generally (see Kalter, 1972b, for review). Usually the only adverse effects have been reduced litter size, associated with increased intrauterine and perinatal death, and some indication of postnatal behavioral changes. One report noted increased central nervous system malformations at high doses in one strain of rats but not in another (Brock and von Kreybig, 1964). Effects on intrauterine development in man are not known.

Reserpine, and its analog deserpidine, has also been extensively studied in laboratory animals (see Kalter, 1972b). Only one group of investigators has reported low incidences of malformations in rats (Tuchmann-Duplessis and Mercier-Parot, 1956, 1961). Two clinical studies examined the question of possible effects during human pregnancy and have found none (Gaunt *et al.*, 1959; Ravina, 1964).

Prochlorperazine is reported to have a low level of teratogenicity in mice and possibly in rats (Bertelli *et al.*, 1968; Roux, 1959; Vichi, 1969), but no adverse effects on development are known in man.

Other tranquilizers, e.g., orphenadrine, perphenazine, and metiapine, have been studied in rodents and generally found to have little or no teratogenic effect although most caused fetal weight reduction and some increase in intrauterine death at high dosage (Beall, 1972; Gibson and Newberne, 1973; Hoffeld *et al.*, 1967). Trifluoroperazine was suspected to be a common factor in 8 cases of human limb malformation within a localized area and time in Canada a few years ago, but this seems not to have been confirmed in further epidemiologic and laboratory studies (Morearity and Nance, 1963).

VI. CONCLUDING REMARKS

Many drugs have never been mentioned in association with embryotoxicity. Prospective surveys have failed to indict particular drugs as causally related to developmental defects; moreover, they have failed to show that the taking of drugs in general during pregnancy predisposes to a higher incidence of defects. Mellin (1964) made particular effort to ascertain the drugs taken by 3200 women during the first 13 weeks of their pregnancies. A·total of 266 (8.3%) of these pregnancies yielded structurally or functionally defective offspring. When they were compared with two sets of control pregnancies yielding normal offspring, born immediately before and immediately after those yielding abnormal infants, it was found that the mothers of abnormal children did not take more or significantly different drugs than the mothers in the control groups. The number was relatively small but the care with which this study was done supports the opinion that the mere taking of drugs during the first trimester does not predispose to abnormal development. In view of the observations presented in this chapter, one is inclined to agree with Mellin who concluded that ". . .drugs in general are not an important factor in congenital malformations in general." Other than the teratological catastrophe caused by thalidomide, the relatively few drugs that have been identified as having some teratogenic potential in man account for only a small fraction of either the drugs used or of the defects produced. It should be noted, however, that two important questions remain unanswered: (1) Is it possible to devise tests and surveillance systems that will prevent another teratological catastrophe such as that caused by thalidomide? (2) How are interactions between two or more drugs, between drugs and other environmental factors, or either of these and a predisposing genotype, to be identified, assuming that such interactions contribute to the large "unknown" segment of causation in human teratology? These questions should serve to dispell any feeling of complacency about use of all but the most essential drugs during human pregnancy.

REFERENCES

Aarskog, D., 1975, Association between maternal intake of diazepam and oral cleft, *Lancet* **2**:921.

Aeppli, V. L., 1969, Teratologische Studien mit Imipramin an Ratte und Kaninchen, *Arzneim. Forsch.* **19**:1617–1640.

Andrew, F. D., and Staples, R. E., 1974, Comparative embryotoxicity of medroxyprogesterone acetate, *Teratology* **9**:A-13.

Andrew, F. D., Williams, T. L., Gidley, J. T., and Wall, M. E., 1972, Teratogenicity of contraceptive steroids in mice, *Teratology* **5**:249.

Andrew, F. D., Christensen, H. D., Williams, T. L., Thompson, M. G., and Wall, M. E., 1973, Comparative teratogenicity of contraceptive steroids in mice and rats, *Teratology* **7**:A-11–A-12.

Anon., 1975, Some anticoagulants can have teratogenic effect, *J. Am. Med. Assoc.* **234**:1015.

Appelt, H., Gerken, H., and Lenz, W., 1966, Tetraphocomelie mit Lippen-Kiefer-gauminspalte und Clitorishypertrophie ein Syndrom, *Paediatr. Paedol.* **2**:119–124.

Askrog, V. F., and Harvald, B., 1970, Teratogen effekt af inhalations anaestetika, *Nord. Med.* **83**:498–500.

Bačič, M., Wesselius de Casparis, A., and Diczfalusy, E., 1970, Failure of large doses of ethinyl estradiol to interfere with early embryonic development in human species, *Am. J. Obstet. Gynecol.* **107**:531–534.

Balci, S., Say, B., Pirnar, T., and Hiesönmez, A., 1973, Birth defects and oral contraceptives, *Lancet* **2**:1098.

Banerjee, B. N., Galbreath, C., and Sofia, R. D., 1975, Teratologic evaluation of synthetic Δ^9-tetrahydrocannabinol in rats, *Teratology* **11**:99–102.

Barr, M., Poznanski, A. K., and Schmickel, R. D., 1974, Digital hypoplasia and anticonvulsants during gestation: A teratogenic syndrome, *J. Pediatr.* **84**:254–256.

Barry, J. E., and Danks, D. M., 1974, Anticonvulsants and congenital abnormalities, *Lancet* **2**:48–49.

Barson, A. J., 1972, Malformed infant, *Br. Med. J.* **2**:45.

Basford, A. B., and Fink, B. R., 1968, The teratogenicity of halothane in the rat, *Anesthesiology* **29**:1167–1173.

Bateman, A. J., and Epstein, S. S., 1971, Dominant lethal mutations in mammals, *in: Chemical Mutagens. Principles and Methods for Their Detection* (A. Hollaender, ed.), pp. 541–568, Plenum Press, New York.

Beall, J. R., 1972, Teratogenic study of chlorpromazine, orphenadrine, perphenazine, and LSD-25 in rats, *Toxicol. Appl. Pharmacol.* **21**:230–236.

Becker, M. H., Genieser, N. B., Finegold, M., Miranda, D., and Spackman, T., 1975, Chondrodysplasia punctata. Is maternal warfarin therapy a factor?, *Am. J. Dis. Child.* **129**:356–359.

Bertelli, A., Polani, P. E., Spector, R., Seller, M. J., Tuchmann-Duplessis, H., and Mercier-Parot, L., 1968, Retentissement d'un neuroleptique, l'halopéridol, sur la gestation et le développment prénatal des rongeurs, *Arzneim. Forsch.* **18**:1420–1424.

Bird, A. V., 1969, Anticonvulsant drugs and congenital abnormalities, *Lancet* **1**:311.

Boletti, M., and Croatto, L., 1958, Deafness in a 5-year old girl resulting from streptomycin therapy during pregnancy, *Acta Pediatr.* **11**:1.

Bongiovanni, A. M., and McFadden, A. J., 1960, Steroids during pregnancy and possible fetal consequences, *Fertil. Steril.* **11**:181–186.

Bongiovanni, A. M., DiGeorge, A. M., and Grumbach, M. M., 1959, Masculinzation of the female infant associated with estrogenic therapy alone during gestation, *J. Clin. Endocrinol. Metab.* **19**:1004–1010.

Borgen, L. A., Davis, W. M., and Pace, H. B., 1971, Effects of synthetic Δ^9-tetrahydrocannabinol on pregnancy and offspring in the rat, *Toxicol. Appl. Pharmacol.* **20**:480–486.

Boué, J., Boué, A., and Lazar, P., 1975, Retrospective and prospective epidemiological studies of 1500 karyotyped spontaneous human abortions, *Teratology* **12**:11–26.

Brandner, M., and Nusslé, D., 1969, Foetopathie due a l'aminopterine avec sténose congénitale de l'espace médullaire des os tubulaires longs, *Ann. Radiol.* **12**:703–710.

Briggs, G. G., Samson, J. H., and Crawford, D. J., 1975, Lack of abnormalities in a newborn exposed to amphetamines during gestation, *Am. J. Dis. Child.* **129**:249–250.

Brock, N., and von Kreybig, T., 1964, Experimenteller Beitrag zur Prüfung teratogener Wirkungen von Arzneimitteln an der Laboratoriumsratte, *Naunyn Schmiedebergs Arch. Pharmakol.* **249**:117–145.

Cahen, R. L., 1964, Evaluation of the teratogenicity of drugs, *Clin. Pharmacol. Ther.* **5**:480–514.

Cahen, R. L., 1966, Experimental and clinical chemoteratogenesis, *in: Advances in Pharmacology* (S. Garottini and P. A. Shore, eds.), pp. 263–334, Academic Press, New York.

Cahen, R., Sautai, M., Montagne, J., and Pessonnier, J., 1964, Recherche de l'effet teratogene de la 2-diethylaminopropiophenone, *Med. Exp.* **10**:201–224.

Caldera, R., 1970, Carbutamide et malformations chez l'enfant, *Ann. Pediatr.* **17**:432–436.

Campbell, G. D., 1961, Possible teratogenic effect of tolbutamide in pregnancy, *Lancet* **1**:891–892.

Campbell, G. D., 1963, Chlorpropamide and foetal damage, *Br. Med. J.* **1**:59–60.

Carr, D. H., 1970, Chromosome studies in selected spontaneous abortions. 1. Conception after oral contraceptives, *Can. Med. Assoc. J.* **103**:343–348.

Carter, M. P., 1965, Discussion of a paper by B. C. S. Slater, *in: Embryopathic Activity of Drugs* (J. M. Robson, F. M. Sullivan, and R. L. Smith, eds.), pp. 253–260, Little, Brown, Boston.

Carter, M. P., and Wilson, F. W., 1963, Antibiotics and congenital malformations, *Lancet* **1**:1267.

Carter, M. P. and Wilson, F., 1965, Antibiotics in early pregnancy and congenital malformations, *Dev. Med. Child. Neurol.* **7**:353–359.

Cervantes, A., Monter, H. M., and Campos, J. L., 1973, Clinical and genetic studies of children born of hormonal contraceptives users, *in: 4th International Conference on Birth Defects* (A. G. Motulsky and F. J. G. Ebling, eds.), p. 53, Excerpta Medica, Amsterdam.

Chang, L. W., Dudley, A. W., Katz, J., and Martin, A. H., 1974, Nervous system development following in utero exposure to trace amounts of halothane, *Teratology* **9**:A-15.

Chaube, S., and Murphy, M. L., 1968, The teratogenic effects of the recent drugs active in cancer chemotherapy, *in: Advances in Teratology* (D. H. M. Woollam, ed.), pp. 181–237, Academic Press, New York.

Chernoff, G. F., 1975, A mouse model of the fetal alcohol syndrome, *Teratology* **11**:14A.

Christenson, R. K., Schafer, J. H., Teague, H. S., and Grifo, A. P., 1971, Influence of methallibure on early pregnancy in swine, *J. Anim. Sci.* **33:**1156.

Christoffel, K. K., and Salafsky, I., 1975, Fetal alcohol syndrome in dizygous twins, *J. Pediatr.* **87:**963–967.

Cohen, E. N., Bellville, J. W., and Brown, B. W., 1971, Anesthesia, pregnancy and miscarriage: A study of operating room nurses and anesthetists, *Anesthesiology* **35:**343–347.

Cohlan, S. Q., 1964, Fetal and neonatal hazards from drugs administered during pregnancy, *N.Y. Med. J.* **64:**493–499.

Coopersmith, H., and Kerbel, N. C., 1962, Drugs and congenital anomalies, *Can. Med. Assoc. J.* **87:**193.

Corbett, T. H., 1972, Anesthetics as a cause of abortion, *Fertil. Steril.* **23:**866–869.

Corbett, T. H., Cornell, R. G., Endres, J. L., and Lieding, K., 1974a, Birth defects among children of nurse-anesthetists, *Anesthesiology* **41:**341–344.

Corbett, T. H., Beaudoin, A. R., Cornell, R. G., Endres, J. L., and Page, A., 1974b, Effects of low concentrations of methoxyflurane on rat pregnancy, *Teratology* **9:**A-15–A-16.

Coté, C. J., Meuwissen, H. J., and Pickering, R. J., 1974, Effects on the neonate of prednisone and azathioprine administered to the mother during pregnancy, *J. Pediatr.* **85:**324–328.

Courtney, K. D., and Valerio, D. A., 1968, Teratology in the *Macaca mulatta*, *Teratology* **1:**163–172.

Crombie, D. L., Pinsent, R. J. F. H., Slater, B. C., Fleming, D., and Cross, K. W., 1970, Teratogenic drugs—R. C. G. P. Survey, *Br. Med. J.* **4:**178–179.

Crombie, D. L., Pinsent, R. J. F. H., and Fleming, D., 1972, Imipramine in pregnancy, *Br. Med. J.* **1:**745.

de Alvarez, R., 1962, Discussion of a paper by C. Goetsch, *Am. J. Obstet. Gynecol.* **83:**1474–1477.

De Costa, E. J., and Abelman, M. A., 1952, Cortisone and pregnancy. An experimental and clinical study of the effects of cortisone on gestation, *Am. J. Obstet. Gynecol.* **64:**746–767.

de Rothschild, B., and Levy, G., 1950, Action de la quinacrine sur la gestation chez le rate, *C.R. Soc. Biol.* **144:**1350–1352.

Diamond, I., Anderson, M. M., and McCreadie, S. R., 1960, Transplacental transmission of busulfan (Myleran) in a mother with leukemia: Production of fetal malformation and cytomegaly, *Pediatrics* **25:**85–90.

Dieulangard, P., Coignet, J., and Vical, J. C., 1966, Sur un cas d'ectro-phocomelia, peut-etre d'origine medicamenteuse, *Bull. Fed. Gynecol. Obstet.* **18:**85–87.

DiPaolo, J. A., and Kotin, P., 1966, Teratogenesis-oncogenesis: A study of possible relationships, *Arch. Pathol.* **81:**3–23.

DiSaia, P. J., 1966, Pregnancy and delivery of a patient with a Starr-Edwards mitral valve prosthesis: Report of a case, *Obstet. Gynecol.* **28:**469–472.

Dishotsky, N. I., Loughman, W. D., Mogar, R. E., and Lipscomb, W. R., 1971, LSD and genetic damage. Is LSD chromosome damaging, carcinogenic, mutagenic, or teratogenic?, *Science* **172:**431–440.

Dluzniewski, A., and Gastol-Lewinska, K., 1971, Search for teratogenic activity of some tuberculostatic drugs, *Diss. Pharm. Pharmacol.* **23:**383–392.

Doig, R. K., and Coltman, O. McK., 1956, Cleft palate following cortisone therapy in early pregnancy, *Lancet* **2:**730.

Dolger, H., Baskman, J. J., and Nechemias, C., 1969, Tolbutamide in pregnancy and diabetes, *J. Mt. Sinai Hosp.* **36:**471–474.

Duke, D. I., 1975, Prenatal effects of the cancer chemotherapeutic drug ICRF 159 in mice, rats, and rabbits, *Teratology* **11:**119–126.

Dumars, K. W., 1971, Parental drugs usage: Effect upon chromosomes of progeny, *Pediatrics* **47:**1037–1941.

Einer-Jensen, N., and Secher, N. J., 1970, Diminished weight of rat foetuses after treatment of pregnant rats with haloperidol, *J. Reprod. Fertil.* **22:**591–594.

Elshove, J., 1969, Cleft palate in the offspring of female mice treated with phenytoin, *Lancet* **2:**1074.

Elshove, J., and van Eck, J. H. M., 1971, Aangeboren misvormingen, met name gespleten lip met of zonder gespleten verhemelte, bij kinderen van moeders met epilepsie, *Ned. Tijdschr. Geneeskd.* **115:**1371–1375.

Emerson, D. J., 1962, Congenital malformations due to attemped abortion with aminopterin, *Am. J. Obstet. Gynecol.* **84:**356–357.

Filippi, B., 1967, Antibiotics and congenitcal malformations: Evaluation of the teratogenicity of antibiotics, *in: Advances in Teratology* (D. H. M. Woollam, ed.), pp. 239–256, Logos Press and Academic Press, London and New York.

Fillmore, S. J., and McDevitt, E., 1970, Effects of coumarin compounds on the fetus, *Ann. Intern. Med.* **73:**731–735.

Finnerty, J. J., and Mackay, B. R., 1962, Antepartum thrombophlebitis and pulmonary embolism, *Obstet. Gynecol.* **19:**405–410.

Fishbein, L., Flamm, W. G., and Falk, H. L., 1970, *Chemical Mutagens. Environmental Effects on Biological Systems,* Academic Press, New York.

Fleischman, R. W., Hayden, D. W., Rosenkrantz, H., and Braude, M. C., 1975, Teratologic evaluation of Δ^9-tetrahydrocannabinol in mice, including a review of the literature, *Teratology* **12:**47–50.

Forfar, J. O., and Nelson, M. M., 1973, Epidemiology of drugs taken by pregnant women: Drugs that may affect the fetus adversely, *Clin. Pharmacol. Ther.* **14:**632–642.

Fraser, F. C., and Fainstat, T. D., 1951, Production of congenital defects in the offspring of pregnant mice treated with cortisone, *Pediatrics* **8:**527–533.

Gabler, W. L., 1968, The effect of 5,5-diphenylhydantoin on the rat uterus and its fetuses, *Arch. Int. Pharmacodyn. Ther.* **175:**141–152.

Gal, I., Kirman, B., and Stern, J., 1967, Hormonal pregnancy tests and congenital malformations, *Nature* **216:**83.

Ganguin, G., 1971, Auswirkungen einer antituberkulösen Chemotherapie bei tuberkulösen Schwangeren auf die Frucht, *Z. Erkr. Atmungsorgane* **134:**95–103.

Gardner, L. I., Assemany, S. R., and Neu, R. R., 1970, 46XY female: Antiandrogenic effect of oral contraceptive?, *Lancet* **2:**667–668.

Gaunt, R., Renzi, A. A., Antonchak, N., Miller, G. J., and Gilman, M., 1959, Endocrine aspects of the pharmacology of reserpine, *Ann. N.Y. Acad. Sci.* **59:**22–35.

Geber, W. F., and Schramm, L. C., 1969a, Effects of marihuana extracts on fetal hamsters and rabbits, *Toxicol. Appl. Pharmacol.* **14:**247.

Geber, W. F., and Schramm, L. C., 1969b, Heroin teratogenic and behavioral activity in hamsters, *Fed. Proc.* **28:**262.

Geber, W. F., and Schramm, L. C., 1969c, Comparative teratogenicity of morphine, heroin and methadone in the hamster, *Pharmacologist* **11:**248.

German, J., Kowal, A., and Ehlers, K. H., 1970, Trimethadione and teratogenesis, *Teratology* **3:**349–362.

Ghanem, M. H., 1961, Possible teratogenic effect of tolbutamide in the pregnant prediabetic, *Lancet* **1:**1227.

Gibson, J. E., and Becker, B. A., 1968, Teratogenic effects of diphenylhydantoin in Swiss-Webster and A/J mice, *Proc. Soc. Exp. Biol. Med.* **128:**905–909.

Gibson, J. P., and Newberne, J. W., 1973, Teratology and reproduction studies with metiapine, *Toxicol. Appl. Pharmacol.* **25:**212–219.

Girotti, M., and Schaer, A.-E., 1973, Malformations after anti-ovulatory drugs, *in: 4th International Conference on Birth Defects,* (A. G. Motulsky and F. J. G. Ebling, eds.), pp. 52–53, Excerpta Medica, Amsterdam.

Giroud, A., and Martinet, M., 1950, Action de la sulfaguanidine sur le développement de l'embryon, *Arch. Fr. Pediatr.* **7:**180–184.

Giroud, A., and Tuchmann-Duplessis, H., 1962, Malformations congénitales. Role des facteurs exogènes, *Pathol. Biol.* **10:**119–151.

Goetsch, C., 1962, An evaluation of aminopterin as an abortifacient, *Am. J. Obstet. Gynecol.* **83:**1474–1477.

Goldfield, M. D., and Weinstein, M. R., 1973, Lithium carbonate in obstetrics: Guidelines for clinical use, *Am. J. Obstet. Gynecol.* **116:**15–22.

Grabowsky, J. R., 1966, Ação do Tofranil sôbre o embrião humano, *Rev. Brasil. Med.* **23:**220.

Gralla, E. J., and McIlhenny, H. M., 1972, Studies in pregnant rats, rabbits and monkeys with lithium carbonate, *Toxicol. Appl. Pharmacol.* **21:**428–433.

Grebe, H., 1952 Können Abtreibungsversuche zu Missbildungen fuhren?, *Geburtshilfe Frauenheilkd.* **12:**333–339.

Green, H. G., 1974, Infants of alcoholic mothers, *Am. J. Obstet. Gynecol.* **118:**713–716.

Greenberg, G., Inman, W. H. W., Weatherall, J. A. C., and Adelstein, A. M., 1975, Hormonal pregnancy tests and congenital malformations, *Br. Med. J.* **2:**191–192.

Greenberg, L. H., and Tanaka, K. R., 1964, Congenital anomalies probably induced by cyclophosphamide, *J. Am. Med. Assoc.* **188:**423–426.

Greenwald, P., Barlow, J. J., Nasca, P. C., and Burnett, W. S., 1971, Vaginal cancer after maternal treatment with synthetic estrogens, *N. Engl. J. Med.* **285:**390–392.

Grosse, K.-P., Schwanitz, G., Rott, H.-D., and Wissmüller, H. F., 1972, Chromosomenuntersuchungen bei Behandlung mit Anticonvulsiva, *Humangenetik* **16:**209–216.

Grumbach, M. M., and Ducharme, J. R., 1960, The effects of androgens on fetal sexual development; androgen-induced female pseudohermaphroidism, *Fertil. Steril.* **11:**157–180.

Guilbeau, J. A., 1953, Effects of cortisone on the fetus, *Am. J. Obstet. Gynecol.* **65:**227.

Hanson, J. W., and Smith, D. W., 1975, The fetal hydantoin syndrome, *J. Pediatr.* **87:**285–290.

Harbison, R. D., and Mantilla-Plata, B., 1972, Prenatal toxicity, maternal distribution, and placental transfer of tetrahydrocannabinol, *J. Pharmacol. Exp. Ther.* **180:**446–453.

Harlap, S., Prywes, R., and Davies, A. M., 1975, Birth defects and oestogens and progesterones in pregnancy, *Lancet* **1:**682–683.

Harpel, H. S., and Gautieri, R. F., 1968, Morphine-induced fetal malformations, *J. Pharm. Sci.* **57:**1590–1597.

Harper, K. H., Palmer, A. K., and Davies, R. E., 1965, Effect of imipramine upon the pregnancy of laboratory animals, *Arzneim. Forsch.* **15:**1218–1221.

Harris, J. W., and Ross, I. P., 1956, Cortisone therapy in early pregnancy: Relation to cleft palate, *Lancet* **1:**1045–1047.

Hartz, S. C., Heinonen, O. P., Shapiro, S., Siskind, V., and Slone, D., 1975, Antenatal exposure to meprobamate and chlordiazepoxide in relation to malformations, mental development, and childhood mortality, *N. Engl. J. Med.* **292:**726–728.

Hendrickx, A. G., 1975, Teratologic evaluation of imipramine hydrochloride in bonnet (*Macaca radiata*) and rhesus monkeys (*Macaca mulatta*), *Teratology* **11:**219–222.

Herbst, A. L., and Scully, R. E., 1970, Adenocarcinoma of the vagina in adolescents—A report of 7 cases including 6 clear-cell carcinomas (so-called mesonephromas), *Cancer* **25:**745–761.

Herbst, A. L., Ulfelder, H., and Poskanzer, D. C., 1971, Adenocarcinoma of the vagina, *N. Engl. J. Med.* **284:**878–881.

Herbst, A. L., Robboy, S. J., Scully, R. E., and Poskanzer, D. C., 1974, Clear-cell adenocarcinoma of the vagina and cervix in girls: Analysis of 170 registry cases, *Am. J. Obstet. Gynecol.* **119:**713–724.

Herbst, A. L., Poskanzer, D. C., Robboy, S. J., Friedlander, L., and Scully, R. E., 1975, Prenatal exposure to stilbestrol. A prospective comparison of exposed female offspring with unexposed controls, *N. Engl. J. Med.* **292:**334–339.

Herrmann, J., Feingold, M., Tuffli, G. A., and Opitz, J. M., 1969, A familial dysmorphogenetic syndrome of limb deformities, characteristic facial appearance and associated anomalies: The "pseudothalidomide" or "SC-syndrome", *Birth Defects: Orig. Art. Ser.* **5:**81–89.

Hirsh, J., Cade, J. F., and O'Sullivan, E. F., 1970, Clinical experience with anticoagulant therapy during pregnancy, *Br. Med. J.* **1:**270–273.

Hoffeld, D. R., Webster, R. L., and McNew, J., 1967, Adverse effects on offspring of tranquilizing drugs during pregnancy, *Nature* **215:**182–183.

Holmes, B., Moser, H. W., Haldórsson, S., *et al.*, 1972, *Mental Retardation: An Atlas of Disease with Associated Physical Abnormalities*, Macmillan, New York.

Holmes, L. B., and Borden, S., 1974, Phocomelia, flexion deformities and absent thumbs: A new hereditary upper limb malformation, *Pediatrics* **54**:461–465.

Hook, E. B., Heinonen, O. P., Shapiro, S., and Slone, D., 1974, Maternal exposure to oral contraceptives and other female sex hormones: Relation to birth defects in a prospectively ascertained cohort of 50,282 pregnancies, *Teratology* **9**:A-21–A-22.

Hoskins, D., 1948, Some effects of nitrogen mustard on the development of external body form in the rat, *Anat. Rec.* **102**:493–512.

Iuliucci, J. D., and Gautieri, R. F., 1971, Morphine-induced fetal malformations. II. Influence of hystamine and diphenylhydramine, *J. Pharm. Sci.* **60**:420–425.

Jackson, A. V., 1948, Toxic effects of salicylate on the foetus and mother, *J. Pathol. Bacterial.* **60**:587–593.

Janerich, D. T., Piper, J. M., and Glebatis, D. M., 1974, Oral contraceptives and congenital limb-reduction defects, *N. Engl. J. Med.* **291**:697–700.

Janz, D., and Fuchs, U., 1964, Sind antiepileptische Medikamente während der Schwangerschaft schädlich?, *Dtsch. Med. Wochenschr.* **89**:241–243.

Jentgens, H., 1968, Ethionamid und teratogene Wirkung, *Prax. Pneumol.* **22**:699–704.

Johansen, K. T., 1971, Lithium teratogenicity, *Lancet* **1**:1026–1027.

Joneja, M. G., 1975, Effects of Δ⁹-tetrahydrocannabinol (THC) on fetuses of Swiss Webster mice, *Anat. Rec.* **181**:387–388.

Jones, K. L., 1974, Fetal alcohol syndrome, Paper presented at 14th Annual Teratology Society Meeting, Vancouver, British Columbia.

Jones, K. L., and Smith, D. W., 1973, Recognition of the fetal alcohol syndrome in early infancy, *Lancet* **2**:999–1001.

Jones, K. L., and Smith, D. W., 1975, The fetal alcohol syndrome, *Teratology* **12**:1–10.

Jones, K. L., Smith, D. W., Ulleland, C. N., and Streissguth, A. P., 1973, Pattern of malformation in offspring of chronic alcoholic mothers, *Lancet* **1**:1267.

Jones, K. L., Smith, D. W., Streissguth, A. P., and Myrianthopoulos, N. C., 1974, Outcome in offspring of chronic alcoholic women, *Lancet* **1**:1076–1078.

Jordan, R. L., and Shiel, F. O'M., 1972, Preliminary observations on the response of the rat embryo to dextroamphetamine sulfate, *Anat. Rec.* **172**:338.

Jurand, A., 1973, Teratogenic activity of methadone hydrochloride in mouse and chick embryos, *J. Embryol. Exp. Morphol.* **30**:449–458.

Kalter, H., 1972a, Nonteratogenicity of isoniazid in mice, *Teratology* **5**:259.

Kalter, H., 1972b, Teratogenicity, embryolethality and mutagenicity of drugs of dependence, *in: Chemical and Biological Aspects of Drug Dependence* (S. J. Mulé and H. Brill, eds.), pp. 413–445, CRC Press, Cleveland, Ohio.

Kaplan, N. M., 1959, Male pseudohermaphrodism. Report of a case with observations and pathogenesis, *N. Engl. J. Med.* **261**:641.

Kasirsky, G., and Tansy, M. T., 1971, Teratogenic effects of methamphetamine in mice and rabbits, *Teratology* **4**:131–134.

Kato, T., Jarvik, L. F., Roizin, L., and Moralishvili, E., 1970, Chromosome studies in pregnant rhesus macaque given LSD-25, *Dis. Nerv. Syst.* **31**:245–250.

Kato, T., and Kitagawa, S., 1973, Production of congenital skeletal anomalies in the fetuses of pregnant rats and mice treated with various sulfonamides, *Congenital Anomalies* **13**:17–23.

Kato, T., and Kitagawa, S., 1974, Effects of a new antibacterial sulfonamide (CS-61) on mouse and rat fetuses, *Toxicol. Appl. Pharmacol.* **27**:20–27.

Kaufman, R. L., 1973, Birth defects and oral contraceptives, *Lancet* **1**:1396.

Kerber, I. J., Warr, O. S., and Richardson, C., 1968, Pregnancy in a patient with a prosthetic mitral valve: Associated with a fetal anomaly attributed to warfarin sodium, *J. Am. Med. Assoc.* **203**:223–225.

Kernis, M. M., Pashayan, H. M., and Pruzansky, S., 1973, Dilantin-induced teratogenicity and folic acid deficiency, *Teratology* **7**:A-19–A-20.

Kimmel, C. A., Wilson, J. G., and Schumacher, H. J., 1971, Studies on metabolism and identification of the causative agent in aspirin teratogenesis in the rat, *Teratology* **4**:15–24.

King, C. T. G., Weaver, S. A., and Narrod, S. A., 1965, Antihistamines and teratogenicity in the rat, *J. Pharmacol. Exp. Ther.* **147:**391–398.

Knill-Jones, R. P., Moir, D. D., Rodrigues, L. V., and Spence, A. A., 1972, Anaesthetic practice and pregnancy: a controlled survey of women anaesthetists in the United Kingdom, *Lancet* **2:**1326.

Kopelman, A. E., McCullar, F. W., and Heggeness, L., 1975, Limb malformations following maternal use of haloperidol, *J. Am. Med. Assoc.* **231:**62–64.

Koppe, J. G., Bosman, W., Oppers, V. M., Spaans, F., Kloosterman, G. J., and De Bruijne, J. I., 1973, Epilepsy and congenital malformations, *in: 4th International Conference on Birth Defects* (A. G. Motulsky and F. J. G. Ebling, eds.), p. 52, Excerpta Medica, Amsterdam.

Kronick, J., Phelps, N. E., McCallion, D. J., and Hirsch, J., 1974, Effects of sodium warfarin administered during pregnancy in mice, *Am. J. Obstet. Gynecol.* **118:**819–823.

Kuenssberg, E. V., and Knox, J. D. E., 1972, Imipramine in pregnancy, *Br. Med. J.* **2:**1972.

Kuenssberg, E. V., and Knox, J. D. E., 1973, Teratogenic effect of anticonvulsants, *Lancet* **1:**198.

Landauer, W., and Wakasugi, N., 1968, Teratological studies with sulphonamides and their implications, *J. Embryol. Exp. Morphol.* **20:**261–284.

Larsson, K. S., Bostrom, H., and Erickson, B., 1963, Salicylate-induced malformations in mouse embryos, *Acta Pediatr.* **52:**36.

Larsson, Y., and Sterky, G., 1961, Possible teratogenic effect of tolbutamide in a pregnant pre-diabetic, *Lancet* **2:**1424.

Laurence, M., Miller, M., Vowles, M., Evans, K., and Carter, C., 1971, Hormonal pregnancy tests and neural tube malformations, *Nature* **233:**495–496.

Lejour-Jeanty, M., 1966, Becs-de-lièvre provoqués chez le rat par un dérivé de la pénicilline, l'hadacidine, *J. Embryol. Exp. Morphol.* **15:**193–211.

Lemoine, P., Harousseau, H., and Borteyru, J., 1968, Les enfants de parents alcooliques. Anomalies observées, *Quest. Med.* **21:**476–482.

Lenz, W., 1962, Drugs and congenital abnormalities, *Lancet* **1:**45.

Lenz, W., 1964, Chemicals and malformations in man, *in: Congenital Malformations* (M. Fishbein, ed.), pp. 263–276, International Medical Congress, New York.

Lenz W., 1966, Malformations caused by drugs in pregnancy, *Am. J. Dis. Child.* **112:**99–106.

Lenz, W., 1972, Genetic diagnosis: molecular diseases and others, *in: Human Genetics* (J. de Grouchy, F. J. G. Ebling, and I. W. Henderson, eds.), pp. 2–5, Excerpta Medica, Amsterdam.

Levin, J. N., 1971, Amphetamine ingestion with biliary atresia, *J. Pediatr* **79:**130–131.

Levy, E. P., Cohen, A., and Fraser, F. C., 1973, Hormone treatment during pregnancy and congenital heart defects, *Lancet* **1:**611.

Levy, G., 1965, Pharmacokinetics of salicylate elimination in man, *J. Pharm. Sci.* **54:**959–967.

Little, W. A., 1966, Personal communication.

Long, S. Y., 1972, Does LSD induce chromosomal damage and malformations? A review of the literature, *Teratology* **6:**75–90.

Loughnan, P. M., Gold, H., and Vance, J. C., 1973, Phenytoin teratogenicity in man, *Lancet* **1:**70–72.

Lowe, C. R., 1964, Congenital defects among children born to women under supervision or treatment of pulmonary tuberculosis, *Br. J. Prev. Soc. Med.* **18:**14–16.

Lowe, C. R., 1973, Congenital malformations among infants born to epileptic women, *Lancet* **1:**9–10.

Malius, J. M., Cooke, A. M., Pyke, D. A., and Fitzgerald, M. G., 1964, Sulphonylurea drugs in pregnancy, *Br. Med. J.* **2:**187.

Markham, J. K., Emmerson, J. L., and Owen, N. V., 1971, Teratogenicity studies of methadone HCl in rats and rabbits, *Nature* **233:**342–343.

Marsh, L., and Fraser, F. C., 1973, Studies on dilantin-induced cleft palate in mice, *Teratology* **7:**A-23.

Martin, P. A., Thorburn, M. I., and Bryant, S. A., 1974, *In vivo* and *in vitro* studies of the cytogenetic effects of *Cannabis sativa* in rats and men, *Teratology* **9:**81–86.

Massey, K. M., 1966, Teratogenic effects of diphenylhydantoin sodium, *J. Oral Ther. Pharmacol.* **2:**380–385.

Matera, R. F., Zabala, H., and Jiminez, A. P., 1967, Bifid exencephalia, teratogenic action of amphetamine, *Int. Surg.* **50**:79–85.

Matsumoto, N., Iijima, S., and Katsunuma, H., 1974, Placental transfer of LiCl and its effects on foetal growth development in mice, *Teratology* **10**:89.

McBride, W. G., 1972, The teratogenic effects of imipramine, *Teratology* **5**:262.

McDonald, A. D., 1961, Maternal health in early pregnancy and congenital defect. Final report on a prospective injury, *Br. J. Prev. Soc. Med.* **15**:154–166.

McNiel, J. R., 1973, The possible teratogenic effect of salicylates on the developing fetus, *Clin. Pediatr.* **12**:347–350.

Meadow, S. R., 1968, Anticonvulsant drugs and congenital abnormalities, *Lancet* **2**:1296.

Meadow, S. R., 1970, Congenital abnormalities and anticonvulsant drugs, *Proc. R. Soc. Med.* **63**:12–13.

Melchoir, J. C., Svensmark, O., and Trolle, D., 1967, Placental transfer of phenobarbitone in epileptic women, and elimination in newborns, *Lancet* **2**:860–861.

Mellin, G. W., 1963, Fetal life study: A prospective epidemiologic study of prenatal influences on fetal development: meclizine and other drugs, *Bull. Soc. R. Belge Gynecol. Obstet.* **33**:79.

Mellin, G. W., 1964, Drugs in the first trimester of pregnancy and the fetal life of *Homo sapiens*, *Am. J. Obstet. Gynecol.* **90**:1169–1180.

Mellin, G. W., and Katzenstein, M., 1964, Increased incidence of malformations-chance or change?, *J. Am. Med. Assoc.* **187**:570–573.

Meltzer, H. J., 1956, Congenital anomalies due to attempted abortion with 4-aminopteroylglutamic acid, *J. Am. Med. Assoc.* **161**:1253.

Mercier-Parot, L., and Tuchmann-Duplessis, H., 1974, Action of diphenylhydantoin in pregnant rats and mice, *Teratology* **9**:A-28.

Milkovich, L., and van den Berg, B. J., 1974, Effects of prenatal meprobamate and chlordiazepoxide hydrochloride on human embryonic and fetal development, *N. Engl. J. Med.* **291**:1268–1271.

Miller, R. P., and Becker, B. A., 1975, Teratogenicity of oral diazepam and diphenylhydantoin in mice, *Toxicol. Appl. Pharmacol.* **32**:53–61.

Milunsky, A., Graef, J. W., and Gaynor, M. T., 1968, Methotrexate-induced congenital malformations, with a review of the literature, *J. Pediatr.* **72**:790–795.

Mirkin, B. L., 1971, Placental transfer and neonatal elimination of diphenylhydantoin, *Am. J. Obstet. Gynecol.* **109**:930–933.

Monnet, P., Kalb, J.-C., and Pujol, M., 1967, De l'influence nocive de l'isoniazide sur le produit de conception, *Lyon Med.* **217**:431–455.

Monson, R. R., Rosenberg, L., Hartz, S. C., Shapiro, S., Heinonen, O. P., and Slone, D., 1973, Diphenylhydantoin and selected congenital malformations, *N. Engl. J. Med.* **289**:1049–1052.

Morearity, A. J., and Nance, H. R., 1963, Trifluoperazine and congenital malformations, *Can. Med. Assoc. J.* **88**:375.

Moss, P. D., 1962, Phenmetrazine and foetal abnormality, *Br. Med. J.* **2**:1610.

Mulvihill, J. J., Mulvihill, C. G., and Neill, C. A., 1974, Prenatal sex-hormone exposure and cardiac defects in man, *Teratology* **9**:A-30.

Murphy, M. L., Dagg, C. P., and Karnofsky, D., 1957, Comparison of teratogenic chemicals in the rat and chick embryos, *Pediatrics* **19**:701–714.

Murray, F. J., 1974, Implications for teratogenic hazard to epileptic women taking diphenylhydantoin as indicated by animal studies, Ph. D. dissertation, University of Cincinnati College of Medicine, Cincinnati, Ohio.

Naranjo, P., and de Naranjo, E., 1968, Embryotoxic effects of antihistamines, *Arzneim. Forsch.* **18**:188–195.

Nelson, M. M., 1963, Teratogenic effects of pteroylglutamic acid deficiency in the rat, *in: Ciba Foundation Symposium on Congenital Malformations* (G. E. W. Wolstenholme and C. M. O'Connor, eds.), pp. 134–151, Little, Brown, Boston.

Nelson, M. M., and Forfar, J. O., 1971, Associations between drugs administered during pregnancy and congenital abnormalities of the fetus, *Br. Med. J.* **1**:523–527.

Nelson, M. M., Wright, H. V., Asling, C. W., and Evans, H. M., 1955, Multiple congenital abnormalities resulting from transitory deficiency of pteroylglutamic acid during gestation in the rat, *J. Nutr.* **56:**349–370.

Nichols, M. M., 1973, Fetal anomalies following maternal trimethadione ingestion, *J. Pediatr.* **82:**885–886.

Nicholson, H. O., 1968, Cytotoxic drugs in pregnancy, *J. Obstet. Gynaecol. Br. Commonw.* **75:**307–312.

Nishimura, H., 1963, Interstrain differences in susceptibility to the teratogenic effects of Mitomycin C in mice, Abstracts of papers presented at 3rd Annual Teratology Society Meeting, Ste. Adele, Quebec.

Nishimura, H., 1964, *Chemistry and Prevention of Congenital Anomalies,* Charles C. Thomas, Springfield, Illinois.

Nishimura, H., and Kuginuki, M., 1958, Congenital malformations induced by ethyl-urethan in mouse embryos, *Okajimas Folia Anat. Jpn.* **31:**1.

Nishimura, H., Uwabe, C., and Semba, R., 1974, Examination of teratogenicity of progestogens and/or estrogens by observation of the induced abortuses, *Teratology* **10:**93.

Niswander, J. D., and Wertelecki, W., 1973, Congenital malformations among offspring of epileptic women, *Lancet* **1:**1062.

Nora, A. H., and Nora, J. J., 1975, A syndrome of multiple congenital anomalies associated with teratogenic exposure, *Arch Environ. Health* **30:**17–21.

Nora, J. J., and Nora, A. H., 1973, Preliminary evidence for a possible association between oral contraceptives and birth defects, *Teratology* **7:**A-24.

Nora, J. J., and Nora, A. H., 1974, Can the pill cause birth defects?, *N. Engl. J. Med.* **291:**731–732.

Nora, J., Trasler, D. G., and Fraser, F. C., 1965, Malformations in mice induced by dexamphetamine sulphate, *Lancet* **2:**1021–1022.

Nora, J. J., Nora, A. H., Sommerville, R. J., Hill, R. M., and McNamara, D. G., 1967a, Maternal exposure to potential teratogens, *J. Am. Med. Assoc.* **202:**1065–1069.

Nora, J. J., McNamara, D. G., and Fraser, F. C., 1967b, Dexamphetamine sulphate and human malformations, *Lancet* **1:**570–571.

Nora, J. J., Sommerville, R. J., and Fraser, F. C., 1968, Homologies for congenital heart diseases: Murine models, influenced by dextroamphetamine, *Teratology* **1:**413–416.

Nora, J. J., Vargo, T. A., Nora, A. H., Love, K. E., and McNamara, D. G., 1970, Dexamphetamine: A possible environmental trigger in cardiovascular malformations, *Lancet* **1:**1290–1291.

Nora, J. J., Nora, A. H., and Toews, W. H., 1974, Lithium, Ebstein's anomaly, and other congenital heart defects, *Lancet* **2:**594–595.

Notter, A., and Delande, M. F., 1962, Prophylaxie des excès de poids chez les gestantes d'apparence normale, *Gynecol. Obstet.* **61:**359–377.

Olwin, J. H., and Koppel, J. L., 1969, Anticoagulant therapy during pregnancy, *Obstet. Gynecol.* **34:**847–852.

Opitz, J. M., Grosse, F. R., and Haneberg, B., 1972, Congenital effects of bromism, *Lancet* **1:**91–92.

Pace, H. B., Davis, M. W., and Borgen, L. A., 1971, Teratogenesis and marijuana, *Ann. N.Y. Acad. Sci.* **191:**123–131.

Paget, G. E., and Thorpe, E., 1964, A teratogenic effect of a sulphonamide in experimental animals, *Br. J. Pharmacol.* **23:**305–312.

Palacios-Macedo, X., Díaz-Devis, C., and Escudero, J., 1969, Fetal risk with the use of coumarin anticoagulant agents in pregnant patients with intracardiac ball valve prosthesis, *Am. J. Cardiol.* **24:**853–856.

Palmer, R. H., Oullett, E. M., Warner, L., and Leichtman, S. R., 1974, Congenital malformations in offspring of a chronic alcoholic mother, *Pediatrics* **53:**490–494.

Palmisano, P. A., and Polhill, R. B., 1972, Fetal pharmacology, *Pediatr. Clin. North AM.* **19:**3–20.

Papara-Nicholson, D., and Telford, I. R., 1957, Effects of alcohol on reproduction and fetal development in guinea pigs, *Anat. Rec.* **127:**438.

Papp, Z., and Gardo, S., 1971, Effect of exogenous hormones on the fetus, *Lancet* **1:**753.

Pashayan, H., Pruzanksy, D., and Pruzansky, S., 1971, Are anticonvulsants teratogenic?, *Lancet* **2:**702–703.

Persaud, T. V. N., and Ellington, A. C., 1967, Ganja (*Cannabis sativa*) in early pregnancy, *Lancet* **2:**1306.

Persaud, T. V. N., and Ellington, A. C., 1968, Teratogenic activity of cannabis resin, *Lancet* **2:**406.

Persaud, T. V. N., and Henderson, W. M., 1969, The teratogenicity of barbital sodium in mice, *Arzneim. Forsch.* **19:**1309–1310.

Peterson, W. F., 1969, Pregnancy following oral contraceptive therapy, *Obstet. Gynecol.* **34:**363–367.

Pettersson, F., Olding, L., and Gustavson, K. H., 1970, Multiple severe malformations in a child of a diabetic mother treated with insulin and dibein during pregnancy, *Acta Obstet. Gynecol. Scand.* **49:**385–387.

Pettifor, J. M., and Benson, R., 1975, Congenital malformations associated with the administration of oral anticoagulants during pregnancy, *J. Pediatr.* **86:**459–462.

Poland, B. J., and Ash, K. A., 1973, The influence of recent use of an oral contraceptive on early intrauterine development, *Am. J. Obstet. Gynecol.* **116:**1138–1142.

Popert, A. J., 1962, Pregnancy and adrenocortical hormones. Some aspects of their action in rheumatic disease, *Br. Med. J.* **1:**967–972.

Poswillo, D. E., 1972, Tridione and paradione as suspected teratogens, *Ann. R. Coll. Surg. Engl.* **50:**367–370.

Potworowska, M., Sianozecka, E., and Szufladowicz, R., 1966, Ethionamide treatment and pregnancy (including drug induced abnormalities), *Pol. Med. J.* **5:**1152–1158.

Powell, P. D., and Johnstone, J. M., 1962, Phenmetrazine and foetal abnormalities, *Br. Med. J.* **2:**1327.

Ravina, J. H., 1964, Les thérapeutiques dangereuses chez la femme enceinte, *Presse Med.* **72:**3057.

Reilly, W. A., 1958, Hormone therapy during pregnancy: Effects on the fetus and newborn, *Q. Rev. Pediatr.* **13:**198–202.

Report of Committee on Adverse Drug Reactions, 1971, Report (6th annual) of the New Zealand Committee on Adverse Drug Reactions, *N.Z. Med. J.* **74:**184–191.

Report of Metropolitan Atlanta Congenital Malformations Surveillance, 1972, Reduction deformities and tricyclic antidepressants, p. 9, (Jan.–Feb.).

Richards, L. D. G., 1969, Congenital malformations and environmental influences in pregnancy, *Br. J. Prev. Soc. Med.* **23:**218–225.

Robertson-Rintoul, J., 1974, Oral contraception: Potential hazards of hormone therapy during pregnancy, *Lancet* **2:**515–516.

Robinson, G. C., and Cambon, K. G., 1964, Hearing loss in infants of tuberculous mothers treated with streptomycin during pregnancy, *N. Engl. J. Med.* **271:**949–951.

Robinson, S. C., 1971, Pregnancy outcome following oral contraceptives, *Am. J. Obstet. Gynecol.* **109:**354–358.

Robson, J. M., and Sullivan, F. M., 1963, The production of foetal abnormalities in rabbits by imipramine, *Lancet* **1:**638–639.

Rossiter, E. J. R., and Rendle-Short, J. T., 1972, Congenital effects of bromism, *Lancet* **2:**705.

Roux, C., 1959, Action tératogène de la prochlorpémazine, *Arch. Fr. Pediatr.* **16:**1–4.

Rutman, J. Y., 1973, Anticonvulsants and fetal damage, *N. Engl. J. Med.* **289:**696–697.

Safra, M. J., and Oakley, G. P., 1975, Association between cleft lip with or without cleft palate and prenatal exposure to diazepam, *Lancet* **2:**478–480.

Sandor, S., and Amels, D., 1971, The action of aethanol on the prenatal development of albino rats, *Rev. Roum. Embryol. Cytol.* **8:**37.

Savini, E. C., Moulin, M. A., and Herrou, M. F., 1971, Experimental study of the effects of quinine on the rat, rabbit and dog fetus, *Therapie* **26:**563–574.

Saxén, I., 1975, Associations between oral clefts and drugs taken during pregnancy, *Int. J. Epidemiol.* **4:**37–44.

Saxén, I., and Saxén, L., 1975, Association between maternal intake of diazepam and oral clefts, *Lancet* **2:**498.

Saxén, L., 1974, Cleft palate and maternal diphenhydramine intake, *Lancet* **1**:407–408.

Scanlon, F. J., 1969, Use of antidepressant drugs during the first trimester, *Med. J. Aust.* **2**:1077.

Schardein, J. L., Hentz, D. L., Petrere, J. A., and Kurtz, S. M., 1971, Teratogenesis studies with diphenhydramine HCl, *Toxicol. Appl. Pharmacol.* **18**:971–976.

Schiff, D., Aranda, J. V., and Stern, L., 1970, Neonatal thrombocytopenia and congenital malformations associated with administration of tolbutamide to the mother, *J. Pediatr.* **77**:457–458.

Schou, M., Goldfield, M. D., Weinstein, M. R., and Villeneuve, A., 1973, Lithium and pregnancy. I. Report from the register of lithium babies, *Br. Med. J.* **2**:135–136.

Seay, P. H., and Field, W. E., 1963, Toxicological studies on haloperidol, *Int. J. Neuropsychiatry* **3**:1–4.

Setälä, K., and Nyyssönen, O., 1964, Hypnotic sodium pentobarbital as a teratogen for mice, *Naturwissenchaften* **51**:413.

Shaul, W. L., Emery, H., and Hall, J. G., 1975, Chondrodysplasia punctata and maternal warfarin use during pregnancy, *Am. J. Dis. Child.* **129**:360–362.

Shaw, E. B., and Steinbach, H. L., 1968, Aminopterin-induced fetal malformation, *Am. J. Dis. Child.* **115**:477–482.

Shelesnyak, M. C., and Davies, A. M., 1955, Disturbance of pregnancy in mouse and rat by systemic antihistamine treatment, *Proc. Soc. Exp. Biol.* **89**:629–632.

Shepard, T. H., and Fink, B. R., 1968, Teratogenic activity of nitrous oxide in rats, *in: Toxicity of Anesthetics* (B. R. Fink, ed.), pp. 308–323, Williams and Wilkins, Baltimore.

Shotten, D., and Monie, I. W., 1963, Possible teratogenic effect of chlorambucil on a human fetus, *J. Am. Med. Assoc.* **186**:74–75.

Sim, M., 1972, Imipramine and pregnancy, *Br. Med. J.* **2**:45.

Sinclair, J. G., 1950, A specific transplacental effect of urethane in mice, *Tex. Rep. Biol. Med.* **8**:623–632.

Slater, B. C. S., 1965, The investigation of drug embryopathies in man *in: Embryopathic Activity of Drugs* (J. M. Robson, F. M. Sullivan and R. L. Smith, eds.), pp. 241–260, Little, Brown, Boston.

Smith, B. E., 1968, Teratogenic capabilities of surgical anaesthesia, *in: Advances in Teratology* (D. H. M. Woollam, ed.), pp. 128–180, Academic Press, New York.

Smith, B. E., Usubiaga, L. E., and Lehrer, S. B., 1971, Cleft palate induced by halothane anesthesia in C-57 black mice, *Teratology* **4**:242.

Smithells, R. W., and Chinn, E. R., 1964, Meclozine and foetal malformations: A prospective study, *Br. Med. J.* **1**:217–218.

Sobel, D. E., 1960, Fetal damage due to ECT, insulin coma, chlorpromazine or reserpine, *Arch. Gen. Psychiatry* **2**:606–611.

Sokal, J. E., and Lessman, E. M., 1960, Effects of cancer chemotherapeutic agents on human fetus, *J. Am. Med. Assoc.* **172**:1765–1771.

Speidel, B. E., and Meadow, S. R., 1972, Maternal epilepsy and abnormalities of the fetus and newborn, *Lancet* **2**:839.

Spira, N., Goujard, J., Huel, G., and Rumeau-Rouquette, C., 1972, Etude du role teratogene des hormones sexuelles premiers résultats d'une enquète épidemiologique portant sur 20,000 femmes, *Rev. Med.* **41**:2683–2694.

Stafl, A., and Mattingly, R. F., 1974, Vaginal adenosis: A precancerous lesion?, *Am. J. Obstet. Gynecol.* **120**:666–677.

Stalsberg, H., 1965, Antiemetics and congenital malformations—meclizine, cyclizine and chlorcyclizine, *Tidsskr. Nor. Laegeforen.* **85**:1840–1841.

Starreveld-Zimmerman, A. A. E., van der Kolt, W. J., Meinardi, H., and Elshove, J., 1973, Are anticonvulsants teratogenic? *Lancet* **2**:48–49.

Stenchever, M. A., and Parks, K. J., 1975, Some effects of diazepam on pregnancy in the Balb/C mouse, *Am. J. Obstet. Gynecol.* **121**:765–769.

Stenchever, M. A., Kunysz, T. J., and Allen, M. A., 1974, Chromosome breakage in users of marihuana, *Am. J. Obstet. Gynecol.* **118**:106–113.

Sterne, J., 1963, Antidiabetic drugs and teratogenicity, *Lancet* **1:**1165.

Sylvester, P. E., and Hughes, D. R., 1954, Congenital absence of both kidneys. A report of four cases, *Br. Med. J.* **1:**77–79.

Szabo, K. T., 1970, Teratogenic effect of lithium carbonate in the fetal mouse, *Nature* **225:**73–75.

Takano, K., 1964, The effect of several synthetic progestagens administered to pregnant animals upon their offspring, *Proc. Cong. Anom. Res. Assoc. Jpn.* **4:**3–4.

Takano, K., Tanimura, T., and Nishimura, H., 1963, Effects of some psychoactive drugs administered to pregnant mice upon the development of their offspring, *Proc. Cong. Anom. Res. Assoc. Jpn.* **3:**2–3.

Takano, K., Yamamura, H., Suzuki, M., and Nishimura, H., 1966, Teratogenic effect of chlormadinone acetate in mice and rabbits, *Proc. Soc. Exp. Biol. Med.* **121:**455–457.

Tanimura, T., 1972, Effects on macaque embryos of drugs reported or suspected to be teratogenic in humans, *in: Symposium on the Use of Non-Human Primates for Research in Problems of Human Reproduction* (E. Diczfalusy and C. C. Standley, eds.), pp. 293–308, WHO Research and Training Center on Human Reproduction, Stockholm.

Tenbruick, M. L., and Buchin, S. Y., 1975, Fetal alcohol syndrome. Report of a case, *J. Am. Med. Assoc.* **232:**1144–1147.

Thiersch, J. B., 1952, Therapeutic abortions with a folic acid antagonist, 4-aminopteroylglutamic acid (4-amino P.G.A.) administered by the oral route, *Am. J. Obstet. Gynecol.* **63:**1298–1304.

Thiersch, J. B., 1956, The control of reproduction in rats with the aid of antimetabolites and early experiments with antimetabolites as abortifacient agents in man, *Acta Endocrinol.* **23:**37–45.

Thiersch, J. B., 1957, Effect of 2,4,6, triamino-"S"-triazine (TR) 2,4,6 "tris" (ethyleneimino)-"S"-triazine (TEM) and *N,N',N''*-triethylene-phosphoramide (TEPA) on rat litter *in utero*, *Proc. Soc. Exp. Biol. Med.* **94:**36–40.

Trautner, E. M., Pennycuik, P. R., Morris, R. J. H., Gershon, S., and Shankly, K. H., 1958, The effect of prolonged sub-toxic lithium ingestion on pregnancy in rats, *Aust. J. Exp. Biol.* **36:**305–321.

Tsun-Yu, K. L., and Stiles, Q. R., 1975, Persistent fifth aortic arch in man, *Am. J. Dis. Child.* **129:**1229–1231.

Tuchmann-Duplessis, H., 1965, *Teratogenic Action of Drugs*, Pergamon Press, New York; Czechoslovak Medical Press, Praha.

Tuchmann-Duplessis, H., 1969, The action of anti-tumor drugs on gestation and on embryogenesis, *in: Teratology, Proceedings of International Symposium at Como, Italy* (A. Bertelli, ed.), pp. 75–86, Excerpta Medica, Amsterdam.

Tuchmann-Duplessis, H., 1973, Teratogenic screening methods and their application to contraceptive products, *in: Pharmacological Models in Contraceptive Development* (M. H. Briggs and E. Diczfalusy, eds.), pp. 203–223, WHO Reseach and Training Center on Human Reproduction, Stockholm.

Tuchmann-Duplessis, H., and Mercier-Parot, L., 1956, Rèpercussions de la rèserpine sur la gestation chez la ratte, *C.R. Acad. Sci.* **243:**410–412.

Tuchmann-Duplessis, H., and Mercier-Parot, L., 1960, À propos de l'action tératogène de l'actinomycine, *C.R. Soc. Biol.* **153:**1697–1700.

Tuchmann-Duplessis, H., and Mercier-Parot, L., 1961, Malformations foetales chez le rat traité par de fortes doses de deserpidine, *C.R. Soc. Biol.* **155:**2291.

Tuchmann-Duplessis, H., and Mercier-Parot, L., 1963, Production de malformations chez le souris et le lapin par administration d'un sulfamide hypoglycemiant, la carbutamide, *C.R. Soc. Biol.* **158:**1193–1197.

Tuchmann-Duplessis, H., and Mercier-Parot, L., 1964, Rèpercussions des neuroleptiques et des antitumoraux sur le développement prénatal, *Bull. Acad. Suisse Sci. Med.* **20:**490–526.

Tuchmann-Duplessis, H., and Mercier-Parot, L., 1967, Delayed nidation under the influence of a neuroleptic agent R 1625 (haloperidol), *C.R. Acad. Sci.* **264:**114–117.

Turner, G., and Collins, E., 1975, Fetal effects of regular salicylate ingestion in pregnancy, *Lancet* **2:**338–339.

Tze, W. J., and Lee, M., 1975, Adverse effects of maternal alcohol consumption on pregnancy and foetal growth in rats, *Nature* **257**:479–488.

Uhlig, H., 1957, Abnormalities in undesired children, *Aerztl. Wochenschr.* **12**:61.

Uhlig, H., 1959, Fehlbildungen nach Follikel-hormonen beim Menschen, *Geburtschilfe Frauenheilkd.* **19**:346–352.

Van Waes, A., and Van de Velde, E., 1969, Safety evaluation of haloperidol in the treatment of hyperemesis gravidarum, *J. Clin. Pharmacol.* **9**:224–227.

van Wagenen, G., and Hamilton, J. B., 1943, The experimental production of pseudohermaphroditism in the monkey, *Essays Bio.* 583–607.

Varpela, E., 1964, On the effect exerted by first-line tuberculosis medicines on the foetus, *Acta Tuberc. Scand.* **45**:53–69.

Vevera, J., and Zatloukal, F., 1964, Případ vrozených malformaci, zpusobených pravděpodobně atebrinem, podávanýn vraném těhotenství, *Cesk. Pediatr.* **19**:211–212.

Vichi, F., 1969, Neuroleptic drugs in experimental teratogenesis, *in: Teratology* (A. Bertelli and L. Donati, eds.), pp. 87–101, Excerpta Medica, Amsterdam.

Vichi, F., and Pierleoni, P., 1970, Effetti letali e teratogeni dell'haloperidol in embroni di topo, *Riv. Ital. Stomatol.* **25**:585–596.

Voorhess, M. L., 1967, Masculinization of the female fetus associated with norethindrone mestranol therapy during pregnancy, *J. Pediatr.* **71**:128–131.

Warkany, J., 1975, A warfarin embryopathy?, *Am. J. Dis. Child.* **129**:287–288.

Warkany, J., and Bofinger, M., 1975, Le rôle de la coumadin dans les malformations congénitales, *Med. Hyg.* **23**:1454–1457.

Warkany, J., and Takacs, E., 1959, Experimental production of congenital malformations in rats by salicylate poisoning, *Am. J. Pathol.* **35**:315–331.

Warkany, J., and Takacs, E., 1965, Congenital malformations in rats from streptonigrin, *Arch. Pathol.* **79**:65–79.

Warkany, J., Beaudry, P. H., and Hornstein, S., 1959, Attempted abortion with aminopterin (4-aminopteroylglutamic acid), *Am. J. Dis. Child.* **97**:274–281.

Warrell, D. W., and Taylor, R., 1968, Outcome for the foetus of mothers receiving prednisolone during pregnancy, *Lancet* **1**:117–118.

Wells, C. N., 1953, Treatment of hyperemesis gravidum with cortisone, *Am. J. Obstet. Gynecol.* **66**:598.

Wickes, I. G., 1954, Foetal defects following insulin coma therapy in early pregnancy, *Br. Med. J.* **2**:1029.

Wild, J., Schorah, C. J., and Smithells, R. W., 1974, Vitamin A, pregnancy, and oral contraceptives, *Br. Med. J.* **1**:57–60.

Wilkins, L., 1960, Masculinization of female fetus due to use of orally given progestins, *J. Am. Med. Assoc.* **172**:1028–1032.

Wilson, J. G., 1964, Teratogenic interaction of chemical agents in the rat, *J. Pharmacol. Exp. Ther.* **144**:429–436.

Wilson, J. G., 1966, Effects of acute and chronic treatment with actinomycin D on pregnancy and the fetus in the rat, *Harper Hosp. Bull.* **24**:109–118.

Wilson, J. G., 1971, Use of rhesus monkeys in teratological studies, *Fed. Proc.* **30**:104–109.

Wilson, J. G., 1972, Abnormalities of intrauterine development in nonhuman primates, *in: The Use of Non-Human Primates in Research on Human Reproduction* (E. Diczfalusy and C. C. Standley, eds.), pp. 261–292, WHO Research and Training Center on Human Reproduction, Stockholm.

Wilson, J. G., 1973a, An animal model of human disease: Thalidomide embryopathy in primates, *Comp. Pathol. Bull.* **5**:3–4.

Wilson, J. G., 1973b, *Environment and Birth Defects*, Academic Press, New York.

Wilson, J. G., 1974, Teratologic causation in man and its evaluation in non-human primates, *in: Birth Defects* (A. G. Motulsky and W. Lenz, eds.), pp. 191–203, Excerpta Medica, Amsterdam.

Wilson, J. G., Maren, T. H., Takano, K., and Ellison, A., 1968, Teratogenic action of carbonic anhydrase inhibitors in the rat, *Teratology* **1**:51–60.

Winckel, C. W. F., 1948, Quinine and congenital injuries of ear and eye of the foetus, *J. Trop. Med.* **51:**2.

Windorfer, A., 1953, Zum problem du Missbildungen durch bewusste Keim- und Fruchtschadigung, *Med. Klin. Berl.* **48:**293.

Wright, T. L., Hoffman, L. H., and Davies, J., 1971, Teratogenic effects of lithium in rats, *Teratology* **4:**151–156.

Yalom, I. D., Grau, R., and Fisk, N., 1973, Prenatal exposure to female hormones: Effect on psychosexual development in boys, *Arch. Gen. Psychiatry* **28:**554–561.

Yasuda, M., and Poland, B. J., 1974, A 63-mm human fetus with radial aplasia, tibial polydactyly, rib and vertebral fusions, and right aortic arch, *Teratology* **10:**119–124.

Yerushalmy, J., and Milkovich, L., 1965, Evaluation of the teratogenic effect of meclizine in man, *Am. J. Obstet. Gynecol.* **93:**553–562.

Zackai, E. H., Mellman, W. J., Neiderer, B., and Hanson, J. W., 1975, The fetal trimethadione syndrome, *J. Pediatr.* **87:**280–284.

Zolcinski, A., Heimrath, T., and Ujec, M., 1965, Quinine as a cause of fetal malformation, *Ginekol. Pol.* **36:**935–938.

Environmental Chemicals 9

JAMES G. WILSON

I. INTRODUCTION

The varieties of chemical substances which man has introduced into his environment since the beginning of the Industrial Revolution can no longer be reasonably classified, much less individually cataloged. There are no reliable records on rates at which new chemicals were introduced year by year, but there is every reason to believe that it has increased tremendously in recent years, perhaps approaching an exponential rate. It is roughly estimated that today the new compounds synthesized each year, or produced inadvertently in industry or as a result of reactions within the environment, may number a few thousand. Possibly 10% of these may persist in more than negligible amounts as environmental contaminants, but of this number very few have been proven to exist in toxic amounts under ordinary conditions of exposure to man. Most of those implicated in overt toxicity have involved high-level exposure to industrial workers prior to recognition of their toxic potential. To date only one chemical agent in the nature of an environmental contaminant or pollutant has been established as embryotoxic in man, namely, methyl mercury which causes both prenatal and postnatal toxicity in the form of Minamata disease (see Section VI).

Lack of proven instances, however, does not remove the basis for concern. This chapter is devoted to an enumeration of some of the natural as well as man-made chemicals which have been examined for embryotoxic effects (intrauterine death, growth retardation, and malformation) in domestic or laboratory animals. In the frequent instances in which embryotoxicity was observed, dosage was almost always above that which would be encountered

JAMES G. WILSON · Children's Hospital Research Foundation, Elland and Bethesda Avenues, Cincinnati, Ohio 45229.

under usual environmental conditions, and in many cases it was at maximum tolerated or even at maternal toxic levels. The absence of reported embryotoxicity in animals may mean any of several things: that the dosage was too low, or the route of administration inappropriate, that the experiment was not otherwise conducted in the most exacting way, that the species used had high resistence to the agent tested, or that the agent indeed had low embryotoxic potential. Even the absence of embryotoxicity in a well-conducted animal study, at doses representing reasonable multiples above estimated environmental levels, does not assure the safety of any chemical. Aside from the infrequent occurrence of accidental overdosage and individual hypersensitivity, the possibility of potentiative interaction between two or more chemical agents, or between these and other potentially embryotoxic factors can never be dismissed. Exaggerated embryotoxic effects from combined action of two or more agents given at low doses have often been demonstrated in experimental teratology and must always be considered in evaluating causative factors in the environment.

II. NATURAL SUBSTANCES—PLANTS, FUNGI, BACTERIA, AND VENOMS

There is a general tendency to think of embryotoxic agents as being chemical and, to a lesser extent, physical factors introduced into the environment by man. Certainly a large proportion of the agents now shown to be embryotoxic in the laboratory are in fact chemicals resulting from man's efforts to protect himself from disease, improve his food supply, raise his standard of living with comforts and luxuries, or the consequence of dumping of the wastes from these efforts into his surroundings in the form of pollutants. It thus is something of a surprise to find that a number of naturally occurring substances are quite teratogenic or otherwise embryotoxic, causing intrauterine death or growth retardation, when pregnant animals ingest or are exposed to these materials.

Table 1 presents a reasonably comprehensive but not exhaustive summary of the natural substances that have been shown to be embryotoxic under laboratory or other controlled conditions. The largest category includes plants that may be eaten by domestic animals, although a few were fed or administered to rats or mice to confirm a suspicion of embryotoxicity in larger animals.

An interesting subject for speculation on examining such a list is the influence toxic plants may have had in determining the phylogenetic evolution and subsequently the geographic distribution of herbivorous animals. Also of broad biological interest is the role mycotoxins of fungal origin might have played in the emergence of human civilization, particularly after man and his animal associates came to depend for food upon stores of grain and plant forage which are liable to molding. The recent "blighted potato" episode

(Emanuel and Sever, 1973; Sever, 1973; Ruddick *et al.*, 1974) in which human neural tube malformations were thought to have been epidemiologically associated with the eating of poorly preserved potatoes would, had it been verified, have been the first proven instance of natural toxins having made major impact on human health. Several more or less widely used plant alkaloids and extracts are known to be embryotoxic in large doses, a fact that is not surprising in view of the pharmacological activity of such substances. Venoms and bacterial toxins seem to have been little studied in relation to reproduction.

III. INSECTICIDES

Of the many and diverse chemicals used to kill insect pests, by no means all have been adequately tested for embryotoxic effects in mammals. Some of those that have been administered to pregnant animals are listed in Table 2. These studies were highly variable as regards route, level, and duration of dosage, as well as other aspects of experimental design, therefore, the results are not amenable to direct comparisons even within the same species. Useful generalizations, however, can be drawn from this heterogeneous compilation of observations. In the first place, it is apparent that not all caused embryotoxic manifestations, despite the fact that many were applied at dosage levels that were at or near maternal toxic levels. This is not to say that most of these insecticides could not be shown to cause some embryotoxicity, i.e., intrauterine death, malformation, or growth retardation, if administered in one or a few high doses during the most susceptible period of organogenesis in the species used. The significant fact seems to be that insecticides as a group have surprisingly low embryotoxic potential, occurring mainly only at doses approaching those causing maternal toxicity, especially since the organochlorine and organophosphate insecticides are well-known to cause general toxicity in mammals at relatively low dosage (Khera and Clegg, 1969).

Further generalizations can be drawn from the data in Table 2. Despite its widely publicized detrimental effects on reproduction in certain birds, DDT has not been found to cause embryotoxicity in mammals, although one study (Diechmann *et al.*, 1971) found that prolonged treatment caused some delay of estrus and loss of libido in dogs. Most other organochlorine insecticides seem to have some teratogenic effect, however, and as a class they may be more effective in this regard than are the organophosphates. There appear to be wide species differences in embryotoxic sensitivity to the same compound, as illustrated by carbaryl which is highly effective in dogs and less so in pigs at relatively low doses, but is generally not effective in rodent–rabbit species until quite large doses are used.

Insecticides have rarely been associated with human embryotoxicity. In one report from Japan it was suggested that a few cases of fetal death and malformation may have resulted from intensive exposure of pregnant women

Table 1. Natural Substances Reported to Be Embryotoxic in Animals

Type of substance	Species affected	Reference
Weeds, trees, other plants		
Locoweed (*Astragalus lentiginosus*)	Sheep, cattle	James *et al.*, 1967, 1969
Lupins (*Lupinus sericius* and *caudatus*)	Cattle	Shupe *et al.*, 1967
	Horse	Rooney, 1966
Jimsonweed (*Datura stramonium*)	Swine	Leipold *et al.*, 1973
Wild black cherry (*Prunus serotina*)	Swine	Selby *et al.*, 1971
Poison hemlock (*Conium maculatum*)	Swine	Edmonds *et al.*, 1972
Chicken peas (*Lathyrus cicera*)	Cattle, sheep, rat	Keeler *et al.*, 1969
		Abramovich and DeVoto, 1968
Sweet peas (*Lathyrus odoratus*)	Baboon	Steffek and Hendrickx, 1972
False hellebore (*Veratrum californicum*)	Sheep	Binns *et al.*, 1963
Tobacco stalks (*Nicotinia tabacum*)	Swine	Menges *et al.*, 1970
		Crowe and Swerczek, 1974
Creeping indigo (indicospicine, amino a. from *Indigofera spicata*)	Rat	Pearn and Hegarty, 1970
West Indian tree (hypoglycene A, amino a. from *Blighia sapida*)	Rat	Persaud, 1968
Bracken fern (*Pteridium aquilinum*)	Mouse[a]	Yasuda *et al.*, 1974
Podophyllin (*Podophyllum peltatum*)	Mouse[a]	Joneja and LeLiever, 1974
	Rat	Dwornik and Moore, 1967
(*Leucaena leucocephala*)	Sheep	Little and Hamilton, 1971
Mycotoxins		
Aflotoxin B$_1$ (*Aspergillus*)	Mice,[a] hamster, rats[a]	DiPaolo *et al.*, 1967
Ochratoxin A (*Aspergillus*)	Mouse	Hayes *et al.*, 1974
Rubratoxin B (*Penicillium rubrum* and *P. purpurogenum*)	Mouse	Hood *et al.*, 1973
Ergotamine (*Claviceps purpurea*)	Mice,[a] rats,[a] rabbits[a]	Grauwiler and Schön, 1973
Fungi (unidentified) on pine needles	Mouse	Chow *et al.*, 1974
"Blighted" potato (organisms unknown)	Marmoset	Poswillo *et al.*, 1972
Several antibiotics from moulds	Various	See Table 3, Chapter 8
Minerals		
Selenium, plants containing	Farm animals	Moxon, 1937
	Chicken	Franke and Tully, 1935
Specific alkaloids and extracts		
Caffeine	Rats[a]	Fujii and Nishimura, 1972
	Mice	Nishimura and Nakai, 1960
Nicotine	Mice	Nishimura and Nakai, 1958
Quinine	Several species[a]	Tanimura, 1972
Colchicine	Hamster	Ferm, 1963a
	Mouse	Shoji and Makino, 1966
Vincristine	Hamster	Ferm, 1963b
Vinblastine	Hamster	Ferm, 1963b
Cycasin (from seeds of *Cycas circinales* and *revoluta*)	Mice, hamster	Hirono *et al.*, 1969
Heliotropine (*Heliotropium sp.* and *Senicis sp.*)	Rats	Green and Christie, 1961

Table 1 (cont'd)

Type of substance	Species affected	Reference
Other natural toxins		
Arvin from venom (*Agkistrodon rhodustoma*)	Mouse, rabbit	Penn *et al.,* 1971
Viper venom (*Vispera aspis*)	Mouse	Calvert and Gabriel-Robez, 1974
Naja venom (*Naja sp.*)	Chicken	Ruch and Gabriel-Robez-Kremer, 1962
Tetanus toxin (*Clostridium tetani*)	Chicken	Corliss *et al.,* 1966
Endotoxin (*Escherichia coli*)	Rat	Ornoy and Altshuler, 1975
Lipolysaccharides (gram neg. bacilli)	Rat[a]	Thiersch, 1962a

[a]Little or no increase in malformations but significant intrauterine growth retardation and/or resorptions.

during agricultural use of methyl parathion (Ogi and Hamada, 1965). It has been postulated that organochlorine and some organophosphorus insecticides, in addition to having direct estrogenic effects, might be able to influence general reproductive functions through changes in normal metabolic patterns (Diechmann *et al.,* 1971). Earlier work by the latter investigators demonstrated that some of these insecticides elevated the activity of hepatic microsomal enzymes which speeded the metabolism of steroid hormones. Chronic exposure to pesticides was shown to have modest effects on reproduction in mice and dogs, but comparable effects have not been reported for man. Analyses of blood levels of DDT and DDE from women undergoing spontaneous abortion and from women with normal pregnancies revealed high blood levels, but these were not associated with increased abortion rates (O'Leary *et al.,* 1970). Thus, presently used pesticides at current use levels have not been shown to increase the risk of human embryotoxicity.

IV. HERBICIDES

There are approximately 100 compounds and preparations that have been used to kill undesirable plants on agricultural lands, rights-of-way, lawns, and in forests and park areas. Some of these have been in use for many years but none had been seriously questioned as having significant effects on mammalian reproduction until the phenoxy herbicides 2,4,5-T, 2,4-D, and 2,4-D isooctyl ester were reported to cause various embryotoxic effects in mice in the so-called Bionetics Report (1969, unpublished). (This was a report of studies conducted between 1965–1968 on the teratogenic effects of a number of pesticides, performed under contract with the National Cancer Institute by Bionetics Research Laboratories.) Numerous subsequent studies have confirmed that 2,4,5-T is teratogenic in several rodent species when relatively large doses are used (Courtney *et al.,* 1970; King *et al.,* 1971; Collins and Williams, 1971; Sparschu *et al.,* 1971a; Khera and McKinley, 1972; and others) and that 2,4-D and some of its esters may be mildly embryotoxic, causing

Table 2. Insecticides Reported to Be Embryotoxic When Given During Pregnancy in Mammals

Agent	Species	Treatment (dose, route, gest. days)	Embryotoxic effects[a]	Reference
Aldrin	Hamster	50 mg/kg, oral, 7, 8, or 9	MAL, IDR, IGR	Ottolenghi et al., 1974
	Mouse	25 mg/kg, oral, 9	MAL	Ottolenghi et al., 1974
	Rat	12.5 ppm, diet, 3 gen.	Reduced pregnancies	Treon and Cleveland, 1955
Carbaryl	Guinea pig	300 mg/kg, oral, var. 11–20	MAL, IDR, IGR	Robens, 1969
	Rabbit	50–200 mg/kg, oral, 5–15	None	Robens, 1969
	Hamster	125–250 mg/kg, oral, 6–8	IDR	Robens, 1969
	Rat	200 mg/kg, diet, 3 gen.	None	Weil et al., 1972
	Guinea pig	300 mg/kg, diet, 0–35	None	Weil et al., 1972
	Rat	2000–10000 ppm, diet, 3 gen.	Reduced fert. 2nd gen.	Collins et al., 1971
	Gerbil	2000–10000 ppm, diet, 3 gen.	Reduced fert. 3rd gen.	Collins et al., 1971
	Dog	3–50 mg/kg, diet, thru preg.	MAL, IDR, PGR	Smalley et al., 1968
	Pig	4–32 mg/kg, diet, various	±MAL, IDR	Earl et al., 1973
	Monkey	2–20 mg/kg, oral, thru preg.	IDR	Dougherty et al., 1971
Carbofuran	Dog	50 ppm, diet, thru preg.	None	McCarthy et al., 1971
	Rat	50 ppm, diet, thru preg.	None	McCarthy et al., 1971
	Rabbit	50 ppm, diet, thru preg.	None	McCarthy et al., 1971
DDT	Rat	20 ppm, diet, thru preg.	None	Ottoboni, 1972
	Rat	25 ppm, diet, 3 gen.	None	Treon and Cleveland, 1955
Demeton	Mouse	7–10 mg/kg; ip, var. 7–10	MAL, IDR, IGR	Budreau and Singh, 1973
Diazinon	Rat	100–200 mg/kg, ip, 11	±MAL, IDR, IGR	Kimbrough and Gaines, 1968
	Rat	9–95 mg/kg, oral, var. 8–15	±MAL, IDR	Dobbins, 1967
	Dog	1–5 mg/kg, oral, thru preg.	SB	Earl et al., 1973
	Pig	5–10 mg/kg, oral, thru preg.	±MAL	Earl et al., 1973

	Species	Dose	Effect	Reference
	Hamster	0.125–25 mg/kg, oral, 6–8	None	Robens, 1969
	Rabbit	7–30 mg/kg, oral, 5–15	None	Robens, 1969
Dichlorvos	Rat	15 mg/kg, ip, 11	MAL	Kimbrough and Gaines, 1968
Dieldrin	Hamster	30 mg/kg, oral, 7, 8 or 9	MAL, IDR, IGR	Ottolenghi et al., 1974
	Mouse	15 mg/kg, oral, 9	MAL	Ottolenghi et al., 1974
	Mouse	5 ppm, diet, 1 gen.	±IDR	Good and Ware, 1969
	Mouse	1.5–6 mg/kg/d, oral, 7–16	None	Chernoff et al., 1975
	Rat	1.5–6 mg/kg/d, oral, 7–16	None	Chernoff et al., 1975
	Rat	2.5 ppm, diet, 3 gen.	Reduced pregnancies	Treon and Cleveland, 1955
Endrin	Hamster	5 mg/kg, oral, 7, 8 or 9	MAL, IDR, IGR	Ottolenghi et al., 1974
	Mouse	2.5 mg/kg, oral, 9	MAL	Ottolenghi et al., 1974
	Mouse	5 ppm, diet, 1 gen.	±IDR	Good and Ware, 1969
Fenthion	Mouse	40–80 mg/kg, ip, var. 7–10	MAL, IDR, IGR	Budreau and Singh, 1973
Imidan	Rabbit	35 mg/kg, oral, 7–12	None	Fabro et al., 1965
Isoflurophate	Rat	1–4 mg/kg, ip, 8, 9 or 12	IGR	Fish, 1966
Kelthane	Mouse	7 ppm, diet, 3 gen.	MAL, PGR 3rd gen.	An Der Lan, 1964
Lindane	Dog	7.5–15 mg/kg, oral, thru preg.	SB	Earl et al., 1973
Malathion	Rat	240 µg/d, diet, 2 gen.	Reduced fert. 2nd gen.	Kalow and Marton, 1961
	Rat	600–900 mg/kg, oral, 2 gen.	None	Kimbrough and Gaines, 1968
	Rat	105–355 mg/kg, oral, var. 8–15	±MAL, IDR	Dobbins, 1967
Methyl parathion	Rat	5–10 mg/kg, ip, 12	IGR	Tanimura et al., 1967
	Mouse	20–60 mg/kg, ip, 10	MAL, IDR, IGR	Tanimura et al., 1967
Parathion	Rat	3.5 mg/kg, ip, 11	IDR, IGR	Kimbrough and Gaines, 1968
	Rat	0.5–5 mg/kg, ip, var. 8–16	SB, PGR	Fish, 1966
Photodieldrin	Mouse	0.15–0.6 mg/kg/d, oral, 7–16	None	Chernoff et al., 1975
	Rat	0.15–0.6 mg/kg/d, oral, 7–16	None	Chernoff et al., 1975
Tsumacide	Rat	80–4000 ppm, diet, 8–15	None	Yasuda, 1972

[a]MAL = malformations; IDR = intrauterine death and resorption or abortion; IGR = intrauterine growth retardation; SB = stillbirth or perinatal death; PGR = postnatal growth retardation.

some increase in intrauterine death and growth retardation. A recent study using large samples of several strains of mice has shown great variability in susceptibility, with the A/JAX strain being sensitive to as little as 15 mg/kg (Gaines *et al.*, 1975).

The defoliant 2,4,5-T has probably been the subject of more controversy and misunderstanding than any other pesticide, possibly excepting the insecticide DDT. In the first place, an appreciable part of the embryotoxicity originally attributed to 2,4,5-T is now thought to have been caused by the presence of a dioxin impurity in the early test samples of 2,4,5-T (Courtney and Moore, 1971; Neubert and Dillmann, 1972; Sparschu *et al.*, 1971b; Moore, 1973). Some of the commotion surrounding this herbicide, however, was undoubtedly an outgrowth of its extensive use as a defoliant in the Vietnam War. In addition to the indignation about the military use of a chemical agent, as well as about consequent damage to crops and forests, there were newspaper reports from Saigon claiming an association between increased human birth defects and exposure of pregnant women to 2,4,5-T (Cutting *et al.*, 1970; Meselson *et al.*, 1970). Repeated attempts to validate the claims of increased occurrence of birth defects have been unsuccessful. Also unsuccessful were attempts to relate this herbicide to human teratogenesis in other parts of the world, e.g., Globe, Arizona, and Swedish Lapland, where pregnant women were alleged to have been exposed during government sponsored defoliation activities (see Report of the 2,4,5-T Advisory Committee to the Environmental Protection Agency, 1971; Rapport fråm en expert grupp, 1971; Binns *et al.*, 1970). Nevertheless, the unfounded claims of human embryotoxicity and the unrestrained emotionalism surrounding the use of this herbicide during wartime may have contributed to the development of a rational level of concern about indiscriminate distribution of toxic chemicals in the environment.

Table 3 provides information on some of the relatively few herbicides that have been adequately examined for embryotoxicity in pregnant mammals. Most have caused mild to moderate adverse effects on intrauterine development when dosage was at or near maternal toxic levels and considerably above likely environmental levels. It thus appears that herbicides as a group are relatively nonembryotoxic at ordinary use levels. Indiscriminate use of herbicides in the agricultural context is to a considerable degree limited by undesirable effects on such nontarget plants as crops, woodlands, and ornamental plants.

V. FUNGICIDES

These compounds are used to combat fungal growth on living human and animal skin, fabrics, wood and leater products, agricultural and ornamental plants, and to protect a number of commodities such as seed grains. The use of methyl mercury for the latter purpose, followed by human consumption of treated grains, has caused instances of congenital and postnatal mer-

Table 3. Herbicides Reported to be Embryotoxic When Given During Mammalian Pregnancy

Agent	Species	Treatment (dose, route, gest. days)	Embryotoxic effects[a]	Reference
2,4-D	Rat	87.5 mg/kg, oral, 6–15	IGR	Schwetz et al., 1971
	Rat	100–150 mg/kg, oral, 6–15	±MAL	Khera and McKinley, 1972
Diquat	Rat	7–14 mg/kg, i.p., 6–14	IGR, IDR	Khera & Whitta, 1968
MCPA (ethylester)	Rat	60–100 mg/kg, diet, 8–15	MAL, IGR, IDR	Yasuda and Maeda, 1972
Norea	Hamster	2000 mg/kg, oral, 6–8	IDR	Robens, 1969
Paraquat	Rat	6.5–13 mg/kg, i.p., 6–14	±MAL, IDR	Khera & Whitta, 1968
Silvex	Rat	100 mg/kg, s.c., 6–15	None	Moore and Courtney, 1971
2,4,5-T	Mouse	46–103 mg/kg, oral, 6–14	MAL, IDR	Courtney et al., 1970
	Mouse	15–120 mg/kg, oral, 6–14	MAL, IGR, IDR	Gaines et al., 1975
	Hamster	40–100 mg/kg, oral, 6–10	MAL, IGR, IDR	Collins and Williams, 1971
	Rat	10–46 mg/kg, oral, 10–15	MAL, IDR	Courtney et al., 1970
	Rat	60–120 mg/kg, oral, 12–16	None	King et al., 1971
	Rat	1–50 mg/kg, oral, 6–15	None	Thompson et al., 1971
	Rat	100–150 mg/kg, oral, 6–15	±MAL, ±IGR	Khera and McKinley, 1972
	Rabbit	10–40 mg/kg, oral, 6–18	None	Thompson et al., 1971
	Sheep	100–113 mg/kg, oral, 14–36	None	Binns and Balls, 1971
	Monkey	5–40 mg/kg X 12, oral, 20–48	None	Wilson, 1972

[a]MAL = malformations; IDR = intrauterine death and resorption or abortion; IGR = intrauterine growth retardation.

cury poisoning (Snyder, 1971; Amin-Zaki *et al.*, 1974). Although not ordinarily added to food as a preservative, fungicides can occur as residues on food and undoubtedly do have access to human blood streams via both the digestive and respiratory tracts.

Table 4 summarizes some of the reports of embryotoxicity studies in domestic and laboratory animals. Although the design and rigorousness of these texts vary widely, as evidenced when the same compound was used in the same species by different investigators, the general impression emerges that fungicides as a group are not highly embryotoxic. Several of the studies that included doses at or near maternal toxic levels did not produce appreciable embryotoxicity. It should be emphasized, however, that some of the reports of no embryotoxicity involved doses well below those that would have been toxic to the pregnant female, although presumably dosage was still above environmental exposure levels. The limited embryotoxicity observed after repeated tests in several species, with such compounds as captan and folpet, permits some confidence that they do not pose appreciable risks under usual conditions of usage.

VI. METALS AND RELATED ELEMENTS

There has been lingering concern for many years about the effects of lead poisoning during pregnancy on the condition of the human newborn. Exposed women were reported (Cantarow and Trumper, 1944) to produce more structurally and neurologically inferior infants, and a study in Japan indicated that abortions increased among women who took employment involving lead exposure (Angle and McIntire, 1964). These suggestive observations were not confirmed, however, in an epidemic of lead poisoning resulting from burning battery casings in Rotterdam, although exposure was shorter in this instance. There is still no well-documented case of human embryotoxicity attributable directly to lead poisoning, but suggestive associations continue to be reported. For example, mental retardation and various neuromuscular disorders were found in a child born to a woman habituated to moonshine whiskey, which is said often to have high lead content (Palmisano *et al.*, 1969), but alcoholism itself is now suspect (see Chapter 8). In Glasgow lead levels in water were measured in the homes of 77 mentally retarded children and found to be significantly higher than in homes of 77 nonretarded, matched controls (Beattie *et al.*, 1975). The probability of mental retardation was significantly increased when water-lead exceeded 800 μg/liter. In South Wales, however, a significant positive correlation between increased CNS malformation rates and several trace elements in water was noted only for aluminum (Morton and Elwood, 1974). Despite the lack of strong evidence that lead is damaging to human intrauterine development, it is certainly prudent to continue to regard it as a potential hazard because of its widespread environmental presence in industrialized societies.

Table 4. Fungicides Studied for Embryotoxicity in Pregnant Mammals

Agent	Species	Treatment (dose, route, gest. days)	Embryotoxic effects[a]	Reference
Alkyldithio-carbamate salts	Rat	1/2–1/20 LD$_{50}$, oral, various	MAL	Petrova-Vergieva, 1971
Benomyl	Rat	0.01–0.5% diet, 6–15	None	Sherman et al., 1975
Captan	Hamster	125–1000 mg/kg, oral, thru gest.	IGR, IDR	Kennedy et al., 1968
	Hamster	300–1000 mg/kg, oral, 7 or 8	MAL, IDR	Robens, 1970
	Rat	500–2000 mg/kg, oral, 8–10	±MAL	Kennedy et al., 1968
	Rabbit	19–75 mg/kg, oral, 6–18	IDR, IGR	Kennedy et al., 1968
	Rabbit	80 mg/kg, oral, 7–12	None	Fabro et al., 1965
	Rabbit	37.5–75 mg/kg, oral, 6–16	MAL	McLaughlin et al., 1969
	Dog	15–60 mg/kg, oral, thru gest.	MAL, IDR	Earl et al., 1973
	Dog	30–60 mg/kg, diet, thru gest.	None	Kennedy et al., 1975
	Monkey	10–75 mg/kg, oral, 21–34	None	Vondruska et al., 1971
Difolatan	Hamster	200–1000 mg/kg, oral, 7 or 8	MAL, IDR	Robens, 1970
	Rat	100–500 mg/kg, oral, 8–10	None	Kennedy et al., 1968
	Rabbit	37–150 mg/kg, oral, 6–18	±IDR, IGR	Kennedy et al., 1968
	Monkey	6.25–25 mg/kg, oral, 22–32	None	Vondruska et al., 1971
Disulfiram	Hamster	125–500 mg/kg, oral, 7 or 8	±IDR	Robens, 1969
Folpet	Hamster	500–1000 mg/kg, oral, 7 or 8	MAL, IDR	Robens, 1970
	Rat	100–500 mg/kg, oral, 8–10	None	Kennedy et al., 1968
	Rabbit	19–75 mg/kg, oral, 6–18	IGR, IDR	Kennedy et al., 1968
	Rabbit	75–150 mg/kg, oral, 6–16	None	McLaughlin et al., 1969
	Rabbit	80 mg/kg, oral, 7–12	None	Fabro et al., 1965
	Monkey	10–75 mg/kg, oral, 21–34	None	Vondruska et al., 1971
Griseofulvin	Rat	500 mg/kg, oral, 7–10, 11–14	MAL, IDR	Slonitskaya, 1969
	Rat	125–1500 mg/kg, oral, 6–15	MAL, IDR, IGR	Klein and Beall, 1972
	Cat	500–1000 mg/kg, oral, various	MAL, ?IDR	Scott et al., 1975
Pentachlorophenol	Rat	10–30 mg/kg, oral, 6–15	IDR	Schwetz et al., 1974a
Pyrithione	Rabbit	20–50 mg, dermal, 7–18	None	Nolen et al., 1975
	Pig	100–330 mg/kg, dermal, 12–36	None	Jordan and Borzelleca, 1975
Tetrachlorophenol	Rat	5–50 mg/kg, oral, 6–15	MAL, IDR, IGR	Schwetz et al., 1974b
Thiram	Hamster	125–500 mg/kg, oral, 7 or 8	MAL, IDR, IGR	Robens, 1969

[a]MAL = malformations; IDR = intrauterine death and resorption or abortion; IGR = intrauterine growth retardation.

Mercury is the only heavy metal proven to enter and interfere directly with the development of the human embryo or fetus, as was shown at Minamata Bay and elsewhere in Japan between 1954 and 1960. Although mercury, as well as lead, had been regarded with suspicion for a number of years (Butt and Simonsen, 1950), it was not until the birth of infants with severe neurological symptoms resembling cerebral palsy was traced to the eating of mercury-contaminated fish (Matsumoto *et al.,* 1965; Harada, 1968) that this metal was accepted as a risk to human intrauterine development. Additional examples of human embryotoxicity from ingestion of methyl mercury have come from the United States (Synder, 1971) and Iraq (Amin-Zaki *et al.,* 1974).

Thus, mercury is the only environmental metal or trace element that has to date been firmly established as embryotoxic to man. There is substantial evidence that lithium carbonate as used therapeutically in the treatment of psychiatric disorders (see Chapter 8) is associated with the birth of an increased number of defective children when given during the early part of pregnancy. A question was raised about selenium following a report that, of 8 women exposed to this trace element in media-preparation kitchens, 4 and possibly 5 aborted and 1 gave birth to a child with bilateral club foot (Robertson, 1970).

Metals and related elements have frequently been found to cause malformations, intrauterine resorption and death, and intrauterine growth retardation in laboratory animals (Table 5). Although the doses were generally high, by no means all were at maternal toxic levels. It should be emphasized that a number of such elements have been adequately studied without significant effects on reproduction or intrauterine development (Ferm, 1972; Schroeder and Mitchener, 1971; Wilson, unpublished data). Nevertheless, the general reactivity of metallic ions and their known involvement in several important biological functions make it necessary that their levels in the environment be closely monitored.

Several metals and trace elements are essential for normal development, and deficiencies are known to interfere with reproduction or embryonic development (see Chapter 7; Ferm, 1972). This raises the question of whether chelating agents which form cyclic complexes with metals may sometimes create artificial deficiencies of metals *in vivo,* in food and water, or otherwise in the environment. Some drugs and useful household and industrial chemicals, as well as environmental pollutants, are chelating agents and may thus have the capacity to sequester scarce metals. For example, a few years ago a chelating agent, nitrilotriacetic acid (NTA), which was proposed for use in the manufacture of detergents, was reported to increase the teratogenicity and lethality of cadmium chloride and methyl mercury chloride in rats and mice (Chernoff and Courtney, 1970). Several subsequent studies, however, were unable to confirm this (Nolen *et al.,* 1972a,b; Scharpf *et al.,* 1972; Tjälve, 1972). In fact a later report (Scharpf *et al.,* 1973) found that the teratogenicity of methyl mercury was appreciably reduced by NTA.

Table 5. Metals and Related Elements Studied for Embryotoxicity in Pregnant Mammals

Element (form)	Species	Treatment (dose, route, gest. days)	Embryotoxic effects[a]	Reference
Al(chloride)	Rat	40–200 mg/kg, ip, var. 9–18	MAL[b]	Bennett et al., 1974
As(Na$_2$HAsO$_4$)	Hamster	20 mg/kg, iv, 8	MAL, IDR	Ferm and Carpenter, 1968
As(Na$_2$HAsO$_4$)	Mouse	40 mg/kg, ip, 9	MAL, IDR, IGR	Hood and Pike, 1972
As(Na$_2$HAsO$_4$)	Rat	30 mg/kg, ip, 8, 9, 10	MAL, IDR, IGR	Beaudoin, 1974
Cd(sulfate)	Hamster	2 mg/kg, iv, 8	MAL, IDR	Holmberg and Ferm, 1969
Cd(not stated)	Mouse	10 ppm, oral, thru gest.	MAL, IDR, IGR, PGR	Schroeder and Mitchener, 1971
Cd(chloride)	Rat	16 µM/kg, ip, 8, 9, 10	MAL, IDR, IGR	Barr, 1973
Cr(trioxide)	Hamster	5–10 mg/kg, iv, 8	?MAL, IDR, IGR	Gale, 1974
Cu(sulfate)	Hamster	2–7.5 mg/kg, iv, 8	?MAL, IDR	Ferm and Hanlon, 1974
Ga(sulfate)	Hamster	40 mg/kg, iv, 8	?MAL, IDR	Ferm and Carpenter, 1970
Hg(methyl Cl)	Hamster	8 mg/kg, ip, 8	MAL, IDR, IGR	Harris et al., 1972
Hg(methyl DCD[c])	Mouse	8 mg/kg, ip, 9–13	MAL, IDR, IGR	Spyker and Smithberg, 1972
Hg(methyl Cl)	Dog	0.1 mg/kg, oral, thru gest.	?MAL, SB	Earl et al., 1973
Hg(methyl Cl)	Cat	0.03–0.25 mg/kg, oral, 10–58	MAL, IDR, ?IGR	Khera, 1973
In(nitrate)	Hamster	0.5–1 mg/kg, iv, 8	MAL, IDR	Ferm and Carpenter, 1970
Li(carbonate)	Mouse	300–465 mg/kg, oral, 6–15	MAL, IDR	Szabo, 1970
Li(carbonate)	Rat	20–50 mg/kg, ip, 1, 4, 7, 9–10	MAL, IDR	Wright et al., 1971
Li(carbonate)	Rat	1–4 mEg, oral, 5–15	None	Gralla and McIlhenny, 1972
Li(carbonate)	Rabbit	0.67–1 mEg, oral, 5–18	None	Gralla and McIlhenny, 1972
Li(carbonate)	Monkey	0.67 mEg, oral, 14–35	None	Gralla and McIlhenny, 1972
Ni(acetate)	Hamster	10–30 mg/kg, iv, 8	?MAL, IDR	Ferm, 1972
Pb(nitrate)	Hamster	50 mg/kg, iv, 8	MAL, IDR	Ferm and Carpenter, 1967
Pb(not stated)	Mouse	25 ppm, oral, thru gest.	?MAL, IDR, IGR	Schroeder and Mitchener, 1971
Pb(nitrate)	Rat	25–70 mg/kg, iv, 8, 9–10, etc.	MAL, IDR, IGR, PND	McClain and Becker, 1975
Pb(tetra ethyl)	Rat	7.5–30 mg/kg, oral, 9, 10, 11	IDR, IGR	McClain and Becker, 1972
Te(H$_2$TeO$_3$)	Rat	3300 ppm, diet, thru gest.	MAL	Agnew et al., 1968, 1973
Tl(sulfate)	Rat	2.5–10 mg/kg, diet, 12, 13, 14	?MAL, IGR	Gibson and Becker, 1970

[a] MAL = malformation; IDR = intrauterine death and/or resorption; IGR = intrauterine growth retardation; SB = stillborn; PGR = postnatal growth retardation; PND = postnatal death.
[b] Maternal toxic level.
[c] Dicyandiamide.

Other studies have also noted that teratogenic metals were reduced in potency by chelation, e.g., lead nitrate was less teratogenic in rats when given with ethylenediaminetetraacetic acid (EDTA), NTA, iminodiacetic acid (IDA), or penicillamine (PEN) (McClain and Siekierka, 1975), and sodium arsenate was less teratogenic in mice when given with 2,3-dimercaptopropanol (BAL) (Hood and Pike, 1972). In the foregoing examples it is assumed that the teratogenic metals were less available to the embryos by virtue of having complexed with chelating agents. Some of these and related chelating agents, however, have been found to be teratogenic themselves when given alone to pregnant animals, e.g., EDTA (Tuchmann-Duplessis and Mercier-Parot, 1956; Swenerton and Hurley, 1971; Kimmel, 1975), BAL (Nishimura and Takagaki, 1959), and DTPA (Sikov et al., 1975). Swenerton and Hurley (1971) demonstrated that the teratogenicity of the chelating agent was related to induced zinc deficiency and that supplemental zinc ingestion prevented the effect. Similarly, the embryolethal effects of disulfiram, a copper-chelating agent, was thought to create deficiency of this essential metal (Salgo and Oster, 1974). Not so readily explained is the observation of Ferm and Hanlon (1974) that copper citrate, which was already chelated when administered, was more teratogenic than inorganic copper sulfate.

VII. SOLVENTS

Solvents are of concern regarding possible embryotoxicity, not only because of their widespread industrial and household uses, but also because of their generally high rates of absorption via skin, lungs, and digestive tract. High absorption is particularly characteristic of fat solvents owing to their ready passage through cellular and other biological membrances, a feature which also undoubtedly enhances passage across the placenta. In a survey of sacral agenesis in man, Kucera (1968) noted that the mothers of 6 of 9 infants with this malformation had had appreciable exposure to organic solvents during pregnancy. It may be of more than incidental interest that two of the mothers, one with and one without occupational exposure to solvents, had diabetes, a disease involving elevated levels of acetone in blood and urine. It is further noteworthy that sacral agenesis is found more frequently than any other malformation among the offspring of diabetic mothers (Kucera et al., 1965; Rusnak and Driscoll, 1965).

In experimental animals fat solvents have not always been found to be highly teratogenic. Dimethylsulfoxide was reported to cause severe brain abnormalities in 11-day hamster embryos after treatment of pregnant females on day 8 (Ferm, 1966), but somewhat larger doses given to pregnant rats on gestation days 8–10 caused increased intrauterine death but no other embryotoxicity (Juma and Staples, 1967). This species difference was later confirmed, and in addition the mouse was found to be susceptible (Staples and

Pecharo, 1973). Chloroform administered orally to pregnant rats and rabbits at maternal toxic levels during organogenesis caused reduced birth weight but no other significant embryotoxicity (Thompson *et al.,* 1974). Administration of subanesthetic doses of this agent by inhalation (7 hr/day, days 6–15) to pregnant rats resulted in increased intrauterine death and resorption, intrauterine growth retardation, and a modest increase in malformations (Schwetz *et al.,* 1974c). To test the possibility that these effects were the result of maternal hepatotoxicity, two chlorinated solvents, carbon tetrachloride, (which is highly hepatotoxic) and 1,1-dichloroethane (which has little or no such action), were given to pregnant rats by inhalation (7 hr/day, days 6–15). Both caused modest developmental retardation and carbon tetrachloride caused a questionable increase in embryolethality, but neither was teratogenic (Schwetz *et al.,* 1974d). A nonchlorinated solvent, methyl ethyl ketone, similarly studied by the latter authors also caused a low incidence of malformations in addition to slight embryotoxicity of other types. A group of widely used chlorinated industrial solvents, trichloroethylene, perchloroethylene, methyl chloroform, and methylene chloride, was given by inhalation to pregnant rats and mice at twice the maximum allowable limits for human industrial exposure (7 hr/day, days 6–15). None caused significant embryotoxicity of any sort at this concentration, which was below the maternal toxic level (Schwetz *et al.,* 1975a). Other industrial solvents have been reported to cause some embryotoxicity, e.g., benzene (Watanabe *et al.,* 1968), xylene (Kucera, 1968), cyclohexanone (Weller and Griggs, 1973), propylene glycol (Gebhardt, 1968), alkane sulfonates (Hemsworth, 1968), and acetamides and formamides (Thiersch, 1962b). Data available at this writing, however, do not indicate that solvents in general or fat solvents in particular represent special embryotoxicity risks in man under current conditions of industrial use.

VIII. DETERGENTS

The extensive use of these synthetic cleansing and emulsifying agents in industry and in the home has occasioned some concern about their possible effects on several biological systems. A nonionic surfactant Triton W. R. 1339 was shown to be highly embryolethal and moderately teratogenic in mice, rats, and rabbits a number of years ago by Tuchmann-Duplessis and Mercier-Parot (1964). More recently the eutrophication of many streams and lakes, thought by some to be the result of accumulated phosphorus from detergents, led to the banning of phosphorus-containing detergents (see Shepard, 1972). A promising substitute ingredient, nitrilotriacetate (NTA), however, was discredited by an unsubstantiated report (Chernoff and Courtney, 1970) that NTA enhanced the teratogenicity and lethality of cadmium and mercury compounds. Several other investigators have failed to confirm this finding (Nolen *et al.,* 1972a,b; Scharpf *et al.,* 1972; Tjälve, 1972). Although NTA is now used in detergent manufacture in a number of countries, its status in the United States remains uncertain, in part owing to the earlier claims of embryotoxicity attributed to certain of its metal chelates.

Also confusing have been persistent reports from the laboratory of Mikami and associates that neutral detergents of the alkyl benzene sulfonate type have caused significant increases in teratogenicity among mice exposed during pregnancy. In these reports (Mikami and Sakai, 1973; Mikami *et al.*, 1969, 1973; Nagai, 1971; and in various news media) the route of administration, the dosage, and the types of malformations varied considerably. Nevertheless, the repeated claims that treated animals showed a higher rate of malformation than did controls were interpreted by this group to mean that exposure to small or moderate amounts of these detergents might involve human teratogenic risk.

These compounds, alkyl benzene sulfonate and linear alkyl benzene sulfonates in particular, have now been studied in other laboratories in Japan and elsewhere. Single- and multiple-generation studies in rats have revealed no effects on general reproduction attributable to chronic eating or drinking of these compounds (Bornmann and Loeser, 1973; Buehler *et al.*, 1971; Omori *et al.*, 1968; Tusing *et al.*, 1960). More intensive dosage during organogenesis, aimed particularly at teratogenic induction, has failed to show such effects in mice, rats, or rabbits, although dosage at or near maternal toxic levels caused some marginally significant reduction of intrauterine growth and increased resorption rates (Iimori *et al.*, 1973; Masuda *et al.*, 1973; Nolen *et al.*, 1974; Palmer *et al.*, 1975a,b; Charlesworth, 1975; Tanaka, 1966).

Thus, although the nonionic surfactant Triton W. R. 1339 is undoubtedly embryotoxic in mice, rats, and rabbits, other compounds in this general functional class have not been found to be significantly so after intensive study by experienced investigators. Chemicals used for cleansing and emulsifying, therefore, are not as a group suspect, although their extensive use and dissemination in the environment justifies close monitoring for effects on reproduction as well as other biological functions.

IX. FOOD ADDITIVES

Relatively few studies on the embryotoxicity of food additives have been published. The nature of their use as preservatives, colorants, flavor enhancers, sweeteners, etc., in human food naturally dictates that such substances have no more than negligible toxicity of any sort. They are usually tested in animals by administration in food or drinking water at low to moderate doses for extended periods to simulate presumed human consumption. Under such conditions little embryotoxicity could be expected. Attention has been focused on a few additives, however, because of concern about other toxicity, and consequently some of these have been tested more intensively at higher dosage given during gestation.

Artificial Sweeteners. Cyclamate given in the diet to rats was reported to have had some detrimental effects on postnatal development of the offspring in multigeneration studies (Nees and Derse, 1965; Ferrando and Huchet, 1968), but no prenatal effects were observed. High levels (5%) of sodium cyclamate or saccharin in the diet of mice during pregnancy, however, caused increased neonatal mortality and, when treatment was continued after birth, resulted in some postweaning growth retardation, although caloric intake was not adjusted to that of the controls (Lederer and Pottier-Arnould, 1969). A 10:1 cyclamate–saccharin mixture had no significant effect on reproduction or organogenesis at 500–2500 mg/kg in rats or rabbits (Vogin and Oser, 1969). Rats raised and bred on relatively low dosage of cyclamate (20 mg/kg/ day) were reported to have increased hyperactivity on an activity wheel as well as enhanced maze-learning ability (Stone *et al.*, 1969), although the latter was not significantly different from controls. These authors (Stone *et al.*, 1971) undertook to survey by questionnaire whether there were behavioral problems among the children of the users of artificial sweeteners and reported the most common problems to be "hyperactivity, nervous irritability, and extreme nervousness," but in view of the methods used such results must be regarded as no more than suggestive. Another possible association was noted between the use of sweeteners and the occurrence of lobster-claw defects and facial clefts in each of two cases from the same locality in New York state (Hillman and Fraser, 1969), but this has not been substantiated by any additional reports and in view of appreciable use of these additives during human pregnancy the association was probably coincidental. Intensive efforts to induce embryotoxicity in rhesus monkeys with large doses of sodium cyclamate (500 and 2000 mg/kg) at various times during organogenesis were unsuccessful (Wilson, 1972).

A toxic metabolite of cyclamate, cyclohexylamine, has been found to cause intrauterine growth retardation and embryolethality but no malformations when given to pregnant mice at 61–122 mg/kg during organogenesis (Becker and Gibson, 1970). Of 16 pregnant rhesus monkeys given this compound at 25–75 mg/kg-day at various 4-day periods during organogenesis, 1 aborted, and 4 produced moderately growth-retarded fetuses, but none had offspring with significant structural defects (unpublished data, author's laboratory). Dose-related chromosomal breakage was produced in adult rats given 1–50 mg/kg day intraperitoneally for 5 days (Legator *et al.*, 1969). Thus, available evidence does not indicate that saccharine, cyclamate, or a toxic metabolite of the latter has appreciable embryotoxicity at doses well in excess of ordinary use levels.

Sodium Glutamate. This food flavoring agent has received considerable attention because of its established neurotoxicity in fetal and early postnatal mice, pigs, and monkeys (Murakami *et al.*, 1972; Shimizu *et al.*, 1973; Filer, 1974). On careful examination, however, the lesions appear to consist of focal damage and repair of the classical pathologic type rather than to involve arrest or diversion of a developmental process. Therefore, it could be contended that this is pathology rather than teratology or embryotoxicity. In any event, exposure of perinatal animals to this additive does not entail the increased mortality and growth retardation usually associated with embryo–fetotoxic effects. Deficient postnatal functions of the central nervous and endocrine systems (Olney, 1969) have been described in mice after large doses, but these seem to be minimal or transitory in monkeys (Reynolds *et al.*, 1971; Olney and Sharpe, 1969). Thus, at present there is little evidence to indicate that sodium glutamate under current conditions of use entails risk to normal human development.

Red Dye No. 2. This food colorant has been extensively used for a number of years and was generally regarded as safe until reports from the Soviet Union suggested that amaranth might have caused increased intrauterine death in a small number of pregnant rats (Shtenberg and Gavrilenko, 1970). The published report lacked many details of procedure, and it is by no means certain that the compound tested was the same or was of comparable purity as that currently used in the food industry. A study at FDA using FD&C Red No. 2 (amaranth) showed this compound not to be teratogenic in rats, but it did appear to cause some increased fetal death when treated litters were compared with one set of controls which had an unusually low rate of intrauterine death (Collins and McLaughlin, 1972). A more recent report by the same authors (Collins and McLaughlin, 1973), however, showed a more typical intrauterine mortality rate for controls that was twice that previously reported, thus tending to place in doubt the significance of the earlier results. (Incidentally, this experience provides further support of the present author's contention that the litter-unit of data presentation involves more variability, and consequently is less reliable, than summation based on all fetuses in a treatment group.) Additional studies by other laboratories have confirmed the lack of teratogenicity of this colorant and, further, have established that no increase in intrauterine death occurs at doses as high as 200 mg/kg/day (Keplinger *et al.*, 1974; Holson *et al.*, 1974; and others).

X. MISCELLANEOUS ENVIRONMENTAL CHEMICALS

Phthalic Acid Esters. These are widely distributed in the environment because of their use as plasticizers in polyvinyl chloride, from which leaching is known to occur, and as vehicles for pesticides, cosmetics, industrial oils, and insect repellents. There is some indication that relatively large doses (0.3–10 ml/kg) may be embryotoxic in mice (Marx, 1972), and doses approximating $\frac{1}{3}$–$\frac{1}{10}$ of maternal LD_{50} given intraperitoneally caused definite growth retardation, increased intrauterine death, and malformations in the offspring of rats (Singh *et al.*, 1971).

Polybrominated Biphenyls. They are normally used as flame retardants but were accidentally introduced into livestock feed in Michigan (Anon., 1975) and were claimed to have caused deformities in the offspring, among other toxic manifestations. Pregnant mice given these compounds in the diet at 50–1000 ppm on days 7–18 produced growth retarded and malformed young (Corbett et al., 1975).

Polychlorinated Biphenyls. Fed at levels of 20–5000 ppm in diet to pregnant rats, they did not cause major malformations but caused significant intrauterine growth retardation at the higher doses which were maternal toxic (Shiota et al., 1974). Mixtures of these compounds fed to sows at 20 ppm of diet throughout pregnancy resulted in reduced litter size but no malformations (Hansen et al., 1975). A number of "cola-colored" infants with dark brown staining of skin at birth was traced to the consumption by their mothers of rice oil that had been contaminated with tetrachlorobiphenyl (Miller, 1971). Measurable amounts of these compounds were found in human fetal tissues (Shiota et al., 1973). They are regarded as potentially important environmental contaminants owing to their extensive use as lubricants, heat transfer agents, insulators, and ingredients of paints, varnishes, synthetic resins and waxes, although serious effects on reproduction or intrauterine development are not presently known.

Organopolysiloxanes. Mixtures of these, used in cosmetics and industry to impart water repellency and lubricity to surfaces, have been found to cause various reversible changes in reproductive organs in rats and rabbits given moderate to high doses (Palazzolo et al., 1972; LeFevre et al., 1972). When given during pregnancy at doses in excess of 200 mg/kg, embryotoxicity was evident in the form of increased intrauterine death and growth retardation but relatively few malformations were observed.

Aerosol Spray Adhesives. These were abruptly banned from sale in the United States in August 1973 on the basis of reports of human malformations and chromosome breaks associated with use of such products by pregnant women. Subsequent well-controlled epidemiologic and cytogenetic investigations of exposed women, however, showed no adverse effects that could be related to these products (Oakley et al., 1974; Hook et al., 1974a; Cervenka and Thorn, 1974). Experimental studies in pregnant hamsters also showed no untoward effects except some intrauterine growth retardation at maternal toxic doses (Murphy et al., 1975).

Tobacco Smoking. Smoking during pregnancy is now generally accepted as being associated with, if not responsible for, reduced birth weight of the infants (Butler et al., 1972; Hook et al., 1974b; Rush and Kass, 1972; Stevens et al., 1973; Lubs, 1973). Most recent authors have also reported increased neonatal mortality among children of heavy smokers, and some have found this effect and growth retardation to be more pronounced in blacks than in whites. Increased perinatal mortality, however, has been questioned by Yerushalmy (1973).

One study presented data suggesting that the incidence of congenital heart disease may be increased among children born to smoking mothers (Fedrick et al., 1971), but this has been refuted by Yerushalmy (1973). Two follow-up studies of children born to smoking mothers have been conducted.

Butler and Goldstein (1973) reported subsequent physical and mental retardation depending on the number of cigarettes smoked, whereas Hardy and Mellits (1972) reported that the small babies born to smokers were no longer significantly different in physical and intellectual measurements from the children of nonsmokers at 4 and 7 years of age. Lubs (1973) studied the influence of socioeconomic factors on prematurity among infants born to smokers and found that there was an effect but that it had less influence than either smoking or race. Nicotine is assumed to be the main detrimental agent in tobacco smoke because of its demonstrated embryotoxicity in mice (Nishimura and Nakai, 1958; Strudel and Gateau, 1973).

Other Chemicals. Sodium fluoride given orally to pregnant rats at 2–10 mg/kg on days 8–11 or 11–14 produced no malformations but was associated with retarded ossification of the skeleton of term fetuses (Goto, 1968). A negative correlation was found between hardness of water in Wales and the occurrence of human nervous system malformations (Lowe, 1971). Inhalation of carbon monoxide at 250 ppm for 7 or 24 hr/day on days 6–15 of gestation by pregnant mice and rabbits produced no embryotoxicity except minor skeletal variations in some offspring (Schwetz *et al.*, 1975b)

In view of the limited observations summarized in this chapter, it can hardly be maintained that any aspect of human development is seriously threatened at this time by chemical agents in the environment.

REFERENCES

Abramovich, A., and DeVoto, F. C. H., 1968, Anomalous maxillofacial patterns produced by maternal lathyrism in rat foetuses, *Arch. Oral Biol.* **13:**823–826.

Agnew, W. F., Fauvre, F. M., and Pudenz, R. H., 1968, Tellurium hydrocephalus: Distribution of [127]tellurium between maternal fetal and neonatal tissues of the rat, *Exp. Neurol.* **21:**120–132.

Agnew, W. F., Snyder, D., Yuen, T. G. H., and Cheng, J. T., 1973, Tellurium hydrocephalus: Accumulation and metabolic activity of tellurium in the choroid plexus, *Teratology* **7:**A–11.

Amin-Zaki, L., Elhassani, S., Majeed, M. A., Clarkson, T. W., Doherty, R. A., and Greenwood, M., 1974, Intrauterine methylmercury poisoning in Iraq, *Pediatrics* **54:**587–595.

An Der Lan, H., 1964, Toxikologische Probleme moderner Pflanzenschutzmittel, *Umschau Wiss. Tech.* **21:**649–652.

Angle, C. R., and McIntire, M. S., 1964, Lead poisoning during pregnancy, *Am. J. Dis. Child.* **108:**436–439.

Anon., 1975, Michigan's new poison, *The New Republic,* April 26.

Barr, M., 1973, The teratogenicity of cadmium chloride in two stocks of Wistar rats, *Teratology* **7:**237–242.

Beattie, A. D., Moore, M. R., Goldberg, A., Finlayson, M. J. W., Mackie, E. M., Graham, J. F., Main, J. C., McLaren, D. A., Murdock, R. M., and Stewart, G. T., 1975, Role of chronic low-level lead exposure in the aetiology of mental retardation, *Lancet* **1:**589–592.

Beaudoin, A. R., 1974, Teratogenicity of sodium arsenate in rats, *Teratology* **10:**153–158.

Becker, B. A., and Gibson, J. E., 1970, Teratogenicity of cyclohexylamine in mammals, *Toxicol. Appl. Pharmacol.* **17:**551–552.

Bennett, R. W., Persaud, T. V. N., and Moore, K. L., 1974, Teratological studies with aluminum in the rat, *Teratology* **9:**A–14.

Binns, W., and Balls, L., 1971, Nonteratogenic effects of 2,4,5-trichlorophenoxyacetic acid and 2,4,5-T propylene glycol butal esters herbicides in sheep, *Teratology* **4:**245.

Binns, W., James, L. F., Shupe, J. L., and Everett, G., 1963, A congenital cyclopian-type malformation in lambs induced by maternal ingestion of a range plant *Veratrum californicum, Am. J. Vet. Res.* **24:**1164–1175.

Binns, W., Ceuto, C., Eliason, B. C., Heggestad, H. E., Hepting, G. H., Sand, P. F., Stephes, R. F., and Tschirley, F. H., 1970, Investigation of spray project near Globe, Arizona, Investigation conducted February, 1970, U.S. Dept. of Agriculture.

Bornmann, G., and Loeser, A., 1973, Über eine waschaktive Substanz auf Basis Dodecylbenzolsulfonat, *Fette, Seifen, Anstrichm.* **65:**818.

Budreau, C. H., and Singh, R. P., 1973, Teratogenicity and embryotoxicity of demeton and fenthion in CF#1 mouse embryos, *Toxicol. Appl. Pharmacol.* **24:**324–332.

Buehler, E. V., Newmann, E. A., and King, W. F., 1971, Two-year feeding and reproduction study in rats with linear alkylbenzene sulfonate (LAS), *Toxicol. Appl. Pharmacol.* **18:**83–91.

Butler, N. R., and Goldstein, H., 1973, Smoking in pregnancy and subsequent child development, *Br. Med. J.* **4:**573–575.

Butler, N. R., Goldstein, H., and Ross, E. M., 1972, Cigarette smoking in pregnancy: Its influence on birth weight and perinatal mortality, *Br. Med. J.* **2:**127–130.

Butt, E. M., and Simonsen, D. G., 1950, Mercury and lead storage in human tissues, *Am. J. Clin. Pathol.* **20:**716.

Cantarow, A., and Trumper, M., 1944, *Lead Poisoning,* Williams and Wilkins, Baltimore.

Cervenka, J., and Thorn, H. L., 1974, Chromosomes and spray adhesives, *N. Engl. J. Med.* **290:**543–545.

Charlesworth, F. A., 1975, Studies on the teratogenicity of alkylbenzenesulfonates, *Bibra Bulletin* **14:**330–333.

Chernoff, N., and Courtney, K. D., 1970, Maternal and fetal effects of NTA, NTA and cadmium, NTA and mercury, NTA and nutritional imbalance in mice and rats, *in: Progress Report from the National Institute of Environmental Health Sciences,* Dec. 1, 1970, Research Triangle Park, North Carolina.

Chernoff, N., Kavlock, R. J., Kathrein, J. R., Dunn, J. M., and Haseman, J. K., 1975, Prenatal effects of dieldrin and photodieldrin in mice and rats, *Toxicol. Appl. Pharmacol.* **31:**302–308.

Chow, F. C., Hamar, D. W., and Udall, R. H., 1974, Mycotoxic effect on fetal development: Pine needle abortion in mice, *J. Reprod. Fertil.* **40:**203–204.

Clavert, J., and Gabriel-Robez, O., 1974, The effects on mouse gestation and embryo development of an injection of viper venom *(Vispera aspis), Acta Anat.* **88:**11–21.

Collins, T. F. X., and McLaughlin, J., 1972, Teratology studies on food colorings. I. Embryotoxicity of amaranth (FD & C Red No. 2) in rats, *Food Cosmet. Toxicol.* **10:**619–624.

Collins, T. F. X., and McLaughlin, J., 1973, Teratology studies on food colorings. II. Embryotoxicity of R salt and metabolites of amaranth (FD & C Red No. 2) in rats, *Food Cosmet. Toxicol.* **11:**355–365.

Collins, T. F. X., and Williams, C. H., 1971, Teratogenic studies with 2,4,5-T and 2,4-D in the hamster, *Bull. Environ. Contam. Toxicol.* **6:**559–567.

Collins, T. F. X., Hansen, W. H., and Keeler, H. V., 1971, The effects of carbaryl (Sevin) on reproduction of the rat and gerbil, *Toxicol. Appl. Pharmacol.* **19:**202–216.

Corbett, T. H., Beaudoin, A. R., and Cornell, R. G., 1975, Teratogenicity of polybrominated biphenyls, *Teratology* **11:**15A.

Corliss, C. E., Fedinec, A. A., and Robertson, G. G., 1966, Teratogenic effect of tetanus toxin on the CNS of the early chick embryo, *Anat. Rec.* **154:**221–232.

Courtney, K. D., and Moore, J. A., 1971, Teratology studies with 2,4,5-trichlorophenoxyacetic acid and 2,3,7,8-tetrachlorodibenzo-*p*-dioxin, *Toxicol. Appl. Pharmacol.* **20:**396–403.

Courtney, K. D., Gaylor, D. W., Hogan, M. D., Falk, H. L., Bates, R. R., and Mitchell, I., 1970, Teratogenic evaluation of 2,4,5-T, *Science* **168:**864–866.

Crowe, M. W., and Swerczek, T. W., 1974, Congenital arthrogryposis in offspring of sows fed tobacco *(Nicotiana tabacum), Am. J. Vet. Res.* **35:**1071–1073.

Cutting, R. T., Phuoc, T. H., Ballo, J. M., Benenson, M. W., and Evans, C. H., 1970, *Congenital Malformations, Hydatidiform Moles, and Stillbirths in the Republic of Vietnam 1960–1969,* U.S. Govt. Printing Office, Washington, D.C.

Diechmann, W. B., MacDonald, W. E., Beasley, A. G., and Cubit, D., 1971, Subnormal reproduction in beagle dogs induced by DDT and aldrin, *Ind. Med. Surg,* **40:**10–20.

DiPaolo, J. A., Elis, J., and Erwin, H. E., 1967, Teratogenic response by hamsters, rats, and mice to aflatoxin B_1, *Nature* **215:**638–639.

Dobbins, P. K., 1967, Organic phosphate insecticides as teratogens in the rat, *J. Fla. Med. Assoc.* **54:**452–456.

Dougherty, W. J., Goldberg, L., and Coulston, F., 1971, The effect of carbaryl on reproduction in the monkey *(Macaca mulatta), Toxicol. Appl. Pharmacol.* **19:**365.

Dwornik, J. J., and Moore, K. L., 1967, Congenital anomalies produced in the rat by podophyllin, *Anat. Rec.* **157:**237.

Earl, F. L., Miller, E., and Van Loon, E. J., 1973, Reproductive, teratogenic, and neonatal effects of some pesticides and related compounds in beagle dogs and miniature swine, *in: Pesticides and the Environment: A Continuing Controversy,* pp. 253–266, Inter-American Conference on Toxicol. and Occupational Medicine, Symposia Specialists, N. Miami.

Edmonds, L. D., Selby, L. A., and Case, A. A., 1972, Poisoning and congenital malformations associated with consumption of poison hemlock by sows, *J. Am. Vet. Med. Assoc.* **160:**1319–1324.

Emanuel, I., and Sever, L. E., 1973, Questions concerning the possible association of potatoes and neural-tube defects, and an alternative hypothesis relating to maternal growth and development, *Teratology* **8:**325–332.

Fabro, S., Smith, R. L., and Williams, R. T., 1965, Embryotoxic activity of some pesticides and drugs related to phthalimide, *Food Cosmet. Toxicol.* **3:**587–590.

Fedrick, J., Alberman, E. D., and Goldstein, H., 1971, Possible teratogenic effect of cigarette smoking, *Nature* **231:**529–530.

Ferm, V. H., 1963a, Colchicine teratogenesis in hamster embryos, *Proc. Soc. Exp. Biol. Med.* **112:**775–778.

Ferm, V. H., 1963b, Congenital malformations in hamster embryos after treatment with vinblastine and vincristine, *Science* **141:**426.

Ferm, V., H., 1966, Congenital malformations induced by dimethyl sulphoxide in the golden hamster, *J. Embryol. Exp. Morphol.* **16:**49–54.

Ferm, V. H., 1972, The teratogenic effects of metals on mammalian embryos, *in: Advances in Teratology* (D. H. M. Woollam, ed.), Vol. 5, pp. 51–75, Academic Press, New York.

Ferm, V. H., and Carpenter, S. J., 1967, Developmental malformations resulting from the administration of lead salts, *Exp. Mol. Pathol.* **7:**208–213.

Ferm, V. H., and Carpenter, S. J., 1968, Malformations induced by sodium arsenate, *J. Reprod. Fertil.* **17:**199–201.

Ferm, V. H., and Carpenter, S. J., 1970, Teratogenic and embryopathic effects of indium, germanium and gallium, *Toxicol. Appl. Pharmacol.* **16:**166–170.

Ferm, V. H., and Hanlon, D. P., 1974, Toxicity of copper salts in hamster embryonic development, *Biol. Reprod.* **11:**97–101.

Ferrando, R., and Huchet, B., 1968, Etude de l'activité éventuelle du cyclamate de soude sur le rat au cours de trois générations, *Bull. Acad. Natl. Med.* **152:**36–41.

Filer, L. J., 1974, Monosodium glutamate metabolism in the fetus and newborn, *in: Perinatal Pharmacology: Problems and Priorities* (J. Dancis and J. C. Hwang, eds.), pp. 177–193, Raven Press, New York.

Fish, S. A., 1966, Organophosphorus cholinesterase inhibitors and fetal development, *Am. J. Obstet. Gynecol.* **96:**1148–1154.

Franke, K. W., and Tully, W. C., 1935, A new toxicant occurring naturally in certain samples of plant foodstuffs. V. Low hatchability due to deformities in chicks, *Poult. Sci.* **14:**273–279.

Fujii, T., and Nishimura, H., 1972, Adverse effects of prolonged administration of caffeine on rat fetus, *Toxicol. Appl. Pharmacol.* **22:**449–457.

Gaines, T. B., Holson, J. F., Nelson, C. J., and Schumacher, H. J., 1975, Analysis of strain differences in sensitivity and reproducibility of results in assessing 2,4,5-T teratogenicity in mice, *Toxicol. Appl. Pharmacol.* **33:**174–175 (abstract).

Gale, T. F., 1974, Effects of chromium on the hamster embryo, *Teratology* **9:**A–17.

Gebhardt, D. O. E., 1968, The teratogenic action of propylene glycol (propanediol-1,2) and propanediol-1,3 in the chich embryo, *Teratology* **1:**153–162.

Gibson, J. E., and Becker, B. A., 1970, Placental transfer, embryotoxicity and teratogenicity of thallium sulfate in normal and potassium deficient rats, *Toxicol. Appl. Pharmacol.* **16:**120–132.

Good, E. E., and Ware, G. W., 1969, Effects of insecticides on reproduction in the laboratory mouse. IV. Endrin and dieldrin, *Toxicol. Appl. Pharmacol.* **14:**201–203.

Goto, K., 1968, Effects of sodium fluoride on the fetuses and sucklings in Wistar strain albino rats, *Proc. Congenital Anomolies Res. Assoc. Jpn.* **8:**46.

Gralla, E. J., and McIlhenny, H. M., 1972, Studies in pregnant rats, rabbits and monkeys with lithium carbonate, *Toxicol. Appl. Pharmacol.* **21:**428–433.

Grauwiler, J., and Schön, H., 1973, Teratological experiments with ergotamine in mice, rats, and rabbits, *Teratology* **7:**227–236.

Green, C. R., and Christie, G. S., 1961, Malformations in foetal rats induced by the pyrrolizidine alkaloid, heliotrine, *Br. J. Exp. Pathol.* **42:**369–378.

Hansen, L. G., Byerly, C. S., Metcalf, R. L., and Bevill, R. F., 1975, Effect of a polychlorinated biphenyl mixture on swine reproduction and tissue residues, *Am. J. Vet. Res.* **36:**23–26.

Harada, Y., 1968, Congenital Minamata disease, *in: Minamata Disease* pp. 73–91, Kumamota University Study Group of Minamata Disease, Japan.

Hardy, J. B., and Mellits, E. D., 1972, Does maternal smoking during pregnancy have a long-term effect on the child, *Lancet* **2:**1332–1336.

Harris, S. B., Wilson, J. G., and Printz, R. H., 1972, Embryotoxicity of methyl mercuric chloride in golden hamsters, *Teratology* **6:**139–142.

Hayes, A. W., Hood, R. D., and Lee, H. L, 1974, Teratogenic effects of ochratoxin A in mice, *Teratology* **9:**93–98.

Hemsworth, B. N., 1968, Embryopathies in the rat due to alkane sulphonates, *J. Reprod. Fertil.* **17:**325–334.

Hillman, D. A., and Fraser, F. C., 1969, Artificial sweeteners and fetal malformations: A rumored relationship, *Pediatrics* **44:**299–300.

Hirono, I., Shibuya, C., and Hayashi, K., 1969, Induction of a cerebellar disorder with cycasin in newborn mice and hamsters, *Proc. Soc. Exp. Biol. Med.* **131:**593–598.

Holmberg, R. E., and Ferm, V. H., 1969, Interrelationships of selenium, cadmium, and arsenic in mammalian teratogenesis, *Arch. Environ. Health* **18:**873–877.

Holson, J. F., Gaines, T. B., Schumacher, H. J., Nelson, C. J., and Gaylor, D. W., 1974, Teratogenic evaluation of the food coloring agent red dye #2 administered orally to rats, *Teratology* **9:**A-20–A-21.

Hood, R. D., and Pike, C. T., 1972, BAL alleviation of arsenate-induced teratogenesis in mice, *Teratology* **6:**235–238.

Hood, R. D., Innes, J. E., and Hayes, A. W., 1973, Effects of rubratoxin B on prenatal development in mice, *Bull. Environ. Contam. Toxicol.* **10:**200–207.

Hook, E., Hatcher, N. H., Brunson, P. S., Stanecky, O. J., Fisher, L., Feck, G., and Greenwald, P., 1974a, Negative outcome of a blind assessment of the association between spray adhesives exposure and human chromosome breakage, *Nature* **249:**165–166.

Hook, E. B., Selvin, S., Garfinkel, J., and Greenberg, M., 1974b, Maternal smoking during gestation and infant morphologic variation: Preliminary report concerning birth weight and incidence of transverse palmar creases, *in: Congenital Defects—New Directions in Research* (D. T. Janerich, R. G. Skalko and I. H. Porter, eds.), pp. 171–184, Academic Press, New York.

Iimori, M., Inoue, S., and Yano, K., 1973, Effect of detergent on pregnant mice by application on skin (preliminary report), *Oil Chem.* **22:**807.

James, L. F., Shupe, J. L., Binns, W., and Keeler, R. F., 1967, Abortive and teratogenic effects of locoweed on sheep and cattle, *Am. J. Vet. Res.* **28:**1379–1388.

James, L. F., Keeler, R. F., and Binns, W., 1969, Sequence in the abortive and teratogenic effects of locoweed fed to sheep, *Am. J. Vet. Res.* **30:**377–380.

Joneja, M. G., and LeLiever, W. C., 1974, Effects of vinblastine and podophyllin on DBA mouse fetuses, *Toxicol. Appl. Pharmacol.* **27**:408–414.

Jordan, R. L., and Borzelleca, J. F., 1975, Teratogenic studies with zinc omadine in swine, *Anat. Rec.* **181**:388.

Juma, M. B., and Staples, R. E., 1967, Effect of maternal administration of dimethyl sulfoxide on the development of rat fetuses, *Proc. Soc. Exp. Biol. Med.* **125**:567–569.

Kalow, W., and Marton, A., 1961, Second-generation toxicity of malathion in rats, *Nature* **192**:464–465.

Keeler, R. F., Binns, W., James, L. F., and Shupe, J. L, 1969, Preliminary investigation of the relationship between bovine congenital lathyrism induced by aminoacetonitrile and the lupine induced crooked calf disease, *Can. J. Comp. Med.* **33**:89–92.

Kennedy, G., Fancher, O. E., and Calandra, J. C., 1968, An investigation of the teratogenic potential of captan, folpet, and difolatan, *Toxicol. Appl. Pharmacol.* **13**:420–430.

Kennedy, G. L., Fancher, O. E., and Calandra, J. C., 1975, Nonteratogenicity of captan in beagles, *Teratology* **11**:223–226.

Keplinger, M. L., Wright, P. L., Plank, J. B., and Calandra, J. C., 1974, Teratogenic studies with FD & C Red no. 2 in rats and rabbits, *Toxicol. Appl. Pharmacol.* **28**:209–215.

Khera, K. S., 1973, Teratogenic effects of methylmercury in the cat: Note on the use of this species as a model for teratogenicity studies, *Teratology* **8**:293–304.

Khera, K. S., and Clegg, D. J., 1969, Perinatal toxicity of pesticides, *Can. Med. Assoc. J.* **100**:167–172.

Khera, K. S., and McKinley, W. P., 1972, Pre- and postnatal studies on 2,4,5-trichlorophenoxy-acetic acid, 2,4-dichlorophenoxyacetic acid and their derivatives in rats, *Toxicol. Appl. Pharmacol.* **22**:14–28.

Khera, K. S., and Whitta, L. L., 1968, Embryopathic effects of diquat and paraquat in rat, *Ind. Med. Surg.* **37**:553.

Kimbrough, R. D., and Gaines, T. B., 1968, Effect of organic phosphorus compounds and aklylating agents on the rat fetus, *Arch. Environ. Health* **16**:805–808.

Kimmel, C. A., 1975, Fetal gonad dysgenesis following EDTA administration, *Teratology* **11**:26A.

King, C. T. G., Horigan, E. A., and Wilk, A. L., 1971, Screening of the herbicides 2,4,5-T and 2,4-D for cleft palate production, *Teratology* **4**:233.

Klein, M. F., and Beall, J. R., 1972, Griseofulvin: A teratogenic study, *Science* **175**:1483–1484.

Kucera, J., 1968, Exposure to fat solvents: A possible cause of sacral agenesis in man, *J. Pediatr.* **72**:857–859.

Kucera, J., Lenz, W., and Maier, W., 1965, Missbildungen der Beine und der kaudalen Wirbel-säule bei Kindern diabetischer Mütter, *Dtsch. Med. Wochenschr.* **90**:901.

Lederer, J., and Pottier-Arnould, A. M., 1969, Toxicité du cyclamate de sodium et de la sac-charine pour la descendance des souris gestantes, *Diabete* **17**:103–106.

Le Fevre, R., Coulston, F., and Goldberg, L., 1972, Action of a copolymer of mixed phenyl-methylcyclosiloxanes on reproduction in rats and rabbits, *Toxicol. Appl. Pharmacol.* **21**:29–44.

Legator, M. S., Palmer, K. A., Green, S., and Petersen, K. W., 1969, Cytogenetic studies in rats of cyclohexylamine, a metabolite of cyclamate, *Science* **165**:1139–1140.

Leipold, H. W., Oehme, F. W., and Cook, J. E., 1973, Congenital arthrogryposis associated with ingestion of jimsonweed by pregnant sows, *J. Am. Vet. Med. Assoc.* **162**:1059–1060.

Little, D. A., and Hamilton, R. I., 1971, *Leucaena leucocephala* and thyroid function of newborn lambs, *Aust. Vet. J.* **47**:457–458.

Lowe, C. R., 1971, CNS malformations and the water supply, *Teratology* **4**:493.

Lubs, M.-L. E., 1973, Racial differences in maternal smoking effects on the newborn infant, *Am. J. Obstet. Gynecol.* **115**:66–76.

Marx, J. L., 1972, Phthalic acid esters: biological impact uncertain, *Science* **178**:46–47.

Masuda, F., Okamoto, K., and Inoue, K., 1973, Effects of linear alkyl benzene sulphonate applied on mouse skin during pregnancy of foetuses, *J. Food Hyg. Soc.* **14**:580.

Matsumoto, H. G., Goyo, K., and Takevchi, T., 1965, Fetal minamata disease. A neuropathological study of two cases of intrauterine intoxication by a methylmercury compound, *J. Neuropathol. Exp. Neurol.* **24:**563–574.

McCarthy, J. F., Fancher, O. E., Kennedy, G. L., Keplinger, M. L., and Calandra, J. C., 1971, Reproduction and teratology studies with the insecticide carbofuran, *Toxicol. Appl. Pharmacol.* **19:**370.

McClain, R. M., and Becker, B. A., 1972, Effects of organolead compounds on rat embryonic and fetal development, *Toxicol. Appl. Pharmacol.* **21:**265–274.

McClain, R. M., and Becker, B. A., 1975, Teratogenicity, fetal toxicity, and placental transfer of lead nitrate in rats, *Toxicol. Appl. Pharmacol.* **31:**72–82.

McClain, R. M., and Siekierka, J. J., 1975, The effects of various chelating agents on the teratogenicity of lead nitrate in rats, *Toxicol. Appl. Pharmacol.* **31:**434–442.

McLaughlin, J., Reynaldo, E. F., Lamar, J. K., and Marliac, J.-P., 1969, Teratology studies in rabbits with captan, folpet, and thalidomide, *Toxicol. Appl. Pharmacol.* **14:**59.

Menges, R. W., Selby, L. A., Marienfeld, C. J., Aue, W. A., and Greer, D. L., 1970, A tobacco-related epidemic of congenital limb deformities in swine, *Environ. Res.* **3:**285–302.

Meselson, M. S., Westing, A. H., and Constable, J. D., 1970, Background material relevant to presentations at the 1970 annual meeting of the AAAS, Herbicide Assessment Commission of the American Association for the Advancement of Science, revised January 14, 1971.

Mikami, Y., and Sakai, Y, 1973, Teratogenicity of some detergents, *Acta Anat. Nippon.* **48:**99.

Mikami, Y., Nagai, H., Sakai, Y., Fukushima, S., and Nishino, T., 1969, Effects of a detergent upon the development of the mouse embryo, *J. Congenital Anomolies* **9:**230(in Japanese).

Mikami, Y., Sakai, Y., and Miyamoto, I., 1973, Anomalies induced by ABS applied to the skin, *Teratology* **8:**98.

Miller, R. W., 1971, Cola-colored babies. Chlorobiphenyl poisoning in Japan, *Teratology* **4:**211–212.

Moore, J. A., 1973, Characterization and interpretation of kidney anomalies associated with 2,3,7,8-tetrachlorodibenzo-*p*-dioxin (TCDD), *Teratology* **7:**A-24.

Moore, J. A., and Courtney, K. K., 1971, Teratology studies with the trichlorophenoxyacid herbicides, 2,4,5-T and silvex, *Teratology* **4:**236.

Morton, M. S., and Elwood, P. C., 1974, C.N.S. malformations and trace elements in water, *Teratology* **10:**318.

Moxon, A. L., 1937, Alkali disease or selenium poisoning, *S.D. Agric. Exp. Stn. Bull.* **311:**1–91.

Murakami, U., Inouye, M., and Fujii, O., 1972, Brain lesions in the mouse fetus caused by maternal administration of monosodium glutamate (MSG), *Teratology* **6:**114.

Murphy, J. C., Collins, T. F. X., Black, T. N., and Osterberg, R. E., 1975, Evaluation of the teratogenic potential of a spray adhesive in hamsters, *Teratology* **11:**243–246.

Nagai, H., 1971, Doctoral thesis, Mie Municipal University, Mie, Japan.

Nees, P. O., and Derse, P. M., 1965, Feeding and reproduction of rats fed calcium cyclamate, *Nature* **208:**81–82.

Neubert, D., and Dillmann, I., 1972, Embryotoxic effects in mice treated with 2,4,5-trichlorophenoxyacetic-acid and 2,3,7,8-tetrachlorodibenzo-*p*-dioxin, *Naunyn Schmiedebergs Arch. Pharmakol.* **272:**243–264.

Nishimura, H., and Nakai, N., 1958, Developmental anomalies in offspring of pregnant mice treated with nicotine, *Science* **127:**877–878.

Nishimura, H., and Nakai, K., 1960, Congenital malformations in offspring of mice treated with caffeine, *Proc. Soc. Exp. Biol. Med.* **104:**140–142.

Nishimura, H., and Takagaki, S., 1959, Developmental anomalies in mice induced by 2,3-dimercaptopropanol (BAL), *Anat. Rec.* **135:**261–267.

Nolen, G. A., Buehler, E. V., Geil, R. G., and Goldenthal, E. I., 1972a, The effects of trisodium nitrilotriacetate (Na$_3$NTA) on cadmium and methyl-mercury toxicity and teratogenicity in rats, *Teratology* **5:**264.

Nolen, G. A., Buehler, E. V., Geil, R. G., and Goldenthal, E. I., 1972b, Effects of trisodium nitrilotriacetate on cadmium and methyl mercury toxicity and teratogenicity in rats, *Toxicol. Appl. Pharmacol.* **23:**222–237.

Nolen, G. A., Klusman, L. W., and Patrick, L. F., 1974, A teratology study of a mixture of tallow alkyl ethoxylated sulfate and linear alkane sulfonate in rats and rabbits, *Teratology* **10:**93–94.

Nolen, G. A., Patrick, L. F., and Dierckman, T. A., 1975, A percutaneous teratology study of zinc pyrithione in rabbits, *Toxicol. Appl. Pharmacol.* **31:**430–433.

Oakley, G. P., Nissim, J. E., Hanson, J. W., Boyce, J. M., and Roberts, M., 1974, Epidemiologic investigations of possible teratogenicity of spray adhesives, *Teratology* **9:**A-31–A-32.

Ogi, D., and Hamada, A., 1965, Case report on fetal death and malformation of extremities probably related to insecticide poisoning, *J. Jpn. Obstet. Gynecol. Soc.* **17:**569.

O'Leary, J. A., Davies, J. E., and Feldman, M., 1970, Spontaneous abortion and human pesticide residues of DDT and DDE, *Am. J. Obstet. Gynecol.* **108:**1291–1292.

Olney, J. W., 1969, Brain lesions, obesity, and other disturbances in mice treated with monosodium glutamate, *Science* **164:**719–721.

Olney, J. W., and Sharpe, L. G., 1969, Brain lesions in an infant rhesus monkey treated with monosodium glutamate, *Science* **166:**386–388.

Omori, Y., Kuwamura, T., Tanaka, S., Kawashima, K., and Nakaura, S., 1968, Pharmacological studies on alkyl benzene sulfonate (2): Effects of oral administration during pregnancy on foetus and newborn in mice and rats, *Food Hyg. Soc.* **9:**473.

Ornoy, A., and Altshuler, G., 1975, Placental mediated endotoxin rat embryopathy, *Anat. Rec.* **181:**441.

Ottoboni, A., 1972, Effect of DDT on the reproductive life-span in the female rat, *Toxicol. Appl. Pharmacol.* **22:**497–502.

Ottolenghi, A. D., Haseman, J. K., and Suggs, F., 1974, Teratogenic effects of aldrin, dieldrin, and endrin in hamsters and mice, *Teratology* **9:**11–16.

Palazzolo, R. J., McHard, J. A., Hobbs, E. J., Fancher, O. E., and Calandra, J. C., 1972, Investigation of the toxicologic properties of a phenylmethylcyclosiloxane, *Toxicol. Appl. Pharmacol.* **21:**15–28.

Palmer, A. K., Readshaw, M. A., and Neuff, A. M., 1975a, Assessment of the teratogenic potential of surfactants. I. LAS, AS and CLD, *Toxicology* **3:**91–106.

Palmer, A. K., Readshaw, M. A., and Neuff, A. M., 1975b, Assessment of the teratogenic potential of surfactants. II. AOS, *Toxicology* **3:**107–113.

Palmisano, P. A., Sneed, R. C., and Cassady, G., 1969, Untaxed whiskey and fetal lead exposure, *J. Pediatr.* **75:**869.

Pearn, J. H., and Hegarty, M. P., 1970, Indospicine—the teratogenic factor from *Indigofera spicata* extract causing cleft palate, *Br. J. Exp. Pathol.* **51:**34–36.

Penn, G. B., Ross, J. W., and Ashford, A., 1971, The effects of Arvin on pregnancy in the mouse and the rabbit, *Toxicol. Appl. Pharmacol.* **20:**460–473.

Persaud, T. V. N., 1968, Teratogenic effects of hypoglycin-A, *Nature* **217:**471.

Petrova-Vergieva, T., 1971, Anomalies induced by dithiocarbamic fungicides in rats, *Teratology* **4:**497–498.

Poswillo, D. E., Sopher, D., and Mitchell, S., 1972, Experimental induction of foetal malformations with "blighted" potato: A preliminary report, *Nature* **239:**462–464.

Rapport från en expert grupp, 1971, Fenoxisyror, granskuing av aktuell information, Giftnämnden, Stockholm.

Report of the 2,3,5-T Advisory Committee, 1971, Submitted May 7, 1971, to William D. Ruckelshaus, Administrator, Environmental Protection Agency, Washington, D.C.

Reynolds, W. A., Lemkey-Johnston, N., Filer, L. J., and Pitkin, R. M., 1971, Monosodium glutamate: Absence of hypothalamic lesions after ingestion by newborn primates, *Science* **172:**1342–1344.

Robens, J. F., 1969, Teratologic studies of carbaryl, diazinon, norea, disulfiram and thiram in small laboratory animals, *Toxicol. Appl. Pharmacol.* **15:**152–163.

Robens, J. F., 1970, Teratogenic activity of several phthalimide derivatives in the golden hamster, *Toxicol. Appl. Pharmacol.* **16:**24–34.

Robertson, D. S. F., 1970, Selenium—a possible teratogen?, *Lancet* **1:**518–519.

Rooney, J. R., 1966, Contracted foals, *Cornell Vet.* **56:**172–187.

Ruch, J. V., and Gabriel-Robez-Kremer, O., 1962, Action tératogène du venin de Naja sur l'embryon de poulet, *C.R. Soc. Biol.* **156**:1508–1509.

Ruddick, J. A., Harwig, J., and Scott, P.M., 1974, Nonteratogenicity in rats of blighted potatoes and compounds contained in them, *Teratology* **9**:165–168.

Rush, D., and Kass, E. H., 1972, Maternal smoking: a reassessment of the association with perinatal mortality, *Am. J. Epidemiol.* **96**:183–196.

Rusnak, S. L., and Driscoll, S. G., 1965, Congenital spinal anomalies in infants of diabetic mothers, *Pediatrics* **35**:989.

Salgo, M. P., and Oster, G., 1974, Fetal resorption induced by disulfiram in rats, *J. Reprod. Fertil.* **39**:375–377.

Scharpf, L. G., Hill, I. D., Wright, P. L., Plant, J. B., Keplinger, M. L., and Calandra, J. C., 1972, Effects of sodium nitrilotriacetate on toxicity, teratogenicity and tissue distribution of cadmium, *Nature* **239**:231–234.

Scharpf, L. G., Hill, I. D., Wright, P. L., and Keplinger, M. L., 1973, Teratological studies on methylmercury hydroxide and nitrilotriacetate sodium in rats, *Nature* **241**:461–463.

Schroeder, H. A., and Mitchener, M., 1971, Toxic effects of trace elements on the reproduction of mice and rats, *Arch. Environ. Health* **23**:102–106.

Schwetz, B. A., Sparschu, G. L., and Gehring, P. J., 1971, The effect of 2,4-dichlorophenoxyacetic acid (2,4-D) and esters of 2,4-D on rat embryonal, foetal and neonatal growth and development, *Food Cosmet. Toxicol.* **9**:801–817.

Schwetz, B. A., Keeler, P. A., and Gehring, P. J., 1974a, The effect of purified and commercial grade pentachlorophenol on rat embryonal and fetal development, *Toxicol. Appl. Pharmacol.* **28**:151–161.

Schwetz, B. A., Keeler, P. A., and Gehring, P. J., 1974b, Effect of purified and commercial grade tetrachlorophenol on rat embryonal and fetal development, *Toxicol. Appl. Pharmacol.* **28**:146–150.

Schwetz, B. A., Leong, B. K. J., and Gehring, P. J., 1974c, Embryo-and fetotoxicity of inhaled chloroform in rats, *Toxicol. Appl. Pharmacol.* **28**:442–451.

Schwetz, B. A., Leong, B. K. J., and Gehring, P. J., 1974d, Embryo-and fetotoxicity of inhaled carbon tetrachloride, 1,1-dichloroethane and methyl ethyl ketone in rats, *Toxicol. Appl. Pharmacol.* **28**:452–464.

Schwetz, B. A., Leong, B. K. J., and Gehring, P. J., 1975a, The effect of maternally inhaled trichloroethylene, perchloroethylene, methyl chloroform, and methylene chloride on embryonal and fetal development in mice and rats, *Toxicol. Appl. Pharmacol.* **32**:84–96.

Schwetz, B. A., Leong, B. K. J., and Staples, R. E., 1975b, Teratology studies on inhaled carbon monoxide and imbibed ethanol in laboratory animals, *Teratology* **11**:33A.

Scott, F. W., de LaHunta, A., Schultz, R. D., Bistner, S. I., and Riis, R. C., 1975, Teratogenesis in cats associated with griseofulvin therapy, *Teratology* **11**:79–86.

Selby, L. A., Menges, R. W., Houser, E. C., Flatt, R. E., and Case, A. A., 1971, Outbreak of swine malformations associated with the wild black cherry, *Prunus serotina, Arch. Environ. Health* **22**:198–201.

Sever, J. L., 1973, Potatoes and birth defects: Summary, *Teratology* **8**:319–320.

Shepard, T. H., 1972, Nitrilotriacetate (NTA) in detergents and human health, *Teratology* **6**:127–128.

Sherman, H., Culik, R., and Jackson, R. A., 1975, Reproduction, teratogenic, and mutagenic studies with benomyl, *Toxicol. Appl. Pharmacol.* **32**:305–315.

Shimizu, K., Mizutani, A., and Inoue, M., 1973, Electron microscopic studies on the hypothalamic lesions in the mouse fetus caused by monosodium glutamate, *Teratology* **8**:105.

Shiota, K., Tanimura, T., Nishimura, H., Mizutani, T., and Matsumoto, M., 1973, Polychlorinated biphenyls and DDE in human fetal tissues: A preliminary report, *Teratology* **8**:105.

Shiota, K., Tanimura, T., and Nishimura, H., 1974, Effects of polychlorinated biphenyls on pre- and postnatal development in rats, *Teratology* **10**:97.

Shoji, R., and Makino, S., 1966, Preliminary notes on the teratogenic and embryocidal effects of colchicine on mouse embryos, *Proc. Jpn. Acad.* **42**:822–827.

Shtenberg, A. I., and Gavrilenko, Y. V., 1970, Influence of the food dye amaranth upon the reproductive function and development in tests on albino rats, *Vopr. Pitan.* **29**:66–73.

Shupe, J. L., Binns, W., James, L. F., and Keeler, R. F., 1967, Lupine, a cause of crooked calf disease, *J. Am. Vet. Med. Assoc.* **151**:198–201.

Sikov, M. R., Smith, V. H., and Mahlum, D. D., 1975, Embryotoxicity of the calcium and zinc salts of diethylenetriaminepentaacetic acid (DTPA) in Wistar rats, *Teratology* **11**:34A.

Singh, A. R., Lawrence, W. H., and Autian, J., 1971, Teratogenicity of a group of phthalate esters in rats, *Toxicol. Appl. Pharmacol.* **19**:372.

Slonitskaya, N. N., 1969, Teratogenic effects of griseofulvin forte on rat fetus, *Antibiotiki* **14**:44–48.

Smalley, H. E., Curtis, J. M., and Earl, F. L., 1968, Teratogenic action of carbaryl in beagle dogs, *Toxicol. Appl. Pharmacol.* **13**:392–403.

Snyder, R. D., 1971, Congenital mercury poisoning, *N. Engl. J. Med.* **284**:1014–1016.

Sparschu, G. L., Dunn, F. L., Lisowe, R. W., and Rowe, V. K., 1971a, Effects of high levels of 2,4,5-trichlorophenoxyacetic acid on fetal development in the rat, *Food Cosmet. Toxicol.* **9**:527–530.

Sparschu, G. L., Dunn, F. L., and Rowe, V. K., 1971b, Study of the teratogenicity of 2,3,7,8-tetrachlorodibenzo-*p*-dioxin in the rat, *Teratology* **4**:247.

Spyker, J. M., and Smithberg, M., 1972, Effects of methylmercury on prenatal development in mice, *Teratology* **5**:181–189.

Staples, R. E., and Pecharo, M. M., 1973, Species specificity in DMSO teratology, *Teratology* **8**:238.

Steffek, A. J., and Hendrickx, A. G., 1972, Lathyrogen-induced malformations in baboons: A preliminary report, *Teratology* **5**:171–179.

Stevens, L. H., Cope, I., Sutherland, R., Skelsey, S., Vinson, A., and Lancaster, P., 1973, Cigarette smoking and pregnancy outcome, *in: 4th International Conference on Birth Defects* (A. G. Motulsky and F. J. G. Ebling, eds.), p. 55, Excerpta Medica, Amsterdam.

Stone, D., Matalka, E., and Riordan, J., 1969, Hyperactivity in rats bred and raised on relatively low amounts of cyclamates, *Nature* **224**:1326–1328.

Stone, D., Matalka, E., and Pulaski, B., 1971, Do artificial sweeteners ingested in pregnancy affect the offspring, *Nature* **231**:53.

Strudel, G., and Gateau, G., 1973, Action tératogène du sulfate de nicotine sur l'embryon de poulet, *Teratology* **8**:239.

Swenerton, H., and Hurley, L., 1971, Teratogenic effects of a chelating agent and their prevention by zinc, *Science* **173**:62–64.

Szabo, K. T., 1970, Teratogenic effect of lithium carbonate in the fetal mouse, *Nature* **225**:73–75.

Tanaka, R., 1966, The toxicity (FLD$_{50}$) of detergents, *Jpn. J. Public Health* **13**:230.

Tanimura, T., 1972, Effects on macaque embryos of drugs reported or suspected to be teratogenic to humans, *in: The Use of Non-Human Primates in Research on Human Reproduction* (E. Diczfalusy and C. C. Standley, eds.), pp. 293–308, WHO Research and Training Center on Human Reproduction, Stockholm.

Tanimura, T., Katsuya, T., and Nishimura, H., 1967, Embryotoxicity of acute exposure to methyl parathion in rats and mice, *Arch. Environ. Health* **15**:609–613.

Thiersch, J. B., 1962a, Effects of lipopolysaccharides of gram negative bacilli in the rat litter *in utero*, *Proc. Soc. Exp. Biol. Med.* **109**:429–441.

Thiersch, J. B., 1962b, Effects of acetamides and formamides on the rat litter *in utero*, *J. Reprod. Fertil.* **4**:219–220.

Thompson, D. J., Emerson, J. L., and Sparschu, G. L., 1971, Study of the effects of 2,4,5-trichlorophenoxyacetic acid (2,4,5-T) on rat and rabbit fetal development, *Teratology* **4**:243.

Thompson, D. J., Warner, S. D., and Robinson, V. B., 1974, Teratology studies on orally administered chloroform in the rat and rabbit, *Toxicol. Appl. Pharmacol.* **29**:348–357.

Tjälve, H., 1972. A study of the distribution and teratogenicity of nitrilotriacetic acid (NTA) in mice, *Toxicol. Appl. Pharmacol.* **23**:216–221.

Treon, J. F., and Cleveland, F. P., 1955, Toxicity of certain chlorinated hydrocarbon insecticides for laboratory animals, with special reference to aldrin and dieldrin, *J. Agric. Food Chem.* **3**:402–408.

Tusing, T. W., Paynter, O. E., and Opdyke, D. L., 1960, The chronic toxicity of sodium alkyl benzene sulphonate by food and water administration to rats, *Toxicol. Appl. Pharmacol.* **2**:464.

Tuchmann-Duplessis, H., and Mercier-Parot, L., 1956, Influence d'un corps de chelation, l'acide ethylenediaminetetraacétique sur la géstation et le développement foetal du rat, *C.R. Acad. Sci.* **243**:1064–1066.

Tuchmann-Duplessis, H., and Mercier-Parot, L., 1964, Avortements et malformations sans l'effet d'un agent provoquant une hyperlipémie et une hypercholesterolémie, *Bull. Acad. Natl. Med.* **148**:392–398.

Vogin, E. E., and Oser, B. L., 1969. Effects of a cyclamate:saccharin mixture on reproduction and organogenesis in rats and rabbits, *Fed. Proc.* **28**:743.

Vondruska, J. F., Fancher, O. E., and Calandra, J. C., 1971, An investigation into the teratogenic potential of captan, folpet, and difolatan in nonhuman primates, *Toxicol. Appl. Pharmacol.* **18**:619–624.

Watanabe, G., Yoshida, S., and Hirose, K., 1968, Teratogenic effect of benzol in pregnant mice, *Proc. Congenital Anomolies Res. Assoc. Jpn.,* 8th Annual Meeting, Tokyo, p. 45.

Weil, C. S., Woodside, M. D., Carpenter, C. P., and Smyth, H. F., 1972, Current status of tests of carbaryl for reproductive and teratogenic effect, *Toxicol. Appl. Pharmacol.* **21**:390–404.

Weller, E. M., and Griggs, J. H., 1973, The covert embryopathic effect of noxious vapors, *Teratology* **7**:A-30.

Wilson, J. G., 1972, Abnormalities of intrauterine development in non-human primates, *Acta Endocrinol. (Suppl.)* **166**:261–292.

Wright, T. L., Hoffman, L. H., and Davies, J., 1971, Teratogenic effects of lithium in rats, *Teratology* **4**:151–156.

Yasuda, M., 1972, Teratologic evaluation of tsumacide (*m*-tolyl-*N*-methylcarbamate) in the rat, *Botyu-Kagaku* **37**:161–165.

Yasuda, M., and Maeda, H., 1972, Teratogenic effects of 4-chloro-2-methylphenoxyacetic acid ethylester (MCPEE) in rats, *Toxicol. Appl. Pharmacol.* **23**:326–333.

Yasuda, Y., Kehara, T., and Nishimura, H., 1974, Embryotoxic effects of feeding bracken fern *(Pleridium aquilinum)* to pregnant mice, *Toxicol. Appl. Pharmacol.* **28**:264–268.

Yerushalmy, J., 1973, Congenital heart disease and maternal smoking habits, *Nature* **242**:262.

Maternal Metabolic and Endocrine Imbalances

10

THOMAS H. SHEPARD

I. INTRODUCTION

An understanding of how maternal metabolic and endocrine disturbances alter embryonic and fetal development requires a knowledge of the control systems that exist in the mother and in the fetus and how these interact with each other. Physiologic and metabolic conditions in the perinatal period are excluded and emphasis is placed on the first half of gestation. Each topic to be covered will be introduced by a brief discussion of existing information on fetal function, the interrelation between maternal and fetal system, and clinical and experimental observations.

II. ENDOCRINE IMBALANCES

A. Thyroid

1. Onset of Function in the Fetus

The proliferation and evagination of the ventral foregut area into the anlage of the thyroid occurs during the early somite stage when the embryo is 2–3 mm in length (Norris, 1918; Orts-Llorca and Galvez, 1958; Shepard, 1965). This site is later marked by the foramen cecum at the base of the tongue. The early stages of thyroid development are associated with marked cell hyperplasia, and there is no evidence of endocrine function. The weight of the gland in relation to body weight increases sharply until the onset of

THOMAS H. SHEPARD · Central Laboratory for Human Embryology, Department of Pediatrics, University of Washington, Seattle, Washington 98105. Supported in part by NIH grants HD005961, HD00836, and DE02918.

function at about the 65-mm crown–rump stage (Shepard *et al.,* 1964). After the onset of function the ratio of thyroid to body weight remains approximately 0.0004 which is very close to the relative size in the adult.

Function begins in the human thyroid at 70 to 80 days of gestation and is closely associated with the appearance of colloid-containing follicles and the ability to concentrate iodide (Shepard, 1967). Although some evidence exists that the attainment of iodide concentration may precede the synthesis of iodothyronines in the chick (Trunnel and Wade, 1955), this dissociation has not been observed in mammals or man. Iodotyrosines are detectable before thyroxine in the fetal rat (Geloso, 1961; Nataf and Sfez, 1961), but the sequential appearances of the appropriate synthesizing enzymes have not been demonstrated. Using organ culture, Shepard (1967) demonstrated the presence of colloid cavities, iodide concentration, and thyroxine synthesis in a human fetus as small as 68-mm crown–rump length. Olin *et al.* (1970) found no iodide trapping in the early period, and they did not detect thyroglobulin synthesis in three thyroids from fetuses of 30–55 mm crown–rump length.

2. Relation of Maternal to Fetal Feedback Control

The appearance of thyroid-stimulating hormone has been detected as early as the 111-mm crown–rump stage (14.0 gestational weeks) (Gitlin and Biasucci, 1969). It is unlikely that a negative feedback control system exists in the fetus before mid-gestation because Shepard and H. Andersen (unpublished) were unable to observe any fetal thyroid hypertrophy after mothers were treated with propylthiouracil or *l*-goitrin.

Fisher *et al.* (1970) have measured thyroxine and thyroid-stimulating hormone in fetal serum and report an abrupt rise in TSH between 18 and 22 weeks, suggesting a maturation of the pituitary–thyroid axis. The presence of such a system in late gestation has been well demonstrated experimentally by Jost (1957), and thyroid hypertrophy in the newborn of treated hyperthyroid mothers is well known. However, the normal weight at birth of thyroid glands from anencephalic monsters lacking an activating hypotholamic–pituitary axis suggests that the fetal pituitary contributes little toward the normal growth of the thyroid (Kind, 1962). Neither the presence or role of thyrotropin-releasing factor in the fetus is known, but in the rat it may cross the placenta (Kajihara *et al.,* 1972).

3. Athyrotic Cretinism Due to Radioiodine Therapy of the Mother

Iodine-131 if given in millicurie doses can damage or ablate the developing thyroid of the human fetus. Hypothyroidism, either congenital or of late onset, has been reported in at least five children whose mothers were treated with ^{131}I during pregnancy (Russell *et al.,* 1957; Shepard, 1976). Green *et al.,* (1971) reported delayed onset of cretinism in the offspring of a mother who received 9.0 mCi of ^{131}I after the 75th gestational day. Two of these patients developed degenerative types of central nervous system disorder. It should be emphasized that no deleterious fetal effects have been shown following diagnostic radioiodine usage. There is evidence that the fetal thyroid may accumulate higher concentrations of iodine than the mother (Beierwaltes *et al.*, 1976; Pickering and Kontaxis, 1961), and this may have public health implications in the event of large population exposures.

4. Athyrotic Cretinism Due to Unknown Causes

Athyrotic cretinism is associated with complete absence of the thyroid gland (Fig 1A and B). It occurs sporadically in all populations but is less common in the Negro race. This congenital defect is singularly localized to the thyroid gland. When mental retardation is associated usually there has been a delay beyond 4 months in the institution of therapy. During development parathyroid III may be carried into the mediastinum with the thymus while parathyroid IV usually remains in the cervical area. Sporadic cretinism is probably unrelated to the lingual or sublingual aberrant thyroid because the latter is associated predominantly with females while in the former the sex ratio is equal. Another dissimilarity is the incidence of phenylthiocarbamide non-tasting which is not increased in the aberrant thyroid group (Shepard and Andersen, 1965).

Based on the finding of a significant increase in the incidence for non-tasting of phenylthiocarbamide (PTC), a recessively inherited state, it has been suggested that this genetic trait may cause an increased susceptibility of the fetal thyroid to damage by environmental agents such as *l*-goitrin in the diet (Shepard, 1961). *l*-Goitrin, a bitter-tasting antithyroid substance present in certain vegetables is also not tasted by PTC non-tasters. The recorded low incidence of cretinism in the Negro matches an equally low incidence of PTC non-tasting in this race. Unfortunately no further clinical support for this theory has appeared, and the administration of the bitter tasting thioamides have never produced athyrosis in experimental animals.

Another possible cause of athyrotic cretinism was suggested when Blizzard observed a 25% incidence of thyroid antibodies in mothers giving birth to athyrotic infants (Chandler *et al.*, 1962a). Whether the antibodies crossed the placenta and ablated the fetal thyroid or were the secondary result of maternal and fetal thyroid damage has not been determined. Fetal thyroid damage following maternal immunization in the rabbit has not been demonstrated (Chandler *et al.*, 1962b).

5. Endemic Goiter Cretinism

Congenital hypothyroidism associated with deafness and mental retardation is commonly found in the offspring of goitrous mothers residing for long

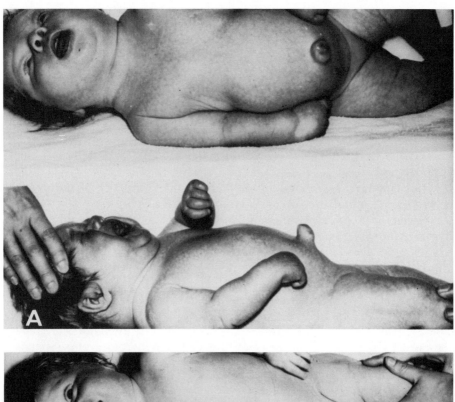

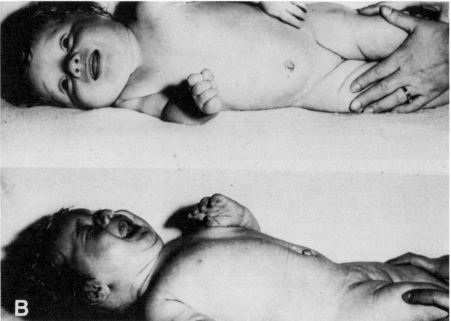

Fig. 1. Athyrotic cretinism due to unknown cause: (A) typical picture at 3 months of age, large tongue, mottled skin, and umbilical hernia. (B) Same infant 10 weeks after dessicated thyroid treatment was started. At age 10 her psychological testing was normal.

periods in iodine-deficient areas. Warkany (1971) has reviewed the subject in detail. The condition differs from athyrotic cretinism in that there is a variable amount of thyroid tissue present and the radioiodine uptake is high. Deafness persists in spite of thyroid replacement therapy. Developmental changes in the brain and cerebellum have been described by Lotmar (1931). Introduction of iodine into the maternal diet initiates prevention of this form of cretinism.

Experimental models of the condition have been produced by use of ^{131}I in pregnant mice and dogs. In the mice treatment after the 17th gestational day produced offspring with thyroids which contained fibrosis and follicular degeneration but no functional changes of hypothyroidism were observed in the animals (Speert et al., 1951). The puppies were hypothyroid, and their thyroids were replaced mostly by fatty tissue (Smith et al., 1951).

6. Maternal Hypothyroidism

Even severely hypothyroid mothers usually deliver normal infants (Hodges et al., 1952). Man (1972) has reported an extensive study on the effects of maternal thyroid function in 1394 women. These women with uncomplicated pregnancies had their butanol-extractable iodine (BEI) measured. In 131 instances the values were below normal (less than 5.5 $\mu g/100$ ml), although few were clinically hypothroid. The group was divided into three subgroups: I, euthyroid; II, hypothyroxinemic treated with dessicated thyroid, and III, hypothyroxinemic not given adequate therapy. The progeny were tested psychologically at 8 months and 4 and 7 years of age; the children born from mothers were low BEI levels and not treated (III) had lower mental scores than the other two groups. Other factors determining whether or not a mother fell into group II or III could have contributed toward these differences.

Experimental hypothyroidism in maternal thyroidectomized dogs (Carr et al., 1959) was found to produce little or not change in thyroid function of the offspring. Hamburgh et al., (1962) hypophysectomized rats on the 12th day of gestation and found no anatomic changes in the pups' thyroids. They also provided some evidence that injected thyroxine could cross the placenta and prevent propylthiouracil goiter in the fetus.

7. Experimental Studies

a. Mechanism of Action. The role, if any, of fetal thyroid hormones in prenatal development is not clear. The athyrotic cretin, except for moderate retardation of secondary epiphyseal ossification, is well developed. Maternal thyroid hormones equilibrate slowly across the human placenta at term, and it is assumed that they are equally available to the embryo and early fetus. Experimental studies in the thyroid-ablated rat have indicated there is a reduction in the size of the fetal cerebellum (Legrand, 1971), and Andersen (1971) has reported clinical signs of cerebellar dysfunction. *In vitro* studies of cerebellum have been used to show that myelinogenesis is stimulated by thyroxine (Hamburgh, 1966). Other tissue culture and organ culture effects of thyroid hormones are reviewed by Lasnitski (1965).

b. Teratogenic Studies. Production of congenital defect of the lens in the offspring of thyroidectomized rats has been reported (Langman and Van Faassen, 1955). Administration of thyroxine was associated by Giroud and Rothschild (1951) with cataracts in rat offspring. Beaudoin (1968) has found some preparations of thyrotropic hormone to be somewhat teratogenic and felt the effect might be related to the method used for preparing the material from bovine serum.

B. Pituitary Hormones

1. Hormone Levels in the Fetus

a. Appearance of Follicle-Stimulating Hormone in the Male and Female Fetus. Some evidence for sexual dimorphism has been found in the human female fetus. Levina (1971), using explanted pituitaries with hypothalamic homogenates, showed that FSH-stimulating ability of the female hypothalmus begins earlier (19–29 weeks of development) than that of the male hypothalamus (24–29 weeks). Grumbach and Kaplan (1973) have shown higher FSH contents in the female pituitary than in the male, although the serum levels did not differ in the two groups. The serum levels approached those of the castrate after 100 gestational days and subsequently decreased.

The time of appearance of gonadotropin-releasing hormone in the fetal hypothalmus and the mechanisms of its delivery and action are of real interest, and hopefully results of such studies will soon be available. Either unrestrained secretion of hypotholamic-releasing factor or autonomous secretion of gonadotropin could account for the high FSH levels in the midterm fetus.

b. Other Pituitary Hormones. Somatotropin has been studied in the fetal pituitary and serum by Kaplan *et al.*, (1973). Physiochemical and immunochemical properties of the growth hormone of fetal pituitary glands were indistinguishable by disk-gel electrophoresis, immunoelectrophoresis, starch-gel electrophoresis, and radioimmunoassay techniques. After about 70 gestational days there was an increasing content in the pituitary and serum. The serum growth hormone content was at acromegalic levels, peaking between 20 and 24 weeks at 119 ng/ml and later dropping gradually to a level of 33 ng/ml at term. The physiologic significance of this high level is not clear, but the possibility has been raised that the fetus lacks somatotropin release-inhibiting factor. Information on the fetal content of somatomedin is not available.

Fetal serum prolactin, in contrast to growth hormone, increases gradually from the 88th gestational day, and at about 180 days shows a sharp and progressive rise (Grumbach and Kaplan, 1973). Luteinizing hormone is present from 97 to 133 gestational days in a concentration of 10 ng/ml but detailed studies of sex differences have not appeared (Kaplan *et al.,* 1969).

2. Relation of Maternal to Fetal Feedback Control

It seems likely that the hypothalamic-releasing factors, even though of smaller molecular weight, are similar to other pituitary trophic hormones in their inability to cross the placenta. An explanation for the very high early fetal values for somatotropin and the gonadotrophins has been made by Grumbach and Kaplan (1973) who postulate that the fetal pituitary at this early stage may not be under control of a feedback mechanism from the hypothalamus and as a result an unlimited release of hormone occurs from the pituitary gland. Choronic somatomammotropin produced by the placenta is thought mainly to ensure the nutritional demands of the fetus by control of the maternal metabolism, especially during starvation (Grumbach *et al.,* 1973). Serum levels in the fetus are constantly at a low level.

a. Evidence For and Against a Somatotropin Effect on the Fetus. Decapitation of fetal rats and subsequent injection of bovine growth hormone into the fetus led Heggestad (1955) to the conclusion that growth hormone had a growth-promoting effect in the fetus. Injection of crude growth hormone into pregnant rats increased the fetal size even when weights were corrected for the observed delay in parturition (Hultquist and Engfeldt, 1949). Subsequently Zamenhof *et al.* (1966, 1971) were able to prevent the brain weight losses in rat fetuses from starved mothers by administering bovine growth hormone. The growth hormone reversed the decrease in fetal body weight caused by maternal starvation. These authors postulated that this growth-hormone effect might be through mobilization of maternal nutrient supplies or be a direct effect on the uterus or placenta, rather than being a specific fetal effect.

C. Virilizing Tumors in the Mother

The effect of androgens administered to the pregnant female will be included in the chapter by Goldman (Vol. 2). Case reports of pregnant women with arrhenoblastoma tumors have indicated masculinization of the external genitalia of the female fetus (Felicissimo and De Abreu Junqueira, 1938; Brentnall, 1945). Verhoeven, *et al.* (1973) have reviewed 44 case reports of pregnant women with masculinizing tumors. In addition to those with arrhenoblastomas, masculinization was found in the fetuses of women with Leydig cell, Brenner, adrenal rest, luteoma, and Krukenberg tumors.

D. Other Endocrine Systems

An effect of maternal parathyroid disease or diabetes insipidus during the first half of pregnancy has not been reported. The pertinent work on pancreatic disease will be discussed under diabetes in a following section.

Parathormone has been detected in the serum of human fetuses (C. Anast, personal communication), but practically nothing is known about parathyroid function in the human fetus. In the experimental animal fetal calcium and phosphorus metabolism differ substantially from that of the mother. Alexander *et al.* (1973), summarizing their work in the lamb fetus, found high serum calcium and virtually no phosphate in the urine. At as early as 80 days of gestation parathormone appeared in the fetal blood after infusion of EDTA. Explants of human fetal parathyroid as early as the 87-mm stage caused bone resorption, giving evidence for parathormone secretion (Scothorne, 1964). Evidence that [^{125}I]parathormone does not cross the placenta has been presented (Garel, 1972). Although calcitonin has been identified in the fetus, its physiologic role is not understood (Alexander *et al.*, 1973).

A vasotocin-like peptide has been obtained from a cell culture system of human pineal gland (Pavel *et al.*, 1973), but the physiologic role of posterior pituitary hormones in the fetus is relatively unexplained. Arginine vasopressin responses have been detected after hyperosmolar loading in the monkey fetus at 140–155 gestational days (Skowsky *et al.*, 1973).

III. METABOLIC IMBALANCES

A. Diabetes Mellitus

1. Incidence of Congenital Defects in Man

There is controversy about whether diabetes contributes to a general increase in human congenital defects. Warkany (1971) has carefully reviewed this evidence. Rubin and Murphy (1958) showed that the increased incidence of defects in the diabetic offspring is the result of the intensified scrutiny these children receive either during hospitalization or at autopsy examination. Farquhar (1965) was unable to show an increased rate of defects, but Hagbard (1961) and Pedersen *et al.*, (1964) reported small increases. Pedersen *et al.* (1964) attempted to control their prospective study, but the observations on the controls were made by different persons during a different period. Their malformation rate in diabetic offspring was 6.4% compared to 2.1% for the controls. In a comprehensive review of autopsies of 95 diabetic offspring, no specific malformation pattern was identified (Driscoll *et al.*, 1960). Comess *et al.* (1969), studying an Indian population, reported a 38% incidence of anomalies in the offspring of diabetic women diagnosed before 25 years of age.

Although there is doubt about the overall incidence of congenital defects, most authors have observed a significant number of offspring with caudal dysplasia or caudal regression syndrome. The defect, appearing in approximately 1% of diabetic offspring, consists of varying amounts of sacral and femoral agenesis, sometimes associated with defects of the palate and branchial arches (Passarge, 1965; Kucera *et al.*, 1965).

a. Correlation of Clinical Type of Diabetes and Therapy with Skeletal Defects. The offspring of mothers with retinopathy or vascular nephropathy in Pedersen's study (1964) accounted for nearly all of the severe limb deformities. There were six among a group of 280, while among offspring from 573 mothers with milder classifications of diabetes only one limb defect occurred. From an autopsy study of 955 Driscoll *et al.* (1960) reported that the four children with skeletal defects were born from mothers with vascular complications. Mothers with mean blood glucose levels below 100 mg/100 ml had significantly fewer offspring with malformations. In the offspring of 439 diabetics treated only with insulin and sex hormones by White (1949), only one skeletal defect was reported (Gellis and Hsia, 1959).

Insulin reactions and hypoglycemia during the first trimester in the mother were not reported by Pedersen *et al.* (1964) to be related to the malformations. After insulin-shock therapy given for psychiatric disease to 17 mothers, Sobel (1960) found two offspring with multiple congenital abnormalities. These defects included blindness, hyperteleorism, mental deficiency, and Hirchsprung's disease. Although tolbutamide is teratogenic in rats, Lazarus and Volk (1963) have reviewed the available negative evidence for teratogenicity in man. Carbutamide has been reported *in rats* to be the most teratogenic of the hypoglycemic sulfonylurea drugs (Tuchmann-Duplessis and Mercier-Parot, 1959). These authors could not produce defects in rats with chlorpropamide. Jackson *et al.* (1962) also observed no congenital defects in the offspring of women treated with chlorpropamide.

2. Experimental Studies

a. Insulin. Attempts to produce congenital malformations by the use of insulin have been reviewed by Kalter (1968a) and Kalter and Warkany (1959). A number of negative reports using rats have appeared (see Kalter, 1968a). Ferrill (1943), using 20–40 units/kg from weaning could find no defects in rat offspring produced during five generations. Love *et al.* (1964), giving 0.5 units every 12 hr during the last two weeks of pregnancy, produced no defects in rats.

The mouse has been used by Smithberg *et al.* (1956) and Smithberg and Runner (1963). They injected 0.1 units of protamine zinc insulin at 8.5 days postcoitum and found a high incidence of exencephaly and rib and vertebral defects. The teratogenic effects of insulin or tolbutamide alone or in conjunction with nicotinamide were compared by Smithberg and Runner (1963) in three strains of mice. The primary role of insulin is hard to judge because the mouse is sensitive also to fasting teratogenesis. In the rabbit Chomette (1955) and Brinsmade *et al.,* (1956) produced microcephaly and other central nervous system defects. Landauer (1947) observed rumplessness, micromelia, and other skeletal defects in chicks injected with 2 units of insulin into the yolk sac at 24 or 120 hr of incubation. Riboflavin given with the insulin caused an increase in the number of deformities (Landauer, 1952), but nicotinamide reduced the incidence (Landauer and Rhodes, 1952). Landauer (1972) has reviewed the evidence that insulin alone is teratogenic.

b. Experimental Diabetic Models. Takano and Nishimura (1967) reported a 6–7% incidence of defects in the offspring of rats and mice made diabetic by alloxan. The defects reported by these and other authors included club feet, microphthalmia, exencephaly, cleft lip and palate, and spina bifida. These authors concluded that the diabetic state and not the alloxan was responsible for the defects. Kalter (1968b) has summarized the extensive experimental literature on defects found after use of alloxan. A defect similar to the caudal regression syndrome has not been reported in mammals, although Landauer's findings (1947) in the chick exposed to insulin might qualify.

A syndrome of giantism resembling that seen in the offspring of diabetic women has been claimed by Hultquist and Engfledt (1949) after administration of crude pituitary hormone to the pregnant rat.

3. Function of Fetal Endocrine Pancreas

Chez (1974), in a review of this subject, points out that the increased plasma insulin level in diabetic offspring is correlated with macrosomia, but a full range of the physiologic derangements is not known. Glucose administration to the normal fetus is associated with either absence or a delayed release of insulin. Organ culture studies of human pancreas have shown insulin release as early as 60 gestational days (Fujimoto and Williams, 1972).

B. Toxemia of Pregnancy and Maternal Hypertension

Since toxemia of pregnancy makes its appearance after major organogenesis is completed, it is unlikely that teratogenic phenomena are associated with the condition. Hypertensive disease of the mother has not been implicated in congenital defects in the offspring. Of possible importance are the observations in the offspring of rats treated during pregnancy with a number of hormones and drugs which could be associated with hypertensive conditions (Grollman and Grollman, 1962). These authors found hypertension in rat offspring one year postnatally when the mother was treated during pregnancy with aldosterone, cortisone, deoxycorticosterone, chlorothiazide, or low-potassium diet. A long-term follow-up of the vascular state of children born of hypertensive mothers seems indicated.

C. Maternal Phenylketonuria

1. Clinical Observations

The offspring of mothers with phenylketonuria have shown a high frequency of mental retardation associated with microcephaly and intrauterine growth retardation (Hsia, 1970; Frankenburg et al., 1968). Those mothers with hyperphenylalaninemia, but with serum phenylalanine levels below 15 mg/dl during pregnancy, however, produced 11 normal infants out of a total of 12 (Hsia, 1970).

Although there are a few normal infants born following successful treatment by maternal phenylalanine restriction, the amount of data is too limited for a conclusion at this time.

Other anomalies found in children from, phenylketonuric mothers include congenital heart disease, dislocation of the hips, and strabismus (Stevenson and Huntley, 1967). In two families they found 8 of 10 children with congenital heart disease, including coarctation of the aorta, patent ductus arteriosus, and other types. Of 26 documented pregnancies, 16 (62%) terminated in abortion. Montenegro and Castro (1965) and Fisch et al. (1969) have reported similar types of defects in the offspring of phenylketonuric mothers.

2. Experimental Studies

Kerr et al. (1968) fed phenylalanine to pregnant rhesus monkeys and reported mental retardation in the offspring, but this was not associated with microcephaly or other congenital defects. They observed higher phenylalanine levels in cord blood than in maternal blood. The feeding of other amino acids to monkey mothers was not associated with mental deficiency in the offspring except for parachlorophenylalanine which inhibits the metabolism of phenylalanine (Chamove et al., 1973).

D. Alcoholism

Maternal alcoholism has been associated with a characteristic clinical picture in about one third of the offspring. (Lemoine et al., 1968; Jones et al., 1973; Jones and Smith, 1973). The mothers were all severe chronic alcoholics.

At birth the infants are undergrown and have facies characterized by shortened palpebral fissures, stub noses, and hirsutism (Fig. 2). Microcephaly and delayed mental development are nearly always present. Cardiac anomalies have been present in over 60% and two of Jones and Smith's (1973) cases had cleft palate. Analysis of the records from the collaborative study of the National Institute of Neurologic Disease and Stroke revealed a 32% incidence of the syndrome in the offspring of women whose chronic alcoholism was ascer-

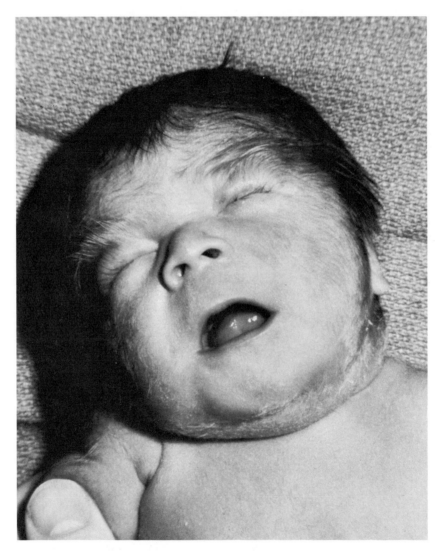

Fig. 2. Fetal alcoholic syndrome. Newborn at birth with small size, hirsutism, short palpebral fissures, and microcephaly. (Courtesy of Doctor David Smith, from Jones and Smith, 1973).

tained during their pregnancy (Jones and Smith, 1973). Other factors besides ethanol, such as poor nutrition (pyridoxine or folic acid deficiency), alcoholic contaminants (lead), or genetic predisposition, may play an important role in the production of this syndrome. In view of the numbers of children involved and the prospect for prevention of the condition, a major investigative effort on all aspects of the problem seems indicated.

1. Experimental Studies

In 1962 Giroud and Tuchmann-Duplessis reviewed the experimental literature on alcoholism and found no evidence for teratogenicity. Sandor and Amels (1971) reported studies on rats given intravenous injections of ethanol at 6, 7, and 8 days of pregnancy. At the highest doses used (2 g/kg of body weight) an increase in abnormal egg cylinders and resorptions was found. At 19.5 days of gestation no increase in congenital malformations was found and the body weights were not altered. Some skeletal retardation was observed at 19.5 days. Sandor (1968) has also reported damage to chick embryos exposed to ethanol at early stages. Chernoff (1975) has briefly reported fetal malformations from alcohol in the mouse. A large body of experimental work dealing with dietary deficiency in experimental animals exists. Both pyridoxine and folic acid deficiency may occur in chronic alcoholism, and both produce teratogenic effects in animals. Nutritional deficiency is discussed in Chapter 7.

E. Other Conditions

1. Acidosis and Electrolyte Disturbances

Acidosis and electrolyte disturbances are nearly always associated with many other changes, such as nutritional deficiency and changes in water content, as well as the disease state that produces them. For this reason it is difficult to analyze their separate effects on development in the human or experimental animal (see chapter by Grabowski, Vol. 2). Information on how these states affect embryonic development in man is not available.

In experimental animals the acidosis produced by salicylate has been studied. Goldman and Yakovac (1964, 1965) have published studies attempting to show the protective role of sodium carbonate and chloride administration on the malformation rate following salicylate treatment. Chebotar (1967) reported that ammonium chloride intensified while sodium bicarbonate reduced the teratogenic effects of sodium salicylate in rats. Acetazolamide administration produces limb defects in rat fetuses, and this treatment is associated with potassium deficiency and respiratory acidosis. Ellison and Maren (1972) found that potassium chloride or bicarbonate replacement resulted in partial or complete protection of the developing embryos.

2. Hypoglycemia

As discussed above hypoglycemic shock due to insulin has not often been associated with an increase in congenital defects in the offspring, but Sobel (1960) found two children with multiple defects among a group of 17 mothers treated with insulin shock for psychiatric disease.

The mouse appears to be the animal most susceptible to fasting teratogenesis (Kalter and Warkany, 1959). A 24–30 hr fast in the mouse during the 7th–10th gestational day produced vertebral and rib defects and occasional exencephaly (Miller, 1962). Fasting on the 8th or 9th day produced the highest incidence of defects. Interruption of the fast with glucose or various amino acids prevented the defects (Runner and Miller, 1956).

3. Pheochromocytoma in the Mother

In 112 pregnancies complicated by pheochromocytoma there were 16 spontaneous abortions, 46 fetal deaths either *in utero* or shortly after birth, and 50 live-born infants without recorded congenital defects (Schenker and Chowers, 1971). This chromaffin tumor thus appears to contribute to intrauterine and perinatal death.

REFERENCES

Alexander, D. P., Britton, H. G., Nixon, D. A., Cameron, E., Foster, C. L., Buckle, R. M., and Smith, F. G., 1973, Calcium, parathyroid hormone and calcitonin in the foetus, *in: Foetal and Neonatal Physiology Proceedings of the Sir Joseph Barcroft Centenary Symposium* (K. W. Cross and P. Nathanielez, eds.), p. 421–429, Cambridge University Press, Cambridge, U. K.

Andersen, H. J., 1971, Prenatal damage in hypothyroidism, *in: Hormones and Development* (M. Hamburgh and E. J. W. Barrington, eds.), p. 559–566, Appleton-Century Crofts, New York.

Beaudoin, A. R., 1968, Other observations on the teratogenic action of the thyroid stimulating hormone, *Teratology* **1**:11.

Beierwaltes, W. H., Hilger, M. T. J., and Wegst, A., 1963, Radioiodine concentration in fetal human thyroid from fallout, *Health Phys.* **9**:1263.

Brentnall, C. P., 1945, Case of arrhenoblastoma complicating pregnancy, *J. Obstet. Gynaecal Br. Emp.* **52**:235.

Brinsmade, A. B., Buchner, F., and Rubsaamen, H., 1956, Missbildungen am Kaninchenembryo durch Insulininjektion beim Muttertier, *Naturwissenschaften* **43**:259.

Carr, E. A., Jr., Beierwaltes, W. H., Raman, G., Dodson, V. N., Tanton, J., Betts, J. S., and Stambaugh, R. A., 1959, The effect of maternal thyroid function on fetal thyroid function and development, *J. Clin. Endocrinol. Metab.* **19**:1.

Chamove, A. S., Kerr, G. R., and Harlow, H. F., 1973, Learning in monkeys fed elevated amino acid diets, *J. Med. Primatol.* **2**:223.

Chandler, R. W., Blizzard, R. M., Hung, W., and Kyle, M., 1962a, Incidence of thyrocytotoxic factor and other anti-thyroid antibodies in the mothers of cretins, *N. Engl. J. Med.* **267**:376.

Chandler, R. W., Kyle, M. A., Hung, W., and Blizzard, R. M., 1962b, Experimentally induced autoimmunization disease of the thyroid, *Pediatrics* **29**:961.

Chebotar, N. A., 1967, Peculiarities of action of sodium salicylate at various stages of embryogenesis in rats and the influence of certain shifts in the female organism on its teratogenic activity, *Pharmacol. Toxicol.* (Russ.) No. **2**:221.

Chernoff, G. F., 1975, A mouse model of the fetal alcoholism syndrome, *Teratology* **11**:14A (abstract).

Chez, R. A., 1974, The development and function of the human endocrine pancreas, *in: The Endocrine Milieu of Pregnancy.* Puerperium, and Childhood, Report of the Third Ross Conference on Obstetric Research (1974), Ross Laboratories, Columbus, Ohio, pp. 43–47.

Chomette, G., 1955, Entwichlungsstorungen nach Insulinschock beim trachtigen Kaninchen, *Beitr. Pathol. Anat.* **115**:439.

Comess, L. J., Bennett, P. H., Man, M. B., Burch, T. A., and Miller, M., 1969, Congenital anomalies and diabetes in the Pima Indians of Arizona, *Diabetes* **18**:471.

Driscoll, S. G., Benirschke, K., and Curtis, G. W., 1960, Neonatal deaths among infants of diabetic mothers, *A.M.A. Am. J. Dis. Child.* **100**:818.

Ellison, A. C., and Maren, T. H., 1972, The effect of potassium metabolism on acetazolamide-induced teratogenesis, *Johns Hopkins Med. J.* **130**:105.

Farquhar, J. W., 1965, The influence of maternal diabetes on the fetus and child, *in: Recent Advances in Pediatrics,* 3rd ed. (D. Gairdner, ed.), pp. 126–129, Little, Brown, Boston.

Felicissimo, P. K. and De Abreu Junqueira, M., 1938, Sobre un caso de arrhenoblastoma de ovario e gravidez topica simultanea. Virilisa cao de gestante e do feto femino, *Rev. Ginecol. Obstet.* **32**:356.

Ferrill, H. W., 1943, Effect of chronic insulin injections on reproduction in white rats, *Endocrinology* **32**:449.

Fisch, R. O., Doeden, D., Lansky, L. L., and Anderson, J. A., 1969, Maternal phenylketonuria: Detrimental effects on embryogenesis and fetal development, *Am. J. Dis. Child.* **118**:847.

Fisher, D. A., Hobel, C. J., Garza, B. S., and Pierce, C. A., 1970, Thyroid function in the preterm fetus, *Pediatrics* **46**:208.

Frankenburg, W. K., Duncan, B. R., Coffelt, R. W., Koch, R., Coldwell, J. G. and Son, C. D., 1968, Maternal phenylketonuria: Implications for growth and development, *J. Pediatr.* **73**:560.

Fujimoto, W. Y., and Williams, R. H., 1972, Insulin release from cultured fetal human pancreas, *Endocrinology* **91**:1133.

Garel, J. M., 1972, Distribution of labeled parathyroid hormone in rat fetus, *Horm. Metab. Res.* **4**:131.

Gellis, S. S., and Hsia, D. Y., 1959, The infant of the diabetic mother, *A.M.A. Am. J. Dis. Child.* **97**:1.

Geloso, J. P., 1961, Date de l'entrée en fonction de la thyroïde chez le foetus du rat, *Compt. Rend. Soc. Biol.* **155**:1239.

Giroud, A., and Rothschild, B., 1951, Repercussions de la thyroxine sur l'oeil du foetus, *C.R. Soc. Biol.* (Paris) **145**:525.

Giroud, A., and Tuchmann-Duplessis, H., 1962, Malformations congenitales. Role des facteurs exogenes, *Pathol. Biol.* **10**:119.

Gitlin, D., and Biasucci, A., 1969, Ontogenesis of immunoreactive growth hormone, follicle-stimulating hormone, thyroid-stimulating hormone, luteinizing hormone, chorionic prolactin and chorionic gonadotropin in the human conceptus, *J. Clin. Endocrinol. Metab.* **29**:926.

Goldman, A. S., and Yakovac, W. C., 1964, Prevention of salicylate teratogenicity in immobilized rats by certain central nervous system depressants, *Proc. Soc. Exp. Biol. Med.* **115**:693.

Goldman, A. S., and Yakovac, W. C., 1965, Teratogenic action in rats of reserpine alone and in combination with salicylate and immobilization, *Proc. Soc. Exp. Biol. Med.* **118**:857.

Green, H. G., Gareis, F. J., Shepard, T. H., and Kelley, V. C., 1971, Cretinism associated with maternal sodium iodide ^{131}I therapy during pregnancy, *Am. J. Dis. Child.* **122**:247.

Grollman, A., and Grollman E. F., 1962, Teratogenic induction of hypertension, *J. Clin. Invest.* **41**:710.

Grumbach, M. M., and Kaplan, S. L., 1973, Ontogenesis of growth hormone, insulin, prolactin and gonadotropin secretion in the human foetus, *in: Barcroft Centenary Symposium in Foetal and Neonatal Physiology* (K. W. Cross and P. Nathanielez, eds.), pp. 462–487, Cambridge University Press, Cambridge, U.K.

Grumbach, M. M., Kaplan, S. L., and Vinik, A., 1973, Human chorionic somatomammotropin (HCS) 5. clinical evaluation of hormonal excess and deficienty: role of HCS in pregnancy, *in: Methods in Investigative and Diagnostic Endocrinology*, Vol. 2B (S. A. Person, ed.) pp. 810–184, North Holland Publishing, Amsterdam.

Hagbard, L., 1961, *Pregnancy and Diabetes*, pp. 21–25, Charles C Thomas, Springfield, Illinois.

Hambaugh, M., 1966, Evidence for a direct effect of temperature and thyroid hormone on myelinogenesis *in vitor, Dev. Biol.* **13:**15.

Hamburgh, M., Sobel, E. H., Koblin, R., and Rinestone, A., 1962, Passage of thyroid hormone across the placenta in intact and hypophysectomized rats, *Anat. Rec.* **144:**219.

Heggestad, C. B., 1955, Retardation of ponderal growth in hypophysectomized fetal rats and its prevention by means of an injected growth hormone, *Anat. Rec.* **121:**339. (abstract).

Hodges, R. E., Hamilton, H. E., and Keettel, W. C., 1952, Pregnancy in myxedema, *A.M.A. Arch. Intern. Med.* **90:**863.

Hsia, D. Y., 1970, Phenylketonuria and its variants, *in: Progress in Medical Genetics*, Vol. 7 (A. G. Steinberg and E. G. Bearn, eds.) pp. 53–68, Grune and Stratton, New York.

Hultquist, G. T., and Engfeldt, B., 1949, Giant growth of rat fetuses produced experimentally by means of administration of hormones to the mother during pregnancy, *Acta. Endocrinol.* **3:**365.

Jackson, W. P. U., Cambell, G. D., Notelovitz, M., and Blumsohn, D., 1962, Tolbutamide and chlorpropamide during pregnancy in humans, *Diabetes* **11:**98.

Jones K. L., and Smith, D. W., 1973, Recognition of the fetal alcohol syndrome in early infancy, *Lancet* **2:**999.

Jones, K. L., Smith, D. W., Ulleland, C. N., and Streissguth, A. P., 1973, Pattern of malformation in offspring of alcoholic mothers, *Lancet* **1:**7815.

Jost, A., 1957, Le problème des interrelations thyréohypophysaires chez le foetus et l'action du propylthiouracile sur la thyroïde foetale du rat, *Rev. Susse Zool.* **64:**821.

Kalter, H., 1968a, *Teratology of the Central Nervous System*, pp. 148–149, University of Chicago Press, Chicago.

Kalter, H., 1968b, *Teratology of the Central Nervous System*, pp. 150–152, University of Chicago Press, Chicago.

Kalter, H., and Warkany, J., 1959, Experimental production of congenital malformations in mammals by metabolic procedure, *Physiol. Rev.* **39:**69.

Kajihara, A., Kojima, A., Onaya, T., Takemura, Y., and Yawada, T., 1972, Placenta transport of thyrotropin releasing factor in the rat, *Endocrinology* **90:**592.

Kaplan, S. L., Grumbach, M. M., and Shepard, T. H., 1969, Gonadotropins in the serum and pituitary of human fetuses and infants, *Pediatr. Res.* **3:**512 (abstract).

Kaplan, S. L., Grumbach, M. M., and Shepard, T. H., 1973, The ontogenesis of human fetal hormones I. Growth hormone and insulin, *J. Clin. Invest.* **51:**3080.

Kerr, G. R., Chamove, A. S. Harlow, H. F., and Waisman, H. A., 1968, Fetal PKU The effect of maternal hyperphenylalaninemia during pregnancy in the rhesus monkey *(Macaca mulatta)*, *Pediatrics* **42:**27.

Kind, C., 1962, Das endokrine System der Anencephalen mit besonderer Berucksichtigung der Schilddruse, *Helv. Paediatr. Acta* **17:**244.

Kucera, J., Lenz, W., and Maier, W., 1965, Missbildungen der Beine und der kaudalen Wirbelsaule bie Kindern diabetischer Mutter, *Dtsch. Med. Wochenschr.* **90:**901.

Landauer, W., 1947, Insulin-induced abnormalities of the beak extremities and eyes in chickens, *J. Exp. Zool.* **105:**145.

Landauer, W., 1952, Malformations of chicken embryos produced by boric acid and the probable role of riboflavin in their origin, *J. Exp. Zool.* **120:**469.

Landauer, W., 1972, Is insulin a teratogen?, *Teratology* **5:**129.

Landauer, W., and Rhodes, M. B., 1952, Further observations on the teratogenic nature of insulin and its modification by supplementary treatment, *J. Exp. Zool.* **119:**221.

Langman, J., and Van Faassen, F., 1955, Congenital defects in the rat embryo after partial thyroidectomy of the mother animal: A preliminary report of eye defects, *Am. J. Ophthalmol.* **40**:65.

Lasnitski, I., 1965, The action of hormones on cell and organ cultures *in: Cells and Tissues in Culture*, Vol. 1 (E. N. Willmer, ed.) pp. 639–658, Academic Press, New York.

Lazarus, S. S., and Volk, B. W., 1963, Absence of teratogenic effect of tolbutamide in rabbits, *J. Clin. Endocrinol.* **23**:597.

Legrand, J., 1971, Comparative effects of thyroid deficiency and undernutrition on maturation of the nervous system and particulary on myelination on the young rat, *in: Hormones and Development*, (M. Hamburgh and E. J. W. Barrington, eds.) pp. 381–390, Appleton-Century Crofts, New York.

Lemoine, P., Harousseau, H., Borteyru, J. P., and Menuet, J. C., 1968, Les enfants de parents alcooliques: Anomalies observées, a propos de 127 cas, *Arch. Fr. Pediatr.* **25**:830.

Levina, S. E., 1971, Sexual development in epiphysial and hypothalamic regulation of the secretion of follicle-stimulating hormone and luteinizing hormone in fetal hypophyses cultured *in vitro, in: Hormones and Development* (M. Hamburgh and E. J. W. Barrington, eds.), pp. 547–552, Appleton-Century Crofts, New York.

Lotmar, F., 1931, Untwicklungsstorungen in der Kleinhirnrinde biem endemischer Kretinismus, *Z. Neurol. Psychiatrie* **136**:412.

Love, E. J., Kind, R. A. H., and Stevenson, J. A. F., 1964, The effect of protamine zinc insulin on the outcome of pregnancy in the normal rat, *Diabetes* **13**:44.

Man, E. G., 1972, Thyroid function in pregnancy and infancy, *C.R.C. Crit. Rev. Clin. Lab. Sci.* **3**:203.

Miller, J. R., 1962, A strain difference in response to the teratogenic effect of maternal fasting in the mouse, *Can. J. Genet. Cytol.* **4**:69.

Montenegro, J. E., and Castro, G. L., 1965, Fenilcetonuria materna: Anomalieas en la descendencia, *Acta Med. Vene.* **12**:233.

Nataf, B., and Sfez, M., 1961, Début de fonctionnement de la thyroïde foetale du rat, *Compt. Rend. Soc. Biol.* **155**:1235.

Norris, E. H., 1918, The early morphogenesis of the human thyroid gland, *Am. J. Anat.* **24**:433.

Olin, P., Vecchio, G., Ekholm, R., and Almqvist, S., 1970, Human fetal throglobulin: Characterization and *in vitro* biosynthesis studies, *Endocrinology* **86**:1041.

Orts-Llorca, F., and Galvez, J. M. G, 1958, On the morphology of the primordium of the thyroid gland in the human embryo, *Acta Anat.* **33**:110.

Passarge, E., 1965, Congenital malformations and maternal diabetes, *Lancet* **1**:324.

Pavel, S., Dorcescu, M., Petrescu-Holban, R., and Ghinea, E., 1973, Biosynthesis of a vasotocin-like peptide in cell cultures from pineal glands of human fetuses, *Science* **181**:1252.

Pedersen, L. M., Tygstrup, I., and Pedersen, J., 1964, Congenital malformations in newborn infants of diabetic women. Correlation with maternal diabetic vascular complications, *Lancet* **1**:1124.

Pickering, D. E., and Kontaxis, N. E., 1961, Thyroid function in the fetus of the macaque monkey. II. chemical and morphological characteristics of the foetal thyroid gland, *J. Endocrinol.* **23**:267.

Rubin, A., and Murphy, D. P., 1958, Studies in human reproduction 3. The frequency of congenital malformations in the offspring of nondiabetic and diabetic individuals, *J. Pediatr.* **53**:579.

Runner, M. N., and Miller, J. R., 1956, Congenital deformity in the mouse as a consequence of fasting, *Anat. Rec.* **124**:437 (abstract).

Russel, K. P., Rose, H. and Starr, P., 1957, The effects of radioactive iodide on maternal and fetal thyroid function during pregnancy, *Surg. Gynecol. Obstet.* **104**:560.

Sandor, S., 1968, The influence of aethyl alcohol on the developing chick embryo, *Rev. Roum. Embryol. Cytol.* **5**:167.

Sandor, S., and Amels, D., 1971, The action of aethanol on the prenatal development of albino rats, *Rev. Roum. Embryol. Cytol.* **8**:37.

Schenker, J. G., and Chowers, I., 1971, Pheochromocytoma and pregnancy: Review of 89 cases, *Obstet. Gynecol. Sur.* **26:**739.

Scothorne, R. J., 1964, Functional capacity of fetal parathyroid glands with reference to their clinical use as homografts, *N.Y. Acad. Sci.* **120:**669.

Shepard, T. H., 1961, Phenylthiocarbamide non-tasting among congenital athyrotic cretins: Further studies in an attempt to explain the increased incidence, *J. Clin. Invest.* **40:**1751.

Shepard, T. H., 1965, The thyroid, *in: Organogenesis* (R. DeHaan and H. Ursprung, eds.) pp. 493–512, Holt, Rinehart and Winston, New York.

Shepard, T. H., 1967, Onset of function in the human fetal thyroid: Biochemical and radioautographic studies from organ culture, *J. Clin. Endocrinol. Metab.* **27:**945.

Shepard, T. H., 1976, *A Catalog of Teratogenic Agents,* 2nd edition, Johns Hopkins University Press, Baltimore.

Shepard, T. H., and Andersen, H. J., 1965, Phenylthiocarbamide non-tasting among different types of cretinism and thyroid disorders, *Acta. Endocrinol. Suppl. 89* **45:**43.

Shepard, T. H., Andersen, H. J., and Andersen, H., 1964, The human fetal thyroid I. Its weight in relation to body weight, crown-rump length, foot length and estimated gestation age, *Anat. Rec.* **148:**123.

Skowsky, W. R., Bashore, R. A., Smith, F. G., and Fisher, D. A., 1973, Vasopressin metabolism in the foetus and newborn, *in: Fetal and Neonatal Physiology Proceedings of the Sir Joseph Barcroft Centenary Symposium* (K. W. Cross and P. Nathanielez, eds.), pp. 439–447, Cambridge University Press, Cambridge, U. K.

Smith, C. A., Oberhelman, H. A., Jr., Storer, E. H., Woodward, E. R., and Dragstedt, L. R., 1951, Production of experimental cretinism in dogs by the administration of radioactive iodine, *Arch. Surg.* **63:**807.

Smithberg, M., and Runner, M. N., 1963, Teratogenic effects of hypoglycemic treatment in inbred strains of mice, *Am. J. Anat.* **113:**479.

Smithberg, M., Sanchez, H. W., and Runner, M. N., 1956, Congenital deformity in the mouse induced by insulin, *Anat. Rec.* **124:**441 (abstract).

Sobel, D. E., 1960, Fetal damage due to ECT, insulin coma, chlorpromazine or reserpine, *A.M.A. Arch. Gen. Psychol.* **2:**606.

Speert, H., Quimby, E. H., and Werner, S. C., 1951, Radioiodine uptake by fetal mouse thyroid and resultant effects in later life, *Surg. Gynecol. Obstet.* **93:**230.

Stevenson, R. E., and Huntley, C. C., 1967, Congenital malformations in offspring of phenylketonuric mothers, *Pediatrics* **40:**33.

Takano, K., and Nishimura, H., 1967, Congenital malformations induced by alloxan diabetes in mice and rats, *Anat. Rec.* **158:**303.

Trunnell, J. B., and Wade, P., 1955, Factors governing the development of the chick embryo thyroid. II. Chronology of the synthesis of iodinated compounds studied by chromatographic analysis, *J. Clin. Endocrinol.* **15:**107.

Tuchmann-Duplessis, H., and Mercier-Parot, L., 1959, Sur l'action teratogène d'un sulfamide hypoglycemiant. Etude experimentale chez la ratte, *J. Physiol.* **51:**65.

Verhoeven, A. T. M., Mastboom, H. A., VanLeusden, I. M., and Van Der Velden, W. H. M., 1973, Virulizing in pregnancy coexisting with an ovarian cystadenoma, case report and review of virilizing ovarian tumors of pregnancy, *Obstet. Gynecol. Sur.* **28:**597.

Warkany, J., 1971, *Congenital Malformations: Notes and Comments,* pp. 102–125, Year Book Publishers, Chicago.

White, P., 1949, Pregnancy complicating diabetes, *Am. J. Med.* **7:**616.

Zamenhof, S., Mosley, J., and Schuller, E., 1966, Stimulation of the proliferation of cortical neurons by prenatal treatment with growth hormone, *Science* **152:**1396.

Zamenhof, S., Van Marthens, E., and Gravel, L., 1971, Prenatal cerebral development: Effect of restricted diet, reversal by growth hormone, *Science* **174:**954.

Atmospheric Gases　　11

VARIATIONS IN CONCENTRATION AND SOME COMMON POLLUTANTS

CASIMER T. GRABOWSKI

I. INTRODUCTION

Pure food, pure water, and pure air are matters of interest to all of us. We can, to a large extent, control what we put into our mouths, but we can control the air we breathe, 500 cc about 15 times every minute, only by the use of elaborate artificial systems or by geographically removing ourselves from a situation which might be harmful. We are beginning to recognize the full extent of the dangers of air pollutants to adult health, but our understanding of the teratogenic potential of the commonest air pollutants is negligible. Our constant need for air and the ubiquitousness of the additives from our industrial society complicate attempts to evaluate these factors precisely. Perhaps the unsatisfactory state of knowledge in this area is at least partly due to a fatalistic dismissal with such attitudes as: "We can't control it, we are constantly exposed to it, so it probably isn't too bad anyhow."

　　There are many components of the atmosphere that can have unfavorable effects on the developing individual. Even normal components, such as oxygen and carbon dioxide, are dangerous when present in too low or too high a concentration. The commonest atmospheric pollutants are carbon monoxide, ozone, nitrogen dioxide, sulfur dioxide, and lead. Though all are known to be toxic to adults, they have been poorly studied from a teratological standpoint. They are of teratogenic concern, not simply because of their toxicological properties, but because of their widespread distribution and the ever-present dangers of low-level chronic and occasional acute exposure. A

CASIMER T. GRABOWSKI · Department of Biology, University of Miami, Coral Gables, Florida.

compelling question is, how much for how long is too dangerous? There is also the important question of whether they can act synergistically with other environmental and genetic factors to induce prenatal damage.

Of necessity this chapter will deal as much with unsolved problems as with demonstrated facts. Emphasis will be placed on mammalian experiments. The current trend to define teratology broadly as a study of the causes of functional as well as structural disorders, and to include postnatal as well as prenatal inception (see Wilson, 1972), is fully accepted in this chapter. For example, at least one agent, hypoxia, has been demonstrated to be dangerous to at least one organ system—the CNS—in preimplantation, early embryonic, fetal, perinatal, and postnatal periods.

II. NORMAL ATMOSPHERIC COMPONENTS

A. Oxygen Deficiency

1. History

Oxygen was discovered as an element in 1774 by Joseph Priestley, and its vital role in animal respiration was demonstrated shortly thereafter by Lavoisier. The teratogenic effects of lack of oxygen was observed as early as 1820 by Geoffrey St.-Hilaire (1820), and the first comprehensive study in experimental teratology was performed by Camille Dareste (summarized in his monographs of 1887 and 1891) on oxygen-deficient chicks. These excellent pioneering experiments rightfully give Dareste the title "father of experimental teratology," and they also make hypoxia the oldest known environmental teratogen.

These early experiments were followed by many other investigations on many species with a wide variety of agents and from diverse viewpoints (see Grabowski, 1970, for review). Although teratogenicity caused by oxygen deficiency has been clearly demonstrated in many ways and in several species, there is reluctance to regard it as a significant human teratogen.

2. Kinds of Hypoxia and Methods of Induction in Mammals

The words anoxia and hypoxia are often used synonymously, but anoxia means total lack of oxygen and is probably rarely achieved biologically outside of frankly lethal situations. Therefore, hypoxia, a general term which means a relative lack of oxygen, is preferred. The major kinds of hypoxia, according to the simplified classification of Van Liere and Stickney (1963) are: (1) anoxic hypoxia (hypoxemia), in which there is a physical lack of oxygen in arterial blood; (2) anemic (hemic) hypoxia, in which arterial oxygen is present at normal tension but there is a deficiency of functioning hemoglobin; (3) stagnant (circulatory) hypoxia, in which arterial blood has a normal oxygen level and hemoglobin is not impaired, but the oxygen supply to tissues is not adequate; and (4) histotoxic hypoxia, in which cellular respiration is inhibited and thus cells and tissues are unable to utilize available oxygen.

There are many ways in which the various forms of hypoxia can be induced in pregnant mammals. Hypoxemia can be produced by exposure to high altitudes or otherwise breathing air at reduced oxygen concentrations at either normal or reduced pressure. It can also occur during periods of maternal respiratory distress, with some types of cardiac dysfunction, at birth by premature separation of the placenta or prolapse of the umbilical cord, or by failure to establish proper breathing after birth. Any serious impairment of the embryonic or fetal cardiovascular system, such as vascular collapse, bradycardia, and/or hypotension, can induce hypoxemia to either localized areas or to all of the developing individual.

Anemic hypoxia can be caused by hemoglobin deficiency in the mother or fetus, whether chronic or acute, as well as by agents which convert hemoglobin to a relatively inactive form. The most familiar type of altered hemoglobin is carboxyhemoglobin, produced by the inhalation of carbon monoxide and to some extent by fluorinated hydrocarbons. It is also a normal by-product of red cell destruction. Equally dangerous is methemoglobin which is formed by the interaction of hemoglobin with the common food additive, inorganic nitrite. Methemoglobin can, in addition, be produced by inorganic nitrate, also a food additive and common contaminant of heavily fertilized vegetables and well water (because it can be converted by bacteria to nitrite), by potassium chlorate, and by a variety of organic compounds such as sulfonamide drugs, methylene blue, phenylhydrazine, pyrogallol, nitrobenzine, and other nitro and amide compounds (Bodansky, 1951).

Stagnant hypoxia can be induced by circulatory problems in a pregnant woman, such as acute bradycardia and/or suddenly lowered blood pressure, either of which could restrict the blood flow to the uterus, or by a circulatory impairment in the uterus or umbilical arteries. There are also numerous drugs, e.g., epinephrine and vasopressin, which can have an acute vasoconstrictive effect on the uterine blood vessels and thereby induce fetal hypoxia (i.e., lowered fetal Po_2, as demonstrated in both rats (Chernoff and Grabowski, 1971) and monkeys (Adamsons *et al.*, 1971). It is possible that the well-documented teratogenic effects of stress may be due to stagnant hypoxia resulting from epinephrine release. In experimental animals stagnant hypoxia has also been induced by clamping the uterine and/or ovarian arteries.

Some anaesthetics and other inhibitors of cellular respiration such as cyanide, azide, dinitrophenol, sodium selenite, sodium malonate, and carbon monoxide can induce histotoxic anemia. Virtually all of these methods for inducing hypoxia have been used in experimental animals, and some have been reported to be effective after accidental application in humans. The possibilities of synergistic combinations for inducing hypoxia are numerous.

3. Abnormalities Produced by Hypoxia in Animals

Hypoxia has been used by numerous investigators to induce abnormal development in a wide variety of lower vertebrates such as fish, amphibians, and chicks (see Grabowski, 1970, for review). Abnormal development due to hypoxia has also been obtained in several species of mammals such as the mouse, rat, rabbit, hamster, cat, and monkey (see reviews by Kalter and Warkany, 1959; Kalter, 1968; Grabowski, 1970). In most of these studies the classical approach was emphasized, namely, exposure of the maternal organism to near-acute conditions of hypoxia followed by examination of fetuses a week or two later for visible abnormalities. Three generalizations are apparent from this approach: (1) Various mammals show different susceptibilities to exposure to O_2-deficient atmospheres (rats are quite resistant, mice very susceptible, rabbits and hamsters intermediate). (2) Relatively extreme conditions are required to produce abnormalities (usually about 5–8% O_2 for periods of up to 48 hr). (3) The most sensitive period is that of major organogenesis. Abnormal development of sensory organs, neural tube, teeth, heart, blood vessels, and urogenital system have been reported by most workers. The axial skeleton is especially sensitive in the mouse (Ingalls and Curley, 1957; Murakami and Kameyama, 1963), rabbit (Degenhardt, 1954, 1958), and hamster (Ericson, 1960). These skeletal defects fit the critical-period concept of susceptibility since progressively more posterior defects are obtained as the age of the embryo at the time of treatment is increased. Since extreme conditions of reduced O_2 supply were generally required to produce significant frequency of abnormalities, these experiments lend some support to the notion, implied if not explicitly stated (Kalter, 1968; Wilson, 1972), that hypoxia is not a strong teratogen.

Other forms of inducing hypoxia are more effective. Although the rat is quite resistant to O_2-deficient atmospheres, stagnant hypoxia induced by uterine artery clamping can produce up to 20–26% externally visible abnormalities (Brent and Franklin, 1960; Franklin and Brent, 1964; Leist and Grauwiler, 1974), depending on time of treatment. Chemical means of uterine vasoconstriction, such as intraperitoneal injection of epinephrine and vasopressin into pregnant rats between 15½ and 17½ days, not only induces severe bradycardia and lowered blood Po_2 but also eventually leads to abnormalities of tail, extremities, liver, and CNS (Chernoff, 1969; Chernoff and Grabowski, 1971).

Relatively few gross terata of the brain and spinal cord have been described subsequent to hypoxic exposure during organogenesis (Kalter, 1968), although paralysis and various behavioral abnormalities have been reported in mice and rats exposed to various kinds of hypoxia during several periods of development; for example, Meier *et al.* (1960) and Vierck and Meier (1963) using an oxygen-deficient atmosphere in mice; Thompson and Olian (1961) using epinephrine injection in mice; and Crist and Hulka (1970) and Chernoff (1969) using epinephrine in rats. Windle and Becker (1943) demonstrated a high incidence of histological damage to the central nervous system subsequent to asphyxia at birth in guinea pigs. Later Windle (1963, 1966), working with rhesus monkeys subjected to asphyxia during birth, demonstrated a high incidence of cerebral palsy and other neurological defects in treated individuals. Ginsberg and Myers (1974), exposing near-term pregnant rhesus monkeys to CO, found neuromotor problems in over 50% of the offspring after birth and clearly visible evidence of extensive brain damage at autopsy. Timiras and Woolley (1966) and Petropoulos *et al.* (1969) exposed rats during the perinatal period to high altitudes and found biochemical indications of retarded differentiation and lowered neuronal populations in the brain. A detailed review of the effects of hypoxia on the CNS of postnatal animals of several species was made by Jílek *et al.* (1970). The CNS of newborns is generally less sensitive to hypoxic damage than that of adults, but the peculiar hazards of birth and early postnatal life make this agent a potentially dangerous one to this vulnerable organ.

Another indication that classical criteria (i.e., grossly visible abnormalities) may not be adequate to evaluate to teratogenic effects of hypoxia is demonstrated by the experiments of Ezrin (unpublished results) in which 10-day rat embryos were exposed to moderate carbon monoxide intoxication. These treated individuals showed a 50% reduction in the rate of heartbeat and ECG abnormalities when examined 9 days later, suggesting some type of subtle damage to the cardiovascular system by the CO.

In conclusion, it is evident that although exposure of pregnant animals to oxygen-deficient atmospheres during the period of organogenesis does not always produce conspicuous abnormal development, there is abundant evidence that other methods of induction of hypoxia, over a wide range of developmental periods, can produce a significant frequency of structural and functional abnormalities.

4. Problem of Evaluating Hypoxic Effects

One of the big problems in evaluating hypoxia as a mammalian teratogen is that, unlike most drugs or metabolites for which dosage, blood level, or total body concentration can be controlled or determined, oxygen deficiency cannot be easily defined in terms of any one or two parameters. Oxygen supply to an organ depends on many factors, both local and those involving the total organism. For example, in the developing mammal it depends on the partial pressure of oxygen in inspired air, oxygen-carrying capacity of the maternal circulatory system, adequacy of uterine circulation and the placental interchange, adequacy of both the bloodstream and the cardiovascular system of the fetus or embryo to deliver oxygen to various parts of the organism, as well as the ability of the developing cells to utilize that oxygen. Sophisticated equipment is available for measuring the total oxygen, bound oxygen, free oxygen, and Po_2 (escaping tendency of oxygen) in blood. Measuring one or two of these factors alone, however, is not adequate and may even give false confidence in data simply because a measurement had been done. Moll (1973) feels that some measurement of the acid–base balance in fetal blood is a more reliable indicator of hypoxia than is any type of oxygen measurement.

Another important consideration in defining degree of hypoxia is the duration of exposure. An animal that has had several hours of prior exposure to hypoxia does not react in the same way to an additional hour of oxygen deficiency as does an animal having no prior exposure. A simple experiment with rats in which fetal heart rate at $17\frac{1}{2}$ days of gestation was monitored during maternal exposure to 5% O_2 (Grabowski, unpublished results) clearly illustrates this point. Several hours of such exposure are required to induce significant bradycardia; but after 5 hours of 5% O_2, the fetal heart rate returns to near normal levels in less than 2 mins after the mother is returned to normal air. However, returning a previously hypoxia-treated rat to atmosphere of 5% O_2 will reduce the fetal heart rate to one third of normal within 30 sec. This experiment dramatically illustrates the importance of the time factor in determining the severity of hypoxic exposure.

Such observations indicate the difficulty of evaluating precisely the degree of hypoxia to which the fetus was exposed. The only experiments known to the author in which the most pertinent factors were monitored with some degree of adequacy are a few on large mammals such as sheep, pigs, large primates, and humans, during late fetal and birth periods (Dawes, 1968; Ginsberg and Myers, 1974). Even these experiments are now being subject to scrutiny because of recent evidence that the elaborate surgical techniques required to exteriorize a fetus and prepare it for monitoring may result in significantly disturbed physiology (Rudolph, 1974).

Another question concerns the oxygen requirements of the mammalian embryo, especially at the early stages of development. Although much is known of the metabolism of late fetal stages, particularly of large mammals, the young embryo remains an enigma. Does it truly develop "in an Everest *in utero*" (Boell, 1955), and if so does this explain its seeming resistance to hypoxia during early stages? To resolve this question there is need for more sophisticated experiments such as those of Shepard *et al.* (1970), who found in carefully explanted rat embryos a big difference in the energy metabolism between 11-somite (10 $\frac{1}{3}$ days) and 35-somite (12 days) stages. The younger embryos converted relatively large amounts of glucose to lactate, and high oxygen concentrations actually inhibited the embryonic heart rate. By the 35-somite stage an oxygen dependency was clearly established. More of such studies are needed to help evaluate the teratogenic potential of hypoxia. It is interesting to note in this respect that in a recent review entitled "Oxygen Supply to the Human Fetus," Moll (1973) could quote only one study in which the oxygen requirements of a human fetus was actually determined (Assali *et al.*, 1960).

5. Mechanisms of Action of Hypoxia

Simple oxygen deficiency is probably rarely, if ever, the only causal factor in a teratogenic situation. Depending on method of induction and the animal used, hypoxia can be accompanied by nutritional restriction, hypercapnia, acid–base disturbances, lactic acid accumulation, hyperkalemia, bradycardia, hypotension, and fluid disturbnaces. Although it is often assumed that the teratogenic action of hypoxia is due to the simple fact that oxygen-deficient cells or organs are damaged (see Stockard, 1921), the action of hypoxia could be enhanced by one or more of the above associated factors.

In some cases the teratogenic action of hypoxia has been demonstrated, or at least suspected, to be directly mediated by one of the associated effects mentioned. Hypercapnia accompanies most forms of mammalian hypoxia and has been shown to be teratogenic in itself (see Section II.C). Lactic acid accumulation has been demonstrated to occur during hypoxia in the chick, and some teratogenic effects were duplicated by injection of this compound (Grabowski, 1961). Most of the teratogenic and lethal effects of moderate hypoxia in the chick embryo have been demonstrated to be caused, not by lack of oxygen itself, but by a disturbance in osmoregulation, leading to a complex edema syndrome (see chapter by Grabowski, Vol. 2). A comparable syndrome has been described in hypoxic rats in which uterine clamping led to edema, venous dilation, blood vessel rupture, necrosis, and malformations (Leist and Grauwiler, 1974). Hypoxia in rats induced by epinephrine (Chernoff and Grabowski, 1971) or sodium nitrite (Grabowski and Chernoff, 1970) has been shown to result in fetal bradycardia, hypotension, and collapse of peripheral fetal vessels. Furthermore these changes are accompanied by an increase in fetal serum potassium far greater than that tolerated by the adult mammalian heart (Grabowski, 1973). The precise relationship between these complex cardiovascular responses of the fetus to hypoxia and the eventual development of abnormalities remains to be worked out, but it is obvious that some of them at least must potentiate the action of oxygen deficiency.

6. Hypoxia as a Human Teratogen

Oxygen deficiency has been frequently suggested as a cause of human abnormalities (Dareste, 1877; Büchner, 1955; Rübsaamen, 1950; Ingalls, 1956; Windle, 1966), but such claims were based primarily on extrapolation of animal data to humans and on accumulated case histories involving possible associations. More direct evidence, particularly with respect to the central nervous system, was provided by Towbin (1969), who examined the brains of numerous individuals dying in the perinatal period and described hypoxic damage as occurring before and during birth. In some of the individuals who lived into the postnatal period, the damage could be correlated with neurological disorders. Jílek *et al.* (1970), using statistics obtained from around the world, estimate that hypoxia may be the cause of perinatal death and permanent disorders of the central nervous system in as much as 70–80% of injured newborns.

How significant, then, is hypoxia as a human teratogen? Consider (1) that the broader definition of teratology now used includes functional as well as structural abnormalities induced prenatally and perinatally; (2) that many different types of abnormalities, some in high frequency, are produced by a variety of forms of hypoxia other than that induced in a low-oxygen chamber; (3) that there are many situations in which pregnant women may be exposed to hypoxia; and (4) that there are many possibilities for synergistic action involving hypoxic agents. The question of human teratogenic potential, then, should be rephrased as, how much hypoxia, at which stage of development, is necessary to induce irreparable damage? Data to answer this question for any species including man are insufficient.

Three examples are particularly relevant to the question of human toxicity:

1. Ephinephrine and other uterine vasoconstrictive drugs are naturally occurring agents which are concentrated in the blood stream as a result of stress; they have been used to treat asthma and acute allergic reactions. They have long been known to induce significant bradycardia in human fetuses (Beard, 1962; Donarin, 1964), as well as in monkeys (Adamsons, 1971), rats and mice (Chernoff and Grabowski, 1971), and rabbits and guinea pigs (Dornhorst and Young, 1962). How much fetal bradycardia and resulting hypotension (Chernoff and Grabowski, 1971) for how long will induce sufficient stagnant hypoxia to be dangerous to the development of human tissues? Is this link between epinephrine, uterine vasoconstriction, and hypoxia a possible explanation for the lethal and teratogenic effects of generalized stress (Archer and Blackman, 1970; Geber, 1966; Geber and Anderson, 1967; Hamburgh et al., 1974; Ward, 1972)? If so, these observations could be extremely important for human teratology.

2. Much has been written on the effect of carbon monoxide on the adult organism, and there is growing concern that exposure in urban and industrial situations is greater than had been suspected previously (Stewart et al., 1974). Despite this concern the prenatal effects have been little studied (see Section III. A).

3. Sodium nitrite is a common food additive as well as an accidental contaminant of both food and well water. The general population is constantly exposed to low levels of this substance; epidemics of poisoning due to overdoses, leading to acute symptoms and deaths, particularly in infants, have been reported. This substance had never been studied as a teratogen until recently (Food and Drug Administration, 1972). It was given tentative clearance as far as these studies were concerned even though some retardation of the skeletal system was obtained at the maximum dosage and this dosage was only about 1/10th of that required to produce symptoms of acute toxicity. In

the author's laboratory it has been found that a single injection of $NaNO_2$ can induce acute bradycardia and hypotension in rat fetuses for periods of up to 5 hr (Grabowski and Chernoff, 1970). Every pregnant woman is liable to frequent exposure to some concentration of epinephrine, carbon monoxide, and sodium nitrite. Granted that there is a safety margin with respect to oxygen requirements of the fetus, the question remains, how much hypoxia is too much, especially as regards the vulnerable CNS?

B. Toxicity of Oxygen Excess

If 21% oxygen in the atmosphere is good, is not 30 or 40% better, particularly in times of physiological stress? This is an old question, asked not only by teratologists but also by obstetricians and such practical men as poultry breeders. The latter in particular are interested in very pragmatic matters such as improving the hatchability of incubators containing thousands of developing embryos. This question was definitively settled by Barrott (1937), who found that hatchability was best at precisely 21% oxygen, and decreased sharply when greater or lesser concentrations of oxygen were used. This was recently confirmed by Taylor et al. (1956) and Taylor and Kreutziger (1966).

The effects of excess oxygen at normal pressures has not been studied in pregnant mammals, but an excess of oxygen after birth is known to have a striking morphogenetic effect on newborns, especially prematures. Before the toxic effect of oxygen was understood, retrolental fibroplasia developed in about 10–15% of all premature infants (those weighing less than 3 lb at birth) and in about 1% of infants weighing 3–5 lb at birth. This condition is now known to be caused by exposure of premature infants to high concentrations of oxygen (above 40% for periods of more than a day), a practice once used in an effort to alleviate respiratory problems. Today premature infants of this size are given oxygen for no more than 24 hr, and the concentration is kept below 40% unless necessary to maintain life (Hellman and Pritchard, 1971).

The effects of hyperbaric oxygen have been studied in both chicks and hamsters. Lehrer et al. (1968) found abnormalities in 30–50% of chick survivors after treatment with 100% O_2 at 30 lb of pressure for 6 hr on days 3–5 of incubation. Ferm (1964) found increased resorptions and some gross abnormalities when pregnant hamsters were treated on days 6–8 of gestation with hyperbaric oxygen at 3–4 atmosphere for 2–3 hr.

C. Carbon Dioxide

The effects of excess carbon dioxide on developing chick embryos have been well documented. Hatchability was significantly reduced with as little as 1.5% carbon dioxide in the atmosphere at the early stages of development, although older embryos tolerate 5–8% without significantly decreasing the percent hatching (Barrott, 1937; Taylor *et al.,* 1956, 1971). Hypercapnia has also been shown to be teratogenic in the chick (Gallera, 1936; Jaffee, 1974).

In mammals hypercapnia is somewhat more difficult to induce, but it has been shown to be teratogenic in the rat. A significant frequency of cardiac malformations due to exposure to excessive CO_2 was found by Haring and Polli (1957) and Haring (1960). In an interesting observation King *et al.* (1962) have shown that exposure of rats to hypercapnia early in development induces postnatal susceptibility to dental caries.

D. Nitrogen

Although nitrogen is the most common constituent in the atmosphere, and apparently also the most innocuous, there is an observation which is of interest to scuba diving enthusiasts. Exposure to excess air pressure for prolonged periods of time causes nitrogen to dissolve in the bloodstream. If ascension from significant depths of water occurs too rapidly, the dissolved nitrogen is released as bubbles, causing the familiar and extremely painful condition known as the bends. Recent information indicates that prolonged submersion, even at nominal scuba depths, can also cause nitrogen solubility and the release of some small bubbles, with or without pain (U. S. Navy Diving Manual, 1970). Considering the teratogenic effects of small hematomas in an embryo (Grabowski, 1964, 1970), it is permissible to wonder what a small bubble of nitrogen could do to the fragile tissues of a young developing embryo. No studies are known on the teratogenic effects of nitrogen under pressure, but in view of the popularity of scuba diving, this would be a suitable subject for investigation.

III. COMMON ATMOSPHERIC POLLUTANTS

A. Carbon Monoxide

Carbon monoxide is the most common as well as the deadliest of atmospheric pollutants. It so effectively inactivates the oxygen-carrying capacity of the hemoglobin molecule that it produces anemic hypoxia. There is a possibility that this effect may be directly enhanced at the cellular level because carbon monoxide also inhibits cytochrome oxidase, but this is usually thought to be more significant *in vitro* than in *in vivo* systems. In this country 1400 deaths a year are caused by carbon monoxide poisoning (not counting 2300 suicides) and, 10,000 people are estimated to need medical attention annually due to the effects of carbon monoxide (DuBois, 1973). Current federal air-quality standards have been set to protect individuals from building up a carboxyhemoglobin level above about 2% (Anon., *Federal Register,* 1971). This is based on indications that as little as 2–6% of carboxyhemoglobin can impair alertness (Schulte, 1963; Beard and Wertheim, 1967; Horvath *et al.,* 1971) and 3% can be dangerous in other ways to the cardiovascular system, especially in people with other health problems (Ayres *et al.,* 1969; DuBois, 1973). A recent survey (Stewart *et al.,* 1974), in which venous blood samples were obtained for carboxyhemoglobin analysis from 29,000 blood donors across the United States, demonstrated that 45% of all the nonsmoking blood donors tested had a carboxyhemoglobin saturation of more than 1.5% whereas the mean for cigarette smokers was approximately 5%, including some values as high as 13%. Heavy smoking, and to a lesser degree, exposure to high vehicular traffic and certain industrial occupations were the greatest factors in increasing the mean carboxyhemoglobin content in the blood.

Carbon monoxide is probably the most thoroughly studied pollutant at the biological as well as chemical levels, yet it is surprising to realize, despite the volumes of other work, that virtually nothing has been done from the teratogenic standpoint. There are few articles describing clinical case histories of accidental carbon monoxide intoxication of pregnant women (Mueller and Graham, 1955; Goldstein, 1965; Longo, 1970; Beaudoing *et al.,* 1969). These usually resulted in fetal mortality, with surviving offspring showing physical abnormalities and neuromotor impairment. Only three studies are known to the writer of experimental teratogenesis using this agent. Astrup *et al.* (1972) and Astrup (1972) found increased mortality and lowered birth weight and a few abnormal extremities in the offspring of a group of 36 pregnant rabbits exposed to a moderate concentration of CO (180 ppm) for 30 days. Tumasomis and Baker (1972) reported hypertrophied hearts and enzymatic changes in chick embryos exposed to a high concentration of CO (425 ppm) for several days. Ginsberg and Myers (1974) exposed 9 near-term rhesus monkeys to acute (0.1–0.3%) CO intoxication for 1–3 hr. Of the newborn animals 4 appeared normal, but the other 5 demonstrated severe neurological symptoms and on autopsy had gross evidence of brain damage. Recent studies in the author's laboratory (Ezrin and Grabowski, unpublished results) have demonstrated skeletal abnormalities, cardiovascular impairment, and postnatal behavioral abnormalities in rats exposed to moderate carbon monoxide levels (1000 ppm for 6 hr and 300 ppm for 1–3 days) at various gestational stages.

Recent controversies about the validity of the new federal regulations on acceptable carbon monoxide levels (Heuss *et al.,* 1971; Barth *et al.,* 1971) have centered on the question of whether for adults there is a threshold level for CO below which no adverse effects can occur. From a teratological standpoint, information is too limited to permit evaluation of the effects of either long-term or short-term exposure. The well-known phenomenon of reduced birth weight in children of mothers who are heavy smokers, and the more recent observation of significant physical and mental retardation in children of such smokers at ages 7–11 years (Butler and Goldstein, 1973), could be related to elevated carboxyhemoglobin levels in the blood of these mothers. Finally, there is the recurring question of to what extent variable degrees of exposure to carbon monoxide can have synergistic effects with other hypoxic agents and other teratogens.

B. Ozone

Ozone is another common atmospheric pollutant and is a dominant constituent of photochemical smog. Its toxic effect is recognized not only as an acute, fast-acting pollutant but also one with potential for long-term exposure at levels as low as 0.05–0.1 ppm. In man it can produce changes in visual acuity, headache, and pulmonary congestion. The physiological responses in mammals are quite varied and include interference with breathing as well as radiomimetic toxicity due to the release of free radicals (Stockinger and Coffin, 1968). Ozone concentrations in Los Angeles smog are reported frequently to reach 0.5 ppm. During September the mean daily level is about 0.2 ppm (Mosher *et al.*, 1970). Since this exceeds the level at which toxicity can be recognized, there is obvious concern over the level of atmospheric pollution with this substance. Only a single report on prenatal effects is known in which increased mortality was reported C57 mice subsequent to exposure of pregnant females to 0.1–0.2 ppm for 7 hr/day, 5 days a week, for 3 weeks (Brinkman *et al.*, 1964).

C. Sulfur Dioxide, Nitrogen Oxides, and Lead

Sulfur dioxides come from the burning of fossil fuels and also are present in automobile exhaust fumes. Several nitrogen oxides are present in auto exhaust, but NO_2 is the predominant form. It, like SO_2, is primarily a pulmonary irritant. The effects of chronic and acute exposure to either of these are varied and fairly well documented (Stockinger and Coffin, 1968). No teratogenic studies are known for either of these pollutants; however, Freeman *et al.* (1974) found some delay in the maturation of lungs of newborn rats exposed for 75 days to air containing 15 ppm of NO_2.

The teratogenicity of lead is discussed in Chapter 9 (see metals), but a few comments on pollution by atmospheric lead are in order here. Approximately 40% of the lead used in the United States is used as the gasoline additives tetraethyl lead and tetramethyl lead; it is released from auto exhaust primarily as particulate lead halides. These can enter the circulatory system by phagocytosis by lung macrophages and by a complex cycle of pulmonary inhalation, ciliary transport of particles to the glottis, and thence into the digestive tract where absorption may occur (Bingham *et al.*, 1968).

D. Miscellaneous Atmospheric Pollutants

To the list of common pollutants discussed above can be added a tremendous number of agents of either known or possible teratogenic danger which are occasionally distributed via the atmosphere. These are mostly of industrial origin and are therefore of regional rather than general significance. Included in this category can be large variety of organic compounds such as vinyl chloride as well as inorganics such as fluoride and arsenic. The latter two, for example, have been released into the air in sufficient quantity to cause well-documented toxicity in livestock (Stockinger and Coffin, 1968). The addition of large numbers of women to the labor force increases the danger of exposure of pregnant women to potentially dangerous air-born pollutants.

REFERENCES

Adamsons, K., Mueller-Heuback, E., and Myers, R. E., 1971, Production of fetal asphyxia in the rhesus monkey by administration of catecholamines to the mother, *Am. J. Obstet. Gynecol.* **109**:248–262.

Anon., 1971, National primary and secondary ambient air quality standards, *Fed. Regist.* **36**:8186–8201.

Archer, J. E., and Blackman, D. E., 1971, Prenatal psychological stress and offspring behavior in rats and mice, *Dev. Psychobiol.* **4**(3):193–248.

Assali, N. S., Rauramo, L., and Peltonen, T., 1960, Measurements of uterine blood flow and uterine metabolism, VIII Uterine and fetal blood flow and oxygen consumption in early human pregnancy, *Am. J. Obstet. Gynecol.* **79**:86–98.

Astrup, P., 1972, Some physiological and pathological effects of moderate carbon monoxide exposure, *Br. Med. J.* **1972**:447–452.

Astrup, P., Trolle, D., Olsen, H. M., and Kjeldsen, K., 1972, Effect of carbon monoxide exposure on fetal development. *Lancet* **1972**:1220–1222.

Ayres, S. M., Mueller, H. S., Gregory, J. J., Gianelli, S., and Penny, J. L., 1969, Systemic and myocardial hemodynamic response to relatively small concentrations of carboxyhemoglobin (COHB), *Arch. Environ. Health* **18**:699–709.

Barrott, H. G., 1937, Effects of temperature, humidity, and other factors on hatch of hen's eggs and on energy metabolism of chick embryos, *U.S. Dept. Agric. Bull. No. 553*, pp. 1–45.

Barth, D. S., Romanovsky, J. C., Knelson, J. H., Altshuller, A. P., and Horton, R. J. M., 1971, Discussion of the report of Heuss *et al.*. (see below). *J. Air Pollut. Control Assoc.* **21**:544–548.

Beard, R. W., 1962, Response of the fetal heart and maternal circulation to adrenalin and nonadrenalin, *Br. Med. J.* **1962**:443–446.

Beard, R. R., and Wertheim, G. A., 1967, Behavioral impairment associated with small doses of carbon monoxide, *Am. J. Public Health* **57**:2012–2022.

Beaudoing, A., Gachon, J., Butin, L. P., and Bost, M., 1969, Les conséquences foetales de l'intoxication oxycarbonée de la mère, *Pèdiatrie* **24**:539–553.

Bingham, E., Pfitzer, E. A., Barkley, W., and Radford, E. P., 1968, Alveolar macrophages: Reduced number in rats after prolonged inhalation of lead sequioxide, *Science* **162**:1297–1299.

Bodansky, O., 1951, Methemoglobinemia and methemoglobin-producing compounds, *Pharmacol. Rev.* **3**:144–196.

Boell, E. J., 1955, Energy exchange and enzyme development during embryogenesis, *in: Analysis of Development* (B. H. Willier, P. A. Weiss, and V. Hamburger, eds.) pp. 520–555, W. B. Saunders, Philadelphia.

Brent, R. L., and Franklin, J. B., 1960, Uterine vascular clamping: New procedure for the study of congenital malformations, *Science* **132**:89–91.

Brinkman, R., Lamberts, H. B., and Veninga, T. S., 1964, Radiomimetic toxicity of ozonised air, *Lancet* **1964**:133–136.

Büchner, F., 1955, Experimentelle Entwicklungsstörungen durch allgemeinen Sauerstoffmangel, *Klin. Wochenschr.* **26**:38–42.

Butler, N. R., and Goldstein, H., 1973, Smoking in pregnancy and subsequent child development, *Br. Med. J.* **4**:573–575.

Chernoff, N., 1969, Physiological and teratological effects of epinephrine and vasopressin on the fetal rat, Ph. D. dissertation, University of Miami, Coral Gables, Florida.

Chernoff, N., and Grabowski, C. T., 1971, Responses of the rat foetus to maternal injections of adrenaline and vasopressin, *Br. J. Pharmacol.* **43**:270–278.

Crist, T., and Hulka, J. F., 1970, Influence of maternal epinephrine on behavior of offspring, *Am. J. Obstet. Gynecol.* **106**:687–691.

Dareste, M. C., 1877 (2nd ed. 1891), *La production artificielle des monstruosités,* C. Reinwald et Cie, Paris.

Dawes, G. S., 1968, *Foetal and Neonatal Physiology: A Comparative Study of the Changes at Birth,* Year Book Medical Publishers, Chicago.

Degenhardt, K. H., 1954, Durch O_2—Mangel induzierte Fehlbildungen der Axialgradienten bei Kaninchen, *Z. Naturforsch.* **9**:530–536.

Degenhardt, K. H., 1958, Analysis of intra-uterine malformations on the vertebral column induced by oxygen deficiency in rabbits, *Proc. Xth Int. Congr. Genet.,* Program 2.

Donarin, A., 1964, Noradrenaline and the foetal heart, *Lancet* **1964**:756.

Dornhorst, A. C., and Young, I. M., 1952, The action of adrenaline and noradrenaline on the placental and foetal circulation in the rabbit and guinea pig, *J. Physiol.* **118**:282–288.

DuBois, A. B., 1973, Reprot of Ad Hoc Committee on Carbon Monoxide Poisoning, Bureau of Community Environmental Management, Rockville, Maryland.

Ericson, A. E., 1960, The effect of maternal hypoxia on metanephric development in the hamster, Ph. D. dissertation, Boston University, Boston, Massachusetts.

Ferm, V. H., 1964, Teratogenic effects of hyperbaric oxygen, *Proc. Soc. Exp. Biol. Med.* **116**:975–976.

Food and Drug Administration, 1972, Teratologic evaluation of FDA 71-9 (sodium nitrite), *Bulletin PB-221 794,* distributed by National Technical Information Service.

Franklin, J. B., and Brent, R. L., 1964, The effect of uterine vascular clamping on the development of rat embryos three to fourteen days old, *J. Morphol.* **115**:273–290.

Freeman, G., Julos, L. T., Furiosi, N. J., Mussenden, R., and Weiss, T. A., 1974, Delayed maturation of rat lung in an environment containing nitrogen dioxide, *Am. Rev. Respir. Dis.* **110**:754–759.

Gallera, J., 1936, Production artificielle des monstres platyneuriques, *Folia Morphol.* **6**:203–251.

Geber, W. F., 1966, Developmental effects of chronic maternal audiovisual stress on the rat fetus, *J. Embryol. Exp. Morphol.* **16**:1–16.

Geber, W. F., and Anderson, T. A., 1967, Abnormal fetal growth in the albino rat and rabbit induced by maternal stress, *Biol. Neonat.* **11**:209–215.

Ginsberg, M. D., and Myers, R. E., 1974, Fetal brain damage following maternal carbon monoxide intoxication: An experimental study, *Acta Obstet. Gynecol. Scand.* **53**:309–317.

Goldstein, D. P., 1965, Carbon monoxide poisoning in pregnancy, *Am. J. Obstet. Gynecol.* **92**:526.

Grabowski, C. T., 1961, Lactic acid accumulation as a cause of hypoxia-induced malformations in the chick embryo, *Science* **134**:1359–1360.

Grabowski, C. T., 1964, The etiology of hypoxia-induced malformations in the chick embryo, *J. Exp. Zool.* **157**:307–326.

Grabowski, C. T., 1970, Embryonic oxygen deficiency—a physiological approach to analysis of teratological mechanisms, *Adv. Teratol.* **4**:125–169.

Grabowski, C. T., 1973, Fetal cardiac physiology and hypoxia-induced hyperkalemia, *Teratology* **7**:A-16.

Grabowski, C. T., and Chernoff, N., 1970, Effects of hypoxia on the cardiovascular physiology of mammalian embryos, *Teratology* **3**:201.

Hamburgh, M., Mendoza, L. A., Rader, M., Lang, A., Silverstein, H., and Hoffman, K., 1974, Malformations induced in offspring of crowded and parabiotically stressed mice, *Teratology* **10**:31–37.

Haring, O. M., 1960, Cardiac malformations in rats induced by exposure of the mother to carbon dioxide during pregnancy, *Cir. Res.* **8**:1218–1227.

Haring, O. M., and Polli, J. F., 1957, Experimental production of cardiac malformations, *A.M.A. Arch. Pathol.* **64**:290–296.

Hellman, L. M., and Pritchard, J. A., 1971, *Williams Obstetrics*, Appleton-Century, Crofts, New York.

Heuss, J. M., Nebel, G. J., and Colucci, J. M., 1971, National air quality standards of automotive pollutants—a critical review, *J. Air Pollut. Control Assoc.* **21**:535–544.

Horvath, S. M., Dahms, T. E., and O'Hanlon, J. F., 1971, Carbon monoxide and human vigilance, *Arch. Environ. Health* **23**:343–347.

Ingalls, T. H., 1956, Causes and prevention of developmental defects, *J. Am. Med. Assoc.* **161**:1047–1051.

Ingalls, T. H., and Curley, F. J., 1957, Principles governing the genesis of congenital malformations induced in mice by hypoxia. *N. Engl. J. Med.* **257**:1121–1127.

Jaffee, O. C., 1974, The effects of moderate hypoxia and moderate hypoxia plus hypercapnea on cardiac development in chick embryos, *Teratology* **10**:275–282.

Jílek, L., Trávníčková, E., and Trojan, S., 1970, Characteristic metabolic and functional responses to oxygen deficiency in the central nervous system, *in: Physiology of the Perinatal Period*, Vol. II, (U. Stave, ed.), Appleton-Century, Crofts, New York.

Kalter, H., 1968, *Teratology of the Central Nervous System*, University of Chicago Press, Chicago.

Kalter, H., and Warkany, J., 1959, Experimental production of congenital malformations in mammals by metabolic procedures, *Physiol. Rev.* **39**:69–115.

King, C. T. G., Wilk, A., and McClure, F. J., 1962, Carbon dioxide-induced acidosis in pregnant rats and carries susceptibility of their progeny, *Proc. Soc. Exp. Biol. Med.* **111**:486–489.

Lehrer, S. B., Usubiaga, L. E., and Smith, B. E., 1968, Hybaroxia, a useful new tool in the study of teratogenesis, *Teratology* **1**:218.

Leist, K. H., and Grauwiler, J., 1974, Fetal pathology in rats following uterine-vessel clamping on day 14 of gestation, *Teratology* **10**:55–61.

Longo, L., 1970, Carbon monoxide in the pregnant mother and fetus and its exchange across the placenta, *Ann. N.Y. Acad. Sci.* **174**:313–341.

Meier, G. W., Bunch, M. E., Nolan, C. Y., and Scheidler, C. H., 1960, Anoxia, behavioral development, and learning ability: A comparative–experimental approach, *Psychol. Monogr.* **74**:1–48.

Moll, W., 1973, Oxygen supply to the human fetus, *Bull. Physio.-Pathol. Respir.* **9**:1345–1364.

Mosher, J. C., MacBeth, W. G., Leonard, M. J., Mullins, T. P., and Brunelle, M. F., 1970, The distribution of contaminants in the Los Angeles basin resulting from atmospheric reaction and transport, *J. Air Pollut. Control Assoc.* **20**:35–42.

Mueller, G. L., and Graham, S., 1955, Intrauterine death of the fetus due to carbon monoxide poisoning, *N. Engl. J. Med.* **252**:1075–1078.

Murakami, U., and Kameyama, Y., 1963, Vertebral malformation in the mouse foetus caused by maternal hypoxia during early stages of pregnancy. *J. embryol. Exp. Morphol.* **11**:107–118.

Petropoulos, E. A., Vernadakis, A., and Timiras, P. S., 1969, Nucleic acid content in developing rat brain after prenatal and/or neonatal exposure to high altitude, *Fed. Proc.* **28**:1001–1005.

Rübasaamen, H., 1950, Uber die teratogenetische Wirkung des Missbildungen bei Mensch und Tier, *Beitr. Pathol. Anat.* **112:**336–379.

Rudolph, A. M., 1974, quoted by Hafez, E. S. E., The mammalian fetus: A dedicational symposium, a report of a metting, *Teratology* **10:**9–12.

St.-Hilaire, G., 1820, Des différents états de pesanteur des oeufs au commencement et à la fin de l'incubation, *J. Complentaire Sci. Méd.* **7:**271 (quoted in Dareste, 1877).

Schulte, J. H., 1963, Effects of mild carbon monoxide intoxication, *Arch. Environ. Health* **7:**524–530.

Shepard, T. H., Tanimura, T., and Robkin, M. A., 1970, Energy metabolism in early mammalian embryos, *in: Changing Syntheses in Development,* 29th Symposium of the Society for Developmental Biology (M. Runner, ed.), pp. 42–58.

Stewart, R. D., Baretta, E. D., Platte, L. R., Stewart, E., Kalbfleisch, J. H., van Yserloo, B., and Rimm, A. A., 1974, Carboxyhemoglobin levels in American blood donors, *J. Am. Med. Assoc.* **229:**1187–1195.

Stockard, C. R., 1921, Developmental rate and structural expression, An experimental study of twins, "double monsters," and single deformities, and the interaction among embryonic organs during their origin and development. *Am. J. Anat.* **28:**115–278.

Stockinger, H. E., and Coffin, D. L., 1968, Biologic effects of air pollution, *Air Pollut.* **1:**445–546.

Taylor, L. W., and Kruetziger, G. O., 1966, The gaseous environment of the chick embryo in relation to its development and hatchability, 3. Effect of carbon dioxide and oxygen levels during the period of the ninth through the twelfth days of incubation, *Poult. Sci.* **45:**867–884.

Taylor, L. W., Sjodin, R. A., and Gunns, C. A., 1956, The gaseous environment of the chick embryo in relation to its development and hatchability, *Poult. Sci.* **35:**1206–1215.

Taylor, L. W. Kreutziger, G. O., and Abercrombie, G. L., 1971, The gaseous environment of the chick embryo in relation to its development and hatchability, 5. Effect of carbon dioxide and oxygen levels during the terminal days of incubation, *Poult. Sci.* **50:**66–78.

Thompson, W. R., and Olian, S., 1961, Some effects on offspring behavior of maternal adrenalin injection during pregnancy in three inbred mouse strains, *Psychol. Rep.* **8:**87–90.

Timiras, P., and Woolley, D. E., 1966, Functional and morphologic development of brain and other organs of rats at high altitudes, *Fed. Proc.* **25:**1312–1320.

Towbin, A., 1969, Mental retardation due to germinal matrix infarction, *Science* **164:**156–160.

Tumasonomis, C. F., and Baker, F. D., 1972, Influence of carbon monoxide upon some respiratory enzymes of the chick embryo. *Bull. Environ. Contam. Toxicol.* **8:**113–119.

U.S. Navy Diving Manual, 1970, NAVSHIPS 0994-001-9010, Navy Dept., Washington, D.C.

Van Liere, E. J., and Stickney, J. C., 1963, *Hypoxia, A Detailed Review of the Effects of Oxygen Want on the Body,* University of Chicago Press, Chicago.

Vierck, C. J., and Meier, G. W., 1963, Effects of prenatal hypoxia upon locomotor activity of the mouse, *Exp. Neurol.* **7:**418–425.

Ward, T. L., 1972, Prenatal stress feminizes and demasculinizes the behavior of males. *Science* **175:**82–84.

Wilson, J. G., 1972, Environmental effects on development, *in: Pathophysiology of Gestation,* Vol. 2 (N. S. Assali, ed.), Academic Press, New York.

Windle, W. F., 1963. Neuropathology of certain forms of mental retardation, *Science* **140:**1186–1189.

Windle, W. F., 1966, Role of respiratory distress in asphyxial brain damage of the newborn, *Cerebral Palsy J.* **27:**3–7.

Windle, W. F., and Becker, R. F., 1943, Asphyxia neonatorium, an experimental study in the guinea pig, *Am. J. Obstet. Gynecol.* **45:**183–200.

Extremes of Temperature 12

MARSHALL J. EDWARDS and
RAOUL A. WANNER

I. INTRODUCTION

For many years excess heat has been recognized as a powerful inhibitor of normal cellular proliferation in such widely differing systems as the testes and cell cultures. Also it has been recognized as an agent which can cause heavy mortality in preimplantation embryos. It is surprising that until recently it has not been examined in any detail as an agent capable of causing disturbances of prenatal development. The effects of hypothermia have attracted even less attention. Both hyperthermia and hypothermia have produced severe developmental defects, particularly of the central nervous system, as well as of the skeleton and soft tissue. Although this review will concentrate on extremes of temperature and prenatal malformations, some reference will be made to the embryonic wastage associated with changes of temperature during development.

II. HYPERTHERMIA

A. Methods of Inducing Hyperthermia

1. Involving Whole-Body Hyperthermia

a. Exposure to High Environmental Temperatures. Incubators of various types have been used for chickens, laboratory animals, and monkeys (Alsop, 1919; Hsu, 1948; Kreshover and Clough, 1953; Pennycuik, 1965; Edwards, 1967, Lecyk, 1969; Poswillo *et al.*, 1974). A bacteriological incubator set at 44–45°C without positive air circulation was used by Edwards (1967) to

MARSHALL J. EDWARDS and RAOUL A. WANNER · Department of Veterinary Medicine, The University of Sydney, Sydney, Australia.

heat guinea pigs and rats. Subsequently animals were stressed for 1 hr in a forced-draught egg incubator set at 42–43.5°C dry bulb (DB) and 23–27.5°C wet bulb (WB) (Edwards, 1968, 1969a–c, 1971a,b). During the process animals were caged singly or in pairs in plastic boxes with wire tops, and the deep rectal temperatures were recorded at the beginning and end of the period. Temperatures of some animals were recorded continuously using an indwelling rectal thermistor with the animal enclosed snugly in an open-mesh wire cage to prevent removal of the device. By varying the incubator temperature to suit individual animals it is usually possible to achieve a wide range of body temperatures.

In the bigger species (pigs, sheep) large environmental chambers set at high temperatures (35–44°C) and with controlled humidity have been used (Yeates, 1953; Tompkins *et al.*, 1967; Alexander and Williams, 1971; Omtvedt *et al.*, 1971.)

The temperature of rabbits has been elevated by immersion in water at 45°C for 20 min (Cameron, 1943).

b. Shortwave Diathermy. Body temperature elevations have been induced by shortwave diathermy in rats and rabbits (Hofmann and Dietzel, 1966; Dietzel and Kern, 1970). Temperatures of 42.7°C have been achieved by applying the electromagnetic field to the lower abdomen.

c. The Use of Pyrogens. Intramuscular injections of boiled milk elevated the body temperature of pregnant rabbits by 1–2.2°C (Brinsmade and Rubsaamen, 1957). Although purified and more specific pyrogens are now available, their use does not seem to have been reported in the literature of teratology.

2. Involving Localized Heating of or About the Uterus

By applying heat directly to the uterus and its contents, or to a localized area including the uterus, many of the complicating effects and dangers of whole-body heating are eliminated.

a. Microwave Radiation. Brent and Wallace (1972) adapted a microwave oven to elevate the temperature of the rat uterus and its contents.

b. Local Heat Applied to the Exteriorized Uterus. The pregnant uterus of the rat has been exteriorized and immersed in saline at 40–41°C or 38–43°C for 40–60 min (Skreb and Frank, 1963; Skreb, 1965).

B. Species Studied and Abnormalities Produced

1. Sheep

Yeates (1953) reported a decline in the number and birth weight of lambs born to ewes exposed for 6–7 hr daily to environmental temperatures of 40.5°C DB and 30.3°C WB or 41.7°C DB and 30.3°C WB during the last third or two thirds of gestation. In a more detailed experiment Alexander and Williams (1971) exposed Merino ewes for 9 hr daily to an air termperature of 44°C and 47% relative humidity during the middle third, last third, or last two thirds of gestation, which resulted in 34.4% reduction in birth weight of lambs and a general disproportion between brain weight and body weight. Cavitation of the white matter of brains was observed in a very large proportion of heated lambs (Hartley *et al.*, 1974).

2. Pigs

Elevated environmental temperatures have reduced embryonic survival in pigs (Tompkins *et al.*, 1967; Edwards *et al.*, 1968; Omtvedt *et al.*, 1971). Details of these experiments are given later.

3. Chickens

Alsop (1919) produced abnormalities in 90% of 1–3-day-old chick embryos by incubation at 39.1–42.2°C. The majority of defects were in the head or neural tube regions. High mortality and abnormalities including hernias and defects of the head, beak, eyes, and spine followed incubation of chick eggs at 39–43°C (normal incubation temperature 37°C) for 1, 2, or 3 days during the first 6 days of embryogenesis (Nielsen, 1969).

4. Rats

Hsu (1948) reported microphthalmia in pups from a rat subjected to heat on days 1–11 of gestation. However, this mother also developed a "spontaneous" fever at 9.5 days of gestation, and Kalter and Warkany (1959) found that this abnormality was not uncommon in the strain of rat used. Kreshover and Clough (1953) induced hyperthermia in pregnant rats in an incubator set at 37.2–38.9°C for 12, 24, or 48 hr. The rectal temperatures, recorded hourly, reached levels between 38.9 and 40.6°C. Hyperthermia induced after the 9th day of pregnancy regularly resulted in severe tooth defects in both the mother and young. Complete cellular degeneration with arrest of enamel matrix formation was observed in some animals. Exposure prior to the 9th day of gestation was associated with a lower incidence of dental damage.

Wilson (1959) failed to produce congenital abnormalities in the rat with elevated maternal temperatures. Skreb and Frank (1963) and Skreb (1965) also failed to produce malformations by raising maternal body temperature to 42°C and believed that temperature elevations of this magnitude were lethal to the mother. However, they induced a high incidence of abnormalities of gestation and fetal development by immersion of the pregnant uterus in saline heated to 40–41°C for 40–60 min on days 7.5–15.5 of gestation. Using more precise equipment Skreb (1965) demonstrated that abnormalities occurred only when the temperature of the saline exceeded 42°C. The incidence of resorption was highest after heat on days 9–12 and 14–15. Anophthalmia, microphthalmia, and small prosencephalon followed heat on days 8 and 10, and anencephaly also on day 10. Rosette formation in prosencephalon and short tail followed heat on day 11. Rosettes in retina, thin cortex and rosettes in prosencephalon, and four toes on forefeet followed heat on day 12. Runting, cleft palate, micromelia, syndactyly, and ectrodactyly followed heat on day 13.

Whole-body hyperthermia in pregnant rats was used by Edwards (1968) and resulted in similar abnormalities. Rats were stressed for 40–60 min twice on one day between days 9 and 14 of gestation; this elevated rectal temperature from 38.1 to 42.7°C, resulting in general retardation of growth in offspring of mothers heated on days 9–13. The abnormalities included microphthalmia after heat on day 10; edema on day 11; micromelia and short bodies and tails on day 13; hydrocephalus, cleft palate, four toes on forefeet, and short or absent tails on day 13; and microphthalmia, short digits, and short tails after heat on day 14. Hofmann and Dietzel (1966) raised the rectal temperature of pregnant rats 1–4°C by shortwave diathermy over the abdomen. Although a few malformations of the head were observed following treatment on days 9 and 10 of pregnancy, the majority of defects involved the tail and extremities and followed treatment on days 13 and 14.

5. Mice

Pennycuik (1965) elevated rectal temperatures of pregnant mice to 41–42°C by exposure for 1 hr in an incubator at 43°C. The anomalies following heat on day 7 were agnathia and reduced maxillary region; day 8, microphthalmia and subcutaneous cyst over the eye; day 11, reduction in number of toes, length of toes, and posterior paralysis; day 12, change in composition of hair coat and in number of vibrissae; and on day 15, tips of toes rotated, usually medially, but sometimes laterally. The incidence of developmental abnormalities other than changes in vibrissa numbers was not particularly high (1–10%). Lecyk (1966, 1969) elevated the body temperatures of pregnant mice to 40–41°C on 1 day between days 7 and 12 of pregnancy by exposing them to 43°C in an incubator for 20 hr. Over 50% of the mothers died, resorbed, or aborted their fetuses. Vertebral anomalies occurred in over 30% of young and were highest following hyperthermia on day 10 of pregnancy (51%), including fusion and wedging of vertebrae, scoliosis, and an increase or decrease in the numbers of vertebrae.

6. Rabbits

Embryonic resorption and congenital abnormalities were reported by Brinsmade and Rubsaamen (1957) to follow the injection of cooked cow's milk into does on days 6 and 7, 7 and 8, or 6, 7, and 8 of gestation. The treatment elevated maternal temperatures by 1–2.2°C on the day of injection, and fetal abnormalities included defects of the limbs and liver, subcutaneous and meningeal hemorrhages, encephalocele, and microcephaly.

7. Guinea Pigs

Hyperthermia during pregnancy causes fetal resorptions, abortions, or abnormalities of development (Edwards, 1967, 1969a–c, 1974). The litter size is often reduced, indicating increased embryonic resorption, and occasionally one or more fetuses are aborted while others retained to term (Edwards, 1969a). In both instances there is a high incidence of defects in the fetuses surviving to term. Abortion following hyperthermia given early in gestation usually occurs at about days 32–38 and appears to be independent of the number of exposures (2, 4, or 8) or when in pregnancy the exposure occurred.

a. Ebryonic Resorption. Resorption has followed heat stress on 2, 4, or 8 occasions between days 11 and 23 but appears most common between days 11 and 15.

b. Stillbirth. The incidence of stillbirths is increased amongst heated litters and is particularly high following heat on days 18–25.

c. Defects of the Eyes. Abnormalities of the eyes most frequently following heat between days 11 and 25 of gestation. Cataract, the commonest defect, occurred in 24% of young heated on days 18–25 and often affects the weaker animals which show a cloudy white central opacity and denser Y-shaped opacities of the capsule. Blindness, strabismus, microphthalmia, and coloboma are uncommon defects which follow heat exposure between days 11 and 18 and occasionally between days 18 and 25.

d. Defects of the Limbs. Talipes affected the hindlimbs of 28% of young from mothers heated on days 18–25 (Fig. 1). All these animals were micrencephalic. The forelimbs were involved in 9% of young heated on days 22 and 23 (Edwards, 1971a).

Arthrogryposis (Fig. 2) was seen in 35% of young from mothers heated on days 35–42 (Edwards, 1971b). The muscles and joints of the neck, forelimbs, thorax, and abdomen were involved following heat at earlier stages (days 26–31), and the abdomen, lumbar region, and hindlimbs followed heat during the later stages (days 30–46) of gestation. The condition may affect a local or an extensive area of the body in which the muscles are atrophic and the joints fixed.

e. Defects of the Brain. Disturbances of brain development (micrencephaly and rarely hydrocephalus and hydranencephaly) are the most common abnormalities of guinea pigs following prenatal hyperthermia. Microencephaly (Fig. 3) has been produced by maternal heating between days 15 and 32, 39 and 46, and 53 and 60. The most vulnerable period was between days 18 and 25, and days 20 and 23 appeared particularly sensitive. The deficit in

Fig. 1. Newborn guinea pigs with talipes of left hindlimb. Mother exposed to hyperthermia on days 20–24 of pregnancy.

brain weight following heat on days 18–25 included deficits in the cerebral hemispheres, brainstem, and cerebellum. The amount of DNA in the smaller brains was less than in controls, indicating a deficit in cell numbers (Edwards *et al.*, 1971). A smaller proportion of micrencephalics followed heat on days 39–46 and 53–60 and their brain weight deficit was less severe. In animals heated on days 40–44 the brainstem and cerebellum were reduced in size (Edwards *et al.*, 1974b).

The growth of the brain can be modified by the stage of development at which the heating is applied, the level of hyperthermia, and the number of exposures given. In early pregnancy increasing doses of heat result in increasing deficits at birth both in wet brain weight and usually in body weight. However, after 1–8 exposures the relative brain weight deficit is about 3–5 times the body weight deficit. The retardation of growth of both the body and brain falls off with increasing exposures.

Fig. 2. Litter of newborn guinea pigs with arthrogryposis of the hindlimbs. Mother exposed on days 39–46.

f. Other Abnormalities. Hypoplasia or agenesis of the incisors occurred in 76% of young heated on days 18–25 of gestation and 5% heated on days 25–32 (Edwards, 1969a).

Absence of hypoplasia of the lateral digits of the forelimbs occurred in 76% of young heated on days 18–25 and 25% heated on days 25–32. Generally the second phalanx was reduced or absent.

Renal agenesis was found in 15% of newborn animals heated *in utero* on days 18–25. All littermates were usually affected, and about 50% were stillborn.

Exomphalos (Fig. 4) is the commonest spontaneous abnormality occurring in our colony, the incidence in nonexperimental young being about 0.5–1%. An incidence exceeding 50% was recorded following heat on days 18–25 of pregnancy.

Multiple abnormalities most commonly followed four or more exposures

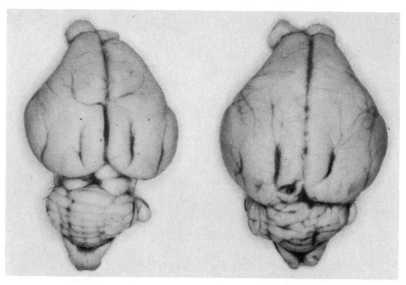

Fig. 3. Micrencephalic brain (left) from newborn guinea pig after exposure to hyper-thermia on days 20–23. Brain from control guinea pig of similar body weight on right.

between days 18 and 25 of gestation. When one or two exposures were given during this period, micrencephaly was usually the only abnormality detected. Over 25% of arthrogrypotic guinea pigs were also micrencephalic, but all those with talipes, short digits, renal agenesis, or hypoplastic teeth also had micrencephaly.

8. Monkeys

Cotton-eared marmosets (*Callithrix jacchus*) were exposed under seda-tion to 42°C for 1 hr daily on 5 consecutive days between days 25 and 50 of gestation. The core temperature of the mothers was elevated from 37.0 to 41.5°C. Of 10 liveborn experimental neonates, 6 died in the neonatal period and 1 died at 7 months. Survivors were clumsy, showed permanent growth retardation, behavioral abnormalities, and skeletal defects (Poswillo *et al.*, 1974).

C. Critical Temperature–Time Relationships

In guinea pigs a general pattern of response follows exposure to heat at susceptible stages of gestation. Maternal deaths follow extreme elevations of temperature; abortions follow less extreme elevations; fetal malformations usually occur at more moderate levels. This response probably occurs in some other species also.

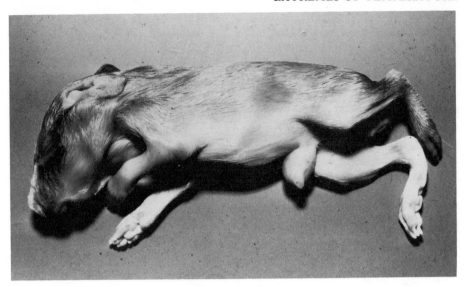

Fig. 4. Exomphalos.

1. Maternal Death

a. Pigs Tompkins *et al.* (1967) found that sows survived 35°C for 1 and 2 days but 8 of 22 sows died when held for up to 5 days at 36.7°C and 50% relative humidity. Omtvedt *et al.* (1971) lost 2 of 14 gilts on days 102–110 of pregnancy when exposed to 37.8°C and 42–50% relative humidity for 17 hr and 32.2°C for the remaining 7 hr daily. Similarly Edwards *et al.* (1968) lost 7 of 44 gilts exposed to 38.9°C and 35% relative humidity for 17 hr and 32.2°C and 65% humidity for the remaining 7 hr for 15 days in early pregnancy.

b. Rats. Skreb (1965) lost all 3 pregnant rats exposed in an incubator at 37°C and whose body temperatures reached 42°C. He did not state the duration of exposure. Nine others whose body temperatures ranged between 40 and 41°C survived. Kreshover and Clough (1953) recorded a mortaility of 9 of 12 rats in late gestation which were exposed for 12, 24, and 48 hr to 37.2–39°C, during which rectal temperatures were elevated to 39–40.6°C. However, Edwards (1968) found a mortality of only 12% following two exposures of 40–60 min on the one day to 43°C. The mothers were watched carefully and removed from the incubator if exhausted.

c. Mice. Pennycuik (1965) found that 10 min in an incubator at 44.4°C caused the death of many mice, while a high proportion exposed for 60 min at 43°C survived. Lecyk (1969) used an incubator set at 43°C with 60% relative humidity which elevated body temperature by 3.4–4.4°C (to 40–41°C) for 20 hr. The total mortality was 37% and reached almost 100% in mothers with a temperature exceeding 42°C.

d. Guinea Pigs. Edwards (1969a) heated pregnant guinea pigs for 1 hr daily in an incubator set at 43–43.5°C. The loss after 2, 4, and 8 exposures was 13, 14, and 15% respectively. In our experiments incubator temperatures used have been reduced progressively from about 44° to 42–43°C. Neither the maternal poststressing temperature nor the incidence of fetal malformations have been reduced greatly but the maternal mortality was. With the incubator set at 44°C the mean poststressing temperatures were 43.1–43.5°C; at 43°C in the incubator, maternal temperatures reached about 43.2°C. The pattern of maternal elevation is illustrated in Fig. 5.

The lethal maternal elevation of temperature is usually over 44°C, but individual susceptibility is variable. Most maternal deaths occur after the first exposure to heat, and animals surviving this usually show about the same temperature elevation at each succeeding exposure (Edwards, 1974).

2. Embryonic Death, Resorption, or Abortion

It is generally recognized that maternal exposure to high temperatures during the preimplantation stages of pregnancy causes increased embryonic losses in most species of animals. As there is an extensive literature on this aspect of the subject, we will only give examples of what might be expected.

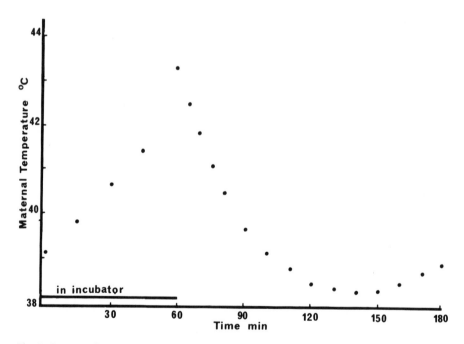

Fig. 5. Pattern of maternal temperature changes during and after exposure to 43°C DB, and 27°C WB for 1 hr.

a. Sheep. Dutt *et al.* (1959) showed that ewes exposed to 32.3°C and a relative humidity of 60%, starting from 5 days before mating and continuing for 16 days, lost an estimated 91.7% of embryos compared with 4% in controls. The loss was highest (77.8–100%) on the first and second days after mating and was heavy (53.8–69.2%) when the heat was given on days 3–5 (Dutt, 1963).

b. Pigs. Tompkins *et al.* (1967) held sows at 35 or 36.7°C for 1 or 5 days between days 1–25 of pregnancy, during which their rectal temperatures were elevated from about 39°C to a maximum of 40.8°C. Embryonic morality was significant with treatment between days 1 and 5 of pregnancy. Omtvedt *et al.* (1971) held gilts in rooms at 37.8°C for 17 hr daily on days 0–8, 8–16, 53–61, or 102–110 of gestation. Rectal temperatures were elevated by 1–1.5°C. Treatment during days 8–16 resulted in higher embryonic losses, and gilts heated in late gestation had more stillborn young. Edwards *et al.* (1968) found 83.2% and 59.7% embryonic survival when sows were exposed to 38.9°C and 35% relative humidity on days 15–30 and 1–15, respectively, of pregnancy.

c. Chickens. Chicken eggs incubated at a temperature over 42.2°C for 22–48 hr during the first few days of embryogenesis had a mortality of 25% (Alsop, 1919). Nielsen (1969) found 100% mortality in eggs incubated at 43°C for 1 day or at 42°C for 2 days between days 2 and 6 of development. A high mortality followed incubation at 42°C for 1 day, but there was good survival at 41°C on the second and third days of development.

d. Rats Embryonic resorptions following maternal exposure to 39°C for 1 or 4 hr reached 49% when the stress was given between days 0.5 and 1.5, 44% between days 2.5 and 4.5, 21% between days 5.5 and 7.5, and 13% between days 8.5 and 10.5. The maternal temperature was elevated from 37°C to 39.75°C at 1 hr and 39.45°C at 4 hr (Hsu, 1948). Macfarlane *et al.* (1957) showed a loss of 58% of implanted fetuses in rats kept at 35°C and 62% relative humidity compared with 7% in controls at 22–28°C. Skreb and Frank (1963) immersed the exteriorized uterine horn in saline at 40–41°C for 40 min and showed that the resorption rate of postimplantation embryos increased from 14.5% when exposed on day 7.5 and 57.4% on day 11.5. Longer exposures resulted in higher rates of resorption. There were resorption rates of 50 and 55% in rats following two 40–60-min exposures to 43°C on one day at days 10 and 12, respectively, of pregnancy (Edwards, 1968). The mean temperature at which total fetal resorption occurred was 42.9±0.64°C (±S.D.).

e. Mice. Lecyk (1969) recorded a loss of 31 of 150 litters (20.7%) in mice exposed to 43°C and 60% relative humidity. Pennycuik (1965) found 45% and 55% of mice to produce litters following exposure of 60 min at 42–43°C on days 7 and 9, respectively.

f. Rabbits. Rabbits, at 72–80 hr after mating, were immersed for 20 min in water at 45°C which elevated the rectal temperature to 42.2–42.8°C, and 17 of 18 does failed to produce litters after this treatment. Immersion in water for 20 min at 37.8°C, 41.7°C, or for only 10 min at 45°C did not affect pregnancy (Cameron, 1943).

g. Guinea Pigs. Edwards (1969a) recorded losses of 10 of 16 litters (62.5%) in guinea pigs given exposures of 1 hr to 43–43.5°C DB and 23–27.5°C WB on days 11–18 of pregnancy. Abortions occurred in 30.8% of mothers given two exposures at daily intervals between days 14 and 17, and 46.2% aborted after exposure on days 18–19. The temperatures of females which aborted were higher than those which produced malformed young (43.5 and 43.2°C, respectively) (Edwards, 1969c).

h. Monkeys. No losses were recorded by Poswillo *et al.* (1974) following 5 daily exposures of 1 hr to 42°C of pregnant marmosets sedated with 0.5% phencyclidine.

3. Congenital Abnormality

Congenital abnormalities following heat stress during embryogenesis appear to have been recorded only in sheep, chickens, rats, mice, rabbits, monkeys, and guinea pigs. The exposure conditions and the abnormalities produced are outlined in preceding sections of this review. In most reports there has been no attempt to analyze the conditions of heating which might result in prenatal death, abortions, or the type of congenital abnormality.

a. Chickens. Both Alsop (1919) and Nielsen (1969) used temperature elevations early in the incubation period (days 0–3 and 0–7, respectively), and at temperatures of 40–42.2°C (1.2–2.3°C above normal) this produced abnormalities especially of the head and ventral body wall. Ventral body-wall defects were most common following excess heat on days 5–7 (Nielsen, 1969).

b. Rats. Kreshover and Clough (1953) maintained the rectal temperatures of rats between 38.9 and 40.8°C for 12, 24, or 48 hr which caused dental enamel abnormalities in the majority of young heated after the 9th day of pregnancy.

Immersion of the uterine horn of rats between days 8.5 and 15.5 of pregnancy in saline at 40–41°C for 40 min produced a higher incidence of abnormalities than at 60 min (Skreb and Frank, 1963). However, by using more precise control of the temperature of the saline, Skreb (1965) showed that malformations followed 30, 45, or 60 min of immersion only when the temperature of the saline was 42°C or over. Again the incidence of malformations did not increase with increasing duration of exposure. Whole-body exposure of rats to 43°C DB and 27°C WB resulted in deep rectal temperatures of 42.7±0.62°C at 60 min (Edwards, 1968). The temperature reached 40°C at 15 min, 41.5°C at 45 min, and returned to normal by about 30 min after the end of exposure. The individual susceptibility to heat was variable, and some rats were removed from the incubator at 40 min. The mean temperatures of mothers which resorbed fetuses or produced malformed young were both 42.9°C, indicating that there is a narrow teratogenic range in the rat.

c. Mice. Lecyk (1969) produced vertebral anomalies by elevation of body temperature to 40–41°C (an elevation of 3.4–4.4°C) for 20 hr between days 8 and 13 of pregnancy. The cervical vertebrae were affected by heat during day 8, the thoracic and lumbar during days 9 and 10, and the caudal vertebrae during days 11–13. Pennycuik (1965) used an exposure of 1 hr and obtained rectal temperatures at intervals of 15 min. Temperatures of the mothers ranged between 40.8 and 41.9°C and were higher at 15 min than at 30, 45, or 60 min. The severity of the defects was less with heat at later stages of development. There was no comparison in either study between the temperatures of mothers which died, produced malformed young, or resorbed their litters.

d. Gunea Pigs. Micrencephaly is usually the only abnormality after one or two exposures of 1 hr to heat between days 18 and 25 of pregnancy. The more bizarre and obvious malformations usually follow four or more daily exposures. The lowest mean elevation of maternal temperature associated with both talipes and arthrogryposis was 3.4°C, and two or more exposures have been required. As it is common to find affected and normal young in the same litter, there appears to be a considerable individual susceptibility.

After one exposure to heat on day 21 of pregnancy the incidence of micrencephaly increases if maternal temperature exceeds 41.3–41.6°C (an elevation of 2–2.3°C). The critical embryonic temperature would probably be slightly less than this. The deficit in brain weight after 1, 2, 4, or 8 exposures between days 18 and 25 is about 8, 14, 20, and 26%, respectively (Edwards, 1974). In addition to the number of exposures, the deficit in brain weight also depends on the level of hyperthermia above the cricial point, and after one exposure for each 1°C elevation above this point, brain weight is reduced by about 8% (Edwards, 1969b). A smaller deficit in brain weight in a smaller proportion of newborn follows maternal exposure between days 39 and 46 and 53 and 60 of pregnancy.

e. Monkeys. Maternal temperature elevation from 37 to 41.5°C in marmosets exposed to heat for 1 hr on 5 days between days 25 and 50 of gestation resulted in behavioral and skeletal abnormalities in their young (Poswillo *et al.*, 1974).

D. Methods of Examination

In the studies reviewed, the examination of fetal and newborn animals has been by standard methods of gross and histological pathology. The interpretation of the effects of heat or other agents on brain development can be difficult, so the methods used in our laboratory for guinea pigs will be outlined in some detail.

The mean brain weight of control newborn guinea pigs usually varies between 2.67 and 2.83 g, and body weight varies between 88 and 99 g. Differences in body weight account for approximately 65% of the variation in brain weight. The brain weights of the stillborn are usually heavier than the surviving guinea pigs, and males have slightly heavier brains than females.

The use of a brain weight:body weight ratio alone in examination of the effects of agents on brain growth can be very misleading. The ratio itself varies with body weight (Edwards, 1974). Small animals have relatively larger brains than larger animals. A newborn control guinea pig weighing 50 g has a brain of about 2.438 g {brain weight (g) = 2.061 + [0.00754 × birth weight, (g)] r = 0.733, n = 178 control guinea pigs]. In this animal the brain makes up 4.88% of the total body weight. A newborn guinea pig weighing 150 g has a brain of about 3.192 g, making up 2.13% of the total body weight. The variation in brain weight:body weight ratios of controls shows that the prenatal influences causing reduced body weight at birth (e.g., large litters) produce a small animal with a relatively large brain. This principle implies that the brain is normally less liable to reduction in size by some prenatal influences than are other organs and tissues.

There are some agents which retard brain growth more selectively than other organs, and this action results in a reduction in the brain weight:body weight ratio for animals of their particular body weight. Heat applied at susceptible stages of development in guinea pigs usually results in a moderate (1.7–8.2%) retardation of body growth and a more severe retardation of brain growth (8.1–25.6%, with 1–8 exposures, respectively). Analysis of covariance may be used to eliminate the effect of differences in body weight on brain size between treated and control groups. Alternatively a rough screening test can be used. The regression of brain weight on body weight of control animals is calculated and plotted on a graph together with the lower 2-S.D. interval. Points corresponding to the brain weights and body weights of treated newborn are plotted on the graph and those falling below the 2-S.D. limit are regarded as micrencephalic (Fig. 6). Differences in the numbers of micrencephalics between control and treated groups are compared with a χ^2 test.

The excised brain may be divided into cerebral hemispheres, brainstem, and cerebellum to determine any differential effects of growth on the different segments.

Additional methods which might be applied where brain development is known to be retarded include the many tests of learning behavior used in psychology laboratories. In our studies we have used serial discrimination reversal tasks (Lyle *et al.*, 1973).

E. Pathogenesis of the Effects of Hyperthermia

The pathogenesis of malformations produced by heat falls into two general categories: the malformations which appear to be related to dysfunction of the central nervous system and malformations which appear to result from disorders of cellular proliferation. Dysfunction of the central nervous system might itself result from disorders of cellular proliferation.

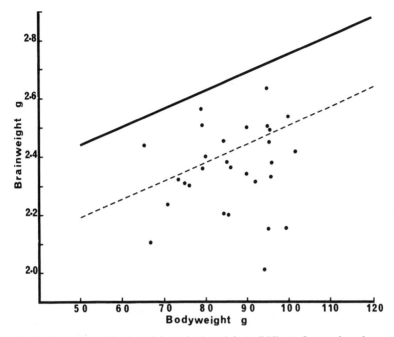

Fig. 6. Regression of brain weight on body weight (solid line) of control newborn guinea pigs with lower 2 standard-deviation interval (broken line). The points represent brain weight and body weight of newborn guinea pigs following one exposure to hyperthermia on day 21 of gestation; those below the 2 standard-deviation interval considered micrencephalic.

1. Malformations Related to Dysfunction of the Central Nervous System

a. Talipes. Areas of myelodysplasia are found in newborn guinea pigs with talipes. Duplication, displacement, and dilatation or absence of the central canal or excessively wide gray commissures may be found singly or in combinations in a cross section at some point along the length of the cord. The gastrocnemius muscle appears hypertonic, and the bones of the tarsus disorientated by tension on the calcaneus (Fig. 1). At times the tibia is bent and the fibula is usually hypoplastic. It was suggested that the skeletal deformity was caused by defective innervation resulting in excess tone in the muscles attached to the affected joint (Edwards, 1971a).

b. Arthrogryposis. The atrophy or fiber-fatty replacement of muscles of the limbs, abdomen, or back are related to reduction in numbers or absence of motor neurones in parts of the cord which innervate affected areas and probably represent a denervation atrophy (Edwards, 1971b). With involvement of the foreparts of the body, the central gray matter of the cord in the cervical area is cavitated (congenital syringomyelia). It is believed that the joints of the affected parts become fixed because of the prenatal immobilization resulting from defective motor innervation (Whittem, 1957).

2. Malformations Related to Disturbed Cellular Proliferation

Many developmental abnormalities which follow hyperthermia during gestation involve reduction in size or aplasia of organs. In guinea pigs these include micrencephaly, exomphalos, hypoplasia of the teeth and digits, and renal agenesis. It seems likely that many of these conditions result from a direct effect of heat on cellular proliferation at the particular site. Micrencephaly following hyperthermia on days 20–23 of gestation was shown to be associated with a deficit in brain cells (Edwards *et al.*, 1971).

The pathological changes occurring in the 21-day guinea pig fetus after hyperthermia were described by Edwards *et al.* (1974a). The developing central nervous system was severely affected, and damaged mitotic cells were apparent immediately after heating. The cells adjacent to the ventricles and central canal showed clumping of nuclear chromatin followed by pycnosis and karyorrhexis. The resumption of mitosis was delayed for 6–8 hr (Figs. 7–11). Similar, though less dramatic, changes were observed in the mesenchyme. The relative significance of the mitotic cell death and the mitotic delay is not clear, and the possibility exists that other mechanisms may contribute to the

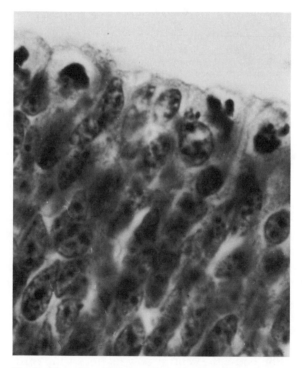

Fig. 7. Clumping of chromatin and vacuolation of cytoplasm in cells adjacent to lateral ventricle. Section from 21-day guinea pig embryo 2 hr after exposure to heat. (From M. J. Edwards *et al.*, 1974a.)

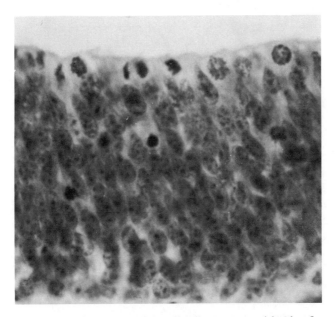

Fig. 8. Mitosis recommenced in cells adjacent to ventricle 6 hr after heating in 21-day guinea pig embryo. Some chromatin fragments deep in ependyma. (From M. J. Edwards *et al.*, 1974a.)

Fig. 9. Brain squash preparation showing normal metaphase chromosomes of control 21-day guinea pig embryo. (From M. J. Edwards *et al.*, 1974a.)

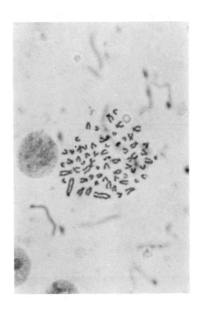

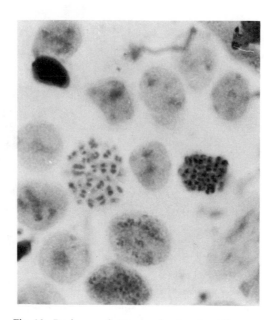

Fig. 10. Brain squash 30 min after heating. Chromosomes shortened and thickened. (From M. J. Edwards *et al.*, 1974a.)

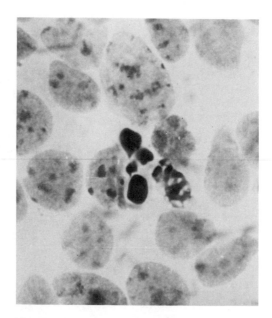

Fig. 11. Chromatin clumps fragmenting 2 hr after heating. (From M. J. Edwards *et al.*, 1974a.)

cell deficit in the brain at birth. The other abnormalities might result from similar localized disturbances of cellular proliferation. However, the effects of hyperthermia on embryonic tissues other than the developing central nervous system appeared minor, suggesting that other factors might be involved. It should be noted, however, that these lesions follow acute hyperthermia and might not represent the pathogenesis in chronic hyperthermia.

Recently McCredie (1974) proposed that thalidomide-induced reduction deformities resulted from damage to neural crest derivatives and the consequent loss of the trophic effect of peripheral sensory neurones essential for normal development. Considering the susceptibility of developing neural elements to hyperthermia, this theory represents an attractive alternative.

III. HYPOTHERMIA

Subjecting chickens and laboratory animals to hypothermia during incubation or gestation has resulted in reproductive wastage and developmental abnormalities.

A. Methods of Inducing Hypothermia

Lecyk (1965) lowered the body temperature of pregnant mice by 11–16°C after placing them in a refrigerator following injections of chlorpromazine. Vidovic (1952) used sealed glass jars immersed in ice water to cool the body temperatures of pregnant rats to 12–15°C. A similar method was used with rats by Brizzee and Brannon (1972). Smith (1956a) reduced the temerpatures of golden hamsters to zero and subzero for up to 70 min by immersion in crushed ice following initial cooling in air-tight glass jars at 2°C. Animals were revived using diathermy or immersion in warm water and artificial ventilation (Smith, 1956b). Lowering of egg-incubator temperatures to 35.3–36.9°C (Alsop, 1919) and 35.83°C (de la Cruz *et al.,* 1966) produced abnormalities in chick embryos.

B. Species Studied and Abnormalities Produced

Delayed implantation, retarded fetal development, and uterine hemorrhage were reported in rats subjected to hypothermia during pregnancy (Courrier and Marios, 1953, 1954), and Barnett and Manly (1954) described reduced litter size in mice living at 3°C.

Vidovic (1952, 1956) lowered the deep-body temperature of pregnant rats to 15–20°C. The litter size was reduced, and the proportion of stillbirths increased in animals cooled after the day 14 of gestation.

White mice cooled to 25°C for 24 hr on day 7.5–12.5 of pregnancy had increased embryonic losses and 18% abnormal young, the majority having reduced numbers of vertebrae (Lecyk, 1965, 1968). In a group cooled to 20°C, 52% of young were abnormal with reduced numbers of vertebrae, wedge and split vertebrae, and vertebral fusions leading to scoliosis.

Smith (1957) cooled hamsters at 1.5–12.5 days of pregnancy to body temperatures below 0°C for 25–35 min. Fetal deaths, resorptions, and uterine hemorrhage were observed when animals were cooled on days 9.5, 10.5, and 11.5. When the hypothermia was maintained for 45 min at 1.5–8.5 days, severe developmental abnormalities including hydrocephalus, anencephaly, anophthalmia, palatoschisis, and deformities of the feet were observed. Freezing of tissue fluids was believed to cause the abnormalities.

In chickens, de la Cruz et al. (1966) reported septal defects in hearts of 18.75% of embryos surviving incubation at 35.83°C, and Alsop (1919) found brain and neural tube abnormalities in 67% of early embryos incubated at 34.4–38°C.

Hypothermia may have a synergistic or additive effect on the action of certain teratogens. Reduction of body temperature to 18–19°C (with hypercapnia) exacerbated the effects of split doses of gamma radiation on the brain of the fetal rat at 13 days of age (Brizzee and Brannon, 1972). Ferm (1967) reported the levels of resorption and malformation in 7- and 8-day pregnant rats treated with vitamin A to be higher when the rats were housed at 0.2°C than when housed at room temperature. Housing in the cold alone did not produce abnormalities.

IV. DISCUSSION

Hyperthermia acts as a typical teratogen. It produces specific malformations at specific stages of development; the abnormalities, however, are more typical of the species affected than of heat as a teratogen. The severity of its action depends both on the level of heat per exposure and the number of exposures given. There appear to be differing species susceptibilities to its teratogenic potential, but in this respect it is not unique. The rat appears to be more likely to resorb its heat-damaged embryos than to give birth to malformed young, and for this reason it is not a very suitable species to study. The heat-damaged guinea pig embryo survives well, and for it the range between the lethal dose and the teratogenic dose appears wider. With short exposures a high proportion of pregnant rats and guinea pigs survive sharp elevations of body temperature. As there is a wide range of individual susceptibility to hot environments, the animals must be watched or monitored carefully. Maternal mortalities can be minimized by frequent or continuous recording of body temperatures. If removed from the source of heat before body temperatures reach 43.5°C, the animals recover quickly especially if they are allowed or assisted to cool quickly.

Our current work indicates that the extreme vulnerability of the central nervous system in early development appears to be due to the rapid rate at which its cells are proliferating. This feature is common to many cellular systems, both *in vivo* and *in vitro,* in which proliferation is rapid.

The possible relevance of these findings in man is of some interest. Pregnant women might be at risk via exposure to hot environmental temperatures or through febrile illnesses.

Exposure to hot summer temperatures might be more likely to raise body temperatures in temperate zones. In these areas seasonal changes are more marked, and the population would be relatively unacclimatized compared with populations in tropical zones. Exposure to hot environments need not raise the body temperatures of individuals at rest, but exercise could cause deep-core temperatures to elevate rapidly. In surveys of intelligence there is a universal trend toward a slightly lower mean among individuals born in winter (early gestation during summer) compared with individuals born in summer (early gestation during winter) (Pintner and Forlano, 1943). In addition Knobloch and Pasamanick (1958) found that significantly more mentally defective children admitted to a state school were born in winter (January–March). They also found that significantly more mental defectives were born following hot, compared with cool, summers.

Congenital abnormalities, especially those affecting the central nervous system, have been reported following influenza epidemics (Coffey and Jessop, 1959; Hakosalo and Saxen, 1971). It was suggested by Hakosalo and Saxen (1971) that some of the drugs used in treatment might be responsible for the malformations. An alternative suggestion was that the associated fever alone might be responsible (Edwards, 1972). Saxen and Klemetti (1973) have shown associations between both influenza and fever and malformations of the central nervous system, skeleton, and palate. These studies have recorded the incidence of bizarre, readily recognizable malformations. Our experience with animal experiments suggests that if the central nervous system is the major target organ of hyperthermia, simple retardation of brain growth would be the major, and usually the only, manifestation. Reduced brain size might not be regarded as a malformation, but it is an important abnormality which could be simply detected by measuring head circumference.

The extent to which the growth of the brain can be retarded without obvious change in shape indicates that a more quantitative approach should be adopted in the study of prenatal disturbances of organ development and the monitoring of the safety of drugs.

REFERENCES

Alexander, G., and Williams, D., 1971, Heat stress and development of the conceptus in domestic sheep, *J. Agric. Sci.* **76**:53.

Alsop, F. M., 1919, The effect of abnormal temperatures upon the developing nervous system in the chick embryos, *Anat. Rec.* **15:**307.

Barnett, S. A., and Manly, B. M., 1954, Breeding mice at $-3°C$, *Nature* **173:**355.

Brent, R. L., and Wallace, J., 1972, The utilization of microwave radiation in developmental biology research, *Teratology* **5:**251.

Brinsmade, A. B., and Rubsaamen, H., 1957, Zur teratogenetischen Wirkung von unspezifischem Fieber auf den sich entwickelnden Kaninchemembryo, *Beitr. Pathol. Anat.* **117:**154.

Brizzee, K. R., and Brannon, R. B., 1972, Cell recovery in the fetal brain after ionising radiation, *Int. J. Radiat. Biol.* **21:**375.

Cameron, J. A., 1943, Termination of early pregnancy by artificial fever, *Proc. Soc. Exp. Biol. Med.* **52:**76.

Coffey, N. P., and Jessop, W. J. E., 1959, Maternal influenza and congenital deformities, a prospective study, *Lancet* **2:**935.

Courrier, R., and Marois, M., 1953, Action de l'hypothermie experimentale sur le gestation chez le rat, *C. R. Soc. Biol.* **147:**1922.

Courrier, R., and Marois, M., 1954, Retard de la nidation et du développement foetal chez la rate en hypothermia, *Ann. Endocrinol.* **15:**738.

de la Cruz, M. V., Campillo-Sainz, C., and Munoz-Armas, S., 1966, Congenital heart defects in chick embryos subjected to temperature variations, *Circ. Res.* **18:**257.

Dietzel, F., and Kern, W., 1970, Fehlgeburt nach Kurzwellenbehandlung; tierexperimentelle Untersuchungen, *Arch. Gynaekol.* **209:**237.

Dutt, R. H., 1963, Critical periods for early embryo mortality in ewes exposed to high ambient temperature, *J. Anim. Sci.* **22:**713.

Dutt, R. H., Ellington, E. F., and Carlton, W. W., 1959, Fertilization rate and early embryo survival in sheared and unsheared ewes following exposure to elevated air temperature, *J. Anim. Sci.* **18:**1308.

Edwards, M. J., 1967, Congenital defects in guinea-pigs following induced hyperthermia during gestation, *Arch. Pathol.* **84:**42.

Edwards, M. J., 1968, Congenital malformations in the rat following induced hyperthermia during gestation, *Teratology* **1:**173.

Edwards, M J., 1969a, Congenital defects in guinea-pigs: Fetal resorptions, abortions and malformations following induced hyperthermia during early gestation, *Teratology* **2:**313.

Edwards, M. J., 1969b, Congenital defects in guinea-pigs: Prenatal retardation of brain growth of guinea-pigs following hyperthermia during gestation, *Teratology* **2:**329.

Edwards, M. J., 1969c, A study of some factors affecting fertility of animals with particular reference to the effects of hyperthermia on gestation and prenatal development of the guinea-pig, Ph. D. thesis, the University of Sydney, Sydney, Australia.

Edwards, M. J., 1971a, The experimental production of clubfoot in guinea-pigs by maternal hyperthermia during gestation, *J. Pathol.* **103:**49.

Edwards, M. J., 1971b, The experimental production of *arthrogryposis multiplex congenita* in guinea-pigs by maternal hyperthermia during gestation, *J. Pathol.* **104:**221.

Edwards, M. J., 1972, Influenza, hyperthermia, and congenital malformation, *Lancet* **1:**320.

Edwards, M. J., 1974, The effects of hyperthermia on pregnancy and prenatal developmental, *in: Experimental embryology and Teratology* 1 (D. H. M. Woollam, and G. M. Morriss, eds.), pp. 90–133, Elek Science, London.

Edwards, M. J., Penny, R. H. C., and Zevnik, I., 1971, A brain cell deficit in newborn guinea-pigs following prenatal hyperthermia, *Brain Res.* **28:**341.

Edwards, M. J., Mulley, R., Ring, S., and Wanner, R. A., 1974a, Mitotic cell death and delay of mitotic activity in guinea-pig embryos following brief maternal hyperthermia, *J. Embryol. Exp. Morphol.* **32:**593.

Edwards, M. J., Penny, R. H. C., Lyle, J., and Jonson, K., 1974b, Brain growth and learning behaviour of the guinea-pig following prenatal hyperthermia, *Experientia* **30:**406.

Edwards, R. L., Omtvedt, I. T., Turman, E. J., Stephens, D. F., and Mahoney, G. W. A., 1968,

Reproductive performance of gilts following heat stress prior to breeding and in early gestation, *J. Anim. Sci.* **27:**1634.

Ferm, V. H., 1967, Potentiation of the teratogenic effect of vitamin A with exposure to low environmental temperatures, *Life Sci.* **6:**493.

Hakosalo, J., and Saxen, L., 1971, Influenza epidemic and congenital defects, *Lancet* **2:**1346.

Hartley, W. J., Alexander, G., and Edwards, M. J., 1974, Brain cavitation and micrencephaly in lambs exposed to prenatal hyperthermia, *Teratology* **9:**299.

Hofmann, D., and Dietzel, F., 1966, Aborte und Missbildungen nach Kurzwellendurchflutung in der Schwangerschaft, *Geburtsh. Frauenheilkd.* **26:**378.

Hsu, C. Y., 1948, Influence of temperature on development of rat embryos, *Anat. Rec.* **100:**79.

Kalter, H., and Warkany, J., 1959, Experimental production of congenital malformations in mammals by metabolic procedure, *Physiol. Rev.* **39:**69.

Knobloch, H., and Pasamanick, B., 1958, Seasonal variation in the births of the mentally deficient, *Am. J. Public Health* **48:**1201.

Kreshover, S. J., and Clough, O. W., 1953, Prenatal influences on tooth development. 11. Artifically induced fever in rats, *J. Dent. Res.* **32:**565.

Lecyk, M., 1965, The effect of hypothermia applied in the given stages of pregnancy on the number and form of vertebrae in the offspring of white mice, *Experientia* **21:**452.

Lecyk, M., 1966, The effect of hyperthermia applied in the given stages of pregnancy on the number and form of vertebrae in the offspring of white mice, *Experientia* **22:**254.

Lecyk, M., 1968, The effect of abnormal temperatures applied during the pregnancy on the structure of the vertebral column in the offspring of mammals showed by the example of white mouse. Part 1. Hypothermia, *Zool. Pol.* **18:**391.

Lecyk, M., 1969, The effect of abnormal temperatures applied during the pregnancy on the structure of the vertebral column in the offspring of mammals showed by the example of white mouse. Part 11. Hyperthermia, *Zool. Pol.* **19:**97.

Lyle, J. G., Jonson, K. M., Edwards, M. J., and Penny, R. H., 1973, Effect of prenatal heat stress at mid- and late gestation on the learning of mature guinea-pigs, *Dev. Psychobiol.* **6:**483.

Macfarlane, W. V., Pennycuik, P. R., and Thrift, E., 1957, Resorption and loss of foetuses in rats living at 35°C, *J. Physiol.* **135:**451.

McCredie, J., 1974, Embryonic neuropathy. A hypothesis of neural crest injury as the pathogenesis of congenital malformation, *Med. J. Aust.* **1:**159.

Nielsen, N. O., 1969, Teratogenic effects of hyperthermia, in: *Teratology* (A. Bertelli and L. Donati, eds.), pp. 102–107, Exerpta Medica, Amsterdam.

Omtvedt, I. T., Nelson, R. E., Edwards, R. L., Stephens, D. F., and Turman, E. J., 1971, Influence of heat stress during early, mid and late pregnancy of gilts, *J. Anim. Sci.* **32:**312.

Pennycuik, P. R., 1965, The effects of acute exposure to high temperatures on prenatal development in the mouse with particular reference to secondary vibrissae, *Aust. J. Biol. Sci.* **18:**97.

Pintner, R., and Forlano, G., 1943, Season of birth and mental differences, *Psychol. Bull.* **40:**25.

Poswillo, D., Nunnerley, H., Sopher, D., and Keith, J., 1974, Hyperthermia as a teratogenic agent, *Ann. R. Coll. Surg. Engl.* **55:**171.

Saxen, L., and Klemetti, A., 1973, The Finnish register of congenital malformation, *Teratology* **8:**210.

Skreb, N., 1965, Température critique provoquant des malformations pendant le développement embryonnaire du rat et ses effets immediats, *C. R. Acad. Sci. (Paris)* **261:**3214.

Skreb, N., and Frank, Z., 1963, Developmental abnormalities in the rat induced by heat-shock, *J. Embryol. Exp. Morphol.* **11:**445.

Smith, A. U., 1956a, Studies on golden hamsters during cooling to and rewarming from body temperatures below 0°C, 1. Observations during chilling, freezing and supercooling, *Proc. R. Soc. London, Ser. B* **145:**391.

Smith, A. U., 1956b, Studies on golden hamsters during cooling to and rewarming from body temperatures below 0°C, 11. Observations during and after rescuscitation, *Proc. R. Soc. London, Ser. B* **145:**407.

Smith, A. U., 1957, The effects of foetal development of freezing pregnant hamsters, *J. Embryol. Exp. Morphol.* **5:**311.

Tompkins, E. C., Heidenreich, C. J., and Stobb, M., 1967, Effect of postbreeding thermal stress on embryonic mortality in swine, *J. Anim. Sci.* **26:**377.

Vidovic, V., 1952, Kalte und Hypoxietoleranz von Ratten-Embryonen, *Experentia* **8:**304.

Vidovic, V., 1956, Effect of hypothermia on the oestrous cycle, the mature egg, pregnancy and lactation of the white rat, Thesis, University of Belgrade, cited by A. U. Smith, 1957, *J. Embryol. Exp. Morphol.* **5:**311.

Whittem, J. H., 1957, Congenital abnormalities in calves: Arthrogryposis and hydranencephaly, *J. Pathol. Bacteriol.* **73:**375.

Wilson, J. G., 1959, Experimental studies on congenital malformations, *J. Chron. Dis.* **10:**11.

Yeates, N. T. M., 1953, The effect of high air temperature on reproduction in the ewe, *J. Agric. Sci.* **43:**199.

Interactions and Multiple Causes 13

F. CLARKE FRASER

I. INTRODUCTION

The truth of Donne's famous adage is nowhere better illustrated than in the field of teratology. No mother, embryo, gene, or teratogen is "an island unto himself," and the literature is replete with examples of interactions between two or more of these that influence the frequency of malformations.

Interactions can be regarded as foes, insofar as they complicate statistical analysis, increase the need for rigor in controls, and decrease the validity of comparisons between sets of results obtained at different times or in different laboratories. They can also be allies, for they may illuminate teratogenic mechanisms and explain otherwise puzzling inconsistencies.

In this discussion we cannot hope to review, exhaustively, the literature on teratogenic interactions and multiple causes, but will present a number of illustrative examples of the various kinds of interactions that may occur and their possible significance. We will consider interactions between teratogens and environmental variables, between teratogens and pharmacological agents (including other teratogens), between teratogens and genes, and between teratogens, genes, and environmental factors. Space will not permit a consideration of interactions between the interactions.

F. CLARKE FRASER · Department of Biology, McGill University, Department of Medical Genetics, The Montreal Children's Hospital, Montreal, Canada. *This chapter is dedicated to Walter Landauer, one of the first pioneers, and still a leader, in the field of teratologic interactions.*

II. INTERACTIONS BETWEEN TERATOGENS AND ENVIRONMENTAL VARIABLES

Interactions, of teratogens and environmental variables, both external and maternal, are important to recognize (Kalter, 1965; and the chapter by Kalter, Vol. 3) as they may complicate experiments, invalidate controls and comparisons between different experiments, and render drugs, pesticides, or various environmental agents that are usually harmless to the embryo teratogenic under certain special circumstances. References to the kind of factors that the teratologist should be aware of will be found scattered throughout these volumes. This chapter will consider the following examples.

The environment in the animal-rearing quarters may vary in such things as the concentration of pesticides, disinfectants, and constituents of the bedding material (Vesell, 1967), which in turn may alter the animals' response to a teratogen by altering the microsomal enzymes or other subtle variables. Variation with season has been noted (Kalter, 1965; Jaworska, 1965; Jacobson, 1968); the cause of this has not been established although possibly variations in the diet may play a role. Interactions with ambient temperature have been reported (Ferm, 1967; Nomura, 1969). Immobilization can act as a teratogen (Hartel and Hartel, 1960; Rozenzweig et al., 1970) and can also alter the effectiveness of other teratogens such as food and water deprivation (Rozenzweig et al., 1970); Brown et al., 1974), salicylates (Goldman and Yakovac, 1963), or vitamin A (Hartel and Hartel, 1960; Rozenzweig et al., 1970). This could complicate studies where the experimental procedure involves immobilization, for instance when measuring body temperature in studies on hyperthermia (Edwards, 1969). It has been assumed that restricted activity acts as a form of "stress," although the term has rarely been precisely defined (Warkany and Kalter, 1962; Hamburgh et al., 1974), and whether the interaction occurs through varying adrenal cortical secretions has not been established (Goldman and Yakovac, 1964). The same applies to other "stressful" procedures that have been shown to be teratogenic, such as loud noises (Ishii et al., 1962; Geber, 1966; Peters and Strasburg, 1969; Ward et al., 1970), handling (Ader and Plant, 1968), and crowding or parabiosis (Hamburgh et al., 1974). Immobilization could alter metabolism or excretion of the pharmacoteratogen, for instance. Incidentally, it is difficult to produce malformations in the mouse by maternal treatment with ACTH (Nakasato, 1953; Heiberg et al., 1959), suggesting that the maternal adrenal can produce teratogenic amounts of glucocorticoid only under maximum stimulation, if at all. The temptation to extrapolate from animal experiments the conclusion that human malformations may result from maternal emotional stress (Strean and Peer, 1956) has not been justified by the data, at least in the case of cleft lip and cleft palate (Fraser and Warburton, 1964).

Several examples of the effects of a teratogen being influenced by the mother's diet have been reported, although the specific nutritional factors related to the differences in malformation frequency have seldom been identified (Kalter, 1965). Decreased dietary protein increased the frequency of malformations produced by treating pregnant rats with trypan blue (Gunberg, 1955) or retinoic acid (Nolen, 1972), and maternal pyridoxine deficiency increased the frequency of cortisone-induced cleft palate in mice (Miller, 1972). The complex interactions between diet and the teratogenicity of cortisone and of 6-aminonicotinamide will be dealt with in a later section (p. 459). The extraordinary difficulty of assuring that the diet is constant throughout the course of a series of experiments should be pointed out here. Probably freezing enough diet for the whole experiment and thawing out portions as needed is the only feasible way.

Maternal food restriction may increase the effect of several teratogens, including cortisone as used to induce cleft palate (Kalter, 1960) and hypoxia and trypan blue to induce vertebral anomalies (Runner and Dagg, 1960). This kind of interaction may complicate the interpretation of experiments in which the teratogenic treatment causes the mother to reduce her food intake. For instance, Szabo and Brent (1975) reported that tranquilizers given to pregnant mice resulted in cleft palate in the offspring but also noted that the tranquilized mice ate and drank less. Forced feeding of the treated mothers strikingly reduced the frequency of cleft palate, but not to zero, demonstrating that the dietary restriction was indeed exacerbating the teratogenic effects of the tranquilizers. The mechanism of this interaction merits further investigation.

In addition the possible interaction of a teratogen with its vehicle should be mentioned. A striking example is dimethylsulfoxide (DMSO). In an elegant study Landauer (1972) compared the teratogenicity in chick embryos of several agents dissolved in either DMSO or water. Compared to the aqueous solution, DMSO decreased the teratogenicity of 3-acetylpyridine, 6-aminonicotinamide and 3-amino-1,2,4-triazole; had no effect on that of physostigmine and nicotine; increased that of sulfanilamide, and changed the spectrum of malformations caused by the insecticide Bidrin. Propylene glycol interacted similarly with these teratogens. DMSO and other solvents, such as acetamides, formamides, benzene and alkyl sulfonates (Wilson, 1973), can be teratogenic in mammals, but at relatively large doses. Other solvents that might be suspected of interacting with teratogenic agents include alkalis or acids and ethyl alcohol.

Finally the interaction of teratogens with maternal variables should be considered. Many examples are known of the effect of maternal characteristics (other than genetic) on the frequency of teratogen-induced malformations. Few conclusions can be drawn, other than that they are complex, unpredictable, and poorly understood. Separation of the effects of maternal age, parity, and maternal weight is complicated by their correlation with one another. Maternal weight seems to be the most important with respect to the effects of cortisone in mice, at least in the situation tested (Kalter, 1956). Heavier mothers have lower frequencies of offspring with cleft palate; the same is true for cleft palate induced by 6-aminonicotinamide (Pinsky and

Fraser, 1959). Conversely, Dagg (1963) found that the frequency of hindfoot malformations induced by 5-fluorouracil given on a per gram body weight basis, increased with increasing maternal weight. This suggests that the teratogenic effect was here related to the absolute amount of drug given, rather than the proportion per unit of maternal weight, and warns us that to adjust the dose of teratogen to the maternal weight will not necessarily reduce variability.

Teratogenic effects may also be influenced by variations in the microenvironment, as reflected by such indicators as littersize (Kalter, 1965) and position in the uterine horn (Trasler, 1960; Woollam and Millen, 1961; Kalter, 1965). Although the specific factors underlying these variables may be important in understanding—and perhaps reducing—the frequency of malformations, they remain obstinately covert.

Further discussion of these questions will be found in several reviews (Kalter, 1965; Dagg, 1966; chapter by Kalter, Vol. 3). Suffice it to say here that the complex interactions of teratogens with environmental variables, whether extrinsic to the mother or related to maternal constitution or to the uterine microenvironment, are important to be aware of, since they emphasize the necessity for rigorous controls and the hazards of comparisons between data from different experiments. So far they have contributed little to our understanding of the mechanisms by which malformations arise. Furthermore, to anticipate the following sections, it is not unlikely that many of the environment–teratogen interactions mentioned above will differ according to the genetic constitution of the responding organism.

III. INTERACTIONS OF TERATOGENS WITH PHARMACOLOGICAL AGENTS

Interactions between drugs are a matter of increasing therapeutic importance, as polypharmacy is still commonplace and it is no longer true that most drugs used are virtually inactive (Anon., editorial, 1975). The resulting changes in potency or toxicity may have teratogenic as well as therapeutic implications.

The editorial cited above classifies the interaction of drugs as either pharmacokinetic (acting during the processes of absorption, distribution, or elimination) or pharmacodynamic (modification by one drug of the action of another on a receptor or physiological control mechanism). In our present state of knowledge, the former class are more familiar to teratologists than the latter. Drugs that decrease gastric emptying and absorption (anticholinergics, strong analgesics), or increase it, may influence the absorption of other agents given concurrently by mouth. The effects will vary according to the site of maximal absorptive capacity. Of greater teratogenic interest are the metabolic interactions. Certain drugs will induce production of enzymes that increase the rate of metabolism of other drugs as well as their own. This is the probable explanation for the fact that acute administration of a drug is often more teratogenic than chronic administration—the drug interacts with itself, by inducing the microsomal enzymes that degrade it, as in the case of antihistamines in the rat (King *et al.,* 1965). Barbiturates, other hypnotics, analgesics, anticonvulsants, and various other compounds act in this way. Thus phenobarbital reduced the teratogenic effects of cyclophosphamide in mice, whereas SKF 525-A (an inhibitor of the drug-metabolizing liver enzymes) potentiated them (Gibson and Becker, 1968); chronic pretreatment with caffeine reduced the teratogenicity of caffeine administered during pregnancy (Terada and Nishimura, 1975). This kind of effect may be the explanation for the reduction in teratogenicity of 6-aminonicotinamide resulting from light etherization at the time of injection (Hamly *et al.,* 1970), a possible source of confusion in teratogenic experiments involving sedation or anesthesia in the experimental procedure.

Other possible means of drug interactions in metabolism include effects of one drug (e.g., hexachlorobenzene) on metabolism of another by the intestinal flora and effects of one drug on the renal secretion of another. More detailed discussion of these problems would be beyond the competence of the author. The main points to be made are that drug interactions can complicate teratogenic experiments, that the interactions are complicated, and that interpretation of the results should preferably be supported by measurement of the metabolism and distribution of the drugs in question.

Rather than being considered a nuisance, the interactions between teratogens and pharmacological agents may be regarded as tools with which to analyze their modes of action. This approach first seems to have been recognized by Landauer (1948), who studied in chicks the mode of action of a teratogen, insulin, by observing whether the teratogenic effects were prevented by various metabolically significant compounds. Nicotinamide and α-ketoglutaric acid had a protective effect but thiamine did not, from which Landauer concluded that the teratogenicity of insulin was probably mediated by disturbance of the embryo's carbohydrate metabolism. A long series of experiments using this approach and others confirmed this conclusion and also showed that the teratogenic effect of the insulin was on the cells of the affected organ primordia, not a secondary effect of hypoglycemia. A review of the subject by Landauer (1972) is recommended for its philosophy as well as its scientific value.

A series of experiments, in which this approach was extended to other teratogens (sulfanilamide, 3-acetylpyridine, 6-aminonicotinamide) and other antiteratogens (glucose, pyruvate, tryptophan, 3-hydroxyanthranilic acid, ADP), led Landauer to the interesting conclusion that these (and probably other) teratogens act by interference with mitochondrial energy production in the responding tissue and that the protective compounds may act by providing bursts of cellular energy that promote the energy-dependent steps of protein and nucleic acid synthesis (Landauer and Wakasugi, 1968; Landauer and Sopher, 1970).

The fact that ascorbate protected against 3-acetylpyridine, 6-aminonicotinamide, and sulfanilamide, but potentiated the teratogenicity of insulin, whereas nicotinamide protected against all of these teratogens, showed that there are separate ways by which teratogens interfere with energy production and by which their effects can be mitigated. Furthermore, since succinate and ascorbate, which become active at different points in the respiratory chain, protect against the same teratogens, it would seem that teratogens producing similar end effects can act along the same pathway at different points (Landauer and Sopher, 1970). These studies are an elegant model of how the modes of action of teratogens can be analyzed by study of their interactions.

A number of workers have used the same approach in mammals. Nicotinamide protects against the teratogenicity of 6-aminonicotinamide in the mouse (Pinsky and Fraser, 1960), rat (Chamberlain and Nelson, 1963), and rabbit (Grote et al., 1971) and of thiadiazole in the rat (Beaudoin, 1973). However, no analytical comparisons like those of Landauer have been reported, perhaps because of the complication that the mammalian embryo is surrounded by its mother. For instance, the paradoxical finding that the pyrimidines uridine and thymine potentiated, rather than reduced, the teratogenic effects of their analog 5-fluorouracil (5-FU) in the rat was shown to result from inhibition, by the pyrimidines, of the metabolism of 5-FU in the maternal liver, thus exposing the embryo to higher concentrations (Schumacher et al., 1969). Similarly, the fact that DNA reduced both the teratogenic effects of actinomycin D and its embryolethal effect in the hamster, whereas progesterone–estrone treatment reduced only the embryolethality, suggested that the teratogenic effects resulted from direct effects on the embryo and the lethality from indirect effects via the mother (Elis and di Paolo, 1970).

The maternal complication may be circumvented in part by intraamniotic injection. For instance, nicotinamide and NAD, but not ATP, protected against the effects of 6-aminonicotinamide by this route in the rat, whereas all three protected against fetal and placental weight loss, suggesting that 6-AN-induced malformation and mortality may involve separate mechanisms in the embryo itself (Chamberlain and Goldyne, 1970).

Other examples of the use of pharmacological agents to analyze the mode of action of teratogens include the demonstration that the teratogenic effects of the diuretic, acetazolamide (a carbonic anhydrate inhibitor), were reduced by agents that prevented maternal loss of potassium. This suggested that potassium deficiency, as well as carbonic anhydrase inhibition, plays a role in the teratogenicity, although the precise cause remains elusive (Ellison and Maren, 1972). Attempts to determine whether the teratogenic effects of diphenylhydantoin are mediated through its action on folate metabolism by attempting to protect with appropriate intermediates of folate metabolism (Kernis et al., 1973; Marsh and Fraser, 1973a) have been disappointingly equivocal, which probably indicates that the hypothesis is wrong.

Interactions have also been used to study whether some teratogens act by virtue of their chelating properties; for instance, chelation of a metal may inhibit an enzyme dependent on that metal. Selenium protects against the teratogenic effects of cadmium and arsenic in the hamster, possibly by chelation (Holmberg, 1969), and zinc protects rat embryos against the teratogenic effects of a chelating agent, EDTA (Swenerton and Hurley, 1971). However, the interactions are complex. Selenium and zinc are themselves mildly teratogenic in rodents, and the teratogenic effects of sodium salicylate (a chelating agent) are potentiated, rather than alleviated, by manganese, possibly by chelation in the maternal serum forming a complex less easily conjugated or excreted or more easily transported to the fetus (Kimmel *et al.,* 1974). Several other teratogens are chelators and may produce their teratogenic effects in this way (Marsh and Fraser, 1973b), and in spite of the complexities involved this approach deserves further exploration.

Finally we would like to consider the question of interactions of teratogens with other teratogens and the conclusions that can be drawn from how they interact. This approach seems to have been exploited first by Runner and Dagg (1960; Runner, 1967), who studied the teratogenic effects of maternal fasting in the mouse in combination with various other teratogenic treatments. They characterized the possible types of interaction as interfering, nonadditive, additive, and reinforcing; they then drew certain inferences about the sites of action of the teratogens concerned. When two teratogens interact nonadditively (that is, their effect in combination is the same as that of the one having the larger effect singly), it is reasonable to suppose that they act on the same pathway; if one teratogen blocks the pathway, adding a second block should not increase the effect. For example, maternal fasting and treatment with iodoacetic acid in mice were nonadditive in combination, with respect to vertebral defects (Runner and Dagg, 1960) and cleft palate (Miller, 1973).

However, the interpretation of interfering, additive, and reinforcing effects is not so simple. Examples include the synergistic interaction of maternal fasting and cortisone in mice (Kalter, 1960), insulin and nicotinamide in mice (Smithberg and Runner, 1963) contrary to the situation in the chick, and trypan blue with hypoxia (Runner and Dagg, 1960). Landauer in a further elaboration of his concept that certain teratogens exert their effect by starving specific sites of developing embryos of energy required for normal functions states (Landauer and Salam, 1974):

> Do the receptors of teratogens which produce homologous results belong to the same chain of developmental events, or are they quite independent, yet converging toward the same end-effect? We do not know.... The synergistic activities of the two interacting teratogens may in the instance of our material all have been based on competition of components for sites of loss, but details of the nature of these sites remain to be clarified. In some instances ... the synergists may simply block cellular components which, unhindered, would have opposed teratogenic activity; in other situations, as is likely to be true for teratogens with homologous end-effects, the supplementing compound way, in addition to occupying sites of loss, add actively to

the synergistic result, as is certainly true in the event of parallel blocking of an alternative pathway.

On a more organismal level, synergism of teratogens can also be interpreted in terms of multifactorial, threshold systems, as mentioned in Chapter 2. If two teratogens act additively in terms of shifting the distribution of the developmental variable relative to the threshold, the result, as measured in terms of frequency of malformations, may appear to be either synergistic (Fraser, 1965) or approximately additive, depending on the magnitude of the shifts and the position of distribution to threshold. Figure 1 illustrates a hypothetical model. The important point is that the effect of the teratogen is properly considered in terms of the amount by which the distribution is shifted, but what is actually measured in a threshold system is the proportion of the population that falls beyond the threshold—that is, the percent affected (Wright, 1926). Thus, if we have a control population that is assumed to be normally distributed, with a mean of three standard deviations (S.D.) from the threshold, only about 1 per thousand will fall beyond the threshold and be affected. If teratogen A shifts the distribution by 1.5 S.D.'s, then (by the properties of the normal curve) 5% of the population will fall beyond the threshold and be affected. Assume that teratogen B has the same effect. Then teratogens A and B together, if they act additively, will shift the distribution by 3 S.D.'s and 50% (not 10%) of the population will be affected. Thus the interaction will appear to be synergistic. If each teratogen moved the distribution by 3 S.D's (50% affected), together they would move it 6 S.D.'s (100% affected), so the effect would appear to be additive, as it actually is.

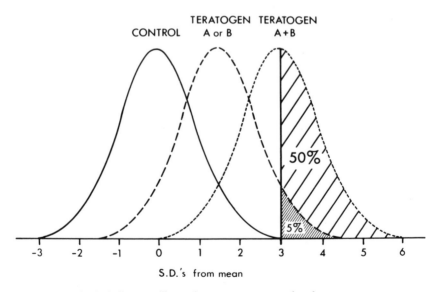

Fig. 1. Hypothetical diagram illustrating apparent synergism between two teratogens that interact additively. For explanation see text.

This does not require the assumption [contrary to Runner's (1967) statement] that similar steps in a common enzymatic pathway are involved. To cite the palate-closure model yet again, one teratogen could delay shelf closure by acting on the tongue, and another on the shelf. If they both delayed palate closure by the same amount, relative to the threshold, the effect on cleft palate frequency could be synergistic. Thus it should have been no surprise to Miller (1972) to find that pyridoxine deficiency and cortisone acted synergistically at a low dose of cortisone and additively at a high dose. Deductions about the synergism of teratogens should be made only if the synergism is still present after a statistical transformation of the data that results in linear dose–response curves (Bliss, 1957; Finney, 1971).

We have seen how the interactions between teratogens and of teratogens with other pharmacological agents may illumine mechanisms of teratogenesis, and we have also considered some of the pitfalls of interpretation. Such interactions also carry the implication that "altogether harmless amounts of potentially teratogenic substances may in combination produce powerful synergism and thereby a high incidence of severe malformations. Moreover ... non-teratogenic compounds may also act as highly effective synergists. The cautionary consequences of these facts should not require emphasis" (Landauer and Clark, 1964).

IV. INTERACTIONS BETWEEN TERATOGENS AND GENOTYPES

The interactions between teratogens and genes provide one of the most interesting aspects of teratology. They are important not only because of their significance in screening for teratogenicity (Wilson, 1973), but because of their contribution to understanding the nature of congenital malformations. (See Chapter 2, this volume, and chapter by Trasler and Fraser, Vol. 2, for examples.)

A. Species Differences

Species differences in response to teratogens gained dramatic prominence from the thalidomide problem but had been recognized previously. Reliable data are still conspicuous by their absence (Runner, 1967). Study of the basis for such a difference should be valuable, as this knowledge could be an aid in extrapolating findings from experimental animals to man. For instance, if one species proved resistant to a drug because it lacked the enzyme that, in other species, converted the drug to a teratogenic metabolite, one could predict whether it would be teratogenic in man by establishing whether man could also make the conversion. An analogous situation has been shown for the carcinogenicity of 2-acetylaminofluorene. Guinea pigs are resistant, since they lack the enzyme that converts the compound to its actively carcinogenic N—OH metabolite (Shellenberger, 1975). Little use has been made of this promising approach to teratology.

One of the earliest examples of species differences in teratogenic response was provided by cortisone-induced cleft palate. Mice (Fraser *et al.*, 1954) and rabbits (Fainstat, 1954) are relatively susceptible, and most strains of rats are highly resistant (Curry and Beaton, 1958). The reasons for the difference are still not clear. The presence of the placenta in the female rat seems to protect it from many of the metabolic effects of cortisone (Curry and Beaton, 1958), and this suggestion is supported by the fact that intraamniotic injection of hydrocortisone can produce cleft palate in rats (Dostal and Jelinek, 1971). Alternatively Harris (1964) has suggested that the species difference in teratogenicity results from a species difference in degree of flexion of the cranial base, influencing the ease with which the tongue is displaced from between the palatal shelves. Explaining species differences is much harder than finding them.

B. Strain Differences

The interaction between environmental teratogens and specific genotypes and their value in providing insight into developmental mechanisms has been discussed at length in several other chapters of this book (Chapter 2, this volume; chapters by Trasler and Fraser and Biddle and Fraser, Vol. 2; and Kalter and others in Vol. 3). We will simply reiterate here that these interactions may be complex and nonlinear, which complicates the problem of statistical analysis and interpretation (Lewontin, 1974).

That the same teratogen may produce different frequencies of types of malformation when imposed on different genotypes is now commonplace knowledge among teratologists, although it caused something of a stir when it was first demonstrated (Landauer and Bliss, 1946; Fraser and Fainstat, 1951). Attempts to analyze the genetic basis of strain differences have been relatively scanty, perhaps because investigators have tended to use strain differences in response as tools to investigate mechanisms of malformation rather than to define the number of genes involved, and perhaps also because of the complexity of the necessary genetic analysis required. Nevertheless, the identification of specific genes influencing teratogenic responses could greatly refine the use of genetic differences to analyze mechanisms.

Let us consider briefly a typical approach to the problem. If one exposes a variety of inbred strains of a given species (usually mice) to a teratogen, there will almost certainly be strain differences in the frequencies, and perhaps types of malformations produced. We will assume that the frequencies of malformations induced in the inbred strains are the same in both sexes, indicating that there are no sex-linked genes influencing susceptibility (Francis, 1973). The first step in a genetic analysis would be to make crosses between strains differing widely in their susceptibility to the teratogen and measure the frequency of the chosen malformation in the offspring of treated mothers. The F_1 frequency will either resemble one strain or the other (dominance), be intermediate (additive), or in some cases be lower than that of either parental strain. The basis for the latter result is not well understood; it may be attributed to heterosis, but in any case it complicates the analysis. Often there will be differences in frequency between the reciprocal crosses, and these may be either matroclinous (the F_1 frequency is higher when the mother is from the susceptible strain), or patroclinous (the F_1 frequency tends to resemble that of the paternal strain). Matroclinous differences may result from maternal uterine effects on susceptibility, cytoplasmically transmitted factors, or X-lined genes, and these can be distinguished by the appropriate backcrosses (see chapter by Biddle and Fraser, Vol. 2, for a detailed discussion).

To estimate the number of autosomal genes involved in the strain difference, the next step is to design backcrosses (BC) that hold the cytoplasmic contribution and maternal genotype constant and see how the BC frequencies compare with those of the F_1 and parental strains (see Biddle, 1975, for example). For instance, if the difference between the two strains results from a single allelic difference, which *acts in the embryo,* the backcross to the susceptible strain will consist of embryos that are heterozygous (50%) and homozygous (50%) for the allele for susceptibility. If so, the distribution of frequencies [after appropriate transformation to allow for the quasicontinuous nature of the defect (see Biddle, 1975, and chapter by Biddle and Fraser, Vol. 2)] should be bimodal. If more than one major locus is involved, there will be correspondingly more modes; and if the number is much more than one, the situation will be too complicated to yield to this type of analysis. One may then turn to methods based on estimating the environmental and genotypic variances (Wright, 1968), but the presence of heterosis, maternal effects, and dominance make critical interpretation difficult; in any case the exercise at this point begins to have only academic interest, since the chances of isolating a major difference are slim.

Even if one does succeed in defining a major gene difference influencing the frequency of a malformation produced by a teratogen, one should not leap to the conclusion that this is a "cause" of the malformation. Suppose we had succeeded in establishing two strains differing only at one locus that made one strain relatively resistant and one relatively susceptible to cleft palate induced by, for instance, 6-aminonicotinamide. The next step would be to identify the mode of action of the gene. Perhaps it would be found to influence the metabolism, placental transport, receptor site activity, etc. of the 6-AN. In this case the gene would have nothing to do with palate closure *per se,* and the effect would be limited to cleft palates induced by 6-AN or similar agents. For instance, we know that the frequency of cleft palate induced by 6-AN in the A/J mouse strain is higher than that in the C57BL strain, that it takes less nicotinamide to protect against a given dose of 6-AN in the C57BL strain, and that the "turnover" of 6-AN (as measured by the time after 6-AN administration that nicotinamide can still reduce the frequency) is faster in the A/J strain (Verrusio, 1966; Fraser, 1969). This has nothing to do with the causes of cleft palate, only (possibly) with the causes for the relative difference between strains in response. If we were lucky, we might find that the gene influenced some aspect of palate closure that increased the probability of failure to close. This difference should then extend to other teratogens, and its study could provide insight into the mechanism of palate closure and its failure.

C. Interactions between Teratogens and Mutant Genes

The interactions of teratogens and single mutant genes have been most extensively analyzed in the elegant series of studies by Landauer in the chick, but comparable studies are appearing in mammals. Treatment of chick eggs with nicotine produces a syndrome closely resembling that produced by the homozygous mutant gene for "crooked-neck dwarf." Embryos heterozygous for the mutant gene are normal, if untreated, but are much more susceptible to the teratogenic effects of nicotine than their sibs that do not carry the mutant gene (Landauer, 1960). This might suggest that the mutant gene and the teratogen act on pathways that intersect (i.e., influence the same end-product), and this is supported by the fact that the heterozygous mutants are not more sensitive to other teratogens, such as insulin (Landauer, 1961).

Another example involves three types of recessively inherited micromelia, and their interactions with 6-aminonicotinamide, which also causes micromelia. In one type (mm^A) the heterozygous embryos were again more susceptible to 6-AN-induced micromelia than their genetically normal sibs. No such effect of heterozygosity was seen in the case of an unrelated micromelic mutant mm^H, suggesting that the mm^H gene acted through a different mechanism. The same was true for another recessive mutant, *ch,* with the following additional interesting feature. Two lines had been established, both carrying the *ch* mutant, but one had been selected for high expression (penetrance) of the mutant, and the other for low expression. The high line was susceptible to 6-AN and the low line was resistant, irrespective of the presence of the mutant gene. Thus it would seem that although the mutant gene and teratogen did not act on the same pathway, the modifiers accumulated by selection were nonspecific, affecting them both. This would fit in nicely with the multifactorial distribution/threshold model (see Chapter 2), according to which modifiers could alter the relationship of distribution to threshold by producing small changes in many factors. The mutant gene and teratogen, respectively, would each produce a major change in the distribution by acting at single points in the system. If they involve different pathways, and if the gene is really recessive, producing no change in the distribution in the heterozygote, they would not interact. However, the resulting change in frequency of malformations will be influenced by the effects of the modifiers in both cases.

Some of the early work on the interaction of mutant genes with teratogens was marred by poor controls. Mice from an anophthalmic strain were mated to animals of an inbred colony with normal eyes. The F_1 offspring had normal eyes, but if the mothers were treated with trypan blue some of the F_1 animals were anophthalmic (Barber, 1957), and the frequency in the F_1 was higher than that produced by treatment of the normal parental strain. This was attributed to the teratogen having increased the penetrance of the genes for microphthalmia, but the difference could just as well have resulted from a difference in susceptibility between the two strains quite unrelated to the presence of the genes for microphthalmia. A similar criticism can be made of attempts to confirm and extend this work (Beck, 1963) which did in fact show such subline differences (Beck, 1964).

A clear-cut interaction of mutant gene and teratogen was demonstrated between a recessive mutant gene, *luxoid,* and 5-fluorouracil (5-FU) in the mouse (Dagg, 1967). The homozygous *luxoid* gene and 5-FU each cause tibial hemimelia; 5-FU is effective at a dose of 0.5 mg, but not at 0.25 mg. The low dose of teratogen produced tibial hemimelia in the mutant heterozygote, but not in sibs homozygous for the normal allele. Thus the mutant gene and teratogen, both in doses that would be nonteratogenic by themselves, reached the threshold of teratogenicity when acting together. The same kind of interaction was observed with 5-FU and another similar but nonallelic mutant, *luxate,* and between the two heterozygous mutant genes, Again a threshold system is involved, although Dagg correctly points out one cannot deduce from this whether the teratogen and genes act on the same or different biochemical pathways. We will return to this point later. The interaction disappeared when the treatment was postponed from day 10 to day 11 of gestation. A similar interaction was shown for another mutant, *Strong's luxoid* and 5-FU, and in this case (as with Landauer's micromelia–6-AN experiments) this action of 5-FU was affected by modifying genes selected by their effects on the expression of the mutant (Forsthoefel, 1972).

Sometimes studies of gene–teratogen interactions lead to unexpected results. In an attempt to increase the penetrance of *lid-gap,* a mutant gene which causes failure of eyelid fusion in the mouse, pregnant mothers were treated with cortisone, in the hope that this teratogenic stimulus would exaggerate the effects of the gene. Surprisingly, the treatment suppressed, rather than exacerbated, the mutant phenotype (Watney and Miller, 1964), possibly by preventing, or removing, adhesions between the corneal epithelium and lid epidermis that may have interfered with normal lid closure (Watney-Harris and Fraser, 1968). In this case there is no reason to assume that the cortisone is acting on the same pathway as the mutant gene, but it does demonstrate the possibility of prenatal correction of developmental errors. Another example in which the deleterious effects of a mutant gene could be prevented by an environmental agent is the *pallid* mouse, where defects of the otolith leading to ataxia could be prevented by maternal treatment with manganese (Erway *et al.*, 1971); in this case there is evidence that the mutant gene and the environmental agent do affect the same developmental pathway, since maternal manganese deficiency causes a similar otolith abnormality in nonmutant rats and in mice.

Another interesting case of interaction of a specific mutant gene with a specific teratogen involved the dominant brachyury gene, *T*, and actinomycin D, an inhibitor of DNA-dependent RNA synthesis (Winfield and Bennett, 1971). The *T* locus presents many fascinating developmental problems (Dunn and Bennett, 1964; Gluecksohn-Waelsch and Erickson, 1970; Erickson, 1975; chapter by Gluecksohn-Waelsch, Vol. 2). Heterozygotes for *T* and the normal allele have short tails. There are series of recessive alleles, each interacting with *T* to produce taillessness and usually lethal in the homozygote. Each *t* homozygote possesses a characteristic phenotype expressing failure of the central nervous and developmentally related systems to pass through a particular stage of development (chapter by Gluecksohn-Waelsch, Vol. 2). The earliest-acting combination, t^{12}/t^{12} fails to progress from morula to blastula and is deficient in ribosomal RNA. When $T/+$ mothers mated to $+/+$ fathers were treated with actinomycin D on day 6¼ of development about 50% of the embryos showed abnormalities, suggesting disturbances of cell and tissue interaction similar to those seen in homozygotes or compound heterozygotes for various *t* alleles (Winfield and Bennett, 1971). No such changes were seen in $+/+ \times +/+$ matings, suggesting that they resulted from the action of actinomycin D on $T/+$ embryos. Trypan blue, which does not appear to inhibit RNA synthesis, produced different malformations and did not appear to interact with the *T* gene in this experiment. When given 2 days later, to mice of another genotype, it appeared to convert the short tail of $T/+$ embryos to taillessness, possibly by imposing an additional nonspecific retardation on a process already jeopardized by the *T* mutant (Hamburgh *et al.*, 1970).

Winfield and Bennett (1971) state that the interaction of actinomycin with the *T* mutant may suggest routes of action similar to those of the *t* alleles (perhaps disturbance of RNA synthesis), but they caution that neither the specificity of the interaction nor the primary site of actinomycin D in the embryo has been firmly established. Nevertheless, this seems a promising approach to the analysis of developmental errors, which is not being widely exploited. By studying the interactions of groups of teratogens with a given mutant gene and of various mutant genes with the same teratogen, one ought to be able to illuminate some of the developmental mechanisms involved.

Finally one may speculate on what role may be placed in human teratology by the interaction of environmental factors with heterozygous recessive mutant genes. Such interactions have been invoked with respect to cancer, diabetes mellitus, and mental illness (reviewed by Swift *et al.*, 1974)—why not with malformations? There is no doubt that such interactions can occur, as the above samples from experimental teratology have shown. Since heterozygotes for rare recessive deleterious genes are comparatively common they might, collectively, contribute significantly to susceptibility. There is some suggestion that a genetically determined predisposition to hearing loss can be exacerbated by treatment with streptomycin (Tsuike and Murai, 1971) or infection with rubella. I am not aware of any other evidence or, for that matter, any attempt to look for it.

D. Interactions between Teratogens, Genes, and Environment

We have reviewed in previous sections numerous examples where the effect of a teratogen varies according to the genotype it is acting upon, and depending on a number of environmental variables. It should be no surprise to learn that all three can interact. The effect of an environmental variable on the response of an embryo to a teratogen may be quite different in animals of different genotypes. Two examples from our own experience will be given. The effects of cortisone on cleft palate frequency, at two dose levels, were measured in two strains of mice, raised on two different commercial diets and treated at varying maternal weights (Warburton et al., 1962). Strain A/J was more susceptible (72% CP) to the teratogenic effect of cortisone than strain C57BL/6 (6%) with the lower dose of cortisone on diet A in "heavy" (over 22 g) mothers, but the reverse was true on diet B (26% and 52%, respectively). Also, the effect of dose varied greatly in heavy A/J mothers, but much less in heavy C57 mothers. "Thus both genetic and nutritional factors influence the embryo's response to cortisone, and genetic factors influence the influence of the nutritional factors"—but differently at different doses and different maternal weights.

Another example involves the teratogenic response, as measured by cleft palate frequency, of two inbred mouse strains to 6-aminonicotinamide. A higher frequency of cleft palate was induced in the A/J than in the C57BL/6J strain. A higher frequency of cleft palate was produced in A × C57 F_1 hybrids than in C57 × A hybrids (mother written first) when the mothers were maintained on Purina Lab Chow but not on Breeder Chow (Verrusio et al., 1968; Fraser, 1969). Backcrosses suggested cytoplasmic transmission. The same difference was then found in the parental strains: the frequency of 6-AN-induced cleft palate was lower in mothers raised on Lab Chow than on Breeder Chow in the C57BL/6J but not the A/J strain. A more extensive discussion will be found in the chapter by Biddle and Fraser (Vol. 2). The point to be emphasized here is that the response to a teratogen varies with diet, but the nature of the variation depends on the genotype.

Complexities such as these may tempt one to despair of ever elucidating the environmental causes of malformations in man, but they should not. Let them serve as cautions in the interpretation of data and their extrapolation from one situation to another, but also as challenges to unravel them. Our present state of confusion may represent the state of conflict, paradox, and uncertainty that often precedes an integrating insight.

REFERENCES

Ader, R., and Plant, S. M., 1968, Effects of prenatal maternal handling and differential housing on offspring emotionality, plasma corticosterone levels, and susceptibility to gastric erosions, *Psychosom. Med.* **30**:277.

Anon., 1975, Drug interactions, *Lancet* **1**:904 (editorial).

Barber, A. N., 1957, The effects of maternal hypoxia on inheritance of recessive blindness in mice, *Am. J. Ophthalmol.* **44**(4):94.

Beaudoin, A. R., 1973, Teratogenic Activity of 2-amino-1,3,4-thiadiazole hydrocholoride in Wistar rats and the protection afforded by nicotinamide, *Teratology* **7**:65.

Beck, S. L., 1963, Frequencies of teratologies among homozygous normal mice compared with those heterozygous for anophthalmia, *Nature* **200**:810.

Beck, S. L., 1964, Subline differences among C57 Black mice in response to trypan blue and outcross, *Nature* **204**:403.

Beck, S. L., 1967, Effects of position in the uterus on fetal mortality and on response to trypan blue, *J. Embryol. Exp. Morphol.* **17**:617.

Biddle, F. G., 1975, Teratogenesis of acetazolamide in the CBA/J and SWV strains of mice II. Genetic control of the teratogenetic response, *Teratology* **11**:37.

Bliss, C. I., 1957, Some principles of bioassay, *Am. Sci.* **45**:449.

Brown, K. S., Johnston, M. C., and Murphy, P. F., 1974, Isolated cleft palate in A/J mice after transitory exposure to drinking-water deprivation and low humidity in pregnancy, *Teratology* **9**:151.

Chamberlain, J. G., and Goldyne, M. E., 1970, Intra-amniotic injection of pyridine nucleotides or adenosine triphosphate as countertherapy for 6-aminonicotinamide (6-AN) teratogenesis, *Teratology* **3**:11.

Chamberlain, J. G., and Nelson, M. M., 1963, Congenital abnormalities in the rat resulting from single injections of 6-aminonicotinamide during pregnancy, *J. Exp. Zool.* **153**:(3):285.

Conney, A. H., 1969, Drug metabolism and therapeutics, *N. Engl. J. Med.* **280**:653.

Curry, D. M., and Beaton, G. H., 1958, Cortisone resistance in pregnant rats, *Endocrinology* **63**:155.

Dagg, C. P., 1963, The interaction of environmental stimuli and inherited susceptibility to congenital deformity, *Am. Zool.* **3**:223.

Dagg, C. P., 1966, Teratogenesis, *in: Biology of the Laboratory Mouse* (E. L. Greene, ed.), pp. 309–328, McGraw-Hill, New York.

Dagg, C. P., 1967, Combined action of fluorouracil and two mutant genes on limb development in the mouse, *J. Exp. Zool.* **164**:479.

Dagg, C. P., 1972, Independent effects of fluorouracil and two mutant genes in mouse embryos, *Teratology* **5**:377.

Dagg, C. P., Schlager, G., and Doerr, A., 1965, Polygenic control of the teratogenicity of 5-fluorouracil in mice, *Genetics* **53**:1101.

Dostal, M., and Jelinek, R., 1971, Induction of cleft palate in rats with intraamniotic corticoids, *Nature* **230**:464.

Dunn, L. C., and Bennett, D., 1964, Abnormalities associated with a chromosome region in the mouse, *Science* **144**:260.

Edwards, M. J., 1969, Congenital defects in guinea pigs: Fetal resorption, abortions, and malformations following hyperthermia during early gestation, *Teratology* **2**:313.

Elis, J., and di Paolo, J. A., 1970, The alteration of actinomycin D teratogenicity by hormones and nucleic acids, *Teratology* **3**:33.

Ellison, A. C., and Maren, T. H., 1972, The effect of potassium metabolism on acetazolamide-induced teratogenesis, *Johns Hopkins Med. J.* **130**:105.

Erickson, R. P., 1975, Differentiation, alloantigens, histocompatibility loci and birth defects, *Am. J. Hum. Genet.* **27**:554.

Erway, L. C., Fraser, A. S., and Hurley, L. C., 1971, Prevention of congenital otolith defect in pallid mutant mice by manganese supplementation, *Genetics* **67**:97.

Fainstat, T., 1954, Cortisone-induced congenital cleft palate in rabbits, *Endocrinology* **55:**502.

Ferm, V. H., 1967, Potentiation of the teratogenic effect of vitamin A with exposure to low environmental temperature, *Life Sci.* **6:**493.

Finney, D. J., 1971, *Probit Analysis,* 3rd ed., Cambridge University Press, Cambridge, U.K.

Forsthoefel, P. F., 1972, The effects on mouse development of interactions of 5-fluorouracil with Strong's luxoid gene and its plus and minus modifiers, *Teratology* **6:**1.

Francis, B. M., 1973, Influence of sex-linked genes on embryonic sensitivity to cortisone in three strains of mice, *Teratology* **7:**119.

Fraser, F. C., 1965, Some genetic aspects of teratology, *in: Teratology Principles and Techniques* (J. G. Wilson and J. Warkany, eds.), pp. 21–38, University of Chicago Press, Chicago.

Fraser, F. C., 1969, Gene–environment interactions in the production of cleft palate, *in: Methods for Teratological Studies in Experimental Animals and Man* (H. Nishimura and J. R. Miller, eds.), pp. 34–49, Igaku Shoin, Tokyo.

Fraser, F. C., and Fainstat, T. D., 1951, The production of congenital defects in the offspring of pregnant mice treated with cortisone. A progress report, *Pediatrics* **8:**527.

Fraser, F. C., and Warburton, D., 1964, No association of emotional stress or vitamin supplement during pregnancy to cleft lip or palate in man. *J. Plast. Reconstr. Surg.* **33:**395.

Fraser, F. C., Kalter, H., Walker, B. E., and Fainstat, T. D., 1954, The experimental production of cleft palate with cortisone and other hormones, *J. Cell. Comp. Physiol.* **43:**237.

Geber, W. F., 1966, Developmental effects of chronic maternal audio visual stress on the rat fetus, *J. Embryol. Exp. Morphol.* **16:**1.

Gibson, J. E., and Becker, B. A., 1968, Effect of phenobarbital and SKF 525-A on the teratogenicity of cyclophosphamide in mice, *Teratology* **1:**393.

Glueckaohn-Waelsch, S., and Erickson, R. P., 1970, The T-locus of the mouse: Implications for mechanisms of development, *Curr. Top. Dev. Biol.* **5:**281.

Goldman, A. G., and Yakovac, W. C., 1963, The enhancement of salicylate teratogenicity of maternal immobilization in the rat, *J. Pharmacol. Exp. Ther.* **142:**351.

Goldman, A. G., and Yakovac, W. C., 1964, Salicylate intoxication and congenital anomalies, *Arch. Environ. Health* **8:**648.

Grote, W., Clausen, U., and Heinz, D., 1971, Verursachung und Verhutung von Entwicklungsstorungen beim Kaninchen durch Verarbreichung von 6-Aminonikotinsauramid und Nikotinsauramide, *Argneimittelforschung* **21:**825.

Gunberg, D. L., 1955, The effect of suboptimal protein diets on the teratogenic properties of trypan blue administered to pregnant rats, *Anat. Rec.* **121:**398.

Hamburgh, M., Herz, R., and Landa, G., 1970, The effect of trypan blue on expressivity of the Brachyury gene "T" in mice, *Teratology* **3:**11.

Hamburgh, M., Mendoza, L. A., Rader, M., Lang, A., Silverstein, H., and Hoffman, K., 1974, Malformations induced in offspring of crowded and parabiotically stressed mice, *Teratology* **10:**31.

Hamly, C.-A., Trasler, D. G., and Fraser, F. C., 1970, Reduction of 6-aminonicotinamide teratogenicity in mice by etherization, *Teratology* **3:**293.

Harris, J. W. S., 1964, Oligohydramnios and cortisone-induced cleft palate, *Nature* **203:**533.

Hartel, A., and Hartel, G., 1960, Experimental study of teratogenic effect of emotional stress in rats, *Science* **132:**1438.

Heiberg, K., Kalter, H., and Fraser, F. C., 1959, Production of cleft palates in the offspring of mice treated with ACTH during pregnancy, *Biol. Neonat.* **1:**33.

Holmberg, R. E., 1969, Interrelationships of selenium, cadmium, and arsenic in mammalian teratogenesis, *Arch. Environ. Health* **18:**873.

Ishii, H., Kamei, T., and Omae, S., 1962, Effects of the concurrent administrations of chondroitin sulfate with cortisone, vitamin A or noise stimulation on fetal development of the mouse, *Gunma J. Med. Sci.* **11:**259.

Jacobsen, L., 1968, *Low Dose X-Irradiation and Teratogenesis,* Munksgaard, Copenhagen.

Jaworska, M., 1965, Cleft palate produced experimentally in C57/B1 strain of mice in two age groups, *Acta Chir. Plast.* **7:**20.

Kalter, H., 1956, Modification of teratogenic action of cortisone in mice by maternal age, maternal weight and litter size, *Am. J. Physiol.* **185**:(1):65.

Kalter, H., 1960, Teratogenic action of a hypocaloric diet and small doses of cortisone, *Proc. Soc. Exp. Biol. Med.* **104**:518.

Kalter, H., 1965, Interplay of intrinsic and extrinsic factors, *in: Teratology: Principles and Techniques* (J. F. Wilson and J. Warkany, eds.), pp. 57–80, University of Chicago Press, Chicago.

Kernis, M. M., Pashayan, H. M., and Pruzansky, S., 1973, Dilantin-induced teratogenicity and folic acid deficiency, *Teratology* **7**:A-19.

Kimmel, C. A., Butcher, R. E., Vorhees, C. V., and Schumacher, H. J., 1974, Metal-salt potentiation of salicylate-induced teratogenesis and behavioral changes in rats, *Teratology* **10**:293.

King, C. T. G., Weaver, S. A., and Narrod, S. A., 1965, Anthihistamines and teratogencity in the rat, *J. Pharmacol. Exp. Ther.* **147**(3):391.

Landauer, W., 1948, The effect of nicotinamide and ketoglutaric acid on the teratogenic actions of insulin, *J. Exp. Zool.* **109**(2):283.

Landauer, W., 1960, Nicotine-induced malformations of chicken embryos and their bearing on the phenocopy problem, *J. Exp. Zool.* **143**(1):107.

Landauer, W., 1961, The interplay of intrinsic and extrinsic factors in the origin of congenital malformations, *Proc. 1st Int. Conf. Congenital Malformations* (M. Fishbein, ed.), pp. 117–122, Lippincott, London.

Landauer, W., 1972, Is insulin a teratogen?, *Teratology* **5**:129.

Landauer, W., and Bliss, C. I., 1946, Insulin-induced rumplessness of chickens III. The relationship of dosage and of developmental stage at time of injection to response, *J. Exp. Zool.* **102**:1.

Landauer, W., and Clark, E. M., 1964, Teratogenic risks of drug synergism, *Nature* **203**:527.

Landauer, W., and Salam, N., 1972, Aspects of dimethyl sulfoxide as solvent for teratogens, *Dev. Biol.* **28**:35.

Landauer W., and Salam, N., 1974, The experimental production in chicken embryos of muscular hypoplasia and associated defects of beak and cervical vertebrae, *Acta Embryol. Exp.* **1**:51.

Landauer, W. and Sopher, D., 1970, Succinate, glycerophosphate and ascorbate as sources of cellular energy and as antiteratogens, *J. Embryol. Exp. Morphol.* **24**(1):187.

Landauer, W., and Wakasugi, N., 1968, Teratological studies with sulfonamides and their implications, *J. Embryol. Exp. Morphol.* **20**:261.

Lewontin, R. C., 1974, Annotation: The analysis of variance and the analysis of causes, *Am. J. Hum. Genet.* **26**(3):400.

Marsh, L., and Fraser, F. C., 1973a, Studies on dilantin-induced cleft palate in mice, *Teratology* **7**:A-23.

Marsh, L., and Fraser, F. C., 1973b, Chelating agents and teratogenesis, *Lancet* **2**:846.

Miller, T. J., 1972, Cleft palate formation: A role for pyridoxine in the closure of the secondary palate in mice, *Teratology* **6**:351.

Miller, T. J., 1973, Cleft palate formation: The effects of fasting and iodoacetic acid on mice, *Teratology* **7**:177.

Nakasato, Y., 1953, Experimental investigation on the influence of abnormal maternal environment (excess of adrenal cortex hormone) upon organ development during the embryonic period, *Nagasaki Igakkai Zassi* **28**(8):825.

Nolen, G. A., 1972, The effects of various levels of dietary protein on retinoic acid-induced teratogenicity in rats, *Teratology* **5**:143.

Nomura, T., 1969, Management of animals for use in teratological experiments, in: *Methods for Teratological Studies in Experimental Animals and Man* (H. Nishimura and J. R. Miller, eds.), pp. 3–15, Igaku Shoin, Tokyo.

Peters, S., and Strasburg, M., 1969, Stress als teratogener Faktor, *Arzneim. Forsch.* **19**:1106.

Pinsky, L., and Fraser, F. C., 1959, Production of skeletal malformations in the offspring of pregnant mice treated with 6-amino-nicotinamide, *Biol. Neonate* **1**(2):106.

Pinsky, L., and Fraser, F. C., 1960, Congenital malformations after a two-hour inactivation of nicotinamide in pregnant mice, *Br. Med. J.* **2**:195.

Rosenzweig, S., Blaustein, F. M., and Anderson, C. D., 1970, An improved apparatus for producing stress-induced cleft palate in mice, *Teratology* **3**(4):311.

Runner, M. N., 1967, Comparative pharmacology in relation to teratogenesis, *Fed. Proc.* **26**:1131.

Runner, M. N., and Dagg, C. P., 1960, Metabolic mechanisms of teratogenic agents during morphogenesis, in: *Symposium on Normal and Abnormal Differentiation and Development,* National Cancer Institute Monograph No. 2, pp. 41–54.

Schumacher, H. J., Wilson, J. G., and Jordan, R. L., 1969, Potentiation of the teratogenic effects of 5-fluorouracil by natural pyrimidines, II. Biochemical aspects, *Teratology* **2**(2):99.

Shellenberger, T. E., 1975, The role of maternal metabolism in developmental pharmacology, *Teratology* **11**(2):34A.

Smithberg, M., and Runner, M. N., 1963, Teratogenic effects of hypoglycemic treatments in inbred strains of mice, *Am. J. Anat.* **113**(3):479.

Strean, L. P., and Peer, L. A., 1956, Stress as an etiological factors in the development of cleft palate, *Plast. Reconstr. Surg.* **18**(1):1.

Swenerton, H., and Hurley, L. S., 1971, Teratogenic effects of a chelating agent and their prevention by zinc, *Science* **173**:62.

Swift, M., Cohen, J., and Pinkham, R., 1974, A maximum-likelihood method for estimating the disease predisposition of heterozygotes, *Am. J. Hum. Genet.* **26**(3):304.

Szabo, K. T., and Brent, R. L., 1975, Reduction of drug-induced cleft palate in mice, *Lancet* **1**(7919):1296.

Terada, M., and Nishimura, H., 1975, Mitigation of caffeine-induced teratogenicity in mice by prior chronic caffeine ingestion, *Teratology* **12**:79.

Trasler, D. G., 1960, Influence of uterine site on occurrence of spontaneous cleft lip in mice, *Science* **132**:420.

Tsuiki, T., and Murai, S., 1971, Familial incidence of streptomycin hearing loss and hereditary weakness of the cochlea, *Audiology* **10**:315.

Verrusio, A. C., 1966, Biochemical basis for a genetically determined difference in response to the teratogenic effects of 6-aminonicotinamide, Ph.D. thesis, McGill University, Montreal, Canada.

Vesell, E., 1967, Induction of drug-metabolizing enzymes in liver microsomes of mice and rats by softwood bedding, *Science* **157**:1057.

Verrusio, A. C., Pollard, D. R., and Fraser, F. C., 1968, A cytoplasmically transmitted, diet-dependent difference in response to the teratogenic effects of 6-aminonicotinamide, *Science* **160**:206.

Warburton, D., Trasler, D. G., Naylor, A., Miller, J. R., and Fraser, F. C., 1962, Pitfalls in tests for teratogenicity, *Lancet* **ii**:1116.

Ward, C. O., Barletta, M. A., and Kaye, T., 1970, Teratogenic effects of audiogenic seizures in albino mice, *J. Pharm. Sci.* **59**:1661.

Warkany, J., and Kalter, H., 1962, Maternal impressions and congenital malformations, *Plast. Reconstr. Surg.* **30**(6):628.

Watney-Harris, M., and Fraser, F. C., 1968, Lid gap in newborn mice: A study of its cause and prevention, *Teratology* **1**(4):417.

Watney-Harris, M., and Miller, J. R., 1964, Prevention of a genetically determined congenital eye anomaly in the mouse by the administration of cortisone during pregnancy, *Nature* **202**:1029.

Wilson, J. G., 1973, *Environment and Birth Defects,* Academic Press, New York.

Winfield, J. B., and Bennett, D., 1971, Gene–teratogen interaction: Potentiation of actinomycin D teratogenesis in the house mouse by the lethal gene Brachyury, *Teratology* **4**(2):157.

Woollam, D. H. M., and Millen, J. W., 1961, Influence of uterine position on the response of the mouse embryo to the teratogenic effects of hypervitaminosis-A, *Nature* **190**:184.

Wright, S., 1962, A frequency curve adapted to variation in percentage occurrence, *J. Am. Stat. Assoc.* **21**:162.

Wright, S., 1968, *Evolution and the Genetics of Populations. I. Genetic and Biometric Foundations,* University of Chicago Press, Chicago.

INDEX